The Correspondence of Charles S. Peirce and the Open Court Publishing Company, 1890–1913

Peirceana

Edited by
Francesco Bellucci and Ahti-Veikko Pietarinen

Volume 5

The Correspondence of Charles S. Peirce and the Open Court Publishing Company, 1890–1913

Edited by
Stetson J. Robinson

DE GRUYTER

ISBN 978-3-11-153194-6
e-ISBN (PDF) 978-3-11-076875-6
ISSN 2698-7155

Library of Congress Control Number: 2022941947

Bibliographic information published by the Deutsche Nationalbibliothek
The Deutsche Nationalbibliothek lists this publication in the Deutsche Nationalbibliografie; detailed bibliographic data are available on the Internet at http://dnb.dnb.de.

This volume is text- and page-identical with the hardback published in 2022.
Typesetting: Jukka Nikulainen

www.degruyter.com

Charles S. Peirce (1839–1914), from *Sun and Shade*, August 1892 (Courtesy of Peirce Edition Project)

Paul Carus (1852–1919), OCP editor, from frontispiece of *The Gospel of Buddha*, 1894 (Courtesy of SIU Special Collections Research Center)

Edward Hegeler (1835–1910), OCP founder, ca. 1900, from frontispiece of *The Open Court*, July 1910 (Courtesy of SIU Special Collections Research Center)

Contents

List of Figures

https://doi.org/10.1515/9783110768756-201

This Edition

Scope and Objective

This edition contains the letters exchanged between Charles S. Peirce and the Open Court Publishing Company (OCP) from 1890 to 1913, roughly the last twenty-three years of Peirce's life. OCP published more of Peirce's philosophical writings than any other publisher during his lifetime, and played a critical role in what little recognition and financial income he received during those difficult yet philosophically rich years. This correspondence is the basis for much of what is known surrounding Peirce's publications in *The Monist* and *The Open Court*, and is referenced often in Peirce editions dealing with his later work. Peirce's OCP correspondents included Paul Carus, editor; Edward C. Hegeler, founder and owner; Thomas J. McCormack, assistant editor and translator; Francis C. Russell, Chicago attorney and OCP editorial contractor; and various other OCP editors and staff members. Also included in this edition (Appendix) are enclosures and other material related to the letters, with some exclusions noted in the text. Not included are letters exchanged within the OCP organization that make reference to Peirce. Significant portions of those letters are quoted or referenced in editorial notes, but their entirety is not necessary for a full account of the Peirce–OCP relationship, and for now falls out of the scope of this edition.

The date range for this correspondence (1890 to 1913) is determined by the first and last extant letters. While the beginning of the correspondence is fairly certain in 1890, the exact end of the correspondence is debatable. The correspondence paused from 1911 to 1913 due to Peirce's preoccupations and declining health, and was revived with a brief exchange between Carus and Peirce in August and September 1913. No other letters are found and no textual evidence suggests that any letters are missing, but it is possible that the exchange continued into 1914 until Peirce's death. This edition determines the end of the formal correspondence to be in 1913 based on the end of the extant material, but assumes that the Peirce–OCP relationship remained intact to some degree until Peirce's death.

The objective of this edition is to provide for the first time a complete and accurate text of this oft-cited correspondence, with contextual annotation and textual apparatus. By so doing, this edition sheds critical light not only on Peirce and OCP, but also on the context, relationships, and concepts that influenced the development of Progressive Era American intellectual history and philosophy.

https://doi.org/10.1515/9783110768756-202

Editorial Method

Text and arrangement

The text for this edition comes from manuscripts, typescripts, and letterpress typescript copies housed in the Peirce papers in the Houghton Library at Harvard University and in OCP records in the Morris Library Special Collections at Southern Illinois University–Carbondale. Most of the material used is what was actually exchanged in correspondence, and the remainder consists of carbon (letterpress) copies of missing typescripts, second-hand OCP copies of missing manuscripts or telegrams, and unsent drafts of letters. The letters are arranged chronologically based on dates provided or estimated (typically the date a letter was written or sent, not necessarily received), except for unsent drafts of letters, which immediately follow the final letter sent even if predated (e.g., an unsent draft of 19 July would follow the sent version of 26 August). Multiple unsent drafts are, however, ordered as chronologically as possible. Letters are divided by correspondent instead of consolidated as a single thread, to highlight the relationships Peirce had with each of his OCP contacts. A headnote preceding each division summarizes Peirce's correspondence with the person and identifies significant themes and events in their relationship. Those with whom he held no regular correspondence, or unknown correspondents with a generic (typed) "Open Court Publishing Company" byline, are grouped under "Peirce and Other OCP Staff." In few cases, letters signed generically but in a known hand are included in a specific correspondence where applicable, such as McCormack to Peirce 9 December 1896 (signed "The Monist" in his hand). Letters typed or written by an OCP amanuensis as dictated by Carus or Hegeler are included in the Carus or Hegeler correspondence—such letters typically contain dictator/transcriber initials following the signature, such as "PC/N" for Paul Carus dictation transcribed by J. G. Nattinger. Missing letters (deduced from the context of extant material), gaps in correspondence, and other critical context are addressed in inter-letter editorial notes enclosed in horizontal rules.

Bibliographical notes, headings, and closings

Bibliographical notes precede each letter and identify the sender and receiver, the type and source of the document(s), stationery or postal information, whether the letter was unsent, external notes found on the document, and any other bibliographical context. Where multiple documents exist for the same letter, such as a typescript and an accompanying letterpress copy, the first listed is given priority and used to establish the text. For example: "TS: RL 77. On OCP Stationery B. LPC: SIU 91.12." The text for this letter comes from a typescript on OCP stationery B found in the Houghton Library, a letterpress copy of which is found in SIU Special Collections. Any variants between excluded letterpress copies or manuscripts and the transcribed document are noted in textual endnotes, as explained in "Contextual annotation and textual apparatus". Images of OCP stationery A–D are provided for reference throughout the Peirce–Carus correspondence at locations indicated in the list of figures in this edition. All other stationeries and letterhead are transcribed directly.

Date and address lines and closing and signature lines are standardized flush right, per common practice for most of the letters, with vertical lines occasionally in place of line breaks as an economy, such as "New York | Aug 7 | 1896". Where closings and signatures are occasionally on one line, such as, "Yours truly, P. Carus," two lines are used. Elements of the date and address line that are part of the stationery, such as "LaSalle,   , 189   " in OCP stationery, are included silently. Other letterhead elements not part of the date line are described in bibliographical notes. Estimated dates for undated letters are enclosed in italicized brackets in the date line. All signed names in the letters, including typescripts, are genuine hand signatures, so as an economy, the editorial note "*[signed]*" has not been used, except in two unclear cases: in McCormack to Peirce 9 December 1896, where McCormack signs "The Monist", and in a draft of McCormack's "Translator's Preface" to Mach's *Science of Mechanics* (Appendix), where he genuinely signs although not a letter.

Unsent letters and drafts of letters are identified accordingly in italicized brackets above the date and address line. An "Unsent" letter is one that was aborted, unsigned, crossed out, or otherwise estimated to be unsent based on textual, contextual, and bibliographical evidence. An "Unsent draft" is an unsent letter, whether aborted or complete, replaced by a sent version of the letter. Whether an unsent letter is truly a draft of another is sometimes debatable (see Carus to Peirce 19 May 1893, for example), but the general rule is that an unsent and a sent letter from the same time and with enough similarities to be considered versions of the same message are identified as drafts.

Formatting and layout

Paragraphing and spatial features are generally retained to convey the physiognomy of the letters, but not necessarily their exact appearance.

Example:

Mr. Charles S. Peirce
    12 West 39th Street,
        New York.

Instead of the more precise:

Mr. Charles S.   Peirce
        12   West   39th   Street,
                        New York.

Diagrams, logical graphs, unique symbols, and other drawings, however, are duplicated more precisely than standard text to ensure that potentially meaningful nuances are portrayed (e.g., see the diagrams in the draft material from Peirce to Russell 10 July 1908). Other examples of spatial elements and variations that are observed include: the greeting on its own line or occasionally part of the first paragraph; extra indentation sometimes used in the first paragraph; unique indentation and spacing in lists, equations, and block quotes within letters; the alignment of tables or table-like text, often including ditto marks (″) referring to text directly above; and extra spacing between words used instead of punctuation in telegrams. Examples of spatial elements typically not observed include: extra spacing between paragraphs (unless used deliberately to divide sections or ideas); exact line breaks in the body of letters; and conventions with heading and closing elements, which are mostly standardized as described previously.

Underlined text is instead italicized in transcription, and double-underlined text is italicized and single-underlined. Underlining is overlooked when used to count words or when added by another hand, typically by the recipient while adding accompanying comments—such cases are often noted in textual endnotes. Superscript and underlined superscript characters are retained.

Spelling and punctuation

In most cases spelling and punctuation have not been polished or normalized, in order to preserve the informal style and texture of the letters, and out of respect for

Peirce's strong defiance of the "tyrant" of unnecessary orthographical convention. In a draft letter to *The Nation* sometime between 1907 and 1914, Peirce wrote:

> I would suggest that every man who thinks that the tyranny of orthography ought to be broken down should regard it as a duty to begin spelling a few words,—not so many as to shock people very badly,—in a rational way; and let every man make his own selection, for the very purpose of disproving the popular prejudice that all educated people spell one way. For instance, I have for some time been asserting my individual liberty so far as to write most words in *ise* by *ize*, to spell *intrinsec*, and to indulge a few other protests against the tyrant. I know very well that I am in consequence of this heroic deed, generally set down as a semi-educated crank and nihilist; but I mean to wear the crown of martyrdom with a smile.[1]

In this light, idiosyncratic and non-standard tendencies that have been retained in the text include: British or non-standard spellings such as Peirce's *favour*, *marvellous*, *develope*, and *proceedure*; Russell's *-ly* for *-lly*, as in *realy*, *totaly*, *especialy*, and *metaphoricaly*; Peirce's *that* for *than*, as in would "rather have him do so that to go"; varying forms of contractions, such as *didnt*, *cant*, etc. with no apostrophe, or archaic *did'nt*, *could'nt*, etc.; varying forms of compounds, such as *anyone* and *any one*, *anyway* and *any way*, *today* and *to day*, etc.; recurring variations in the spelling of proper names, such as *Keppler* for *Kepler* (who himself occasionally added the extra *p*, as discussed in *W* 8:452.), *Schroeder* for *Schröder*, *MacCormack* for *McCormack*, *Halstead* for *Halsted*, *Lobatchewsky*, *Lobatchewski*, or *Lobachewski* for *Lobachevsky* (or *Lobachevski*), and *Gramercy* for *Grammercy*; Russell's period following other end punctuation ("text?.", "text!.", etc.); and omitted final periods occasionally at ends of paragraphs.

However, anomalous misspellings (including of proper names), slips, repeated words, and other typos that fall outside of the special treatment described above, as well as idiosyncrasies that compromise readability or overly distract the reader, have been emended and noted in textual endnotes (denoted with an *[E]*). En dashes and other marks used as dashes (such as Russell's "=", as in "text = text") have been rendered as em dashes throughout the text. Peirce's combination of punctuation and spaced en dash ("text, – text") has been rendered as the given punctuation and em dash ("text,—text"). In some cases additional punctuation, such as necessary commas and closing quotation marks or parentheses, has been inserted silently for clarity, particularly for Russell who tended to omit commas in ways that belabor comprehension. For example (all commas mine):

> I am too sensible of the favor accorded me, too sensible of the value and engrossment of your time to ask any such thing.... I have been in the habit ever since your "Illustrations of the

1 *NEM* 1:x–xi, from R 1204.

> Logic of Science" were published in the Pop. Sci. My. of picking up the numbers containing the six parts, taking out the leaves on which they were printed, binding them together, and then of donating such to ..." (to Peirce 22 Jan 1889).

With this selective approach to emendation, gray areas are bound to arise since what counts as a significant variation versus a mere mistake is not always clear. More concrete alternatives previously considered for this edition were to establish a stricter diplomatic text and note all authorial errors and peculiarities or simply warn the reader of them in advance, or conversely to normalize virtually everything and note all editorial interventions. But such polarized approaches would result in a text that either is excessively rough (the former) for an edition with a larger readership than the intended audience of the original letters, or is excessively smooth (the latter) and neutralizes the personal touches inherent in letters. This edition therefore attempts to strike a balance between these public and private considerations, and employs best editorial judgement in debatable cases.

Equations and special characters

Peirce was a highly technical philosopher with an extensive background in mathematics and science and a passion for formal logic. He was also deliberate with his notation and occasionally used custom symbols. The dense mathematical and logical equations that appear in these letters, primarily between Peirce and Russell, have therefore been transcribed with utmost care and have not been normalized or emended. Three unique logical symbols have been custom designed for this edition: Peirce's cursive claw or sign of inclusion (⤙), often represented as −< in print; his sign of relative addition (ɟ), often represented as the overly "stiff" ⸸ in print; and his sign of logical addition or nonexclusive disjunction (ψ), often represented similarly as Ψ in print. The basis for their design rests largely on Peirce's detailed instructions and illustrations sent to McCormack 2 September 1896 and to Carus early November 1896 during the typesetting of his "The Logic of Relatives," as well as on Peirce's recurring usage of them in his manuscripts. Similar treatment has also been given to Peirce's logical diagrams and graphs.

Contextual annotation and textual apparatus

Numbered footnotes provide background and contextual details, identify manuscripts and works, and cross-reference relevant letters and sources. Works sufficiently cited in the text are not footnoted; for these, readers can refer directly to the Bibliography. Peirce's contemporaries mentioned repeatedly in the letters are identified in the Biographical Register and typically not footnoted in the text, while persons mentioned incidentally or those deemed to be well-known historical figures (e.g., Kepler, Kant, and the like) are not included in the Register and are footnoted in the text as needed. Notes are not intended for in-depth philosophical analysis or other extensive commentary on the letters, although occasionally the temptation has been yielded to.

Referenced line numbers with an identifying lemma serve as the textual apparatus to record authorial alterations, editorial emendations, and other textual features, such as "4 new] *inserted above*" (i.e., the word "new" on line 4 was inserted above the text in the manuscript). If the same lemma appears more than once on the same line, then a superscript numeral indicates which instance of the lemma the note refers to, such as "4 new^{2}] *inserted above*" (i.e., the second instance of the word "new" on line 4 was inserted above the text in the manuscript). As an aid, line numbers are displayed in the margin of each page. The elements of the textual apparatus convey to the reader the state of the manuscript and of the writer's mindset at the time of writing, and in many cases expose substantial content omitted in the letter or in a previous draft. For each variant, to the left of the closing bracket is the final state of the text in question (the lemma) as it appears in the edition, and to the right of the bracket are one or more of these: a notation of the authorial alteration, the uncorrected authorial text before editorial emendation (denoted by *[E]*), or the variant or altered text in another copy or draft not included (denoted by *[MS]*, *[TS2]*, etc.). A combination of descriptive and symbolic notation is used for indicating alterations, which allows the editor to choose one in cases where the other would break down or obscure, or to use a combination of the two for optimal clarity or concision. Typically, descriptive notation is used in simple revisions and symbolic notation in more complex revisions. For example, "new] *inserted above*" instead of "new] \new/" or "new] \~/", but "the new paper, I presume] the ~~two~~ \new/ paper~~s~~$^{\vee}$,$^{\vee}$ I ~~heard~~ paper~~s~~$^{\vee}$,$^{\vee}$ I ~~heard~~ /presume\" instead of several descriptive notes: "new] *above* ~~two~~", "paper,] *final* s *deleted and comma inserted*", etc. A simpler alternative to recording variants that is used in many editions is to note the lemma to the left of the bracket, as usual, but indicate the original pre-revised text to the right of the bracket without tracking revisions. The same example above would then become "the new paper, I presume] the two papers I heard". Readers could deduce many of the revisions that took place and would be spared from notation and symbols, but because this convention does not explain how the revisions were made (inserted above or below, superimposed, etc.), nor reflect re-revision and heavily worked-over content, readers cannot get a full sense of the state of the manuscript or of the writer's thought process. The more detailed approach described here has therefore been used to preserve these elements. Alterations in typescripts are made in pen unless noted otherwise (“his] *over* her” vs. “his] *typed over* her”). Editorial emendations in endnotes are indicated with an *[E]*, and in-text editorial notes, conjectures, and other notations are set off with italicized brackets distinct from Peirce’s own roman brackets. In some cases, substantive alternate portions of a letter are included in footnotes instead of the textual apparatus for greater accessibility to the reader.

Below is a summary of the editorial notation and symbols used in this edition. Occasionally more detailed descriptions than what is here listed are used, with or without a lemma, as in “Alter his … the sponsor.] *across header and down right margin*” or in another case simply “*This paragraph up left margin*”.

inserted above, below, in-line, etc.	Simple insertion
above, below	Insertion above or below another alteration (typically deletion) Example: critic] *above* ~~writer~~
over	Superimposition of whole words or false starts Examples: My] *over* Our all] *over* ev (For superimpositions in complex revisions or in partial-word substitutions, the symbolic form below is used.)
after	Indication of alteration (typically deletion) relative to lemma Examples: critic] *after* ~~writer~~ critic] *after deleted period*
~~text~~	Deletion of text
~~<text>~~	Deletion within deletion in complex revisions Example: Nor do I] *after* ~~I <didn’t> \don’t/~~
<punctuation>	Deletion of punctuation in complex revisions, where strikethrough is unclear Examples: <,> vs. ~~,~~ <—> vs. ~~—~~

*text A*text B	Superimpositions in complex revisions or in partial-word substitutions (text B over text A) Examples: Your] ~~Although~~ *y*Your discoveries] discover*y*ies
\text/	Insertion above in complex revisions
/text\	Insertion below in complex revisions
$^{\vee}$text$^{\vee}$	Insertion in-line in complex revisions
[E]	Endnoted editorial emendation Example: aware] award *[E]* (i.e., the MS reads "award" but has been emended in this edition to "aware".)
[MS], [TS1], [TS2], etc.	Endnoted variant or alteration from an excluded draft or copy Example: sent] \\~~delivered~~/ sent *[TS2]* (i.e., an excluded second TS copy *[TS2]* contains a correction that was retained in the final document transcribed.)
*[*text*]*	In-text editorial note or added word for clarity
[?text*]*	In-text conjecture of illegible word or words (handwriting is unclear or MS is deteriorated or torn)
[?]	Illegible word or words (handwriting is unclear or MS is deteriorated or torn)
[ ]	End of content before page end, likely aborted
[…]	Missing connecting page, before or after
[..]	Editorial ellipsis for partial transcriptions in endnotes

Abbreviations

Below is a list of abbreviations widely adopted in Peirce editions and used in this one. The format chosen is similar to that used in the *Writings* and *Essential Peirce* editions by the Peirce Edition Project. Also included are bibliographic abbreviations specific to this edition.

CBPW [catalog #]	*A Comprehensive Bibliography of the Published Works of Charles Sanders Peirce*, 2nd edition rev., ed. Kenneth L. Ketner. Bowling Green: Philosophy Documentation Center, 1986.
CN [vol #: pg #]	*Contributions to The Nation*, 4 parts, ed. Kenneth L. Ketner and James E. Cook. Lubbock: Texas Tech Press, 1975–1988.
CP [vol #.para #]	*Collected Papers of Charles Sanders Peirce*, vols. 1–6, ed. Charles Hartshorne and Paul Weiss; vols. 7–8, ed. Arthur Burks. Cambridge: Harvard University Press, 1932–35, 1958.
EP [vol #: pg #]	*The Essential Peirce*, vol. 1 ed. Nathan Houser and Christian Kloesel; vol. 2 ed. Peirce Edition Project. Bloomington, IN: Indiana University Press, 1992, 1998.
HP [vol #: pg #]	*Historical Perspectives on Peirce's Logic of Science*, 2 vols., ed. Carolyn Eisele. New York: Mouton, 1985.
LPB	Letterpress copy of a typescript letter found in OCP Letterpress Book in SIU Special Collections.
LPC	Generic letterpress copy of a letter in general OCP correspondence in SIU Special Collections, not in OCP LPB and containing no LPB copy number. LPCs are of typescripts unless noted otherwise (e.g., "LPC (MS)").
MS	Manuscript
NEM [vol #: pg #]	*New Elements of Mathematics*, 4 vols. in 5, ed. Carolyn Eisele. The Hague: Mouton, 1976.
OCP	Open Court Publishing Company

R [ISP #]	A manuscript from the Peirce papers in Houghton Library at Harvard University listed in Richard S. Robin's *Annotated Catalogue of the Papers of Charles S. Peirce* (Amherst: University of Massachusetts Press, 1967). Letters are denoted RL [ISP #]. For the most part, Russell–Peirce MSS are in RL 387 while all other Harvard letters are from RL 77. The full citation to these letters is: Charles Sanders Peirce papers, MS Am 1632 (L[77/387]), Houghton Library, Harvard University.
RLT [pg #]	*Reasoning and the Logic of Things: The Cambridge Conference Lectures of 1898*, ed. Kenneth Laine Ketner. Cambridge, MA: HUP, 1992.
SIU [box #.folder #]	A letter or document from OCP company records in Morris Library Special Collections at Southern Illinois University. The full citation to these letters is: MSS 27, Box [#], Folder [#], Open Court Publishing Company records, Special Collections Research Center, Southern Illinois University–Carbondale.
SIU [box #.folder #. LPB copy #]	An LPB copy of a typescript letter from SIU Special Collections. The LPB copy number is found in the top-right corner of the document.
TS	Typescript
W [vol #: pg #]	*Writings of Charles S. Peirce: A Chronological Edition.* Bloomington, IN: Indiana University Press, 1982–.

De Gruyter publication considerations

In addition to the editorial conventions described in this chapter, this edition also complies with De Gruyter publication conventions that require certain elements of the text to be standardized or otherwise modified for optimal layout. Paragraph indentation and line breaks have occasionally been modified to avoid problematic textual spacing and page breaks. Some diagrams are presented between paragraphs instead of wrapped by surrounding text as shown in the source text to improve readability. Large numbers without commas are spaced out every three orders of magnitude for improved readability, and similarly monetary symbols are spaced from their values, as in "$ 100 000" or "$ 100,000" instead of "$100000" or "$100,000". Mathematical and logical characters, both alone and in formulae, are italicized throughout the edition.

Acknowledgements

During my research for this edition, I came across a small set of typescript letters in Box 91 Folder 31 of SIU Special Collections between Peirce scholar Max Fisch and OCP editor Eugene Freeman 1959–1962. Freeman was then working on a Carus biography and preparing to resurrect *The Monist* after its twenty-six-year hiatus since 1936, while Fisch was working on a much-anticipated Peirce biography. The incomplete correspondence describes their efforts to collect and catalog what Peirce–OCP letters were then available in preparation for their respective projects and for a joint paper on the Peirce–Carus relationship. Sadly, none of the projects were completed due at least partly to a lack of accessibility to all the needed material at the time.[2] However, that valiant undertaking was a critical precursor to this edition, which over sixty years later has finally been made possible by Peirce scholars like Fisch, OCP historians like Freeman, and Hegeler–Carus family members who have devoted themselves to making accessible the extant Peirce–OCP material and to expanding its circle of scholarship.

I am particularly grateful to Andre De Tienne of the Peirce Edition Project (PEP), who suggested this edition as a formidable option for my dissertation and eventually a book, and who provided essential support and resources throughout my research. De Tienne also kindly granted me beta access to the Scholarly Text-Editing Platform (STEP), an open source TEI-compliant transcription and editing software created by the PEP, with which this edition was initially transcribed. I am also very grateful to Joe Kaposta and other PEP staff members for painstakingly authenticating all transcriptions in this edition against the digitized source texts.

I am indebted to Pam Hackbart-Dean, Christina Bleyer, Nicholas Guardiano, and other staff of the SIU Special Collections and Research Center, as well as the staff of Houghton Library at Harvard, for allowing me to digitize all Peirce–OCP material for this project and for the future use of others; the Alwin C. Carus Research Grant committee for funding and making possible my research at SIU; Professor Steven Skaggs of the University of Louisville for his immaculate (and gratuitous) designs of Peirce's unique logical symbols used in the early versions of this

2 Andre De Tienne, Peirce Edition Project, personal communication (5 Nov 2015).

edition, and Delve Withrington of Delve Fonts for the custom New Peircean typeface used initially to implement them; *Peirceana* editors in chief Francesco Bellucci and Ahti-Veikko Pietarinen for welcoming this edition into the series, and especially Pietarinen for sharing pre-press material from his *Logic of the Future* series during my research and design stages of Peirce's logical diagrams in the letters; Mara Weber and André Horn at De Gruyter for orchestrating the final editorial stages and the copyediting for this edition; Jukka Nikulainen for his thoroughness and skill in typesetting this edition; and my Editorial Institute colleagues Amanda Jarvis, Emily Kramer, and Jillian Saucier at Boston University for their help with some of the Greek, Latin, and German transcriptions and translations in the letters.

The final word of appreciation is reserved for Archie Burnett, Christopher Ricks, and the late Marilyn Gaull, professors and mentors at the Boston University Editorial Institute. Their valued instruction, collaboration, and feedback throughout this project have turned an ambitious goal into a rewarding reality.

Chronology

The following chronology provides an overview of significant events, submitted articles, and publications related to the Peirce–OCP correspondence. Additional context is also provided for gaps in the correspondence.

1887	Feb	OCP publishes first issue of *Open Court* under editorship of Benjamin and Sara Underwood, formerly of Boston *Index*. Carus arrives at Chicago per Hegeler's invitation to fill an unclear role with OCP, and is disfavored by Underwoods.
	Apr	Charles and Juliette Peirce move from New York to Milford, PA. Charles maintains correspondence on logic and mathematics with friend and mutual OCP contact Francis Russell, and over the next few years learns of OCP through him.
	early Dec	Underwoods resign from OCP due to disputes between them, Hegeler, and Carus, leaving Carus as editor.
1888	May	Charles and Juliette purchase Quick farmhouse in Milford, which they would expand and rename "Arisbe," and which would be their home the rest of their lives.
1890	Jul 1	Russell invites Peirce on Carus's behalf to write for inaugural issue of *Monist*.
	Jul 2	Carus himself invites Peirce to write for inaugural issue of *Monist*. Peirce–OCP correspondence begins officially.
	Aug 7	Peirce reviews Carus's *Fundamental Problems* in *The Nation*; includes criticism of OCP.
	Aug 28	Hegeler publishes "Religion and Science," *Open Court*, responding to Peirce's critical *Nation* review.
	Aug 30	Peirce submits "The Architecture of Theories" to Carus (too late for the October *Monist*).
	Sept 11	Carus publishes "The Unity of Truth" and "The Superscientific and Pure Reason," *Open Court*, responding to Peirce's critical *Nation* review.
	early Oct	OCP publishes first issue of *Monist*—without Peirce's article.

https://doi.org/10.1515/9783110768756-203

1891	early Jan	Peirce's "Architecture of Theories" published, *Monist.*
		Carus publishes "The Criterion of Truth," *Monist*, responding to Peirce's "Architecture of Theories."
	Nov 5	Peirce submits "The Doctrine of Necessity Examined" to Carus for *Monist.*
1892	early Apr	"Doctrine of Necessity Examined" published, *Monist.*
		Carus publishes "Mr. Charles S. Peirce on Necessity," *Monist*, responding to Peirce's "Doctrine of Necessity Examined" and setting up the debate that would follow.
	Apr 27	Carus invites Peirce to help with the OCP translation of Ernst Mach's second edition of *Science of Mechanics* (1889), led by McCormack.
	May 24	Peirce submits "The Law of Mind" to Carus for *Monist.*
	early Jul	Peirce begins working with McCormack on Mach translation.
		Peirce's "The Law of Mind" published, *Monist.*
		Carus publishes "Mr. Charles S. Peirce's Onslaught of the Doctrine of Necessity," *Monist*, responding further to Peirce's "Doctrine of Necessity Examined."
	Jul 15	Peirce submits "Man's Glassy Essence" to Carus for *Monist.*
	Aug 6	Peirce submits "Pythagorics" to McCormack for *Open Court.*
	Aug 25	Peirce submits "Critic of Arguments. I." to McCormack for *Open Court.*
	Aug 31	Peirce submits "Dmesis" to Carus for *Open Court.*
	Sep 3	Peirce submits "Critic of Arguments. II." to McCormack for *Open Court.*
	Sep 8–29	"Pythagorics," "The Critic of Arguments. I.," and "Dmesis" published, *Open Court.*

	Oct	Carus publishes "The Idea of Necessity, Its Basis and Its Scope," *Monist*, another rejoinder to Peirce's "Doctrine of Necessity Examined."
		Peirce's "Man's Glassy Essence" published, *Monist*.
	Oct 7	Peirce submits "Evolutionary Love" to Carus for *Monist*.
	Oct 13	Peirce's "The Critic of Arguments. II." published, *Open Court*.
1893	early Jan	Peirce's "Evolutionary Love" published, *Monist*.
	Jan 20	Peirce submits "Reply to the Necessitarians" to Carus for *Monist*, responding to Carus's "Onslaught" and "Idea of Necessity."
	bet Jan 28 and Feb 6	Per Hegeler's invitation, Peirce meets with Hegeler and likely Carus at Hegeler's home in La Salle to discuss larger publication projects, including Peirce's "Quest for a Method" (or "Search for a Method") and several math books. Shortly after begins investigating printers to contract with OCP for the books.
	Feb 7	Peirce meets with Russell at Richelieu in Chicago—details undetermined, though an unsatisfactory meeting for Peirce due to Russell's "not showing me the side of him that I wanted to see" (Peirce to Hegeler 11 Feb 1893).
	early Feb	Peirce submits "Marriage of Religion and Science" to Carus for *Open Court*.
	Feb 10	Peirce meets separately with William Appleton of Appleton & Co. and Henry Vail of the American Book Company, the former regarding his arithmetic, the latter regarding two arithmetics and an Elementary Geometry.
	Feb 16	Peirce's "Marriage of Religion and Science" published, *Open Court*.
		Carus publishes "Religion Inseparable from Science," *Open Court*, responding to Peirce's "Marriage."

	Mar 7	Peirce sends Hegeler an outline of his proposed "Quest for a Method" book.
	late Apr	Peirce submits to Carus "What Is Christian Faith?" for *Open Court* and a collection of MSS for his "Quest for a Method" book.
	May 4	Peirce submits "Immortality in the Light of Synechism" to Carus for *Open Court* or *Monist* (never published in either).
	May 18	Peirce submits "Critic of Arguments. That all inferences concerning matters of fact are subject of certain qualifications" to Hegeler—apparently not for publication in "Critic of Arguments" series of *Open Court* but merely "in hopes it may lead you to take a less unfavorable view of my opinions." (However, compare Peirce to Carus 10 May 1893 regarding his intention to publish.)
	May 22	Peirce sends Hegeler a report of MSS ready or in process for his arithmetic book.
	early Jun	Peirce submits "Cogito Ergo Sum" to Carus as a "letter to the editor" for *Open Court.*
	Jun 15	"Cogito Ergo Sum" published, *Open Court.*
	early Jul	Peirce's "Reply to the Necessitarians" published, *Monist.*
		Carus publishes "The Founder of Tychism," *Monist*, responding to Peirce's "Reply to the Necessitarians."
	Jul 27	Peirce's "What Is Christian Faith?" published, *Open Court.*
1894	Jan 11	Peirce's "The Principles of Philosophy" book announced in *Nation.*
	ca. Mar	Peirce is invited by Ginn & Co to publish his "New Elements of Geometry" and "How to Reason" books (never published).
	Jun	Peirce submits MSS for "How to Reason" to Ginn & Co.

	Jun 21	Carus publishes "The Circle-Squarer," *Open Court*, which Peirce interpreted as an unflattering representation of him and his necessitarian debate with Carus the previous year. (Part 2 of the article appeared 28 June.)
	Jul 6	Peirce submits "Rienzi, Last of the Tribunes" (rejected) and "Philosophical Reflections on Table Turning" (never published) to Carus for *Monist*.
	Aug	Due to consistent complications and misunderstandings with agreements, Peirce officially falls out of favor and contact with Hegeler, ending completely their correspondence and postponing all other OCP contact (except with Russell) for the next two years. The remainder of Peirce's dealings with OCP would be under the auspices of Russell and later Carus.
	Sep	Russell attempts to persuade Hegeler to reconsider working with Peirce, which Hegeler refuses to do 19 Sept.
	ca. Oct	Peirce's "The Principles of Philosophy" book announced by Henry Holt & Co. (never published).
1895	Summer	Ernst Schröder publishes *Algebra und Logik der Relative*, which would be the subject of Peirce's 1896–1897 "Regenerated Logic" and "Logic of Relatives" in *Monist*. Charles and Juliette relocate to New York to be closer to Juliette's doctors, but soon after are forced to remain there to evade an arrest warrant in Milford for Charles's alleged assault against a house servant. This hampers his exchange of Logic MSS (sequestered at Arisbe) with Russell.
	Sep–Oct	Carus asks Russell to review Schröder's *Logik*, but Russell suggests instead that Peirce be invited to do so. Carus insists, so Russell secretly asks Peirce for help 25 Sept. Peirce supplies some of his own logic material as an aid 4 Oct and the two remain in collaboration on related matters.
1896	early Jan	Carus reviews Schröder's *Logik* in *Monist*.

	Feb 25	Apparently persuaded by Schröder's insistence to have Peirce be the one to review his *Logik*,[1] Carus asks Russell to elicit a review from Peirce after all.
	Apr 22	Russell invites Peirce to write on Schröder for *Monist*.
	Apr 23	Peirce reviews Schröder's *Logik* in *Nation*.
	Apr 26	Peirce accepts Russell's invitation and submits "The Regenerated Logic" ca. 30 Apr.
	May 9	Carus ends the two-year silence between Peirce and OCP acknowledging the submission of "Regenerated Logic."
	Jun 19	Peirce delays in returning the proofs for "Regenerated Logic" sent him by OCP for review, postponing the article until Oct *Monist*.
	Jul 15	Peirce submits "The Logic of Relatives" to Carus for *Monist*.
	Aug 31	Peirce submits "On Logical Graphs" to Carus for *Monist* (never published).
	Oct	Peirce's "The Regenerated Logic" published, *Monist*.
1897	Jan	Peirce's "The Logic of Relatives" published, *Monist*.
	May 12	Carus informs Peirce of William James's efforts to procure lectureship for him during the upcoming 1898 Cambridge Conferences.
1898	Feb–Mar	Per James's invitation, Peirce delivers Cambridge lectures on "Reasoning and the Logic of Things."
		Peirce is invited by Putnam's Sons to prepare a "History of Science" book (never published).
	May	Charles and Juliette return permanently to Milford from New York following the resolution of the assault case against Charles.

1 Brent, 256; also *Peirce Project Newsletter*, 1994.

	Dec 5	Peirce submits "Knotty Points on the Doctrine of Chance" to Carus for *Monist* (never published).
1900	Mar 17	Carus proposes to Peirce that OCP publish their Necessity–Tychism debate from April 1892–July 1893. (They would never agree on the form the book would take and it was never published.)
1901–1904		For the next three years, with the exception of his unusual, incomplete correspondence with OCP compositor Fred Sigrist, no letters between Peirce and OCP are found. It is likely that Peirce simply maintained little contact with OCP during what was then a very busy time for him: he translated for the *Annual Report of the Board of Regents* of the Smithsonian Institution (1901), wrote definitions for Baldwin's *Dictionary of Philosophy* (1901–1902), submitted his elaborate grant proposal to the Carnegie Institute for "Memoirs on Minute Logic" (1902),[1] delivered his Harvard lectures on pragmatism[2] (an arrangement again made by William James) and his Lowell Institute lectures on logic (1903),[3] and all along continued reviewing books for *The Nation*. He was also, however, gearing up for a new series of *Monist* articles on pragmatism to disentangle his own form of it, *pragmaticism*, from other contemporary variants, such as those of William James, F. C. S. Schiller, and John Dewey.
1904	Sep 12	Peirce submits "What Pragmatism Is" to Carus for *Monist*.
	Oct 3	Peirce submits "Substitution in Logic" to Carus for *Monist*, a review of Taine's *De L'Intelligence* to be altered and signed by Russell.
1905	ca. Apr 1–3	Peirce's "What Pragmatism Is" published, *Monist*. Russell (for Peirce) publishes "Substitution in Logic," *Monist*.

1 Ransdell, "MS L75: Logic,Regarded as Semiotic", and partly in *NEM* 4:13–73.
2 *EP* 2:133–241, also in full in Turrisi, 1997.
3 *EP* 2:258–299.

		Peirce submits to Carus second and third pragmatism articles, at this point likely "Consequences of Pragmaticism" and "Evidences for Pragmaticism."
	Apr 3	Carus proposes again (see 17 Mar 1900) to Peirce that OCP publish their Necessity–Tychism debate from April 1892–July 1893.
	ca. May 20	Peirce submits "Issues of Pragmaticism" to Carus for *Monist*, to replace "Consequences of Pragmaticism" as his second article in the series.
	ca. Jun 28	Peirce submits review of Edward Ross's 1905 *Foundations of Sociology* to Carus for *Monist*.
	early Oct	Peirce's "Issues of Pragmaticism" published, *Monist*.
	ca. Nov 3	Peirce submits "Mr. Peterson's Proposed Discussion" to Carus for *Monist*, responding to James Peterson's "Some Philosophical Terms" from Oct *Monist*.
		Also submits a paper on Josiah Royce's 1905 *The Relation of the Principles of Logic to the Foundations of Geometry* (ultimately rejected). This was likely a version of, if not the very paper, "The Relation of Betweenness and Royce's O-collections" that Peirce would present 14 Nov to the National Academy of Sciences.
1906	early Jan	Peirce's "Mr. Peterson's Proposed Discussion" published, *Monist*.
		Peirce submits a third pragmatism article to Carus for *Monist*, at this point possibly "The Basis of Pragmaticism [in Phaneroscopy]."
	ca. May 25	Peirce submits "Prolegomena to an Apology for Pragmaticism" to Carus for *Monist*, to replace his previously submitted article as the third in the series.
	early Jul	Peirce's review of Ross's *Foundations of Sociology* published, *Monist*.

	early Oct	Peirce's "Prolegomena to an Apology for Pragmaticism" published, *Monist.*
1907	Jan 23	Carus invites Peirce to re-publish his 1877–1878 *Popular Science Monthly* "Illustrations of the Logic of Science" series in book form, and would work with him to that end for the remainder of their correspondence. (Peirce was not able to revise it to his satisfaction and ultimately the book was never finished.)
	Mar–Jul	Peirce prepares a lengthy "letter to the editor" for *The Nation* and *Atlantic Monthly* on pragmatism (rejected; *EP* 2:398–433) and delivers Harvard Philosophy Club lectures on "Logical Methodeutic" 8–13 April (see *W* and *EP* chronologies).
		Peirce pauses his pragmatism series and begins working on his "Some Amazing Mazes" series for *Monist.*
	ca. Sep 28	Peirce submits to Carus "The First Curiosity," "Explanation of Curiosity the First," and "The Second Curiosity" of his "Amazing Mazes" series for *Monist.*
	ca. Oct	Peirce resumes work on pragmatism series, which he is never able to complete.
	early Dec	Peirce submits to Carus "The Third Curiosity" and "The Fourth Curiosity" of his "Amazing Mazes" series for *Monist* (never published by OCP).
1908	early Apr	Peirce's "Some Amazing Mazes: The First Curiosity" published, *Monist.*
	May	Peirce publishes a "Letter from Mr. Peirce," *Open Court*, endorsing Carus's "Problems of Modern Theology" from the April issue.
	early Jul	Russell publishes "Hints for the Elucidation of Mr. Peirce's Logical Work," *Monist.*
		Peirce's "Some Amazing Mazes: Explanation of Curiosity the First" published, *Monist.*

	Dec 9	Carus attempts to re-engage Peirce in work on the "Illustrations" book.
1909	early Jan	Peirce's "Some Amazing Mazes: A Second Curiosity" published, *Monist*.
	Jan 7	Peirce begins revising his "Illustrations" for the book proposed by Carus, initially to take only a few months but ultimately never finished.
1910	early Jan	Carus publishes "The Nature of Logical and Mathematical Thought," *Monist*, quoting Peirce to Russell early Oct 1895 at pp. 45 and 158, on non-Aristotelean Logic.
	Aug 26	Peirce sends Carus a long letter in reply to his repeated requests for "Illustrations" material promised since Jan 1907. Explains the difficulties he has had in revising the work and proposes a new plan to simply add a preface to the original articles stating the alterations he would have made, which he describes in detail.
1911–1913		Peirce re-visits a commitment from 1909 to submit an essay for a volume in honor of Victoria Lady Welby, Peirce's semiotic confidant and correspondent 1903–1908. Prepares several drafts of "A Sketch of Logical Critics" to that end ca. spring and summer 1911, though never published (R 673–677, one version in *EP* 2:451–462). In Nov 1911 presents papers on "A Method of Computation" and "The Reasons of Reasoning" to National Academy of Sciences, his last public address. Likely continues to work on his "System of Logic," though greatly impeded by growing health challenges. Carus, meanwhile, continues preparing for the "Illustrations" book and writes a proposed preface of his own.
1913	Aug 23	Carus sends Peirce his proposed preface for the "Illustrations" book (included in Appendix).
	Sep 10	Carus proposes a third time (see 17 Mar 1900 and 3 Apr 1905) to Peirce that OCP publish their Necessity–Tychism debate from April 1892–July 1893.

1914	Apr 19	Peirce dies of cancer at Arisbe.
	Apr 22	Carus invites Russell to write Peirce's obituary in *Monist*.
	early Jul	Russell publishes "In Memoriam Charles S. Peirce," *Monist*.

1 Peirce and the Open Court

A New Platform

In the winter of 1886–1887, after nearly a decade of attempting to set up a publication for religio-philosophical thought, La Salle zinc miner and entrepreneur Edward C. Hegeler came to an agreement with Boston *Index* editor Benjamin F. Underwood to bail out the then struggling *Index* with a new publisher, The Open Court Publishing Company, new offices in Chicago and La Salle, a new mission to establish a monism of religion and science, and a new name: *The Open Court*.[1] The fortnightly ran its first issue in February 1887, shortly after which German philosopher and scholar Paul Carus arrived to the organization per Hegeler's invitation to fill an uncertain, though heavily involved, role with OCP that led to the unsavory resignation of the Underwoods by year's end. Hegeler and Carus immediately announced their intention to devote the journal to "the work of conciliating Religion with Science ... in Monism, to present and defend which will be the main object of The Open Court,"[2] and began making plans for a broader quarterly publication that would be similarly devoted to "the establishment and illustration of the principles of monism in philosophy, exact science, religion, and sociology."[3] This one, still in print today, would long outlive *The Open Court* and attract renowned philosophical and scientific minds that made it instrumental in professionalizing philosophy in the U.S. By the summer of 1890, the first issue of *The Monist* was underway, and Charles S. Peirce would be among its first and longest-standing contributors.

Invitations for Peirce to contribute to the new quarterly came 1 and 2 July 1890 from Francis Russell, Chicago attorney and mutual friend of Peirce and Carus, and from Carus himself. Russell had already been in correspondence with Peirce at least as early as 1887 regarding Peirce's work in logic and regarding OCP's begin-

---

Much of the context in this introduction, especially in the first two sections, is detailed in Henderson, 21–44 and 125–141, the introduction to *W* 8, the introductions and headnotes to *EP* 1 and 2, and Brent, 203–347. As an economy, individual citations are in many instances not provided to these sources, to which the reader is referred for a fuller account.

1 Hegeler preferred *The Monist*, but acquiesced to this, Sara Underwood's coinage. He would get his way, however, with the quarterly journal that soon followed.

2 Subhead of *Open Court* that began running 23 Dec 1887.

3 *Science*, vol. 6 (19 Sept 1890), pp. 166–167.

https://doi.org/10.1515/9783110768756-001

nings and intentions, and was instrumental in persuading Carus to elicit from him related topics for the inaugural issue of *The Monist*.[4] "I wish that in the new Quarterly," wrote Carus to Peirce 2 July, "our most prominent American authors should be represented and shall be greatly indebted to you for an article from your pen on 'Modern Logic' or some similar topic—perhaps 'Logic and Ethics.' You may choose any theme with which you are engaged at present." Peirce accepted the invitation 3 July and proposed to Russell not one article but a remarkable series known today as Peirce's *Monist* metaphysical series:

> I should be very glad to write for the Open Court. One can profitably put but very little into a single article. I should therefore prefer to write a number. I would write in a general way about the ways in which great ideas become developed, *not* about verification and assurance, to which my Johns Hopkins lectures used chiefly to be directed. If you or Dr. Carus will write and say how many articles the Open Court will take on that subject and how many thousand words per article, I will endeavor to fill the bill.

This was the beginning of a relationship that lasted until Peirce's death and one of the most important he ever made: it would provide the platform for the most philosophically productive years of his life, a fact that led Max Fisch to label these years Peirce's "*Monist* Period."[5]

The Peirce–OCP relationship was not a harmonious one, however, even from the outset. A month after Peirce accepted the call for papers, an unflattering review of Carus's *Fundamental Problems* appeared in *The Nation* criticizing not only Carus's "superscientific" philosophy and underwhelming views that were "the average opinions of thoughtful men," but also his and Hegeler's general mission of reconciling religion with science:

> Is this wise? Is it not an endeavor to reach a foredetermined conclusion? And is not *that* an anti-scientific, anti-philosophical aim? Does not such a struggle imply a defect of intellectual integrity and tend to undermine the whole moral health?… Religion, to be true to itself, should demand the unconditional surrender of free-thinking. Science, true to itself, cannot listen to such a demand for an instant.… Why should not religion and science seek each a self-development in its own interest, and then if, as they approach completion, they are

---

4 Probably Russell's greatest influence on Carus was his introducing him in 1888 to Peirce's "Illustrations of the Logic of Science" series of 1877–1878. For details, see the "Illustrations" section in this introduction.

5 The fifth and final period or "step" (specifically 1891–1914) in Peirce's progress from nominalism to realism. See Fisch, 184–197, for details. This period also corresponds roughly with what Fisch calls Peirce's "Arisbe Period" (1887–1914), the third and final period of a broader three-part division of Peirce's philosophic activity as a whole, initiated by Peirce's move from New York to his Milford, PA, estate "Arisbe." See Fisch, 227–245, for details; also "Arisbe" in the Biographical Register of this edition.

found to come more and more into accord, will not that be a more satisfactory result than forcibly bending them together now in a way which can only disfigure both?[6]

Carus rightly suspected the review to be Peirce's and wrote to him 3 August 1890: "I read your review in the Nation and shall make a brief reply to the single points you raise.... Mr Hegeler discussed your objection ... in the last number." Hegeler's "Religion and Science" and Carus's double-blow of "The Unity of Truth" and "The Superscientific and Pure Reason" responded in August and September to Peirce's remarks. In reply 6 September, a bit on the defensive, Peirce warned Carus that "You should not attribute anonymous articles to me, as you don't know what editorial liberties may have been taken with them.... But," Peirce admits, "I have always declined to identify myself with associations for 'conciliating religion with science.' Best let them work out their own roads to truth." The tiff would not be the last, and others far less innocuous, but Peirce was nevertheless able to publish more of his philosophical writings through OCP than any other medium.

In addition to being the outlet for Peirce's mature philosophy, OCP was also critical to his income during desperate times. After Peirce's forced resignation December 1891 from his 30-year career with the U.S. Coast and Geodetic Survey, Peirce never again held regular employment and relied solely on income from OCP and other freelance projects. Due to bad luck and self-destructive money management, Peirce's final twenty or so years were plagued with intense poverty, illness (particularly with his wife, Juliette), starvation, occasional homelessness, failed investments, swindled business agreements, legal action, and other troubling matters that surface throughout his correspondence with OCP. The generous payments and advances on articles that Carus and Hegeler provided Peirce were at times all that kept Peirce from death and literally bought him just enough time to take a few next steps in expounding some of his greatest work.

Joseph Brent poetically illustrates the juxtaposition of the fruitful yet tragic years Peirce spent in correspondence with OCP, which sets the stage for a closer look to which we turn next:

I imagine him as a confidence man and prestidigitator—a frenzied juggler elegantly dressed in harlequin costume, on a flimsy but deceptively substantial stage of his own devising, gambling everything on keeping too many improbably seductive objects in flight, while trying to decide which one to snare. At the end, he stands there in tatters, surrounded by the melancholy debris of his life, contrite and apologetic, asking our—and especially his dearest friend William James's—indulgence. But all the while, this poor fool, behind the scenes and between the acts, has been building piece by piece the armature of a most marvelously intri-

6 7 Aug 1890, pp. 118–119.

> cate universe, so beautiful it transfigures him amidst the wreck of his afflictions, and we gratefully see the signs around us with new eyes.[7]

Metaphysics, Mechanics, and Mathematics: 1890–1894

These first few years of Peirce's writing for OCP were by far his busiest. During this time he published six articles in *The Monist* and six in *The Open Court*,[8] assisted in the translation of Ernst Mach's second edition of *Science of Mechanics* (1889), and began working in earnest on several math books on arithmetic and geometry. His opening series in *The Monist* 1891–1893 comprised five articles that supplied, for the first time in his career, a systematic account of his metaphysics, largely derived from the evolutionary philosophy Peirce had been engaged in laying out in his intricate "A Guess at the Riddle." The articles included "The Architecture of Theories," "The Doctrine of Necessity Examined," "The Law of Mind," "Man's Glassy Essence," and "Evolutionary Love," and argued most notably for his doctrines of tychism, synechism, and agapism, or the roles of chance, continuity, and love in the universe. The series was cut short of the sixth and final article, however, which "would have been the keystone of the whole."[9] The reason was that shortly after his "Evolutionary Love" was published January 1893, Peirce paused the series to respond to Carus's interjections against Peirce's anti-necessitarian arguments in "The Doctrine of Necessity Examined"—for Peirce, "the strongest piece of argumentation I have ever done"—that included "Mr. Charles S. Peirce's Onslaught of the Doctrine of Necessity" in July 1892 and then "The Idea of Necessity, Its Basis and Its Scope" three months later. Peirce's extensive rejoinder, "Reply to the Necessitarians," appeared July 1893—immediately followed in the same issue by Carus's counter, "The Founder of Tychism"—and, whether due to a miscommunication or breach of agreement, was the final article Hegeler and Carus would accept from Peirce for the series.[10] Peirce expressed his unmistakable disappointment with Hegeler 27 July 1893:

> I consider that a positive wrong has been done to me in not allowing me the six articles promised last summer in the Monist on my philosophy. In sending the rejoinder to Carus, I

---

7 p. 203. Specifically in reference to the years 1890–1900, but applicable to the whole correspondence.

8 Seven if we include his brief "letter to the editor," "Cogito Ergo Sum" (15 June 1893).

9 Peirce to Russell 4 Oct 1895. Preliminarily to be titled "A General Sketch of a Theory of the Universe," as listed in Peirce's outline of "A Quest for a Method" sent to Hegeler 7 March 1893.

10 More details about this Necessity–Tychism dispute are footnoted in Peirce to Carus 3 April 1892.

> distinctly said I did not consider that one of them; and therefore it could have been rejected, if another had been considered excessive. Besides that, I think it unjust that when my preliminary plan proved insufficient, I should have been left with an important philosophy half developed, because you would not give a *little more* space than was promised,—instead of breaking the positive promise made to me.
>
> Another positive promise in reference to the Open Court has been broken by you. It is difficult for me to believe that you are a man to break an explicit promise when the fact is pointed out to you.

The mishap surrounding the metaphysical series would be a recurring sore spot between Peirce, Carus, and Hegeler.

At the same time, between the summers of 1892 and 1893, Peirce was also writing for *The Open Court* on other religious and logical matters, most notably his "Critic of Arguments" series intended "to further the cause of right reasoning" and "to contain remarks, mostly rather special, upon good and bad reasoning"[11]—another series cut short with only two articles that appeared fall 1892—and filling a contract position extended him by Carus to review proof sheets for the OCP translation of Mach's *Science of Mechanics*, led by OCP assistant editor and translator Thomas McCormack. Carus considered Peirce "an authority in this province,"[12] and he was right, to the point that Peirce took on a much more substantial role with the edition. During the year Peirce worked with McCormack, he corrected and supplied translations, footnotes, and an entire section (Chapter III, Section III, Sub-section 8) on the mechanical units in use in the U.S. and Great Britain, which was acknowledged gratefully in McCormack's "Translator's Preface" and included in several subsequent English and German editions.[13]

Peirce's involvement with OCP would become more intimate still when, after submitting his "Reply to the Necessitarians" 20 January 1893, Hegeler invited him Monday, 23 January, to meet with him in his home in La Salle, believing that "it would be a help to our work if we became nearer acquainted." Peirce telegrammed immediately Wednesday accepting the invitation and by that Saturday night arrived by train in La Salle.[14] The meeting with Hegeler, and likely Carus, was held sometime between 28 January and 6 February and addressed

11 Peirce to Carus 3 June 1892. For more on the other *Open Court* articles, see the Chronology or Bibliography of this edition.

12 Carus to Peirce 27 April 1892.

13 For this Preface and Peirce's section 8, see Appendix, "Peirce and McCormack." Mach in his own Preface was also to have recognized Peirce's work, but Peirce seems ultimately to have declined the honorable mention (see McCormack to Peirce 23 May 1893). Details on subsequent editions in *HP* 1:538–539.

14 Hegeler and Peirce 23 and 25 Jan 1893.

larger publication projects that Peirce might undertake with OCP, particularly his "A Quest for a Method" (or "Search for a Method") and an arithmetic textbook.[15] Peirce had begun pursuing more seriously his interests in mathematical textbooks around the time of his forced resignation from the Survey, especially his Arithmetic since 1888, which he believed would be the most profitable.[16] Peirce seems to have persuaded Hegeler of this fact because he was provided a stipend of $1 750 in $250 installments over the next six months to complete the Arithmetic, to be paid back at 6 percent interest from future profits of the book.[17] Following their meeting, Peirce began working, among many other tasks, on his Arithmetic, his "Quest for a Method" logic book, and likely an Elementary Geometry instigated by the American Book Company and later Ginn & Co. of New York.[18] The rest of the year proved personally challenging for Peirce and, despite his sidelining other endeavors per Hegeler's request to focus on the Arithmetic,[19] he regretfully informed Hegeler on New Year's Eve 1893 that

> Several circumstances impossible to foresee prevented the completion of my arithmetics within the time I had set. Among these I may mention the bursting of my boiler and blocking of the roads with snow afterward, rendering my house uninhabitable for a long time, the breaking of a bank in which my wife was a depositor, the breaking of another which indirectly affected me. Other calamities there were, still worse, which I shall not further mention.

By the fall of 1894, Peirce's still-incomplete Arithmetic manuscripts had been seized by Hegeler,[20] and from this point, due to the consistent complications and misunderstandings with agreements, Peirce officially fell out of favor and contact with Hegeler, ending completely their correspondence and postponing all other OCP contact for the next two years—only Russell remained in touch during the lapse.

---

15 See Peirce to Hegeler 24 Feb and 7 March 1893 for details. Also de Waal, 15–17, and Brent, 221–224.

16 *NEM* 1:xxix.

17 Hegeler to Peirce 24 Aug 1893.

18 Peirce to Hegeler 11 Feb 1893 and 11 July 1894. The Geometry later evolved into a revision of Peirce's father's (Benjamin's) 1837 *Elementary Treatise on Plane and Solid Geometry*, to be entitled "New Elements of Geometry."

19 Hegeler to Peirce 15 May 1893.

20 See Peirce to Russell 8 Sept 1894. Exactly when is unclear.

Lonely Logician: 1894–1897

Having gotten wind of the falling out between Hegeler and Peirce, Russell provided Peirce candid yet supportive counsel that the controversy was "partly your own fault (which I am very very willing to attribute to the 'eccentricities of genius' although that does not alter the fact) and partly his fault that 1st he does not adequately appreciate your genius, and 2d that he acts on certain principles good enough for the most part, but not adapted for cases like that one in hand." He offered too "to write for The Monist my own ideas of your philosophy say an article on 'The Philosophy of C. S. Peirce' as a means in the first place of perhaps starting you up again in The Monist and also of clearing up various misconceptions of your ideas (according to me) that have been published."[21] In a lengthy, introspective draft reply ca. 1 September 1894, Peirce provides a succinct and revealing description of his personal and philosophical challenges:

> No person can understand what the nature of my difficulties are. Besides the "eccentricity of genius" which seems to consist in being a perfect idiot in all dealings with human beings, in my case, I have to contend with certain peculiarities in the health & mind of some of those I have to think of, which I certainly ought not to explain, I ought not even to hint at them as I am doing, but which would account for certain things for which I am violently blamed.
>
> The only objection I should have to you or anybody writing about my philosophy is that my philosophy has never been stated even in the most general outline. It is absolutely impossible to get the slightest comprehension of it until certain sides of it not yet touched upon are sketched & the way the different parts fit together to form one unitary conception are explained. I defy the greatest intellect,—I would defy Aristotle himself,—to see in a general way what I am driving at from what I have said so far. As I cannot get any help to say it, I mean to publish it myself.

Peirce thus recognized and in some ways ensured that he was on his own, temporally and philosophically, but was determined to succeed nevertheless in bringing out his system of logic and philosophy. On the outs with OCP, the means whereby Peirce was to do so were less promising, and his needs increasingly great. He desperately relied on Russell to convince Hegeler to reconsider working with him, or to find some other way to provide income for him:

> I guess you can influence Hegeler if anybody can; but his animosity is nasty. Try to get me scholars; to have me invited to give lectures or readings.... Above all *[?dont]* lose time. Starvation dont wait. If you cant raise all at once, raise a part and send in haste. I need it & my

21 Russell to Peirce 29 Aug 1894.

> wife needs it even more. Need cannot be more extreme. My two brothers are both abroad. There is nobody to whom I can make my condition known.[22]

Hegeler's definitive rejection came to Russell 19 September 1894: "I have to positively decline. I will not enter again into any kind of business relation or correspondence with Mr. Peirce."

Among Peirce's troubling letters during his isolation from OCP we find also some of his richest informal writings on logic, largely by way of commentary on Ernst Schröder's 1895 *Algebra und Logik der Relative*, the third volume of his *Algebra der Logik* series. Russell had been asked by Carus to review the new volume in *The Monist*, but suggested instead that Peirce be invited to do so. Carus refused and Russell consented, secretly asking for Peirce's help 25 September 1895 to "find some way to *enable* me to write such a review or something that will pass for it without my making a *fool* of myself." In reply, Peirce sent Russell 4 October several portions of his "How to Reason" (or "Grand Logic")[23] and other logical papers and remained in touch on related matters. By 25 February 1896, apparently influenced by Schröder's insistence that Peirce was the only one suited to review his *Logik*,[24] Carus asked Russell to invite Peirce to write the review after all. "As matters are," Carus adds, "I cannot ask him to write the review, but if he would send it I would not refuse it.... Mr. Hegeler does not wish me to be in any business connections with him, but that would not exclude the acceptance of a contribution which he might offer at a cash price, the bill to be settled at once ... without further complications."[25] Peirce received the news from Russell 26 April with great delight, especially since it would pay at his "*old rates* and as soon as the article is written," and within days submitted his review, "The Regenerated Logic," to appear in the October *Monist*. Due to a miscommunication between Russell and Carus,[26] however, Peirce would not be paid for nearly two agonizing weeks—which he took to be an intentional reminder of OCP's sentiments toward him—captured in several undated draft letters to Russell ca. 12 May noting "the terrible effects which have resulted from the breach of Carus's promise," and pleading, "For God's sake send me *some* money if you can communicate with Carus. It is a matter of life and death!" Peirce was eventually paid and, despite the rocky re-entry, was back in touch with OCP and would submit various other articles on logic, although only

---

22 Peirce to Russell 8 Sept 1894.

23 At that time cycling through various other titles, including "The Art of Reasoning" and "Principles of Strict Reasoning," outlined previously in Peirce to Russell 8 Sept 1894.

24 Brent, 256; also *Peirce Project Newsletter* (1994).

25 SIU 91.7.

26 See footnote to 9 May 1896 letter.

one would be accepted, "The Logic of Relatives"—a second article on Schröder and a watershed in Peirce's move from nominalism to realism.[27] Carus's reluctance to publish Peirce's manuscripts extended his and especially Juliette's difficulties of health and finances into the spring of 1897. Weeks before Juliette was to undergo a life-saving hysterectomy, Peirce wrote to Carus 1 April 1897 of the seriousness of the situation and of his fear of losing her:

> As I appear to be drawing to the end of a long period of misery, I examine myself to see if I have really learned all the lessons of an unusual experience. They have been impressive. The most delightful has been to discover all the virtues of my wife whose life is in such danger at this moment. If I lose her it seems as if every light within my soul must become extinguished. It is a question of days now. An advance for an article would make a great difference in the hopefulness of the case.

No such advance was possible for Carus at the time, though he tried to meet with Peirce in New York late April to discuss his work further. The two failed to cross paths and Peirce was no closer to progressing his logic.

Some relief finally came on 12 May 1897 when Carus informed Peirce of having learned from William James, one of Peirce's most loyal friends, of a series of lectures he was organizing for Peirce to deliver in Cambridge, MA. Whether or not this was the first Peirce had heard of this is undetermined, but by 30 May he wrote excitedly to James, "If I were to have such a charge, I ought to spend every moment from now till October in preparing the first part of my course. Therefore, I beg you to let me know what the prospects are."[28] Once the invitation became concrete, Peirce began dividing his attention between OCP and the Cambridge lectures on logic, as well as other pragmatic affairs that would arise over the next six years.

Preparations for Pragmatism: 1898–1904

Peirce delivered his series of eight Cambridge Conferences lectures on "Reasoning and the Logic of Things" February–March 1898, and received a glowing review from James who wrote to Carus 18 March 1898:

> Chas. Peirce has just been giving a course of lectures here, with great success—abstruse in parts, but always something popular and inspiring, and the whole thing leaving you with a sense that you had just been in the place where ideas are manufactured. They ought in some

27 See *EP* 2:xx for details.

28 Brent, 261–262.

> shape or other to get into print. I am sure there is stuff there for several more Monist articles, and I tip you this hint accordingly.[29]

Six months later, possibly influenced by Peirce's lectures,[30] James delivered before the Philosophical Union at Berkeley 26 August his famous "Philosophical Conceptions and Practical Results," wherein he introduced the term *pragmatism* and credited Peirce as its founder from their Cambridge Metaphysical Club days in the early 1870s.[31] This recognition elevated Peirce to a new prominence that would allow him to enter the soon-to-be international pragmatic debate, and elicited from him distinct views that would constitute his next *Monist* series of 1905–1906.

Meanwhile, between fall 1898 and January 1901, Peirce continued to submit material to OCP on logic, religion, and other miscellaneous topics that Carus, despite James's previous endorsement, found overly dense and unsuitable for general readers, explaining to Peirce that "although your contributions are valuable, from a scientific standpoint, they are not so from a business standpoint,"[32] and reminding him repeatedly to

> bear in mind that what we need mainly is popular articles.... You may not care for popularity, but if you had seen all the letters we received complaining about your articles on "The Logic of Relatives," you would understand that an editor has also to look out for his readers. Even your stanchest friends wrote me that the article did not do you justice.[33]

Whether consequentially or coincidentally, from 1901 to 1904 no letters between Peirce and OCP are found, with the exception of his unusual, incomplete correspondence with OCP compositor Fred Sigrist (see "Peirce and Other OCP Staff" correspondence). It is likely that Peirce simply maintained little contact with OCP during what was then a very busy and pivotal time in his career: he translated for the *Annual Report of the Board of Regents* of the Smithsonian Institution (1901), wrote definitions for Baldwin's *Dictionary of Philosophy* (1901–1902), submitted his elaborate grant proposal to the Carnegie Institute for "Memoirs on Minute Logic" (1902),[34] delivered his Harvard lectures on pragmatism and his Lowell In-

29 *RLT*, 36.
30 This is the belief of Ketner and Putnam in *RLT*, 36.
31 For a full account, see Fisch, 283–302; also Murray, 357–359.
32 Carus to Peirce 13 Oct 1898.
33 2 Dec 1898 and 3 April 1900. Peirce, however, considered his "Logic of Relatives" "the fruit of very great study & labor" that cost him "2 years over and above anything spent in work already published" to that point (Peirce to Carus 13 Aug 1896).
34 Ransdell, "MS L75," and partly in *NEM* 4:13–73.

stitute lectures on logic (1903),[35] and all along continued reviewing books for *The Nation*. His Harvard and Lowell Institute lectures were especially significant in the development of his philosophy and in his imminent return to the pages of *The Monist*. As interest in pragmatism grew from James's 1898 lecture, Peirce had become increasingly uncomfortable with the misrepresentations of his ideas (he being the declared founder) bundled with those of others, including James himself, F. C. S. Schiller, and John Dewey, and with the related movements that began to appear, such as practicalism, humanism, and instrumentalism. The program of his 1903 lectures—one that would guide him the rest of his life—was to establish more definitively the underpinnings and proof of pragmatism, to clarify his distinct conceptions of it, and to position it more as a proposition of logic than a matter of psychology, sociology, or other field to which it had been applied.[36] By the summer of 1904, Peirce was gearing up for a new series of *Monist* articles to further disentangle his pragmatic doctrines from the masses, to provide examples of its application, and to again take up the proof of it—this time adding to the growing list a new name for it, one hopefully "ugly enough to be safe from kidnappers": *pragmaticism*.[37]

Proof of Pragmatism and Some Amazing Mazes: 1904–1908

Peirce's pragmatism series appeared in *The Monist* April 1905 to October 1906, and consisted of three articles: "What Pragmatism Is," "Issues of Pragmaticism," and "Prolegomena to an Apology for Pragmaticism." Peirce's intention was to pro-

35 *EP* 2:133–241, also in full in Turrisi (1997); and *EP* 2:258–299.

36 See *EP* 2:xxv and Turrisi, 8, for more details.

37 "What Pragmatism Is," *Monist* (April 1905), p. 166. It should be noted that Peirce's intention with the new series was not to depreciate or disregard the work of other pragmatists, whose efforts he respected, but to point out what they "do not see," as he later explained to Carus 18 Oct 1907:

> As for W. James, he is one of my dearest friends, and I should dislike extremely to say *all* I think about his way of handling philosophy. Meantime he and Schiller & the Leonards man whose name just now escapes me, in the first place: have as clear a grasp of the leading ideas of pragmatism and of how they are to be applied *as they have of anything in the world*, and in my opinion are doing splendid work beneficial extremely to a wider public, and in the second place, they sincerely want to learn, and are learning notwithstanding certain difficulties under which they labor.... *[M]*y purpose has mainly been to insist upon points that other pragmatists do not see, and thus one does not perceive in what I write the warm sympathy I really have with their work.

duce a series of five or six articles, similar to his metaphysical series from 1891–1893, but the plan metamorphosed several times as he struggled to work out his proof and was ultimately never completed. Unfortunately the correspondence surrounding the series is mysteriously incomplete, especially from January 1905 to January 1907, which is composed almost exclusively of Carus's LPB copies. Nevertheless, the letters we do have during these four years provide valuable insight into Peirce's thought process and obstacles in writing the articles, as well as OCP's challenges and misgivings in editing them.[38]

When Peirce submitted his "What Pragmatism Is" to Carus ca. 12 September 1904, opening the series with an introduction to pragmaticism, it was to be followed by "The Consequences of Pragmaticism" and "The Evidences for Pragmaticism." By ca. May 1905, the second was retitled (if not altogether exchanged for)[39] "Issues of Pragmaticism," which treated pragmatism in semiotic terms and identified Peirce's forms of critical common-sensism and scholastic realism as consequences, or "issues," of pragmaticism. The third by this point was to proceed with Peirce's proof in the context of his theory of signs, six drafts of which he entitled "The Basis of Pragmaticism" (no longer "Evidences") but what Peirce submitted instead to Carus ca. 25 May 1906 was not the expected proof but a "Prolegomena to an Apology for Pragmaticism," an introduction to his system of logical graphs, or Existential Graphs, which he believed would more clearly facilitate the proof that was to appear in subsequent articles. Peirce's "Prolegomena"—and to some degree all of his pragmatism articles—was not well received by Carus, nor were Peirce's last-minute revisions. On 8 September 1906, roughly a week before the October issue was to be sent for printing, Carus wrote frantically to Peirce in light of his recent revisions:

> On my return home from a trip to Lake Michigan I find my office staff in consternation and despair concerning your corrections. The compositors are in rebellion, but I have insisted to let your changes be made as you wanted. I am sorry that there is no more time for reconsideration, nor can I myself give the necessary leisure to reading over and carefully considering your articles and the difficulties which they present. I will this time again follow the advice of our mutual friend, Mr. Holden of West Point to publish your article even though I myself feel rather reluctant in offering it to the public in the shape in which you have written it.... I fear that I shall be again in receipt of a number of letters accusing me of partiality in publishing unintelligible articles. I appreciate your significance in the development of science, and I will publish your article exactly as you want it, but I must confess that if I myself had

38 In addition to the evidence from the correspondence, the discussion of MSS in this section is based on *EP* 2:xxvii–xxix and 331–433; Bisanz (2009); and commentary in Robin's catalogue.
39 See Peirce and Carus May–June 1905.

> written it I would first give it a careful revision and render the diction of it more obvious to the reader.

In a missing reply ca. 15 September, Peirce ensured Carus of the value of his pragmatism writings and that he could "easily publish" them in "some pragmatist journal," to which Carus agreed "whether it would not be advisable for you to have them appear there."[40] As if following the suggestion, Peirce between the end of 1906 and spring 1907 began preparing a separate account of his proof of pragmaticism, entitled simply "Pragmatism," intended to be a "letter to the editor" for *The Nation* and later for *The Atlantic Monthly*.[41] The task was not done so "easily" after all and the heavily worked-over article was never published.

Peirce paused his pragmatism series around the summer of 1907, being "far too ill" (and probably a bit discouraged) to write anything more of substance for it,[42] and turned his attention instead to another *Monist* series on "Some Amazing Mazes" addressing four mathematical "curiosities" and other applications of Existential Graphs. This series only added to Carus's growing frustration and disappointment with Peirce's writings and their "typographical difficulties," and it too was cut short, with only the first three of the proposed five articles—"The First Curiosity," "Explanation of Curiosity the First," and "The Second Curiosity" (to be followed by the "Third" and "Fourth" curiosities)—appearing April 1908 to January 1909. Carus wrote again in frankness to Peirce 2 December 1907 regarding the obliqueness of Peirce's writing:

> While I wish to bring the best of your ideas before the public, and while I am anxious to let the world know what you have done, good and valuable work, I find myself greatly embarrassed by the objections which a great many readers offer to your writings, and I must confess that to some extent their complaints are justified. Though you have really something to say, and although your articles contain valuable propositions, you enwrap them unnecessarily in such difficulties, and sometimes switch off your ideas without explaining why you do so, that you tax the patience even of your most attentive readers. You actually write only for a few men, and even they have to work hard before they can dig out what you are really driving at.... You limit your public by frightening away your readers with your ponderous expressions of thoughts which could very easily be stated in clear language.

He reassured Peirce, however—as if hinting at what he would engage Peirce in next—that his ability to write well was "plainly shown in your former articles on logic in the Pop. Sc. M." (i.e., his 1877–1878 "Illustrations of the Logic of Science"

---

40 Carus to Peirce 17 Sept 1906.

41 Several portions in *EP* 2:398–433. Peirce's proof was again based on his theory of signs and likely related to what he intended for *The Monist* before pursuing instead his Existential Graphs.

42 Reported later in Peirce to Carus 12 Oct 1908.

series) and pled with Peirce to "continue in that same forcible style which is at once instructive and to the point." Perhaps taking Carus's remarks to heart, Peirce a few days later replied 7 December to a previously unattended letter from Russell (not found) and asked for Russell's help with the deficiencies of his first three pragmatism articles (which apparently Russell too had pointed out) as he began around this time to take back up the rest of the series:

> [O]ne purpose of your letter was to remonstrate with me about my style of exposition. I am just about to write out an explanation of the last sentence of my Monist paper of Oct. 1906. It is a very difficult task & I wish you would kindly revert to that and to the two others of April & Oct. 1905 and point out some of the worst passages & especially those in which, as you say, I have made too great leaps in my reasoning, in order that I may profit by the admonitions.

Peirce's plan for the two or three remaining pragmatism articles changed several times over the next few years, so they are difficult to determine with much certainty. For the fourth article, to follow "Prolegomena" in the series, Peirce composed various options, including "Phaneroscopy," "The First Part of an Apology for Pragmaticism," and "The Bed-Rock Beneath Pragmaticism," all following suit from "Prolegomena" to incorporate his Existential Graphs-based proof of pragmaticism instead of a semiotics-based proof from previous attempts. Although it is unclear what the fifth was to be, in the same 7 December letter to Russell above Peirce seems to have in mind "The First Part" for the fourth, leaving a fair chance that the fifth might have been "Apology for Pragmatism." This would also be in line with Robin's commentary to "The First Part" (R 296), which additionally notes a sixth article that Peirce at some point had in mind: "The fourth article was to begin the apology *["The First Part"]*, the fifth to have contained the main argument *["Apology"]*, and the sixth to have provided the subsidiary arguments and illustrations." By 12 October 1908, however, Peirce re-proposed the remaining articles to Carus in a way more closely resembling his Harvard Philosophy Club lectures on "Logical Methodeutic" and his "letter to the editor" for *The Nation* and *The Atlantic Monthly* on pragmatism from the spring and summer of 1907: "This year I am quite well & have the abstracts of the concluding papers of my series on Pragmatism fully elaborated and one of the articles itself nearly written, treating of logical analysis and definition and its *methodeutic.*" But Carus had lost patience with Peirce's pragmatic meanderings and mathematical recreations, and 9 December 1908 drew Peirce's attention back to the "old and cherished plan" to reprint Peirce's 1877–1878 "Illustrations of the Logic of Science" series from *The Popular Science Monthly*, a project that would dominate the remainder of their correspondence.

"Illustrations of the Logic of Science": 1909–1913

Carus's admiration for Peirce's "Illustrations" harked back to the beginning of their correspondence, and it was partly because of these papers that Carus initially invited Peirce to contribute to *The Monist*, as he noted in an initial conversation about the reprinting of the papers 30 January 1907:

> I think that your papers on logic in the Popular Science Monthly are the best introduction to the new movement in logic that have been written in the English tongue—perhaps in all tongues—and they are not sufficiently known among our teachers of logic, and have scarcely reached at all broader circles. It is the first I have read of you, and at the time they made an extraordinarily favorable impression on me so as to create a desire to be better acquainted with you and have articles of yours for The Monist.

Carus was introduced to the "Illustrations" ca. 1888 when he received from Russell—who through the same series also became attracted to Peirce and recognized him as "a master to be followed"[43]—a personally bound set of the articles, Russell later recalling to Peirce "the influence its perusal and study" had upon Carus.[44] Carus had encouraged Peirce to re-publish the series throughout their correspondence at least as early as 1893,[45] but in this 30 January letter Carus proposed more seriously to bring out the book with OCP and even to pay Peirce himself "entirely on my own hook," believing that "nothing will bring the importance of your best thoughts so well before the public as will these essays, to be printed as a book in handy form." A tentative agreement and revision plan seems to have been made,[46] but little more was done for the next two years while Peirce recovered from illness, worked on his "Amazing Mazes," and struggled with his remaining pragmatism articles.

Substantial work on the "Illustrations" project didn't begin until 7 January 1909, when Peirce, in response to Carus's 9 December 1908 reminder of the "old and cherished plan" to print the series, committed to revisit the papers if only to further the work on his remaining pragmatism articles and his proposed "System of Logic" book, upon which he was then earnestly engaged.[47] His "Logic" was to

43 Russell to Peirce 2 Jan 1889.

44 Russell to Peirce 22 Jan 1889. To this booklet (in SIU 91.40) Carus later appended a proposed Preface and table of contents for the (unsuccessful) republication of the papers in book form. See Carus to Peirce 23 Aug 1913 and the accompanying enclosure in Appendix.

45 See Hegeler to Peirce 27 Feb 1893.

46 Peirce to Carus 4 Feb 1907 proposes a tentative new title "Cream," noting that "it will be well understood in LaSalle that milk is one thing, cream quite another."

47 "System of Logic, Considered as Semiotic," or "Logic as the Theory of Signs in General" as reported to William James on Christmas 1909, addressing the classifications of Signs (Book 1), Critic

be his "chief work" that would be "addressed to a quite large class of intelligent, or somewhat intellectual men, and is intended to initiate them into my views of reasoning and to put them into condition to examine them critically & to form opinions of their own on the subject."[48] But given how much Peirce's thought had evolved since the "Illustrations" appeared over thirty years earlier, he explained that "they naturally require much revision" and proposed to provide the revised papers in three months' time. The agreement was made, but Peirce was never able to follow through. Three months turned into well over a year and after repeated requests for the promised revisions, and payments to Peirce now totaling $250 of Carus's own money, Carus wrote Peirce 30 April 1910 in deceit:

> You have promised again and again to send me the revised copy of your logical papers. I wish you would bear your promise in mind and send them at your earliest convenience.... Perhaps it would not be advisable to change too much because the papers are good as they stand, and the little shortcomings which they may have according to your present conception ... are perhaps not very important. Please let me hear from you, and if possible send me the corrected logical papers. You ought to do so not only to make good your promise but in your own interest as an author and a thinker.

And again 1 August 1910, having heard no reply:

> I have been waiting in vain for a review of your papers "Illustrations of the Logic of Science". You had even forgotten that I had paid you for them, and I am afraid you have again forgotten that after some reluctance you had acknowledged your promise.

Finally in a dense and revealing letter 26 Aug 1910 (and in an unsent draft of 19 July 1910), Peirce described his intense struggles in trying to revise the articles and acquiesces to Carus's suggestion to leave them as they are and simply add a preface stating the alterations he would have made:[49]

> I have written a great deal but am satisfied with but the smaller part of it, and the result is that I worked, until,—what with worries, too,—I was in a state of downright nervous exhaustion from which I have now been for a good number of months recovering, but owing to the decay of my powers and being so sick of writing about the same things, I cannot even yet write but a few equivalents of pages of this letter very slowly and for not over three hours and with many days when I can do nothing. Since I got your letter I have—often trying in vain to accelerate my speed,—the slowness of which is largely due to the labor of writing with the extreme precision required, gradually been forced to the conclusion that since you are very

---

and different kinds of reasoning (Book 2), and methods of research or "Methodeutic" (Book 3) (letter in *NEM* 3/2:867–875 and in part in *EP* 2:500–502).

48 Peirce to Carus 6 Jan 1909 (unsent draft) and 9 March 1909.

49 For a draft of Peirce's proposed preface, see Appendix, "Peirce and Carus."

> reasonably impatient, my best course is simply to write a preface in which I state in general terms how what I then say ought to be altered, and I will here (if my strength holds out) try to indicate the points I should make.

The remaining forty-seven pages of Peirce's letter detail the "principal positive error" of the "Illustrations" papers with which he had been relentlessly grappling, above all, their nominalism both in the portrayal of the general theory of inquiry in the first two papers (which he had attempted to rectify in his 1903 Harvard Lectures and his 1905–1906 *Monist* pragmatism series) and in the application of it to probability in the other four. Although Peirce would attempt still as he worked on his "System of Logic" to correct this "positive error" and mold his "Illustrations" around the realism he had embraced in roughly the last decade of his life, his efforts were futile and the results ultimately unsatisfactory to him.[50]

With Peirce's permission to use the articles as they stood, Carus and Russell agreed to move swiftly on the work with little revision, Russell to annotate, and without overly involving Peirce, who was now "old and might break down" and who would surely delay it further.[51] Even without Peirce, progress on the book remained slow and after three years of sparse correspondence,[52] Carus finally reached out to Peirce 23 August 1913 enclosing a preface of his own for Peirce's review.[53] In his response 28 August, Peirce made no mention of Carus's preface, although on the state of his own explained that he had in 1911 been further "puzzling how to get those Pop. Sci. papers into any decent shape" when ca. December of that year decided that rather than the preface proposed previously he would "add a new essay giving an outline of my more mature studies,"[54] but that he had been gravely hampered of any "close thinking" for the following year as he convalesced from a debilitating fall he suffered in his home. Peirce added sorrowfully:

> It is needless to say that we are on the verge of indigence, of course unable to keep a servant, my wife struggling against effects of overwork, and all the old friends dead, Stickney, Garrison, Alex Agassiz, and several others. As I am right in the midst of writing something that looks as if it might be worthy of your acceptance, I will postpone for a few days longer replying further to your letter.

---

50 See Fisch, 196, for more details on this struggle.

51 Carus to Russell 28 Nov 1910 (SIU 91.7).

52 See Chronology for some of Peirce's occupations during this time.

53 Draft included in Appendix, "Peirce and Carus."

54 Possibly related to his "A Sketch of Logical Critics" written the same year for Lady Welby, which discusses logical critics as "the theory of the kinds and degrees of assurance that can be afforded by the different ways of reasoning" and addresses faults in his first two "Illustrations" articles.

Sadly, no follow-up letters from Peirce survive, and Carus's response 10 September 1913 is the last of the correspondence. Carus welcomed Peirce's additions but reiterated, perhaps prophetically, that either way he did "not expect that the Open Court Publishing Company will make any money from your papers," but intended to present them as "an introduction to your more ponderous works" and as "propaganda for a sound study of logic."

Seven months later, Peirce died in his Milford home 19 April 1914 and the "Illustrations" book was never completed. Although the individual articles were republished in various posthumous editions of Peirce's writings—and proved to be highly valued logical "propaganda" indeed, as Carus anticipated—they were not treated at book-length until a century later by Cornelis de Waal (published by OCP), who brings to fruition Carus's long-awaited passion project and who chronicles in detail this chaotic episode of Peirce's "Illustrations" and their extensive revisions.[55]

Straws in the Wind: 1914–

Peirce's life and relationship with OCP ended with much philosophical work unconcluded. All five article series between *The Monist* and *The Open Court* were broken off short, both books under serious consideration by OCP (his Arithmetic and "Illustrations") were never completed, and many other book projects ("How to Reason," "New Elements of Geometry," "Elements of Mathematics," "Memoirs on Minute Logic," "System of Logic") were left unfinished or unpublished. His relentless and at times life-threatening personal struggles, some of which entirely self-inflicted, greatly encumbered what he hoped to achieve in his writings and career, especially in his later years. Yet Peirce endured with that characteristic "Peirce-istence" and "Peirce-everance" in pursuit of his elusive "METHOD of thinking" toward defining which he had dedicated much of his career.[56] This pursuit he described in a poignant self-examination in a letter to Russell 15 November 1904:

> I hold that a man of 65 well read in philosophy & a thinker himself must be a precious fool or be able to place himself better than anybody else can do, and I place myself somewhere about the real rank of Leibniz. Of course, Leibniz had the advantage of coming to a field into which no reapers had come. But what I want to say which is more practical, is that I am by nature most inaccurate, that I am quite exceptional for almost complete deficiency of imaginative power, & whatever I amount to is due to two things, first, a perseverance like that

55 de Waal, *Illustrations of the Logic of Science*, 2014.
56 Brent, 16.

> of a wasp in a bottle & 2nd to the happy accident that I early hit upon a METHOD of thinking, which any intelligent person could master, and which I am so far from having exhausted it that I leave it about where I found it,—a great reservoir from which ideas of a certain kind might be drawn for many generations. It is a pity that necessities have prevented my having a scholar to take up this method. From my point of view, it seems like awful stupidity & waste. But that is the way of things, they get done, but they get done in the least economical way imaginable.

What precisely the "METHOD" was is terrifically—perhaps necessarily—unclear,[57] but the by-products of his teasing it out, including a great lot of material produced for OCP, contain rich insights into an array of logical, mathematical, and metaphysical matters and hints of the underlying philosophical system toward which he was working. Carus very astutely recognized this in Peirce's writings early on. Two articles in to Peirce's *Monist* metaphysical series and with his third in galleys, Carus wrote to Peirce 31 May 1892:

> My impression is that we are working, perhaps not so much in the same line, as on lines converging to a common goal. Before I had special reasons for believing that, I felt it instinctively. And this instinctive belief originated from having found among your writings, certain hints, undeveloped ideas and suggestions, which might have meant little to others, but to me were indicative of the underlying tendencies of your philosophy. They were straws in the wind, which I observed before making my tests with the vane.

Very few appreciated the full merit of Peirce's philosophy during his life, and Carus, Russell, and (to an extent) Hegeler were among them. As with most minds ahead of their time, Peirce's real triumph came only after his passing—truly "the least economical way imaginable." Around Christmas 1914, eight months after Peirce's death, Juliette Peirce had arranged for Charles's papers, books, and letters—including much of this correspondence—to be donated to the Harvard Department of Philosophy, soon after which editions and organizations dedicated to Peirce's work began to appear. Now hailed by many as the "American Aristotle," Peirce has far outlived his initially neglected existence, and the straws scattered throughout his papers and letters have for the last century received their due recognition and thorough investigation, and continue to instigate much testing with the vane.

57 For various speculations, see Brent, 322–347.

2 Letters

2.1 Peirce and Carus

iPeirce's relationship with OCP editor Paul Carus was the most ificant of his OCP connections, rivaled only by his correspondence with Francis Russell, who introduced Carus and Peirce and who often served as their intermediary. Carus was more instrumental in promoting Peirce's philosophy than anyone else in Peirce's lifetime, publishing twenty of his articles between *The Monist* and *The Open Court* and eliciting much unpublished material as well. Their twenty-three years of correspondence serves as an extensive commentary on many of the manuscripts (and criticisms) exchanged between these two great thinkers, and on the strain that Peirce's idiosyncrasies and unpredictability inflicted on Carus and the OCP organization. Despite personal differences and occasional acrimony, Carus maintained an appreciation for Peirce's thought and was drawn repeatedly to two areas of his philosophy in particular: his tychism and logic of science. As detailed in the introduction, it was Peirce's 1877–1878 "Illustrations of the Logic of Science" series in which Carus initially saw great promise in Peirce and which he later went to great lengths to republish after suggesting such for years (though he never succeeded). Even more luring to Carus, if only because of how much he disagreed, was Peirce's anti-deterministic theory of tychism introduced in his 1892 "The Doctrine of Necessity Examined," which resulted in a year-long dispute between the two in *The Monist*.[1] Carus found unusual pleasure in this conflict and proposed several times throughout their correspondence to republish their five articles (two by Peirce and three by Carus) together in book form, to which Peirce never consented. He reassured Peirce that their refutations did not lessen his admiration for Peirce's "spirited temper and ingenious buoyancy," and that in fact his "appreciation should count the higher the more I disagree from your views."[2] "The more you attack me the more I like it," Carus later reiterated, "and I prefer by far to carry on a fight with you than with some incompetent small fellow, who has little or no scientific knowledge or method to back him."[3] Carus's tolerant and elevated perception of Peirce, at the time not shared by many nor much reciprocated by Peirce, brought

1 See Chronology and introduction ("Peirce and the Open Court"). More details about the dispute also footnoted in Peirce to Carus 3 April 1892.

2 Carus to Peirce 6 April 1892 and 1 July 1894.

3 Carus to Peirce 15 March 1909.

https://doi.org/10.1515/9783110768756-002

Peirce's work into a light that it might not have received otherwise and that undoubtedly played a role in Peirce's posthumous following.[4]

1890

Carus to Peirce. MS: RL 77. On OCP Stationery A.

Chicago July 2d 1890

Charles S. Peirce Esq.

Dear sir: The Open Court Publ Co intends to publish a new Quarterly[5] for which I have enlisted as contributors the very best names—among them Max Müller, Prof. Mach of Prague, Prof. Sigwart of Tübingen, Prof Romanes of London and others. I wish that in the new Quarterly our most prominent American authors should be represented and shall be greatly indebted to you for an article from your pen on "Modern Logic" or some similar topic—perhaps "Logic and Ethics." You may choose any theme with which you are engaged at present.

I shall be much obliged to you if you can send your MS. soon so that I can publish your essay in the first number.[6] If you will kindly let me know the price of your contribution, with which you may be pleased to favor us, I shall send remittance at once.

Respectfully yours,
P Carus

Peirce to Carus. MS: SIU 91.9.

Milford Pa. 1890 July 19

Dr. P. Carus
Dear Sir:

I shall be happy to write the article you propose for your new quarterly, and will entitle it "The Architecture of Theories." How will 4 000 words suit you, and when do you want it?

Yours very truly
C. S. Peirce

---

4 For more on the Peirce–Carus relationship, see Henderson, 125–141.

5 *The Monist*.

6 Peirce's maiden *Monist* article, "The Architecture of Theories," not submitted in time for the first number as Carus had hoped, and postponed to the Jan 1891 number (as shown in the letters that follow).

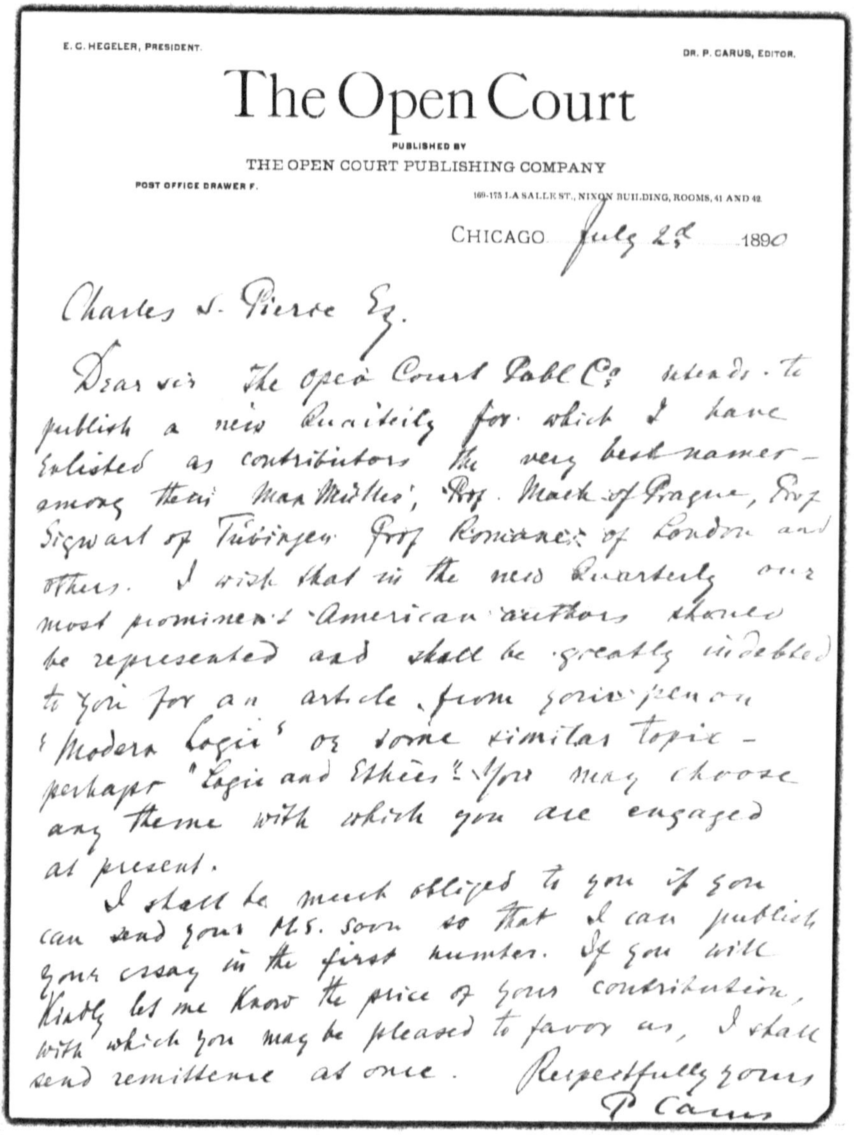

E. C. HEGELER, PRESIDENT. DR. P. CARUS, EDITOR.

The Open Court

PUBLISHED BY

THE OPEN COURT PUBLISHING COMPANY

POST OFFICE DRAWER F. 169-175 LA SALLE ST., NIXON BUILDING, ROOMS, 41 AND 42.

CHICAGO July 2d 1890

Charles S. Pierce Eq.

Dear Sir The Open Court Publ Co intends to publish a new Quarterly for which I have enlisted as contributors the very best names – among them Max Müller, Prof. Mach of Prague, Prof Sigwart of Tübingen Prof Romanes of London and others. I wish that in the new Quarterly our most prominent American authors should be represented and shall be greatly indebted to you for an article from your pen on 'Modern Logic' or some similar topic – perhaps "Logic and Ethics". You may choose any theme with which you are engaged at present.

I shall be much obliged to you if you can send your MS. soon so that I can publish your essay in the first number. If you will kindly let me know the price of your contribution, with which you may be pleased to favor us, I shall send remittence at once. Respectfully yours

P Carus

Figure 1: Carus to Peirce 2 July 1890, on OCP Stationery A. Invitation to contribute to *The Monist*.

Carus to Peirce. MS: RL 77. On OCP Stationery A. LPC: SIU 91.9.

Chicago July 22d 1890

Charles S. Peirce Esq.

Dear sir:

Your letter is at hand. At the same time I was informed by Judge Russell of your kind readiness to write for the Open Court. Both subjects—that proposed in your letter to the Judge as well as that proposed in your letter to me—shall be welcome.[7]

Judge Russell showed me your anonymous article on Mr Spencer. I take pleasure in mailing you my answer to Mr Spencer's criticism on Kant's Ethics.[8] I wish you would write an essay on Mr Spencer's philosophy. It is greatly needed—to open the eyes of people. Perhaps you or somebody else should write first an article as a letter similar to the one I saw, but betraying less knowledge of the subject. Your queries contain more than half of the answers—as was intended for that purpose, I suppose. I would publish a letter demanding information concerning the value of Spencer's philosophy and add in an editorial note, that I had solicited from Mr Charles S. Peirce an article on the subject and hope that he would respond to the request. What do you think of it. Would you be willing to undertake it?

Respectfully yours

PCarus.

Let me have the article for the Quarterly at your earliest convenience.

Peirce to Carus. MS: SIU 91.9.

Milford, Pa. 1890 Aug 30

Dr. Paul Carus

Chicago:

7 Peirce to Russell 3 July 1890 (the letter referenced) proposes not a separate but the same subject as that in his letter to Carus, namely an article ("Architecture of Theories") showing that "a philosophy is not a thing to be compiled item by item, promiscuously" but "should be constructed architectonically." Perhaps what confused Carus were Peirce's two (undetermined) enclosures with his letter to Russell, which were merely writing samples to give Russell an idea of "the kinds of things" he writes.

8 Likely Peirce's "Herbert Spencer's Philosophy. Is it Unscientific and Unsound?," *New York Times* (23 March 1890); and Carus's "Herbert Spencer on the Ethics of Kant," *Open Court* (16 Aug, concluded 23 Aug, 1888), a rejoinder to Spencer's "The Ethics of Kant," *Popular Science Monthly* (Aug 1888).

20 Let me have … convenience.] *This sentence down left margin.*

My dear Sir:

The illness of my wife has prevented my sending you before the Article[9] for your Quarterly which I now enclose. You did not limit me as to length; and I have found it impossible to express myself within the limits to which I proposed to restrict myself, although I have endeavored to be concise. The article has 5 600 words, and I will thank you to send me a check for $ 140.

Yours very truly
C. S. Peirce

Carus to Peirce. TS: RL 77. On OCP Stationery A. LPC: SIU 91.9. Misdated "August" for "September".

Chicago *[September]* 3d. 1890

Charles S. Peirce Esq.
Milford Pa.

Dear Sir:

Your MS. is at hand. I enclose check of $ 140. Whether I shall be able to put it in the first number I can not as yet decide, I fear it is too late for that, in which case I shall announce it for the second number.

I read your review in the Nation[10] and shall make a brief reply to the single points you raise, which will appear in No. 159.[11] Concerning the explanation of life, I refer you to my article "Feeling and Motion"[12] of which I send you a copy. Scattered remarks on the subject are also contained in the Montgomery controversy of recent date.[13]

---

9 "The Architecture of Theories."

10 Of Carus's *Fundamental Problems* (1889), *Nation*, 7 Aug 1890.

11 Carus's "The Unity of Truth," *Open Court* (11 Sept 1890), mentioned in the next paragraph. Among the "single points" raised in the review that Carus addresses is Peirce's argument beginning "Religion, to be true to itself, should demand the unconditional surrender of free thinking." Carus's "The Superscientific and Pure Reason" in the same number (pp. 2509–2511) also responds to Peirce's review, specifically to the paragraph beginning "The Philosophy it [the *Open Court*] advocates is superscientific."

12 *Open Court*, 31 July and 7 Aug ("Concluded."), 1890.

13 Referring to Edmund Montgomery, opponent of the monistic objectives of Hegeler and the *Open Court*, to his criticism in general at the time (not unlike Peirce's criticism of Carus's tolerance of mechanical or necessitarian evolutionary theory), and in particular to his controversial article "The Monism of 'The Open Court' Critically Examined" to appear *Open Court* 21 Aug 1890. (See Biographical Register for more information.)

Mr Hegeler discussed your objection to the conciliation of Religion in the last number,[14] and he would be glad to hear from you concerning this point. I myself shall follow with an article in No. 159 "The Unity of Truth." I confess that I believe you threw these remarks out more as seeds than to represent your opinions. They will set people thinking and point towards the same solution we propose, but I cannot believe that you stand on the standpoint you set forth in the Nation.

With kind regards,

Respectfully yours,
P. Carus
B. C. D.[15]

Peirce to Carus. MS: SIU 91.9.

Milford 1890 Sep 6.

My dear Sir:

Yours dated "August 3'[d]" with check has come safely to hand.

You should not attribute anonymous articles to me, as you don't know what editorial liberties may have been taken with them. As you justly say a brief article in a weekly paper can do little more than scatter suggestions. But I have always declined to identify myself with associations for "conciliating religion with science." Best let them work out their own roads to truth. It is not likely I shall be drawn into any further discussion of this question which being one of wisdom has to be settled by the slow peristalsis of the mind. I shall read with attention your articles and that of Mr. Hegeler.

I have been looking over your article on Kant's Evolutionism.[16] I myself came into philosophy by the road of Kant, of whom I was long an enthusiastic student, & for whom I retain a certain veneration.

14 "Religion and Science," *Open Court* (28 Aug 1890). In response to Peirce's *Nation* review of Carus's *Fundamentals*, primarily (even quotes in full) Peirce's paragraph beginning, "The profession of the *Open Court* is to make an 'effort to conciliate religion with science.' Is this wise? Is it not an effort to reach a foredetermined conclusion? And is not *that* an anti-scientific, anti-philosophical aim?..." See also Peirce's "Notes" in *The Nation*, 23 Oct 1890, criticizing the first number of *The Monist* on similar grounds.

15 B. C. Diesterweg, OCP Chicago office secretary. See Biographical Register.

16 "Kant on Evolution," *Open Court* (4 Sept 1890).

5 same] *typed above*

22 Mr.] *over* Dr.

Your leaded quotation from the Critic of the Pure Reason,[17]—a celebrated passage,—appears to me to have no bearing on the question, being merely a point of logic.

Nor can I admit that Kant either in the passage about the baby's cry or anywhere else anticipates the Darwinian idea in the least.

I have heard too much of Kant's being hard reading. I think he is one of the easiest of philosophers; for he generally knows what he wants to say, which is more than half the battle, and he says it in terms which are very clear. Of course, it is quite absurd to try to read Kant without preliminary studies of Leibnizian & English philosophers, as well as of the terminology of which Kant's is a modification or transmogrification. But there is a way of making out what he meant, while such writers as Hume and J. S. Mill, the more you study them the more they puzzle you.

I wish you would not talk about the "Critique of Pure Reason"; first because the proper English word is *critic* not *critique* (we having two forms where the German has only one) and second because the names of faculties take the definite article. *Reason* is one thing in English, and *the reason*, that is the reasoning faculty, is quite another. It is absurd to talk of criticizing *Reason*, for *reason* means conformity with truth as far as we can see it.

What you say of men's attitude of hostility to those whose views are very nearly their own is quite true. Your views of ethics, by the way, seem to me a good deal nearer those of Spencer than to those of Kant.

Kant was splendid;—but after all he lived a hundred years ago,—and what a hundred years it has been!

Yours faithfully
CSPeirce

Dr Paul Carus
Editor Open Court
Chicago.

Carus to Peirce. TS: RL 77. On OCP Stationery A. LPC: SIU 91.9.

Chicago Sept. 23rd. 1890

17 Likely for Critic "of Judgement," as in Kant's *Critique of Judgement* (1781), which is what Carus quotes and analyzes extensively in his article, and what Carus addresses in the letter that follows.

19 men's] *after* ~~people's~~

Charles S. Peirce Esq.
Milford, Pa.

Dear Sir: Your letter was received some time ago, and I read your remarks with great interest. The leaded passage is a point of logic, nevertheless it is of great importance with reference to the evolution idea. It proves that Kant was not *a priori* averse to the idea of evolution, as Kant directly states. He says: "*A priori* this does not contradict the judgment of pure reason".[18] It seems to me that we would have to give an artificial interpretation to Kant's words, if we maintain that he "nowhere anticipated the Darwinian idea in the least." (You probably mean by "Darwinian" Evolution.)

I agree with you that Kant is not hard reading. Kant is hard reading [as I said p. 2492[19]] to those who attempt to read his works without taking into consideration the historical development of his philosophy, and, in addition, most English scholars are hampered by bad translations. My favorite work is his *Prolegomena.*[20] It is a small pamphlet yet it contains the trend of his philosophy and is the best introduction to his "Critique of Pure Reason."

Your distinction between <u>the</u> *reason* and *reason* is fully justified by logic, and I wish you would introduce it into the English language. But you should not recommend it to me who had trouble in unlearning the German custom. We say in German *Kritik der reinen Vernunft*—not *reiner Vernunft.* I consider the use of the article as logical, and the dropping of the article as a loose habit. But after all it is a habit. I have been told, and I find it true, that the omission of the article is idiomatic. *Usus est tyrannus.* I searched the translations and the Encyclopædia Britannica, yet all translators and the Encyclopædia Britannica agree in translating *Critique of Pure Reason.* I look upon this custom as a similar idiomatic expression as "Man" for mankind. In German we say "*Der* Mensch". Goethe says "Es irrt *der* Mensch so lang er strebt"[21] and the Greek also say ὁ ἄνθρωπος in this sense.

18 *Critique of Judgement*, section 80, footnote 120; entire (translated) note quoted by Carus in his article at p. 2495.
19 Paragraph beginning, "It is, however, fair to state that these misunderstandings appear excusable if the difficulties are borne in mind with which the English student of Kant is confronted."
20 *Prolegomena to Any Future Metaphysics* (1783); translated by Carus 1902, reviewed by Peirce in *The Nation* 18 June 1903.
21 *Faust* (1808), Part 1, Scene 3: "Man errs so long as he strives."

9–10 (You … Evolution.)] *parentheses inserted in-line*

24 Encyclopædia Britannica] Encyclopǣdia Bri˅t˅tan~~n~~ica *(both instances) [E]*

26 Mensch] *m*<u>M</u>ensch *(here and in next two instances)*

To say ἄνθρωπος or Mensch instead of ὁ ἄνθρωπος or Der Mensch would have been considered not only ungrammatical but also illogical according to the logical lessons I received. Similar are my objections to *critic* instead of *critique. Critic* is commonly employed for the German "Kritiker" and *critique* for "Kritik." Webster quotes Lock sub voce "critic" and refers to "critique," defining this meaning of "critic" as "art of criticism". The meaning "critical examination" is confined to *Critique* alone. The Century dictionary is the only authority (and if I am not mistaken the item is written by yourself) which mentions that *critic* should be used, as proposed by Hodgson, in the sense of "the science of criticism" similarly as *logic* and *arithmetic* signify sciences.[22] Yet the same work viz., the Century Dictionary, calls sub voce *critique* Kant's work, *Critique of Pure Reason.* I have no objection to your usage "Critic of the Pure Reason," nay, more so, I like your way, but I shall not give up the customary way until yours has become generally accepted.

You as a logician will think that these distinctions are no trifles; and I think so myself. I have given much valuable time to philological studies, and I know the importance of words and their usage. But I find it often necessary to neglect things for the sake of other things. The things neglected are perhaps no trifles in themselves, but they are trifles when the purpose of the enquiry has nothing to do with them.

My policy is best explained by another instance. I have been requested to introduce the phonetic spelling. Of course I consider the customary spelling as wrong in principle and inconsistent, even from the historical point of view. A correct and practical spelling is of highest importance. I should recommend to introduce a new and better system, but I shall not introduce it first in The Open Court. I do not propose to trifle with attempts at spelling reforms, but I shall not deviate from the present custom until the custom has changed all over the English speaking world. I know for instance that the *l* in *could* is utterly nonsensical and yet I retain it, simply because it is custom, and it would lead my readers astray if I did not follow custom in these things which, because they are not the present object of investigation, must be considered as trifles.

22 *Century Dictionary* (1889), p. 1355, third usage of *critic* as "the art or science of criticism"; reference to Shadworth Hodgson, *The Philosophy of Reflection* (1878), vol. 1, p. 17, quoted in the entry, beginning: "Kant had introduced *Critic*, name and thing: it was a branch of analysis, like Logic,…"

5–6 and refers … of "critic"] *inserted above*
10 Dictionary,] *inserted above*
11–12 objection … Reason,"] objection ~~of~~ to your ~~your~~ usage ˅"˅Critic of Pure Reason˅"˅
14 think[1]] k *inserted in-line*

I read your article[23] and wish that it had been in time to find room in the first number of our new quarterly, but it will be welcome for the second. Should you like to have the proof-sheets sent to some critical mind to write a few comments upon your propositions? Your ideas are new, and it might give some additional interest if they were contrasted with some other ideas. I differ from you as you perhaps may surmise yourself. Nevertheless your propositions have excited not only my keenest interest, but also my admiration. The way you propose them is splendid. I feel tempted myself to write an article on your essay.

With kind regards,

Respectfully yours,
P Carus

Peirce to Carus. MS: SIU 91.9.

Milford, Pa 1890 Sep 25

My dear Sir:

Yours of the 23’$^{\underline{d}}$inst. settles a doubt I had as to whether you were born a German. You write English so much better than many men whose native tongue it is,—and if I may venture to say so, so much better than many Germans write German,—that I was puzzled. It is curious we use no article with *man* when we describe characters of the species; for we always speak of *the* negro, *the* student, *the* horse, etc. But English has been so much influenced by medieval latin, and phrases about *homo* are so familiar, that it may be that is the reason. Certainly, we never could tolerate “*the* man.” But as for “*the* reason,” though it is not idiomatic, I do not think it an unpleasant or unintelligible expression. We almost always speak of *the* understanding, *the* judgment, *the* imagination, *the* heart, etc. I do not see why “*the* reason” should be objected to. “The pure reason” still less.

As for *critic*, if I mistake not, it was a new word in German in the sense of critical science when Kant used it; while in English it is as old as Locke—I inserted *critique* in this sense in the Century Dictionary to record the fact of its use—God forbid I should *approve* of above 1/10 of what I insert.

Thank you for your cordial appreciation of my essay. Of course, I should be glad of any discussion of my ideas, though I hardly expect them to find much

23 “The Architecture of Theories.”

8 I feel … essay.] *inserted in-line*
13 Pa] *inserted above*
22 tolerate] *above* ~~support~~

favor, at first. Whatever you see fit to do in order to bring them into the arena of thought, I shall approve of.

Yours very truly
C. S. Peirce

Dr. P. Carus—

P.S. By the Darwinian idea I do not at all mean the general notion of evolution. That was of course familiar to Kant and to the ancients. But I mean by the Darwinian idea, the conception of accidental small variations, combined with destruction of forms unable to maintain themselves, as the great factors of evolution.

Carus to Peirce. LPC (MS): SIU 91.10.

December 22 0

Charles S. Peirce Esq.

Dear sir:

In case you have another article for a future Number of the Monist, I shall be glad to hear from you again.—I gave order to send you an advance sheet of my article[24] in which I referred to your views set forth in your essay The Architecture of Theories. You will have the second number of the Monist, containing your and my article in a few days.

Respectfully yours
P Carus

1891

Peirce to Carus. MS: SIU 91.10. Stationery headed, "Manhattan Club. | Fifth Avenue."

1891 Jan 12

My dear Doctor Carus:

I beg you to pardon my silence which is due to my being so pressed with work that I could not find time to draw up a list of persons to whom I would like the Monist No 2 sent.

24 "The Criterion of Truth," *Monist* (Jan 1891), addressing Peirce at pp. 241–244.

I should be mad to expect or *desire* people to fall in with my views at once. They are the fruit of long studies & I have held them in reserve,—studying them, testing them, developing them,—for 12 years before venturing before the public with them. If I myself hesitated so long, I cannot ask others to be precipitate in accepting them. I like your article very much. I wonder anew at your graceful English. I don't think you are very far from me & am hopeful of counting you for an ally before very long. As for a *chance*, without any degree of conformity to law, I regard that as nonexistence, a mere germ of being in so far as it may acquire habits. My 1878 article[25] is in no conflict with my present views, which were then in embrion.

I thank you for the invitation to send another article which I shall do as soon as I can get time.

May I trouble you to have copies sent (at my expense—Please send bill) to the following.[26] I will send a further list soon.

Clarence King Esq
Century Club
W. 43'$^{\underline{d}}$St. New York

Century Club.
W 43'$^{\underline{d}}$St. New York

University Club.
E 26$^{\underline{h}}$St & Madison Av
New York

Player's Club.
Gramercy Park
New York

Prof Simon Newcomb LLD
Sup't Nautical Almanac
Washington D.C.

25 Likely "The Doctrine of Chances" from his 1877–1878 "Illustrations" series.
26 See Biographical Register.

7 law] *over* ru

Prof S. P. Langley LL.D.
Smithsonian Institution
Washington D. C.

Prof. T. C. Mendenhall LL.D.
Coast and Geodetic Survey
Washington D.C.

Professor O. N. Rood
Columbia College
E 49$^{\underline{h}}$St & Madison Av
New York

Mrs. Dr. Mary Putnam Jacobi
110 W 34$^{\underline{h}}$St
New York

Prof. O. H. Mitchell
Marietta, Ohio

Prof W. James M.D.
Cambridge Mass

Prof James Mills Peirce
Cambridge Mass

Prof. Joseph Jastrow
Madison, Wisconsin

Professor Dr. Schröder
Karlsruhe Baden
Germany

Monsieur Faye
Membre de l'Institut

---

7 N] *over* C
12 110] *after* ~~E~~
14 Prof. O. H. Mitchell] O. C. Mitchell *[E]; below* ~~Dr A Jacobi~~
23 Karlsruhe] Carlsruhe *[E]*
25 Monsieur] *above* ~~Professor~~

Paris

Professor A. Cayley
Cambridge
England

Professor J. J. Sylvester
University College
Oxford
England

Professor G. G. Stokes PFRS M.P.
Cambridge

A. B. Kempe Esq F. R. S.
Royal Society
Burlington House
Piccadilly London W
England

Professor Marquand
Princeton University
New Jersey

Yours very truly
C. S. Peirce

Carus to Peirce. LPC: SIU 91.10.

Jan. 22d. 1

Charles S. Peirce Esq.
Milford, Pa.
Dear Sir:

Thanks for your letter. I handed the addresses to our clerk who will attend to the matter. I am glad to learn that you will write again for the Monist. Prof. Mach

---

10 Cambridge] *before inserted* Eng. *by OCP*
17 Princeton University] *below* ~~Fran~~
27 the matter] *above* ~~it~~
27 for the Monist] *inserted above*

mentions your article in a communication to me which I intend to publish in the next number.[27]

With kind regards,

Yours respectfully,
P Carus

Peirce to Carus. MS: SIU 91.10. Stationery headed, "Arisbe,| Milford, Pa.". "See over" written by Carus in red pencil in top-right corner, corresponding to this note on the verso of page 1 to M. A. Sacksteder (OCP office manager) regarding Peirce's advertisement mentioned in the letter: "to Mr Sacksteder! I shall leave this point to be answered by you. ~~te Please answer this point.~~ State either our old terms or the terms of the Nation. Let me just see your letter! and state that we make an exception for his sake." See OCP (Sacksteder) to Peirce 15 Nov 1891 addressing this.

1891 Nov 5

My dear Sir:

I enclose herewith an article[28] for The Monist. I consider it the strongest piece of argumentation I have ever done. If accepted, kindly oblige me with a check for the enclosed bill.

In regard to an outrageous thing in the Nation on your Soul of Man,[29] you must know that I have been so busy all this year that I have written scarcely anything. I did send to the Nation to ask them to send me your book, but was too late. It had already been given to somebody else. When the notice appeared I was incensed; but what could I do. I set it down as a gross error of judgment. But recently, I must tell you, a number of little circumstances have led me to suspect it to be a malicious performance. I shall do all in my power to ascertain this, and to punish it.

Would you consent to printing weekly in the Open Court an advertisement of my Instruction in the Art of Reasoning by Correspondence?[30] If so, please let me know what the terms would be.

---

27 Mach, "Some Questions of Psycho-Physics," *Monist* (April 1891); last paragraph at p. 399 discusses Peirce's "Architecture of Theories."

28 "The Doctrine of Necessity Examined," to appear in *Monist* April 1892.

29 Likely the "Recent Philosophical Works" notes in the 25 June 1891 issue, which says briefly of Carus's *Soul of Man* (1891): "The volume by Dr. Carus is remarkable only for 152 illustrations and diagrams; not exactly of the soul, but of everything except that. Such an amount of ill-digested material is seldom put in book form for the philosophic public."

30 Advertisement in *Open Court*, 1 Sept 1892. A similar course attempted 1887 with advertisements distributed in other magazines. See *W* 6:10–60 for circulars and related material.

22–23 suspect it to be] *above* ~~suppose it~~

Yours very truly
C. S. Peirce

Carus to Peirce. LPC (MS): SIU 91.10.

La Salle, Ill. | Nov. 7$^{\underline{th}}$ 91

Charles S. Peirce Esq.

My dear sir: I shall be most happy to use your article and expect to have it appear in the April N$\underline{o}$ of 92. I read the greatest part of your arguments and am much interested in your position but have not as yet been able to give your ideas due consideration.

Yours very truly,
P Carus

Please find enclosed check of $ 160$^{\underline{00}}$

Peirce to Carus. MS: SIU 91.10. Stationery headed, "The Century | 7 West Forty-Third St."

New York | 1891 Nov 14

My dear Sir:

I received your receipt of my registered package, but nothing else.[31] I should like to know whether you are going to use my article, and if not I will send stamps for its return.

I should also like to have a reply to my inquiry about advertisements in The Open Court.

Yours very truly
C. S. Peirce

Please address me here.

Peirce to Carus. MS: SIU 91.11. Stationery headed "The Players | 16 Gramercy Park". Undated, likely sent soon after receiving permission to advertise from OCP 15 Nov 1891.

New York *[ca. 18 Nov 1891]*

My dear Dr. Carus:

31 It is unclear why Peirce says this since Peirce to OCP 8 Nov 1891 acknowledges received check for $ 160.

Please cause the enclosed advertisement to be inserted in the Open Court weekly on the terms given.

My address will be as above for a short time.

Yours very truly
C. S. Peirce

Thanks for cheque.

1892

Peirce to Carus. MS: SIU 91.11. Stationery headed, "Century Club | 7 West Forty-Third St."

New York 1892 Feb 8

My dear Dr. Carus:

My address will be here till further notice.

Yours very faithfully
CSPeirce

Peirce to Carus. MS: SIU 91.11. Stationery headed "The Century | 7 West Forty-Third St." Note by Carus on verso reads, "Send *12* copies Monist | if he sends addresses | we had better make no changes".

Dr Carus

New York, 1892 Apr 3

My dear Sir:

I suppose my article[32] is out in this month's Monist. Will you kindly send me *six* copies to the above address. I shall soon be sending you another article.

32 "The Doctrine of Necessity Examined." In the same number Carus included the notice "Mr. Charles S. Peirce on Necessity" (p. 442), which briefly (though sharply) distinguished the results of Peirce's tychistic investigations as standing "in a strong contrast" to Carus's deterministic or necessitarian beliefs, and set the stage for the ongoing Necessity–Tychism controversy between the two that would persist in the months (and years) to come. For details about this debate, see the general Introduction (especially at p. lxxiii) to *W* 8, individual headnotes to Peirce's "*Monist* Metaphysical Series" articles in *W* 8 and *EP* 1, and Corrington, *An Introduction to C. S. Peirce* (1993), pp. 167–198.

1 be] by *[E]*

6 cheque] checque *[E]*

Yours very faithfully
C. S. Peirce

Carus to Peirce. LPC (MS): SIU 91.11.

April 6th 92

C. S. Peirce Esq.

My dear sir: Your article will be welcome! Indeed I shall be exceedingly glad to have another article of your pen. I can not help admiring the vigor of your thought and my appreciation should count the higher the more I disagree from your views. I have just finished an article in reply to your onslaught on the doctrine of necessity and shall send you advance proofs as soon as I can.[33] Another article of mine on the same subject will follow in the Fall number.[34]

I shall send you copies of the last Monist, and at the same time send as many copies with your card placed inside to as many addresses as you desire. Please do not trouble with paying for them. We shall be glad to be of service to you.

With kind regards
Yours very truly
P Carus

Peirce to Carus. MS and TS copy: RL 77. Undated. A note on the MS estimates ca. 20 July 1890, positioning this as a response to Carus's initial invitation of 2 July 1890 to write for *The Monist*. However, as explained in *W* 8:lxxvi–lxxvii, the letter was more likely written late April 1892, shortly after a spiritual conversion of sorts Peirce had 24 April 1892, recounted in a letter that day to ecclesiastical correspondent and friend Reverend John Wesley Brown (see *W* 8:lxxvi). "The transformational power of the religious feelings Peirce had begun to experience," *W* 8 notes, was also manifest in his subsequent *Monist* articles, "The Law of Mind" and "Man's Glassy Essence," upon which Peirce was working around this time (see Peirce to Carus 8 May 1892).

[End of April 1892]

Dr. Paul Carus
Dear Sir:

I think I could write an acceptable article or even two for the Open Court on the following materials

33 "Mr. Charles S. Peirce's Onslaught of the Doctrine of Necessity," *Monist* (July 1892).
34 "The Idea of Necessity, Its Basis and Its Scope," *Monist* (October 1892).

1. Personal experience has a positive value always. This is greater the more unusual the experience. That which I have to report seems worth mention.

(A.) Of late years I have suffered extreme adversity & affliction, being

(B.) In the somewhat unusual situation of a student of philosophy, laboratory-bred, who holds on essentially to the creed and communion of the church.

(C.) Now, the facts which seem worth reporting are $1^{\underline{st}}$, what kind of reflections I have found really consoling, $2^{\underline{nd}}$, how the different literary works addressed to those in such circumstances sound.

2. Of course, it is nothing but the experience of a single individual; still, it is out of individual experiences that general experience is built. But I wish to say clearly that a single case can have, until verified & supported by others, no importance at all. Still, I write as a means of collecting other testimony.

Under the $1^{\underline{st}}$ head, my experience would, if generally borne out by others, go to support the law of continuity. For first, I find, ideas about heaven of very little or no support, evidently because that life is completely cut off from this. Myself in a life the whole aim, motives, means, problems, of which radically differ from those of this life, does not seem to come within my special interests.

Second, I have found immense help from certain other reflections. Such as this. "If," I would say to myself, "by voluntarily enduring what I am forced to bear I could further certain objects I had at heart, would I not do so and more? And if I could comprehend the purposes of God, would I not give an absolute preference to those purposes over the objects I actually have at heart,—which indeed I only now prefer as being as near as I can make out the objects it is God's will I should pursue? Since then God is doubtless using me, so far as I can be of use, to promote his own purposes, should I not be content?

Why should I not feel particularly honored that I have been selected to undergo all this agony?

Carus to Peirce. TS: RL 77. On OCP Stationery B. LPC: SIU 91.11. MS (draft): SIU 91.11, an initial hand-written draft on scratch envelope. Text comes from final TS. Textual variants and emendations from MS draft are indicated in the textual notes.

---

1 positive] *after* ~~valu~~
2 worth] *after* ~~rather~~
5 essentially] *after* ~~as if~~
14 go] go$^\vee$es$^\vee$ *[E]*
18 this] *l*this
20 by] be *[E]*
21 not] *d*not
24 make] *after* ~~t~~
24 should] shou*d*ld
26 purposes,] purposes which *[E]*

Chicago, April 27, 1892.

Mr. Charles S. Peirce, Esq.
My dear Sir:

We intend to publish a translation of Prof. Ernst Mach's book, "*Die Geschichte der Mechanik*".[35] The translation has been done by Mr. McCormack, who has shown himself competent as a translator on other occasions. Although the translation is made with great care, I should nevertheless like someone who is an authority in this province, or at least in similar fields of investigation, to look over the proof-sheets before they go to press. Of course *I do not mean* that the proof reading should be done by him, but only a revision of the whole, so as to assure us that the translation is good. Would you be kind enough to undertake the work, and if so, what are your charges?

With kindest regards,

Yours truly,
P. Carus

Peirce to Carus. MS: SIU 91.11.

12 W 39$^{\underline{\text{h}}}$ St. New York | 1892 May 8

My dear Dr. Carus

Owing to absence, I only just receive yours of Apr. 27. I shall take pleasure in doing the work you speak of. I will leave the matter of compensation to you. I shall, however, need a copy of the original.

My next two articles[36] for the Monist, upon which I am now engaged, are to be, I think, the most valuable things I have done. Thanks for your kind offer to

---

35 *The Science of Mechanics* (1893), from *Die Mechanik In Ihrer Entwickelung*, second edition (1889). For details about Peirce's work on this edition, see the Peirce and McCormack correspondence.

36 "Law of Mind" and "Man's Glassy Essence," *Monist* (July and October 1892). The former was soon after criticized by George McCrie in "The Issues of 'Synechism'," *Monist* (April 1893).

---

4 Ernst] Er*a*ns*s*t *[TS]*
4 *Geschichte*] *Er*Geschichte *[TS]*
5 *Mechanik*] *before* ~~and have almost~~ *[MS]*
5 done] made ~~with great care~~ *[MS]*
6 occasions.] occasions*,*. ~~and~~ *[MS]*
7–8 someone … investigation,] some ~~authority~~ one who is an authority in this \province, or at least in similar/ field$^{\vee}$s$^{\vee}$ $^{\vee}$of$^{\vee}$ \investigation,/ *[MS]*
9 Of course] *inserted above [MS]*
10 of the whole,] *after* ~~to judge~~ *[MS]*
11 that the] ~~of~~ that ~~pr~~ the *[MS]*

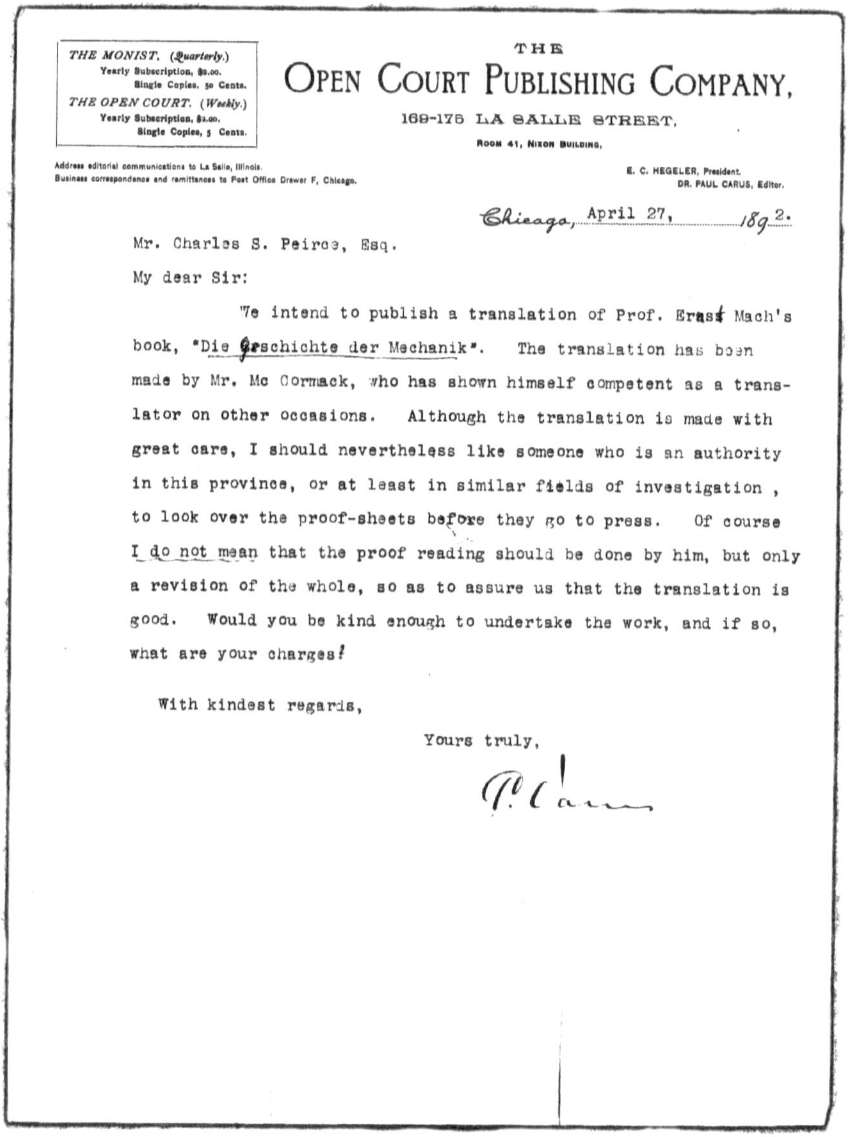

THE MONIST. (Quarterly.)
Yearly Subscription, $2.00.
Single Copies, 50 Cents.
THE OPEN COURT. (Weekly.)
Yearly Subscription, $2.00.
Single Copies, 5 Cents.

THE
OPEN COURT PUBLISHING COMPANY,
169-175 LA SALLE STREET,
ROOM 41, NIXON BUILDING,

Address editorial communications to La Salle, Illinois.
Business correspondence and remittances to Post Office Drawer F, Chicago.

E. C. HEGELER, President.
DR. PAUL CARUS, Editor.

Chicago, April 27, 1892.

Mr. Charles S. Peirce, Esq.

My dear Sir:

We intend to publish a translation of Prof. Ernst Mach's book, "Die Geschichte der Mechanik". The translation has been made by Mr. Mc Cormack, who has shown himself competent as a translator on other occasions. Although the translation is made with great care, I should nevertheless like someone who is an authority in this province, or at least in similar fields of investigation, to look over the proof-sheets before they go to press. Of course I do not mean that the proof reading should be done by him, but only a revision of the whole, so as to assure us that the translation is good. Would you be kind enough to undertake the work, and if so, what are your charges?

With kindest regards,

Yours truly,

P. Carus

Figure 2: Carus to Peirce 27 April 1892, on OCP Stationery B. Invitation to help with Mach translation.

send copies. May I name the following persons to whom I would like the last number sent?

Herbert Spencer,—Johnson who is writing on Logic in Mind,[37] Prof W. James, John Fiske Esq, Prof. Josiah Royce, Rev. F. E. Abbot all of Cambridge, Prof. J. E. Oliver, Ithaca, N.Y., Mrs. Prof. Fabyan Franklin, Johns Hopkins University, Baltimore, Md, Mrs. Mary Putnam Jacobi M.D. 110 W 34$^{\underline{h}}$ St, New York, Albert Stickney Esq 128 W 59$^{\underline{h}}$ St, N.Y.

I have written a story entitled "An Excursion into Thessaly: a Tale."[38] It is interesting and pretty, largely descriptive, and intended to be read aloud. If I could get an audience in Chicago large enough to pay expenses, I should like to go and read it there, and so have an opportunity of meeting you.

Yours very truly,
C. S. Peirce
12 W 39$^{\underline{h}}$ St. New York.

Carus to Peirce. TS: RL 77. On OCP Stationery B. LPC: SIU 91.11.

Chicago, May 10, 1892.

Charles S. Peirce, Esq.,
Dear Sir:

I shall have to reprint the article, Kant, on Evolution, which appeared about a year ago in the OPEN COURT.[39] I had, at that time, a letter from you in which you stated that you had heard tootoo much about the hardness of reading Kant, and you protested against it.[40] I do not quite remember the exact words, but I have the letter on file in Chicago and shall look the passage up. I endorse your view most heartily, and desire to quote your words in a foot-note. I hope that you will make no objection to my doing so.

Before I shall publish the article and said foot-note, proofs will be sent to you.

With kind regards,

Yours very truly,
P. Carus

37 William E. Johnson, "The Logical Calculus," *Mind* (Jan, April, and July 1892). See Biographical Register.
38 Later titled "Embroidered Thessaly," *W* 8:296–340.
39 "Kant on Evolution," *Open Court* (4 Sept 1890), reprinted in *Monist* (July 1892).
40 Peirce to Carus 6 Sept 1890.

17 Charles] *after* ~~Mr.~~ 21 too] *over* so

Carus to Peirce. LPC: SIU 91.11.

May 14, 2.

Mr. Charles S. Peirce,
My dear Sir:

Please find enclosed the foot-note containing the quotation from your letter.[41] You are, of course, welcome to modify the style as you please, but I hope you will not alter anything in it, for I like the remark exactly as it stands.

Concerning your lecture, Mr. Francis C. Russell wrote you from our office in Chicago.[42] He can better attend to the matter than I. He has more connections in Chicago. Moreover, he is constantly there, while I stay in La Salle, a place one hundred miles west of Chicago. I hope you have answered the questions in Mr. Russell's letter; and Mr. Russell is by this time, busily engaged in preparing the way for you, and I need not add that I shall be very much pleased to meet you personally. I am sorry that Mr. Hegeler is at present not at home. He is out West and will not return for many weeks and then only to return after a few days stay.

I remain,

Very truly yours,
P. Carus

Peirce to Carus. MS: SIU 91.11. Undated. Message written on the page proof (below the proposed text to be quoted) enclosed in Carus's of 14 May.

[ca. May 16, 1892]

Dear Doctor Carus:

I have no objection to the above being quoted. My parturition of the next two articles for the Monist on the nature of mind is not yet complete & I have been suffering torments with it.

very faithfully
CSPeirce

41 "The proposition that Kant is no easy reading found an unexpected and strong opposition. Immediately after the publication of this article, Sept. 4th, 1890, Mr. Charles S. Peirce made the following incidental remark in a letter to the author dated Sept. 6th, 1890:..." Carus then quotes the paragraph of Peirce's letter beginning, "I have heard too much of Kant's being hard reading..." (*Monist*, July 1892, p. 44)

42 Russell to Peirce 11 May 1892, regarding a potential audience in Chicago for Peirce's reading of this piece.

Carus to Peirce ca. 18 May 1892 (missing): Eagerly requests these articles that Peirce is "suffering torments with" (i.e., "Law of Mind" and "Man's Glassy Essence") and seems to have threatened their and subsequent articles' inclusion in future numbers.

Peirce to Carus ca. 20 May 1892 (missing): Responds angrily to Carus for apparently breaking what Peirce believed to be a promise "to be allowed to go right on in successive numbers" and boasts that "Few philosophers, if any, have gone to their tasks as well equipped as I in the study of other systems and of various branches of science."

Carus to Peirce 23 May 1892 (missing): Telegram requesting that Peirce "Send manuscript at once. Intend to print next week. Letter follows. P. Carus."

Carus to Peirce. TS2 (final): RL 77. On OCP Stationery B. TS1 (edited): SIU 91.11. On OCP Stationery B. LPC (of TS1): SIU 91.11. Text comes from TS2, except where p. 2 is missing, as indicated in the textual notes. Textual variants and emendations from TS1 are also indicated.

Chicago, May 23, 1892

Charles S. Peirce, Esq.
12 West 39th St.,
N.Y. City.

My dear Sir:

Your letter is received and I telegraphed at once to 12 West 39th Street, N.Y. City:

"Send manuscript at once. Intend to print next week. Letter follows. P. Carus."

I shall do my best to let your article[43] appear in the July number of the MONIST.

You might have known that I am willing to do whatever I can for you, and that I wish to bring your ideas before the public as effectually as possible. So, for instance, I could have had my reply to your first article ready to let it appear in the same number with your article, but I thought it would cause a greater stir if I kept back with it. Being disposed toward you as I am, I was surprised to receive a letter from you which I do not know whether I shall take it as an insult or an

---

43 "Law of Mind."

---

14 Charles S. Peirce, Esq.] ~~Mr.~~ Charles S. Peirce$^{\vee}$, Esq. $^{\vee}$ *[TS1]*
20 manuscript] MS *[TS1]*
23 effectually] ~~a~~\e/ffectually *[TS2]*
23–24 So, for instance,] *inserted above [TS1]*
25 with your article] *inserted above [TS1]*
26 Being … surprised] \Being disposed toward you as I am,/I was ~~therefore~~ /very much\ surprised *[TS1]*
27 I shall take it as] ~~it is intended as~~ /I shall take it as\ *[TS1]*

unwarranted outbreak of temper. If you are displeased with somebody, don't you think that the other party has a right to be heard before you use angry words?

I am willing to publish articles of yours, and am glad to receive contributions as often as possible. But I must have the MSS in time. The MONIST of July 1st has to appear on that date in *London, England*. Therefore we must have finished printing June 10th. Binding takes several days, for the sheets must dry. On June 14th they ought to be on the ocean, so as to arrive a few days before the first of July, in order to leave the dealers time enough for distribution. One number consists of ten forms, and every form takes about one day to print. There is always some delay somewhere, either with the binder or the printer, and I have learned to take this delay in account. So we must begin press work about the 25th of May.

By the bye, I never rely on the promises of contributors, for the very greatest contributors are , as a rule, the most unreliable men in this respect. When arranging the make-up I calculate only with what I have, and never take into consideration what has been promised me. I have found by experience that this is the only way of editing. I trust that you have sufficient reasonableness not to expect an editor to wait for important contributions almost to the very day when he goes to press.

You write, "I understood that I was to be allowed to go right on in successive numbers, and state my philosophy." Although I am willing to grant you as much space as possible, and am even glad to be informed about these your intentions, I cannot find, in any of your letters on file, a word about such an "understanding". Nor can I find in the copies of my letters to you an acceptance of your offer. Although your contributions are welcome, I doubt very much whether I would have made a promise of that kind. I never make promises extending so much in advance without giving Mr. Hegeler due notice thereof. I have no objection to publishing six consecutive articles of yours, but I have an objection to binding the editor of the MONIST, (i.e. myself, or in case any accident would happen to me, a successor of mine) for more than a year ahead.

Perhaps you know that Mr. Hegeler wants the MONIST to represent a definite world view, and I edit the MONIST with the understanding to represent it. I accept

---

1–2 If you are displeased ... angry words?] \If you are displeased with somebody, ~~do~~ \don't/ you ~~ever~~ think that *[. .]* before you ~~jump at him with~~ use angry words?/ *[TS1]*
6–27 On June 14th ... consecutive] *This page of TS2 (p. 2) is missing, so text comes from TS1.*
11 must] *after* ~~have to~~ *[TS1]*
12 By the bye,] *above* ~~I must add that~~ *[TS1]*
13 , as a rule,] *inserted above [TS1]*
26 due] *inserted above [TS1]*
27–29 binding ... mine)] binding \the Editor of The Monist, (i.e.,/ myself, or in case any accident would happen to me, a successor of mine*,*) *[TS1]*
30 Perhaps you] ˅Perhaps˅ *Y*you *[TS1]*

articles from any standpoint. But when I publish articles that differ in some important point considerably from our position, I have to point out the difference. I do not hurry to the rescue of Monism against Mr. Dixon[44] or similar fellows, but it is my duty to say something by way of reply to your first article, not so much in criticism of your position, as in order to state the position of the MONIST. And I shall even take the risk of your thinking me 'to be goose enough in risking to run foul of you before discovering what your position is.'

In venturing to reply to you before you have fully developed your system, I trust that after all, I can rely on your words and that you will say nothing in your subsequent articles inconsistent with the principles laid down in your first article.

I am very well conscious of the fact that, in not waiting till you have explained your whole scheme, I forego a great advantage; but this consideration does not prevent me from speaking out boldly. I have a definite opinion of the problems under discussion and have weighed every pro and con that I could think of. I know very well that all you say of yourself is true. "Few philosophers, if any," you say, "have gone to their tasks as well equipped as I, in the study of other systems and of various branches of science." I know that you do not exaggerate, and if your opinion differs from mine, I think twice before I speak. But I do speak because I

---

44 Edward T. Dixon, involved in a *Monist* dispute with Carus around this time about the Oct 1891 review of Dixon's *Foundations of Geometry* (1891) and about his forthcoming "Future Position of Logical Theory" (July 1892).

1–2 important] *inserted above [TS1]*
2 I have … difference.] inserted above [TS2]
2–4 I do not hurry … my duty] ~~Accordingly,~~ \I do not hurry *[. .]*, but/ *[TS1]* ~~I have to point out the difference.~~ *I*it is my duty *[TS2]*
8–9 In venturing …after all,] ~~But, after all,~~ \In venturing *[. .]* your system,/ I trust that $^{\vee}$after all,$^{\vee}$ *[TS1]*
10 principles … article.] ~~propositions of~~ \principles laid down in/ ~~your first \former/~~ articles$^{\vee}$.$^{\vee}$ ~~containing the principles of your inquiry. \of yours/~~ *[(first revision)* ~~propositions of~~ \principles laid down in ~~your first~~ \former/ articles ~~containing the principles of your inquiry.~~ \of yours/ *(second revision shown in letter)] [TS1]*
12 forego] forgo$^{\vee}$e$^{\vee}$ *[TS1]* for\e/g~~oe~~ *[TS2]*
13 opinion] *inserted above [TS1]*
15 you say,] $^{\vee}$"$^{\vee}$ \you say,/ $^{\vee}$"$^{\vee}$ *[TS1]*
16 equipped] equip$^{\vee}$p$^{\vee}$ed *[TS1]*
17–18 I know that … speak.] \I know that you do not exaggerate, ~~but that is no reason to give my opinion~~/ ~~and~~ /~~although I myself am not exactly~~ unequ~~ipped~~\ ~~if your opinion differs from mine, I think twice before I speak~~. *[(first revision)* \I know that you do not exaggerate, but that is no reason to give my opinion/ ~~and if your opinion differs from mine, I think twice before I speak~~. *(second revision)* \I know that you do not exaggerate, ~~but that is no reason to give my opinion~~/ ~~and~~ /although I myself am not exactly unequipped\ *(third revision shown in letter)] [TS1]*
18 But I do speak because] But I /~~do not intend to give my opinion captive~~\ \do speak because/ *[TS1]*

am not so anxious to come out of a discussion victorious, as to arrive at the truth. If that be reached I am satisfied, and should not be ashamed to alter my opinion.

You write that you wish to contribute to the OPEN COURT. I did not know that you care to write for the OPEN COURT, but I shall be glad to have contributions from your pen to the OPEN COURT also.

Please wire me at once whether at all in the next days, and when, I can expect your MS. I shall try

Yours very truly,
P. Carus

Peirce to Carus. Telegram: SIU 91.11. Form headed, "Night Message. | The Western Union Telegraph Company."

May 23 1892 | Cambridge Mass

Dr Paul Carus
Open Court Company
Nixon Building
169 Lasalle st
Manuscript will be mailed tomorrow

C. S. Peirce

Peirce to Carus. MS: SIU 91.11.

12 W 39$^{\underline{h}}$ St. New York | 1892 May 24

My dear Sir:

---

2 reached] reached, *[TS1]*
6 at all … and when,] \at all in the next days/ and when *[TS1] commas inserted in-line [TS2]*
7 I shall … wait long.] \I shall \try to/ wait with the make-up of/ ~~Also how many pages it will make in the MONIST.~~ \~~If you do not let me wait too long,~~/ ~~I shall not make up~~ the July number until I hear from you*.*;̲ ˅but you must not let me wait long.˅ *[(first revision)* Also how many pages it will make in the MONIST. \If you do not let me wait too long,/ I shall not make up the July number until I hear from you *(second revision shown in letter)] [TS1]*

Also in TS1 is this additional paragraph:

~~I have to add that~~ \If you write that you wish to contribute to The Open Court, I did not know that you care to write for The Open Court, but/ I shall be glad to have contributions from you to the OPEN COURT \also/.

7 try] *after* ~~not~~ *[TS2]* to wait with the make-up of the July number until I hear from you; but you must not let me wait long.

I enclose one[45] of the two articles on Mind in my series on my philosophy. I have not counted the words; but there cannot be far from 8 400; so, I shall be much obliged if you will send me $ 200, on account.

The next article,[46] which will be fully as interesting, will be ready before the first of July.

I went on to Cambridge by invitation of the philosophical club there to read the article I send. It created much interest, as well as I could judge. The last article on necessity[47] had formed the subject of several séances of Royce's seminarium. He, of course, opposes it; and on lines probably about like yours. He is to attack it in Schurmann's Journal.[48] So, between you, that view will get pretty thoroughly set forth.

Yours very truly
C. S. Peirce

I have your telegram, but no letter.

Peirce to Carus. MS: SIU 91.11.

12 W 39$^{\underline{h}}$ St. New York | 1892 May 27

My dear Dr. Carus:

I must plead guilty to the charge of using an offensive expression in regard to your possible reply. I much regret it, & ask your pardon. It was a case of the pen of a tired and hurried man running away & making him say things whose *portée* he did not perceive. One of "Things one would rather have left unsaid."[49]

I do not think I said that there was an understanding *between* us that my articles were to be published in consecutive number. What I said was that *I* had been

45 "Law of Mind."
46 "Man's Glassy Essence."
47 "The Doctrine of Necessity Examined."
48 Jacob Gould Schurman's *The Philosophical Review* of Cornell. Although no such "attack" by Royce appeared in the *Review*, he did argue against Peirce in a paper read to the Philosophical Club at Brown University 23 May 1895: "I do not myself accept this notion that the laws of phenomenal nature … are the evolutionary product of any such cosmical process of acquiring habits, as Mr. Peirce has so ingeniously supposed in his hypothesis of 'Tychism' " (*Studies of Good and Evil*, 237; noted in *W* 8:lxxxi n101). For further discussion of Peirce and Royce on necessity, see Kegley, "Josiah Royce: Classical American Philosopher," 2013.
49 Cartoon series in *Punch*, British magazine founded 1841.

4 be] *after* ~~sure~~
10 Schurmann's Journal.] *after* ~~th~~
14 no] no~~t~~

led so to understand, by your making no comment and giving no warning to the contrary when I talked in that way.

As to my articles not being in harmony with the position of the Monist, you much surprize me. I read the prospectus carefully, and judged the definition there given of Monist would easily take me in. I am particularly surprized that it is Mr. Hegeler who takes the narrow view of the purpose of the journal. I had the idea that he was particularly concerned with the conciliation of Science & Religion. From *that* point of view, he should welcome my views; unless he only desires the conciliation if it takes place in a preconceived way. Anyway, I think the narrow platform *is* a pretty good reason why my writings should not appear in the *Monist*, and makes me wonder why they are welcome to you. On the whole, I am rather puzzled what to say.

I shall not deviate in the least from the principles of my first Paper; but these principles when developed will take on an aspect which will perhaps surprize some persons, and which I should have supposed would have rendered them more agreeable to you. But I am now rather inclined to think you stickle for a special form of monism which I have never seen clearly stated.

Yours very faithfully
CSPeirce

Carus or OCP to Peirce ca. 28 May 1892 (missing): Sends galleys of "Law of Mind" for review.

Peirce to Carus. MS: SIU 91.11.

1892 May 31

Dear Dr Carus

I return galleys 7–12 of my "Law of Mind" within half an hour of their reception. I have not yet received the earlier galleys. They must have gone to Cambridge. But it is not indispensable I should send them, I think, if you will do so carefully, noting that by Kanticity & Aristotelicity,[50] I do not mean Kantianity and Aristotelianity

Very truly
C. S. Peirce.

---

50 In Peirce's personal interleaved copy of *The Century Dictionary* he added the following definitions: "The Kanticity is having a point between any two points. The Aristotelicity is having every point that is a limit to an infinite series of points that belong to the system." (*W* 8:395)

---

3 my] *after* ~~your~~

26 indispensable] indispensible *[E]*

Carus to Peirce. TS: RL 77. Stationery headed, "The Monist. | The Open Court. | Editorial Department. | La Salle, Ill. | Dr. Paul Carus." LPC: SIU 91.11.

La Salle, Ill., May 31, 1892.

Charles S. Peirce, Esq.,
12 West 39th Street,
New York.

My dear Sir:

Your letter[51] is just received. Many thanks for the spirit in which you write. I am greatly pleased that the harmony of our minds is not endangered, and I hope that it never will be. I am glad that you express your surprise at not being in harmony with the MONIST. It is not impossible that I did not make myself sufficiently clear. I *do* mean to say that you *are* in harmony with the monist; but that your first paper[52] contains an important deviation from our conception of necessity. And it is a matter of course that such a fundamental idea as is necessity will affect its superstructures also—and I shall state why in this point I differ from you.

My impression is that we are working, perhaps not so much in the same line, as on lines converging to a common goal. Before I had special reasons for believing that, I felt it instinctively. And this instinctive belief originated from having found among your writings, certain hints, undeveloped ideas and suggestions, which might have meant little to others, but to me were indicative of the underlying tendencies of your philosophy. They were straws in the wind, which I observed before making my tests with the vane.

My discussion of your article on "Necessity" will soon be in your hands, and I am curious what you will think of it.[53] I do not believe that my objections are similar to those made by Prof. Royce; for I glanced at his book, "The Spirit of Modern Philosophy", and apprise that there is as great a gulf between his theoretical views and mine, as there is between Prof. Adler's ethics and mine.[54]

Mr. Hegeler's monism, no less than mine, is broad enough to encompass not only sympathizers and men as you are, who steer for the same goal, but also opponents. We would not hesitate to admit to our columns the bitterest opponents of monistic thought, among them the Roman Catholics, who, by the bye, are in many respects much more monistic and also consistent than their Protestant brothers.

---

51 Of 27 May.

52 "Doctrine of Necessity Examined."

53 "Mr. Charles S. Peirce's Onslaught of the Doctrine of Necessity."

54 See "Adler" in Biographical Register.

---

12 the monist] $^{\vee}$the$^{\vee}$ monis*m*t

15 —and ... you.] *inserted in-line*

Mr. Hegeler is not narrower than I am, but I should say that he is more positive. I would restrain editorial interference to matters of principle only, and leave the working out of our aim to the contributors, admitting, however, the editor as one of the contributors. As such I would often abide my time longer than I do. So, for instance, my proposition was to name the new quarterly “THOUGHT” and not “THE MONIST”; but in Mr. Hegeler's opinion, the name “THOUGHT” was unmeaning. He regarded it as an attempt at being non-committal. He wants to have the solution which he arrived at expressed unequivocally. The name of the journal is to him the flag; while to me it would be an invitation to the class of people who are welcome to contribute. Mr. Hegeler had the same objection to the name, “THE OPEN COURT”. I shall be pleased to have your article for the Fall number of the MONIST.

With kind regards,

Yours very truly,
P. Carus

Peirce to Carus. MS: SIU 91.11.

New York 12 W 39th St. 1892 June 3

My dear Dr. Carus:

Thanks for your kind letter. The name “Thought” would have been a superb one. It would have given a send-off to the thing. It particularly describes the kind of magazine I have always wanted to see, in which articles of deep thought on all sorts of subjects (not too abstruse) should find their natural place.

As for flags and parties in philosophy, I think it is 10 to 1 we all are in the wrong. We should therefore exercize the utmost toleration. Besides, philosophy has little practical value. It is a poor thing to base religion or conduct or politics or business of any kind on. The *study* of philosophy is not altogether useless in elevating the spirit, and we hope it will one day yield some unquestionable result. But little or nothing of that kind has yet been reached. So I for one am not at all disposed to risk any skin upon it or make a party to advocate any particular variety of philosophy. When I enter a philosophical disputation it is in the hopes that I and the audience will learn something, not at all to cause the triumph of any doctrine. In fact, when philosophy becomes partisan, it may be *sophy*, but it ceases to be *philosophy*. I should as soon think of getting up a party to further the case of a proposition in mathematics or chemistry. The scientist, like the philosopher, does

---

4 As such … I do.] *inserted above*
6–7 unmeaning. He regarded it] unmeaning*,*. \He regarded it/
20 describes] *above* ~~expresses~~
21 magazine] *after* ~~journal~~

not busy himself with vindicating doctrines but in searching out truth. He is a student, not a party-leader. A chemist would not think the Church Journal or the Methodist a proper place to print his researches, even if the clergymen should think his conclusions somehow helped their purposes and furthered the cause of righteousness. I as a philosopher have no more to do with the cause of religion than the chemist has. I am just pegging away at my studies and giving the results to the world without any ulterior purpose. I want to publish them in a journal which does not care a straw what the results may be, or what cause they forward or injure.

I have been asked to have the April Monist sent to the following persons.

Monsieur Ch. Renouvier
au Pontet, Avignon
France (Vaucluse)

Monsieur J. Delboeuf
Boulevard Frère Orban 32
Belgium Liège

Mr. A. D. Risteen
Hartford Steam Boiler
Inspection & Insurance Co.
Hartford
Conn—

How would you like a series of short articles by me for the Open Court intended to further the cause of right reasoning, in general? Say, to be entitled "The Critic of Arguments,"[55] to contain remarks, mostly rather special, upon good and bad reasoning.

I was inclined to think your defence of necessity would be something like Royce's. At any rate, his is quite in the Kantian way. I shall be curious to see both, and may respond in the fullness of time.

Yours very truly
CSPeirce

Peirce to Carus. MS: SIU 91.11.

12 W 39$^{\underline{h}}$ St. New York | 1892 June 4

My dear Sir:

55 Published in *Open Court* in two parts: "I. Exact Thinking" (22 Sept 1892) and "II. The Reader is Introduced to Relatives" (13 Oct 1892). Might also refer to his entire *Open Court* series, which would include "Pythagorics" and "Dmesis" (8 and 29 Sept 1892).

In asking you to send a Monist to Delboeuf, I forgot he already would have received it since he had an article in it.

If you like, we *[can]* call the July article 8 400 words, or you can have them counted & send check accordingly.

Yours very truly
C. S. Peirce

Dr. Paul Carus

Peirce to Carus. MS: SIU 91.11. Two notes across header: (Carus) "make out an additional check of $ 100 (one hundred)"; (OCP): "Sent June 14$^{\underline{th}}$/92".[56]

12 W 39$^{\underline{h}}$ St. New York | 1892 June 9

My dear Doctor Carus:

A succession of misfortunes in the last year has put me for the time being into great straits financially. I am on the point of placing two inventions each for a very large sum, but negotiations are slow, and it is essential that I should keep up appearances. Consequently, a few hundred dollars now will be worth to me many thousand a little later. For this reason, I should take it as a very particular kindness if you would send me immediately the money for the article[57] now in your hands. I should also like to have the money for the article[58] I am now engaged in writing as soon after the copy reaches you as may be. Please understand that I only ask this on the presumption that the early payments will put you to no personal inconvenience; for, otherwise, I should far prefer to submit to the disadvantages of my situation and await your convenience.

Yours very truly
C. S. Peirce
12 W 39$^{\underline{h}}$ St. | New York

Peirce to Carus. MS (letter): SIU 91.11. Stationery headed, "Long Beach Hotel | Chas. E. Hitchcock | Manager" with beachfront hotel image at center. MS (list of persons): RL 77. Stationery headed, "En Route | New York & Chicago Limited | Pennsylvania Lines | Pullman Vestibuled | Train". Letter answered by McCormack 20 July 1892.

---

56 See OCP (Sacksteder) to Peirce 16 June 1892.
57 "Law of Mind."
58 "Man's Glassy Essence." Money advanced by OCP (Sacksteder) to Peirce 16 June 1892.

---

3 we *[can]* call] we we call *[E]*
9 $ 100] $ *2*100
21 prefer] *after* ~~rather~~

Long Beach, L.I. July 15 1892

My dear Dr. Carus:

It was most kind in you to send the hundred dollars for the article I now forward.[59] It was particularly handy.

The first pages of the translation of Mach needed a fearful amount of polishing. But when he once reached the main part of the book, it was no longer so.

Your reply to my tychism is very persuasive. I am impatient to see the rest of it. I feel very grateful for the flattering manner in which you speak of me.

I only saw the July number a day or two ago on inquiring for it in a shop. I take the liberty of sending a list of persons to whom I would like to have it sent.

The article I now send has been delayed somewhat by unusually pressing business and by the phenomenal heat, to which my brain succumbed. I have counted the words on every seventh page and find 111.89.105.112.111.111.123.106.101. 104. According to this there are a little over 8 000 words or $ 200 of which I have already received $ 100. The article is not so good as it ought to have been, owing to preoccupations and anxieties; but still, I think it has some value

Yours very faithfully
C. S. Peirce

Address continues
12 W $39^{\underline{h}}$ St. New York

Please send July Monist to[60]

Prof. Josiah Royce Cambridge Mass
Graduate Metaphysical Club,
Cambridge, Mass.
Dr. F. E. Abbot, Cambridge Mass
Prof. J. M. Peirce, Cambridge Mass
Prof. Palmer, Cambridge Mass
Prof. J. E. Oliver, Ithaca, N.Y.
Dr. John Winslow, Ithaca, N.Y.
Century Club. 7 W $43^{\underline{rd}}$ St. New York
The Players 16 Gramercy Park, New York

59 "Man's Glassy Essence."
60 See Biographical Register.

13 112] 1*0*12

Professor S. P. Langley, Smithsonian Inst.
Washington D.C.
Major Powell, U.S. Geological Survey
Washington D.C.
Clarence King Esq
Union League Club
New York
Dr. W. K. Otis, 5 W $50^{\underline{h}}$ St
New York
Rev. W. R. Huntington DD
Grace Church New York
Prof. G. B. Halstead
Austin Texas

very truly
C. S. Peirce

Carus to Peirce late July 1892 (missing): Sends a proof of "Man's Glassy Essence," which Peirce returned to McCormack 5 Aug 1892.

Carus to Peirce mid-August 1892 (missing): Sends a proof of "Pythagorics," the MS of which Peirce submitted to McCormack 6 Aug 1892.

Peirce to Carus. MS: SIU 91.11. Answered by McCormack 29 Aug 1892.

Milford, Pa. 1892 Aug 25

To the Acting Editor of the Open Court
Dear Sir:

I return the proof of my *Pythagorics* and send the copy of the first number of my *Critic of Arguments*.[61] In a few days, I will send a discussion of a question of public ethics.

I should like to have my advertisement discontinued, now that I am to appear as a contributor. I have never got a reply to it.

I should be much obliged for a check if convenient. To excuse my importunacy, let me tell you what recently happened to me. Some three months ago, I

61 "Critic of Arguments. I. Exact Thinking."

2 Washington] *over* Ne

12 Prof. G. B. Halstead] *below* ~~Dr. Edmund Mont~~ *for Edmund D. Montgomery*

was invited by an inventor to examine his process of electrical bleaching. I looked at it and made some criticisms. Shortly after, I was requested by the Wall Street promoter to report upon the chemistry of the process and upon any modifications of which it might be susceptible. A contract was executed by which I was to receive $ 500 for my work and besides, in case of making valuable modifications, a hundred thousand dollars of the stock. I spent a good deal of time on it, doubled the intrinsic economy of the process, and by new chemical discoveries gave a real basis to patents that was wholly wanting before. I received a check. Said check was returned as "no good;" and I now find there is a combination of millionaires to use my work and pay me nothing. That is a severe disappointment; for after I got my check, I naturally thought I was all right. Hence it is that I am obliged to make myself a nuisance by asking for money before I ought. I hope the circumstances will beg pardon for me. I am in hopes soon to have other chemical contracts that will work better.

Yours very truly
C. S. Peirce

Peirce to Carus. MS: SIU 91.11.

1892 Aug 31

Editor Open Court
Dear Sir

I enclose herewith my *third*[62] article for the *Open Court*. I have not yet heard whether or not my second has reached you.

Please oblige me with a check

Yours very truly
C. S. Peirce
Milford | Pa.

Peirce to Carus ca. 3 Sept 1892 (missing): Submits "Critic of Arguments. II." for *Open Court*.

Peirce to Carus. MS: SIU 91.11.

62 "Dmesis."

1 bleaching] bleeching *[E]*

7 economy] *after* ~~value~~

Oct 7 1892

Editor of Open Court

I enclose my article[63] for the January which I regret has unavoidably been delayed.

C. S. Peirce

Carus to Peirce. Postcard: SIU 91.11. Undated, unofficial card with no printed header, verso blank. An immediate reply to Peirce's letter to McCormack 17 Oct and telegram to OCP 19 Oct requesting payment for "Evolutionary Love." Followed by Carus's detailed letter the same day.

[19 October 1892]

Charles S. Peirce
Buckingham Hotel
New York, City.

MS received check will be sent tomorrow to Buckingham Hotel

Carus

Carus to Peirce. MS: RL 77. On OCP Stationery B. LPC: SIU 91.11.

Chicago, October 19th, 1892

Charles S. Peirce Esq.
Buckingham Hotel
New York, City.

Dear sir: Your letter was received today, also your telegram.[64] Since you seem to be in a great hurry to have your bill attended to, I asked Mr Hegeler to be so kind to make out the check, which I herewith enclose.

Your 4 Open Court articles[65] contain about 10.400 words

$= \$\,262^{\underline{50}}$

Your Monist article[66] contains about 10.000 words

$= \$\,250^{\underline{00}}$

You have received on account 250 dollars, which leaves a balance of $262^{\underline{50}}$ dollars in your favor.

Twice, when passing through New York, I tried to see you; but in vain. I am sorry having missed you.

With kind regards from | Mr Hegeler

63 "Evolutionary Love," *Monist* (Jan 1893).
64 See Peirce to McCormack 17 Oct 1892 and Peirce to OCP 19 Oct 1892.
65 "Pythagorics," "Critic of Arguments" I and II, and "Dmesis."
66 "Evolutionary Love"

Yours very truly,
P. Carus

Peirce to Carus (or possibly OCP). MS: SIU 91.12. Undated, bottom half of a torn sheet. Date estimated based on Buckingham Hotel return address, Peirce's address between Oct 1892 and Feb 1893, and likely late October in response to Carus's summary of Peirce's contributions to *The Monist* and *Open Court*.

[late October 1892]

[. . .]
One set of C. S. Peirce's Monist papers to
Dr. Geo. Ferdinand Becker
U.S. Geological Survey
Washington D.C.

Subscription to Open Court for—
Capt. E. Stanton Huntington
Wollaston, Mass.

One bound set of all publications except Soul of Man—F. P.—Phil. of Attention[67] to Prof. C. S. Peirce, Buckingham Hotel New York City.

Carus to Peirce. TS: SIU 91.11. On OCP Stationery B. LPC: SIU 91.11.

La Salle, Ill., Nov. 14, 1892.

Mr. Charles S. Peirce, Esq.,
Milford,
Pike Co., Pa.

My dear Sir:

I am just arranging the contents of the next Monist, which is much overcrowded. Several articles have to lie over until the next number, which in this way will also be overcrowded. Editorially considered, it is quite indifferent which ar-

67 Carus, *The Soul of Man* (1891); "F. P." unknown; and likely for Théodule Ribot, *The Psychology of Attention* (1890).

15 Wollaston, Mass.] *OCP marginal notes regarding the above three lines: [Hand 1, purple pen]* "Send the nos. of *The Monist* containing all Peirce's articles." *[Carus, pencil]* "don't charge as yet."
16 One bound set of all publications] *Marginal note with line pointing here: [Carus, pencil] "including M. & OC."*

ticles are selected for publication in the January number and which for the April number. In case your article should be left over for the April number, I would again give it the first place, which in the January number was reserved for Prof. Lloyd Morgan.[68] May I ask you whether you have any preference as to when it shall appear? I regard the April number as of special importance in so far as it is the first number during the World's Fair.[69]

Will you kindly answer at your earliest convenience?

Yours very truly,
P. Carus

Possibly Peirce to Carus late November (missing): Acknowledges Carus's inquiry to postpone his "Evolutionary Love" to April (which Carus ultimately didn't do) and perhaps discusses the possibility of preparing a "Logic of Relatives" text for *The Open Court*, mentioned in the second 20 Jan letter below.

1893

Peirce to Carus. MS: SIU 91.12. Note by Hegeler across header reads, "Ansd by E.C.H. —Jan 23^{d} 93 | ordering check for $ 200" (see Hegeler to Peirce 23 Jan 1893).

[20 January 1893]

Dear Dr. Carus:

I will write you another note today.

This is merely to say here is my article.[70] It is very irregularly written so that I cannot tell for certain how much there is, but probably there are 10 000 words. If you will send me a check for $ 200 on account, I shall be obliged.

Address, till further notice

C. S. Peirce
Buckingham Hotel | $5^{\underline{h}}$ Avenue & $50^{\underline{h}}$ St | New York

---

68 Conway Lloyd Morgan, "The Doctrine of Auta," *Monist* (Jan 1893).
69 *Chicago World's Fair: Columbian Exposition*, to be held 1 May–30 Oct 1893, celebrating the 400th anniversary the previous year of Christopher Columbus's discovery of the New World.
70 "Reply to the Necessitarians," *Monist* (July 1893).

---

20 here] *after* ~~t~~

Peirce to Carus. MS: SIU 91.12. Stationery headed, "Buckingham Hotel, | Fifth Avenue and 50th Street. | Wetherbee & Fuller, | Proprietors."

Address: C. S. Peirce | 821 Sixth Avenue | New York
New York, Jan 20 1893

My dear Dr. Carus:

You have not found, I trust, that in my rejoinder I have anywhere overstepped the limit of amiable disputation. If anything of that kind did, unconsciously to me, in the heat of composition, slip from my pen, I am most anxious to have it pointed out to me, so that there may be no feeling in the matter of a disagreeable kind. For if you should not mention it, I should at some future time discover it, and it would be a source of real unhappiness to me.

I have to thank you for a pretty book at Christmas time. My wife has been reading it with much pleasure, and I shall do so when I have leisure.

I broke down completely in attempting to put the outlines of the Logic of Relatives into a sufficiently lucid and elementary shape for the *Open Court*. Years ago I had to abandon work on that subject, because it interests me so intensely, as to seize upon me completely; and I found that I must not think of it until I could arrange my affairs so as to do nothing else for a while. But my wife and I both thought that if I put it off longer, it would never get done at all; and so I thought I would just dip into it a little. But it was no use. I wrote enough for nearly a dozen articles; but far too difficult for such a publication, and I struggled and labored to hurry it up, until I was made seriously ill.

Then I was forced to deliver 12 lectures in Boston on the History of Scientific Thought,[71] and they had to be written out in rather elaborate shape. Thus, I have not yet had time to send you anything but that article.

But I shall begin at once with lighter subjects.

I feel I must not end my life without developing my Logic. I have had the most beautiful discoveries lying by me for ten years without being able to publish them; for I *cannot* do half way work in that branch. I had my plans all laid to put everything aside and devote myself to that; but I was so terribly swindled last summer that it sickens me.[72] My wife says, and I feel, that I must take up my logic within a year, or I shall never do it; for I am kept working at a pace which would break

71 "The History of Science," 12 lectures delivered at the Lowell Institute 28 Nov 1892 to 5 Jan 1893 (*HP* 2:139–296).

72 See Peirce to Carus 25 Aug 1892.

down a man in the prime of life before long. I have several chemical processes; one an entirely new process of electrical bleaching which is just ready while the bleaching is about to be revolutionized, and is, the business people tell me worth a great deal. My process is founded on a chemical fact hitherto unknown and is irreplaceable. I have a process certainly destined to revolutionize the distillation of wood-alcohol. This is also capable of a basic patent. I have an unpatentable but highly practical process for preventing scaling in locomotive boilers, which I have tested on several thousand gallons evaporated to dryness, and which railway men tell me there is a living in. But I am decided that if I go on with these things my Logic will be sacrificed; and I ought not to do it.

Everybody has always expressed a mild interest in my logical and philosophical work; but nobody has ever offered me a helping hand. Probably they thought I did not need it.

I am now determined to see if I cannot induce some of my friends to buy out my processes, I care not at what sacrifice, so it is enough to enable me to go on for a couple of years in uninterrupted work on what I was put into this world to do. I will then devote six months to writing a school arithmetic which the largest publishers assure me will give me an income to get along on, after 18 months; and I will add when I can, a geometry and algebra upon which I have been working and experimenting with pupils for many years.

I wonder if Mr. Hegeler would be willing to buy any of my chemical processes. I should rather have him to do so that to go to anybody who understands neither *chemistry* (so as to know the value of what he is buying) nor *philosophy* (so as to appreciate my motive in selling for a song, what is of great value.)

I must and will bring the struggle to get my proper work under weigh to a termination somehow, at once.

Yours faithfully
C. S. Peirce

P.S. All my difficulties have to be kept secret from my wife, or she would be wanting to compromise her prospects in a way I would never consent to

---

Between 28 Jan and 6 Feb 1893: Peirce meets with Hegeler and likely Carus at Hegeler's home in La Salle to discuss larger publication projects, including Peirce's "A Quest for a Method" (or

1 processes] process *[E]*
3 about] *over* go
12 ever] every *[E]*
19 when I can,] *inserted above*

"Search for a Method")[73] and several math books. Hegeler seems also to have asked Peirce to begin investigating into printers to contract with OCP on the books, especially his arithmetic book.

7 Feb 1893: Returning from La Salle, Peirce meets with Russell at Richelieu in Chicago[74]—details undetermined, though an unsatisfactory meeting for Peirce due to Russell's "not showing me the side of him that I wanted to see" (Peirce to Hegeler 11 Feb 1893).

Peirce to Carus early Feb 1893 (missing; or possibly during the La Salle meeting above): Submits "Marriage of Religion and Science" for *Open Court* (published 16 Feb).

Peirce to Carus. MS: SIU 91.12. Stationery headed, "Buckingham Hotel, | Fifth Avenue and 50th Street. | Wetherbee & Fuller, | Proprietors."

New York, Feb 10 1893

My dear Carus:

I sat up till after two the night I left you working over my article,[75] but could not finish it, though I got up at 5 to work more—I therefore brought it along. Last night was the first time I had for it & though I had not slept much in the sleeping cars the two previous nights, I again sat up till half past two on it. On retiring, found my young wifeunable to sleep & spent the time till five in endearments. I write this at 8 while they are getting my breakfast. I am so tired I can hardly speak; but am going to get in a little work at the Aston today notwithstanding various business.

I shall write at once or telegraph to John Wilson & shall try to go on tomorrow. Today I have to see Appleton & after seeing him shall write to Mr. Hegeler.

My article is made much longer, much clearer, & somewhat more interesting. I hope it will be *much* improved before I have done with it. But the points are not essentially changed. There is one part that I may modify the tone of

Yours very faithfully
C. S. Peirce

I shall send messages to the family in my letter to Mr. Hegeler. This is only business

---

73 (R 592–594). For details, see Peirce to Hegeler 24 Feb and 7 Mar 1893. Also de Waal, 15–17, and Brent, 221–224.

74 Carus to Russell telegram 6 Feb 1893: "Peirce will arrive tonight about seven invite you to lunch tomorrow Richlieu" (RL 387).

75 "Reply to the Necessitarians."

Peirce to Carus. MS: SIU 91.12. Stationery headed, "Century Club | 7 West Forty-Third St."

1893 Feb 13

My dear Dr. Carus:

I send herewith my revised article. Only one of the cancelled pages had any pencil-mark of yours and that one I have replaced on the new page by a blue mark just like yours. I wish to add a few words about an article by J. E. Oliver in The Philosophical Review.[76] If you will send me that article.

I have written to my farm to have the house made ready for me, & shall return there day after tomorrow. I shall now at once take up the Mach proofsheets.

Yours very faithfully
C. S. Peirce

Peirce to Carus. MS: SIU 92.12. Stationery headed, "Century Club | 7 West Forty-Third St." First and last paragraphs of letter struck out by OCP.

1893 Feb 15

My dear Carus:

Please strike out of my article all reference to the Century Dictionary. The Century Company have had the incredible folly to prohibit my copying my own definitions even with giving them credit. Only, in place of the last § please substitute the following.[77]

It is true that I wrote many definitions for the *Century Dictionary*. But they were necessarily rather vaguely expressed, in order to describe the popular usage of terms, and in some cases were modified by proofreaders or editors; and for reasons not needful here to explain, they are hardly such as I should give in a Philosophical Dictionary proper. Under the circumstances, I shall not undertake to defend their strict accuracy.

I have read Oliver's paper. I think he has made a mistake. Shall not need any copy of it. Shall send you a bit about it.

Yours very faithfully
C. S. Peircepar

76 "A Mathematical View of Free Will" (May 1892).
77 "Necessitarians," p. 570.

21 describe the] *above* ~~meet,~~

22 proofreaders or editors] proofreaders<,> \or/ editors<,> ~~etc~~

23 hardly] *above* ~~not altogether~~

Carus to Peirce. LPC: SIU 91.12.

February 18, 3.

Mr. Chas. S. Peirce,
New York City, N. Y.,
Dear Sir:—

Your MS. is at hand, but I have not as yet been able to read it through. I fear that your second edition of the article will not be an improvement upon the first, because it is too long. I always prefer short articles. Well, it cannot be helped!

The letter which you wrote me some time ago, and from which I intended to quote a passage in my reply to you, has been misplaced.[78] Will you kindly write those sentences, or some similar sentences, again, in which you state that you would have been sorry to write anything offensive to me, and that you would be willing to change those expressions which I did not like? In case you wish to add something please add it in the shape of a footnote, in which you give your permission to make this quotation from your private letter.

Mr. McCormack is looking out anxiously for the proofs of "Mach" and we shall be glad to see them back as soon as possible, because there is so much type tied up in them.

With kind regards,

Yours very truly,

Peirce to Carus. MS: SIU 91.12. Stationery headed, "Fowler House | J. E. Wickham, Prop."

Port Jervis, N. Y., Feb 24 1893

Dear Dr Carus

There is nothing to prevent that paper of mine from being greatly abridged if you like. All reference to The Century Dictionary had better be omitted in that case.

I have been dissatisfied with those people ever since Roswell Smith[79] died. Scott[80] wrote to me after I wrote you from The Century Club & withdrew his prohibition against quoting & admitted I had a right to make a dictionary of the definitions. But—well, I worked with them so long, I guess I will hold my tongue further

78 Peirce to Carus 20 Jan 1893. Quoted in Carus's "Founder of Tychism," p. 571.
79 (1829–1892), founder of The Century Company 1881 and involved in creating *The Century Dictionary*.
80 Charles P. G. Scott (1853–1936), etymologist for *The Century Dictionary*.

26 have been] *after* ~~am~~

than to say I am not altogether satisfied with them, and I don't mean in future to talk much about the Dictionary.[81]

There are many other things in the article that are perfunctory & uninteresting & which I left in my last revision because you might like to comment on them as I saw you had made some pencillings on the margin.

That about Venn was too severe.[82] The truth is his first edition took up a theory of probability that Mill had taken up in his first edition & subsequently dropped. But Venn in his second edition treats the matter in a way I totally disapprove. My praise of his first edition in The N. A. Review[83] was written before I saw him. When I saw what a namby pamby creature he was, I was rather irritated.

very truly
C S Peirce

Carus to Peirce. LPC: SIU 91.12.

February 27, 3.

Chas. S. Peirce, Esq'r,

My Dear Sir:—

Your kind letter is at hand.

There is no hurry about your reply since we cannot publish it in the next "Monist". I shall have it set, however, as soon as we have type free.

If you desire to revise it a second time you are welcome to it, but I must confess that I do not regard it as advisable. I do not believe in working over articles several times. They lose freshness; and especially controversies should possess this quality. Even in a higher degree than essays. Looking at your articles from your standpoint I should have preferred the first version. It seemed to me a more vigorous criticism and attack upon my position than the second, and I am sure a third version would appear to me again as a weaker edition of the second one.

With kind regards,

Very truly yours,

---

81 Peirce contributed definitions to *The Century Dictionary* throughout much of his career, particularly leading up to the 1889–1891 edition. His *Century* writings between 1883 and 1909 are to comprise the core of *W* 7.

82 "Necessitarians," p. 528. Criticizes John Venn's *Logic of Chance* (1866) as a "blundering little book" and his whole method of logic as "of the weakest."

83 Review in *The North American Review*, 1867.

Carus to Peirce. MS: RL 77. Stationery headed, "The Monist, | Editorial Department, | La Salle, Ill."

Febr. 28th 1893

My dear sir:

I forgot to mention that the pictures from the light and shade Publ Co.[84] arrived and that I sent one of them to Schroeder. Mr Hegeler rc'd another one and you mention in your letter (he told me) that you had sent one to Schroeder too. I am sorry if my neglect to inform you about this receipt has caused you trouble. I also gave one copy to Judge Russell who was very happy to have one.

With kind regards
Yours very truly
P Carus

Peirce to Carus. MS: SIU 91.12.

Port Jervis N.Y. 1893 March 3

My dear Dr. Carus:

I received the last *Open Court*. I trust that means I am now to receive it regularly. Please send a bound copy of Vol VI when you have it.

In my humble opinion you are never likely to say again anything so false as that writings lose their freshness by being worked over. The first page or two of my Theory of Probable Inference[85] was put into more than 90 forms very varied before I was satisfied; yet nobody would suspect any elaborate work on it. I purposely took the snap out of my article. For one thing I have set myself up as a model of deportment & therefore must be careful.—The article is tedious because it discusses small points not worth notice; but it contains some passages worthy of preservation & some of the best were added in the revision.

Very faithfully
C. S. Peirce

Did my portrait come to hand?

84 *Sun and Shade*, which in August 1892 (vol. 4, no. 12) published a brief biographical sketch of Peirce with accompanying photogravure (displayed in Carus to Peirce 17 March 1900, to be used in a reprint of their necessitarian saga), and who apparently had several other photos on file. Peirce might have been considering options for a frontispiece for his arithmetic, logic, or "Illustrations" book.

85 In *Studies in Logic* (1883)

27 Did my portrait come to hand?] *This line down left margin.*

Carus to Peirce. LPC: SIU 91.12.

La Salle, March 6, 3.

Prof. Charles S. Peirce,
Milford, Pa.,
My Dear Sir:—

Having considered the case with Mr. Hegeler, we have come to the conclusion that we cannot return the MS. for a second revision. After I had spent much time in a careful perusal of the first version of your rejoinder, you took the MS. back to make alterations and additions. As we have now again spent much time on it, we have this time to refuse a new revision on your part. I am behind in many things and other work would have to suffer.

I have at once given an order to send you Vol. VI of The Open Court.

In the hope that you are well, and that your wife's health is improving with the returning Spring, which will relieve you of much anxiety, I remain,

Yours very truly,
P Carus

Carus to Peirce. TS: SIU 91.12. On OCP Stationery B. LPC: SIU 91.12. Unsent draft of 6 March letter. Notes along header indicate that this letter was "not sent" (top right) and merely kept on "file" (top left).

[Unsent draft]

La Salle, March 6, 1893.

Prof. Charles S. Peirce,
Milford, Pa.,
My Dear Sir:—

I have ordered to be sent to you a copy of Vol. VI of The Open Court.

As to your proposition that writings do not lose their freshness by being worked over, I have to reply that that is quite true, in cases in which an author is not as yet ready in his own mind. I have sometimes pursued the method myself of working out my own way of salvation to clearness by writing. If I do, I have to work the same material over again and again. If that was your way of writing "The Theory of Probable Inference" I fully believe that you might have had it in

---

14 the] *inserted above*
27–28 reply that … own mind.] reply that \\~~under certain circumstances~~/ that ~~may be~~ \is quite/ true, ~~or rather, that I do not intend to deny that it is true for I believe it myself~~<—>in cases ~~where~~ \in which/ an author is not ~~clear~~ \as yet ready/ in his own mind.

more than ninety very varied forms before you were satisfied. The present case, however, I hope is different and in my humble opinion you are never likely to propose again anything so egotistical as that the articles which you have sent in for reply should be worked over several times. Shall I alone have the benefit of "the snap" in your article and will you take it out after having fired it away on me? Nor do you seem to be aware of the fact that you treble my work. As for my part, I am ready to write a reply to any rejoinder of yours, but every reply takes away from my time. All the time spent on your first rejoinder has been wasted. Shall I have lost another week in which I carefully read over your rejoinder, so as to bring in shape everything that I have to say in reply? Mr. Hegeler told me positively to refuse any further revision of your article, because he says that he wants my time for writing on other subjects with which I have my mind partly filled already. There is additional work of World's Fair Auxiliary lectures besides editorial business.

I must confess that I actually regard your first version as much superior to the second one. That which you have added may be worthy of preservation, but whether it improves your reply to me is quite another question.

As editor, I wish to have articles as perfectly and well written as possible, and so I should like your article, too, to be the best you can do. Taking that ground I personally am not disinclined to admit a third revision, but I think it would be preferable to let you make your alterations in the proof sheets after the article has been set. We might otherwise keep on revising and revising until July.

At present I must let the question rest because I am too busy in other quarters. But I shall attend to the subject as soon as possible and then let us settle it for good.

With kind regards,

Yours very truly,

Peirce to Carus 16 March 1893 (missing): Perhaps discusses further the issue of working over articles. Carus references this letter in his unsent draft 19 May 1893 (critical note for line 19 on page 78) as one he would like to quote in addition to Peirce's 9 April 1893 letter below.

Peirce to Carus. MS: SIU 91.12. Note by Carus in top-left corner reads, "O. C. | *[?Review]*". Header and first paragraph struck out by Carus, and the remainder typeset to be printed as an *Open Court*

4–6 Shall I alone … you seem] \Shall I alone … me? Nor/ ~~You~~ do ∨you∨ not seem *(omitted* "not" *mine)*
10 in shape] *inserted above*
10 reply] *before* ~~together~~
18 perfectly and] *inserted above*
24 But … good.] *inserted in-line*

"Correspondence". Upon receiving the proof, Peirce, although flattered by the gesture, replied 5 May 1893 withholding consent to publish it for reasons there explained.

Milford Pa 1893 Apr 9

My dear Carus:

I send you a Shelley. I won't say "Read, mark, learn, and inwardly digest," as our old phrase goes but keep it and read it sometimes in memory of my friendship for you.

I must tell you much of your homiletic writing for The Open Court does not over well please me. It seems addressed to persons who are supposed to hang upon your lips and take everything you say as gospel truth. That is not quite philosophic I must say. But what is worse, your message to these your faithful draws upon their credulity to an extent to which no clergyman I hear would dare to ask his congregation to degrade their intellects, and still worse, it calls upon their evil passions & excites them by blind prejudice & bigotry.

Take your article on Idolatry for example. No enlightened person can fail to have a high esteem for true idolatry or can fail to think the most narrow element of the Hebrew Scriptures is the condemnation of it. Yet you seize on the word because there is a blind unintelligent prejudice against—one knows not what it is—but whatever is called by that name you use this word for that reason & apply it to religions to which it is in no way properly applicable. To do this you produce a definition which would include your own and all religious ideas. You dont reason about it. You just call on your blind followers to *hate*.

The essence of true religion involves catholicity. It must embrace in its sympathy the Christian, the Buddhist, the Jew, the Pagan,—every discerner of Gods. The pest of religion is emphasizing two penny ha' penny differences. That is what you put all your strength into.

What you say about Prayer is utterly unjustifiable. Tyndall proposed a prayer-gauge. That was scientific. It recognized the necessity of submitting to facts of observation. You without any such facts,—or rather with the whole host of them dead against you,—propose to settle the question in the *high priori* fashion. Your rationale of the matter would sound rather queer in a psychological treatise.

---

12 to which] *inserted above*
14 prejudice] pre~~d~~judice
15–22 Take your ... to *hate*] "no" *written by Carus at the end of this paragraph.*
15–17 No enlightened person can fail to have a high esteem for true idolatry or can fail to think the most narrow element of the Hebrew Scriptures is the condemnation of it.] "?" *written by Carus in the left margin here.*
24 Buddhist] Boudist *[E]*
25 pest] *b*pest *[Peirce]* ~~*b*p~~\r/est *[Carus]*

In some of our churches they use morning & evening the prayer of St. John Chrysostom, as follows: "Almighty God, who hast given us grace at this time with one accord to make our common supplications unto thee, and dost promise that when two or three are gathered together in Thy name Thou wilt grant their requests, fulfil now, O Lord, the desires and petitions of Thy servants as may be most expedient for them, granting us in this world knowledge of Thy truth, and in the world to come life everlasting." This recognizes that all prayer is for two things:—in this world, knowledge of divine things; in the world to come, life. It also recognizes that the supplicants have not attained as yet such knowledge of divine things as may be attained in this world. It therefore solemnly adopts the spirit of inquiry. But you, on the other hand, have no need of this prayer, the Summa of all prayers, because you seem to be so cocksure of everything already.

The idea that people are to pray for the sake of the reflex good it does them, is an idea so odious to any healthy mind, that it is infinitely better they cease all conscious prayer at once than practice such damned affectation.

I remember a gentleman from the rural districts,—though a real gentleman, a little gauche—who being presented to a lady at an evening party where I was, stood silent some minutes & then turned away. The introducer calling him to account, he said: "Well, I had no statement to make to her, and she had no statement to make to me." This is just about your notion of prayer. You have no information to impart to Omniscience! If this is your only view of it, on all accounts, hold your peace!

The difference between the beliefs of a Newman and a Huxley is an utterly trivial thing compared with the agreement in their real religious belief. To make this common ground felt would be the best service a philosopher could do for religion.

very faithfully
CSPeirce

Peirce to Carus late April 1893 (missing): Submits "What Is Christian Faith?" for *Open Court*, announces his "Immortality in the Light of Synechism" (EP 2:1–3, never published by OCP) to be sent shortly, and sends a portion of the MSS for his "Quest for a Method" book as a preview of the work and in hopes that Hegeler would review Peirce's 1868 "Some Consequences of Four Incapacities" in particular (see Peirce to Carus 12 May 1893).

15 damned] *deleted by Carus*
17 an] an~~y~~
18 turned] *after* ~~whe~~
23–26 The difference ... for religion] "C. S. P." *written by Carus following this paragraph.*

Carus to Peirce late April (missing): Sends Peirce a proof of a typeset portion of his 9 April 1893 letter to Carus for a potential *Monist* discussion. Also around this time he or OCP sends Peirce proofs of "Reply to the Necessitarians" (see McCormack to Peirce 24 April 1893).

Peirce to Carus ca. 4 May (missing): Submits "Immortality in the Light of Synechism" for *Open Court* or *Monist.*

Carus to Peirce. TS: RL 77. On OCP Stationery B. LPC: SIU 91.12.

La Salle, May 4, 1893.

Chas. S. Peirce, Esq'r,
Milford, Pa.,
My Dear Sir:—

Your article is duly received.[86]

Mr. Hegeler wishes me to ask you, before its publication, what your conception concerning the honorarium would be; and I should like to know whom you address in the article by "you". If you mean us, viz., the management of The Open Court, would it not be advisable to express it more plainly, so that among our readers there would be no mistake about it?[87]

Your MS.[88] and copy of Shelley announced by you have not as yet arrived.

With kind regards
Yours very truly,
P. Carus

Peirce to Carus. MS: SIU 91.12.

Milford Pa 1893 May 5

My dear Carus:

Your having the enclosed[89] set up is very flattering; but I really cannot consent to the publication of it.

---

86 "What Is Christian Faith?," *Open Court* (27 July 1893); also identified in *Peirce Project Newsletter*, 1994.
87 To cover his bases, Carus noted in the same volume that Peirce's potential allusion to the OCP as one of those "setting up a church for the scientifically educated" was unjustified (p. 3750).
88 Either "Immortality in the Light of Synechism" or the portion of "Quest for a Method."
89 Proof of Peirce's letter to Carus 9 April 1893, beginning "I must tell you …" to the end. Ultimately not published.

---

24 flattering] *over* appr

In the first place, it does not relate to any subject about which I care to say I am competent to instruct the public.

Second, it would convey to the general public & especially to an inattentive reader a very false Idea indeed of my sentiments concerning the Open Court & its management.

Third, it is intolerably exaggerated, & though I indulge in such talk to my friends—perhaps injudiciously,—I endeavor not to do so with the general public.

Fourth, its publication would involve me in the very fault I am principally blaming, namely, a tendency to exaggerate differences.

Oh no! The thing amused you, but I cannot indulge in such flippancy before the public. We have our responsibilities which make us very prosy and unnewspaperial.

I have besides lately sent you two pieces for the *Open Court*. One of them says pretty much the same as this letter; and the other which I sent yesterday contains some valuable matter,—almost worth keeping for the Monist.[90]

I am going to write to ask Mr Hegeler to come & pay me a visit. If he will not come, I wish you would; and Mrs. Peirce would be delighted if Mrs Carus would do us the pleasure of coming too with the baby & nurse. If Hegeler does come why then I hope you & Mrs Carus will favor us later.

Yours very truly
CSPeirce

I hope Trumbull[91] will say a word against the Russian Extradition Treaty, if he has not already done so. I am glad to receive a blank petition to sign against it. I shall send out my wife to get signatures. No doubt you already receive the blank.

Carus to Peirce. TS: RL 77. On OCP Stationery B. LPC: SIU 91.12.

La Salle, May 8, 1893.

Charles S. Peirce, Esq'r,
Milford, Pa.,
My Dear Sir:—

90 "What Is Christian Faith?" and "Immortality in the Light of Synechism," respectively.
91 M. M. Trumbull, Civil War General and contributor to *Monist*. See Biographical Register.

2 public] *over* g
6 intolerably] *above* ~~strongly~~
8 would involve me] *inserted above*
23 blank] *after* ~~petit~~

Shelley, and also the book containing a number of your articles,[92] were duly received Saturday, May 6th. I am very much obliged for the beautiful edition of Shelley, and have noticed with pleasure that it contains more material than other editions.

Will you kindly let me know whether the book containing your articles is intended as copy for your book to be published, or as a personal gift? In case you intend it to be a copy for the printers I should be much obliged to you for further information concerning the title of the whole book, arrangement of the articles, etc., etc.

With kind regards,

Yours very truly,
P. Carus

Carus to Peirce. TS: RL 77. On OCP Stationery B. LPC: SIU 91.12.

La Salle, May 9, 1893.

Chas. S. Peirce, Esq'r,
Milford, Pa.,
My Dear Sir:—

Your article on "Synechism" and your letter of May 5th are duly received. You are right that the article is suited for The Monist, and I will consider what I shall do with it.

I had a hearty laugh when reading your letter, but I must confess that your objection to having your lines on *prayer* published is a disappointment to me. The two articles[93] in which, as you state, you say pretty much the same, are not so plain as you are in your letter, and I dislike very much indirect statements. I would rather you would call me "a philosophical crank" directly than that you speak in a general way about "cranks" of such and such a nature. It always strikes me as a breach of etiquette in a true knight. It amounts to striking an adversary without taking the responsibility.

I should add at the same time that all your critical remarks are evidences that you do not know the aims of our work. After having seen so much of our publica-

92 Portion of his "Quest for a Method" book containing "Some Consequences" article and other pieces.
93 "Christian Faith" and "Synechism."

3 material] *inserted above*
6 to be published] *inserted above*
19 is] *after* ~~might be~~
27 breach of etiquette in] breach ~~in the~~ \of/ etiquette ~~of~~ \in/

tions you should know better. You make imputations especially on our conception of Religion, which are miles off the mark.

I am much obliged for your kind invitation, but I do not think that either I or Mrs. Carus shall find the time to accept it.

With kind regards,

Yours very truly,
P. Carus

As to the duality of nature, I have always insisted upon it very strongly. Yet this duality does not make monism impossible. Again, I have always criticized the Unitarians for their objections to the Trinity. There is a "Three-ism" in nature which is very obvious. Neither the duality nor the Trinity make monism impossible. There is further a polyism etc

Peirce to Carus. MS: SIU 91.12.

Milford Pa 1893 May 10

My dear Dr. Carus

Yours of the $4^{\underline{h}}$ just reaches me. As Mr. Hegeler wishes to know what the honorarium should be, I infer that the terms hitherto agreed upon for the Critic of Arguments no longer meet his views. In that case, it will only be necessary to return to me the two articles sent.[94] I was to have mailed you today two more papers for the Critic of Arguments;[95] but I can print them elsewhere.

I am very much surprized that my indefinite "you" could be supposed at all to refer to the management of the Open Court. The article is not a letter, nor addressed to the Editor but to the readers. But I see that my English meets your approval as little as my logic.

The Shelley has been sent. The MS. of the book has not been sent owing to reasons which I have explained in a letter to Mr. Hegeler.[96] With kind regards to all the family, I remain,

Yours faithfully

94 "Christian Faith" and "Synechism."

95 "Critic of Arguments. III. Synthetical propositions a priori" and "Critic of Arguments. That all inferences concerning matters of fact are subject of certain qualifications," R 590 (neither published).

96 6 May 1893, regarding "Quest for a Method."

1–2 especially … Religion,] *inserted above*

8–12 As to the … polyism etc] *This paragraph added in pen.*

CSPeirce

I trust we may soon have the pleasure of seeing you here.

Peirce to Carus. MS: SIU 91.12.

Milford Pa. 1893 May 12

Dr. Paul Carus
La Salle, Ill
My dear Sir:

The book of my papers was sent on merely that Mr. Hegeler might, if he liked, look over the one called Some Consequences of Four Incapacities, which contains some views somewhat similar to his own and to yours. You may also like to glance at it.

I should not dream of sending copy for printers in such a shape; but as I remarked in my last, I have fully explained to Mr. Hegeler in a long letter[97] the motives of my delay in sending copy, and have inquired whether on the whole he wished to hasten the matter. No doubt the letter reached him if not before very soon after the writing of yours of the 8$^{\underline{h}}$ which I received yesterday evening.

With kind regards

yours faithfully
C. S. Peirce

Peirce to Carus. MS: SIU 91.12.

Milford Pa 1893 May 13

Dr. Paul Carus
My dear Sir:

Yours of the 9$^{\underline{h}}$ comes to hand this evening.

My private letters are, no doubt, habitually too exaggerated in tone. My friends for the most part do me the grace to interpret them in the light of my general behaviour toward them. My friends have always been bright spiritual people.

I do not believe I so utterly mistake your aims as you say. I think, on the contrary, that it is you who mistake criticism on special sentiments, as being out of harmony with the tenor of your ideas, for aspersions upon your purposes. Remember this: I *never* criticise anybody's fundamental aims. If they are such as I do not

---

97 6 May 1893.

13–14 motives] *over* re

approve, I simply drop those persons. But if I do mistake your views, one thing is certain: it is not for lack of systematic study of them. Few doctrines at any time have ever given me so much trouble to understand them; and really, if I, trained to the study of opinions, after all the pains I have been at to make out yours, fail as utterly as you say I do, may it not be owing to some defect of method and clearness in the exposition of them?

You say that "*all*" my "critical remarks are evidences that" I "do not know the aims of" your "work." Would you extend this to all critical remarks upon any *other* work, or only to yours? Is it true only of *my* critical remarks, or of all possible critical remarks? If I mistake your aims my mistake consists in this, as you must clearly see: that I think your aims are such as I approve, while in truth they are not so.

You speak of "imputations." An *imputation* is the charging of blame, when something has gone wrong, against a person (or figuratively, an element of character) against whom it is not justly, or, at least, not naturally chargeable. The nearest to that in any of my letters has been that on several occasions I have attributed views of Mr. Hegeler and you to a narrow and sectarian tendency, very much at variance with your main spirit. The occasions have been, first, last summer when I found that, far from welcoming my articles in the *Monist*—you and Mr. Hegeler thought you were most liberal to print them, so far were they opposed to the "flag you fly." A *flag* is the mark of a sect or party. I hold that all *parties* are contrary to the spirit of science and to the spirit of religion. If I edited a Philosophical Journal, I might and no doubt should decline articles because I believed their doctrine, but never because I did *not* believe it. To do so, or feel like doing so, I call narrow, and in conflict with your deeper purposes. Second, when Mr. Hegeler informed me that he disliked to see my book published by his publishing house, without a warning to readers expressing how far he and you did, and how far you did not, endorse my doctrine, I thought and said that that was remarkably narrow and sectarian. A number of scientific and philosophical counsellors emphatically agreed with me. When, at this moment, Mr. Hegeler, if I understand him, is averse to publishing my thoughts on religion, because they differ too widely from his own, I think that is slightly narrow and sectarian,—only slightly, because I cannot claim much authority for my writing on that subject, although I have thought about it very long and deeply. When I see you, with magisterial waiving of argumentation, laying down the *dicta* of science (or what you say are such) about religion, I fancy, considering the other things, I smell a little sectarian spirit there; and when you apply to those

7 know] *over* fo
11 such] *over* al
23 and no doubt should] *inserted above*
26 he] *over* his
30 my] *after* ~~&~~

who really differ from you very little, epithets intended to do them harm, or, if not so intended by you, really sure to have that effect, calling odious names, I call that a vestige of the devil who animates the theological breast. If you call these cases *imputations*, then (unless you misuse the word) you must mean they are faults not chargeable against your spirit and intention but against something else, say to want of acquaintance with the scientific world and inexperience in dealings about philosophy. But if you mistake such criticisms for attacks upon the "aims of your work," I shall be tempted to call you dull,—or else too microscopic.

Now as to published matter. It would not be possible for me to express myself more candidly and directly, and scarcely to criticize more fully your opinions than I do in next July's Monist.[98] Except that I refrain from indicating the pigeon hole to which your doctrine belongs, I could hardly be more explicit. Controversy, however, is a tedious bore to me. My whole interest is in certain aspects of general truth; and I know not what you mean when you talk of "flags" and "knights" and "adversaries." Crass would be the mind who could see nothing in divine philosophy but a match! When I have to condemn a tendency of thought, I frequently try to show to what enormities it would lead if carried out in a thoroughgoing manner. Such remarks have always been suggested by a lifetime of experience and I have more than one or two parties in my mind's eye. If I do not pick out one of these and accuse him of carrying the fault to its worst extreme, it is mayhap because such sweeping and slashing would seem to me untruthful. I do not remember now where or what I have said about "philosophical cranks," and doubt if it had the slightest reference to you. If it squinted at all at you, it could have been very slightly. If I see a fault of method or principle in your views and choose to illustrate it by showing how it would appear if exaggerated, and if in doing so I speak of a *crank*, you can reply as warmly as the case calls for, but it is absurd to find fault with me for not applying to you an epithet which I dont think applicable to you. That is the Much-Ado-about-Nothing style.

Except in the coming Monist, my animadversions in public upon your philosophy have been few, and upon your aims and conception of religion, I think absolutely none. My private remarks have always been of this nature. You or Mr. Hegeler have said something which struck me as too narrow to fit your general position; and I have inquired, do you really hold this narrow position or this other broader one? The reply has in each case been unequivocal, we embrace the narrow doxy,

98 "Necessitarians."

8 ,—or else too microscopic.] *inserted in-line and between paragraphs*

11 refrain] *above* ~~abstain~~

32 to fit] $^{\vee}$to$^{\vee}$ ~~for~~ \fit/

we repulse the broad one. One of these days, I think you will find out you are not really as narrow as you accuse yourselves of being.

I should have no objection to expressing my ideas about *prayer* for the Open Court, should it be decided that I write for it any more. Mr. Hegeler has to decide whether he wishes to pay the same price as heretofore. If so, and if you want articles from me on religious topics, I will give my views on *prayer*. But my private letters are hastily written and are not intended for the public.

Your PS. about duality is well enough. Of course, I have always been a decided monist, though not of Haeckel's stripe.[99]

very truly
C. S. Peirce

Possibly Peirce to Carus ca. 15 May 1893 (missing): Objects further to his 9 April 1893 letter being published with the address and other header elements omitted, as Carus proposed. (Carus addresses this concern in the unsent draft of the 19 May response below.)

Carus to Peirce. LPC (of missing TS2, final), TS1 (edited) on OCP Stationery B, and LPC (of TS1): SIU 91.12. The text of the letter comes from the LPC of the missing final TS (TS2), which implements emendations made in a previous draft (TS1). The edits from TS1 are indicated below, all but the last honored in the sent version.

Chicago, May 19, 1893.

Chas. S. Peirce, Esq'r,
Milford, Pa.,
My Dear Sir:—

You are right. Your letter better remains unpublished. I confine myself strictly to my work, trying to improve every occasion in its interest, and make all my life subservient to it. Under these conditions I wish to let all critique bear upon the position I have taken. Critique is always welcome to me, but there is no use of replying to private letters which are "habitually too exaggerated in tone" as well as "hastily written". Under these circumstances it is better to confine myself only

99 Ernst Heinrich Haeckel (1834–1919), German zoologist, evolutionary theorist, and monist philosopher. His "stripe" of monism detailed in Carus, "The Monism of 'The Monist,' Compared with Professor Haeckel's Monism," *Monist* (July 1913).

23 Your letter better remains] You$^{\vee}$r$^{\vee}$ letter better remain$^{\vee}$s$^{\vee}$ *[TS1]*
24 in its] *above* ~~of~~ *[TS1]*
27 too exaggerated] to$^{\vee}$o$^{\vee}$ exaggerate$^{\vee}$d$^{\vee}$ *[TS1]*
77.28–78.1 confine … pen] *above* ~~wait until I have something matured~~ *[TS1]*

to the matured productions of your pen. Perhaps I am to be blamed that I have taken your letter too seriously.

Concerning the honorarium for your articles Mr. Hegeler will write you himself.

With kind regards,

Yours very truly,

Carus to Peirce. MS: SIU 91.12. Undated, estimated to be an initial, aborted version of Carus to Peirce 19 May 1893. The sent, acquiescent letter above is in stark contrast to this and others leading up to it, which Peirce notes immediately in his reply of 23 May.

*[*Unsent, ca. 19 May 1893*]*

Dear sir: You must excuse our compositors for altering the addresses of all the correspondence. It is a custom adhered to (so far as I know) in all journals and I am glad of it, for it is not agreeable to have one's name appear too often in one's own journal. However, if I had known that you regard it as a part of the letter itself and of consequence, I should have given order to preserve the address as it stands.

I did not take your letter to be private. Nor do I have any reason for considering it as such. I hope that you are not dissatisfied at its publication. I want to let every view be presented as strongly and concisely as possible; and found that you show a very serious spirit in your belief which always goes far in convincing others.

With kind regards
yours very truly

Peirce to Carus. MS: SIU 91.12. Stationery headed, "Arisbe, | Milford, Pa."

1893 May 23

Dear Sir:

Your anxiety to print what I have to say about prayer appears by yours of the 19$^{\underline{h}}$ to have all evaporated. It is plain you only wanted to print it because it was a particularly private letter. American editors have not a very high esteem for

---

1–2 I have … seriously.] I ~~have taken you~~ \take your letters too/ serious~~ly~~. *[TS1]*
11 all] *after* ~~prin~~
12 journals and] journals<.> ~~I <did not> \neither/ alter~~$^{\vee}$~~ed~~$^{\vee}$ ~~it, <but it is done> nor <take> took any notice of its~~ and
15 stands.] *before (new paragraph)* ~~I published your letter not~~
16 private. Nor] private *,*. ~~but as~~ *n*Nor
19 a very] ~~so much~~ \a ~~gr~~ very/
19 others.] *before* ~~I shall publish your letter of March~~ ~~16th~~ ~~too \which brings out your views/ <if> but as you may regard it as private I do not know whether your last letter is private or not.~~

the itch to print private letters. But the whole thing shows well what comes of repudiating christian sentiments.

Tell me: does the *Open Court* desire my exposition of the philosophy of prayer, or does it not? You know my terms.

Will the *Monist* print on the same terms four articles[100] by me in continuation of the exposition of my philosophy, to begin in January and the four to appear within five numbers? I will supply the articles for four successive numbers if desired. Of course, they will handle objections as I think best. You are free to decline the articles now, but I will not send them unless it is agreed, in advance of sending them, that they shall appear.

Yours very Truly
C. S. Peirce

Carus to Peirce. TS: RL 77. On OCP Stationery B. LPC: SIU 91.12.

Chicago, May 27, 1893.

Chas. S. Peirce, Esq'r,
Milford, Pa.,

Dear Sir:—

Your favor of May 23d is duly received, and I have to answer that I must decline to make such an agreement as you offer for the publication of a series of Monist articles. In reply to your question as to whether or not The Open Court desires your exposition of the philosophy of prayer, I have to state, that this subject was set forth in a private letter of yours,[101] for the publication of which you have withheld your permission. The two articles in my hands ("What is the Christian Faith" and "Immortality in the light of Synechism") do not touch upon the subject of prayer.

Yours truly,
P. Carus

Carus to Peirce. TS: SIU 91.12. On OCP Stationery B. LPC: SIU 91.12. Unsent draft.

[Unsent draft]
Chicago, May 27, 1893.

---

100 Undetermined.
101 Peirce to Carus 9 April 1893.

---

5 on the same terms] *inserted above*
7 numbers¹] *above* ~~months~~
7 for] fo*u*r~~r~~
14 27] 2*3*7

Charles S. Peirce, Esq'r,
Milford, Pa.,
Dear Sir:—

Having come to the conclusion not to publish your articles in The Open Court, I take the liberty to return them to you.

Yours truly,

Peirce to Carus early June 1893 (missing): Submits "Cogito Ergo Sum" in *Open Court* (published 15 June 1893) responding to Eugene Dreher's "Basis of Dualism" from 18 May 1893.

Carus to Peirce. TS2 (final): RL 77. TS2 (edited): SIU 91.12. Both on OCP Stationery B. The two TSS are identical with exception of one variant in the first paragraph.

La Salle, June 10, 1893.

Chas. S. Peirce, Esq'r,
Milford, Pa.,
Dear Sir:—

You ask, in a letter to Mr. Hegeler,[102] whether The Open Court desires further articles from you in the "Critic of Argument" series, and I have to reply that you are welcome to send articles but that unless a special arrangement be made, we preserve the liberty of rejecting them.

There are three articles of yours in my hands: one on "Immortality in the Light of Synechism", which I intend to publish in *The Monist*; another, "What is Christian Faith" is in the hands of the compositors, and you may expect proofs in a few days. The third one, entitled "The Critic of Arguments"[103] was handed me by Mr. Hegeler some time ago for inspection only. It was not sent as a contribution. If you mean it as a contribution, I should suggest (*at least* for its republication as a part of the whole series in book or pamphlet form) the omission of one or two passages. I think, mainly of the footnote on "uniformity of nature". We shall be ready to publish this third article (viz., on the Critic of Arguments) in about two months.

---

102 5 June 1893.

103 "Critic of Arguments. That all inferences concerning matters of fact are subject of certain qualifications," R 590 (never published). See Peirce to Hegeler 18 May 1893.

17–18 but that … rejecting them.] but that \unless a spe/ ~~I cannot bind myself to accept them~~. *[TS1]*

23 inspection] *after* ~~my~~

I enclose the two articles which are not as yet in the hands of the compositors.[104]

Please return them in case you are satisfied with this arrangement.

Yours very truly,
P. Carus

Carus to Peirce. TS: RL 77. Stationery headed, "The Monist, | Editorial Department, | La Salle, Ill." LPB: SIU 29.2.37.

La Salle, Ill., June 21, 1893.

Chas. S. Peirce, Esq'r,
Milford, Pa.,
Dear Sir:—

The new number of The Monist, I am sorry to say, will be late, and the composition work of my reply[105] to you is just finished. I shall send you a copy of The Monist, or advance sheets, as soon as I can conveniently do so. I trust that you will find no reason to complain of me. I have, of course, stated my objections without reserve, but I have endeavored to do it in such a way as to encourage our readers to study your philosophy. Should you desire to continue your discussion with me, I have to say that any further articles on the subject must be limited in size.

I have not as yet received the proof of your article "The Christian Faith" which was sent you about a week ago. I intended to publish it in the last number, but thought it wiser to await the return of the proof.

With kind regards,

Yours truly,
P. Carus

Carus to Peirce. LPB: SIU 29.3.47.

La Salle, July 6, 1893.

Chas. S. Peirce, Esq'r,
Milford, Pa.,
Dear Sir:—

Your proofs are on hand.[106]

---

104 "Christian Faith" and "Synechism."

105 "Founder of Tychism," responding to Peirce's "Reply to the Necessitarians."

106 Of "Christian Faith," although in a dispute during July 1894, both Carus and Peirce very enigmatically deny having heard from the other regarding the MSS Carus sent 10 June 1893. Yet Carus seems here to have received return-proofs for "Christian Faith" (though oddly not

Your article was destined for the number of two weeks ago, and I shall now publish it as soon as possible. The next two numbers are, I am sorry to say, already overcrowded. The number for next week will contain articles on questions of the day which I cannot hold back, and the next succeeding number will be almost entirely taken up with my lecture which is to be delivered before the World's Fair Auxiliary on "The Philosophy of the Tool".[107]

With kind regards,

Yours very truly,
P. Carus

1894

21 June 1894: Carus publishes in *The Open Court* "The Circle-Squarer," a fictitious mathematical dispute between two mathematicians—an esteemed Professor Newman and a little-known, abrasive Mr. Charles Gorner—that shows striking similarities (which Carus would deny) with Carus and Peirce and their necessitarian feud the previous year. Peirce found the representation of him unflattering and an indication of Carus's "rancorous sentiments" toward him, opening correspondence as follows.

Peirce to Carus. MS: SIU 91.13.

Confidential

Milford Pa 1894 June 25

My dear Carus:

I do not comprehend why you entertain rancorous sentiments toward me; although I feel sure they arise from some misapprehension. But the fact that I do not comprehend the origin of your enmity makes it impossible for me to do anything to bring about a better state of feeling. But though I cannot do this, I request you to postpone personal attacks upon me until next winter, when they will be just as effective in injuring *me* as they can be now, and when they will no longer harm third parties toward whom I do not think you have any resentment. I am at this time straining all my energy, with every prospect of success, to pay off *all*

of "Synechism"), which did go on to appear in *Open Court* 27 July 1893. See Peirce and Carus beginning 1 July 1894.

107 Delivered 18 July 1893, published by OCP the same year. His other World's Fair lectures included "Our Need of Philosophy" and "Science: A Religious Revelation."

21 comprehend] compr*h*ehend   28 straining] train-*[line break]* all *[E]*

my creditors, by means of some remunerative books. These books do not relate to any subject of controversy between you and me. I do not *count* a treatise on logic, the MS of which has already left my hands, and which will be printed anyway.[108] That, of course, has an indirect bearing upon our philosophical differences; although as the MS stands, owing to an oversight, no reference is made to you or your opinions. It is, however, and has been all along, my intention to make some allusion to your different conception of "explanation," and illustrate its bearing upon practical logic. I do not mean to have the book marred by any display of personal irritation against any philosophical writer. In doing so I should violate my own code of philosophical conduct.

Upon the only book, then, which I am going to print before winter which has any particular interest for you, your attacks can have no influence whatever; for it is completed, and, with the exception of inserting something accidentally omitted, the copy cannot be altered.

But in several ways your hostility might operate to prevent my paying my debts. I beg you will not extend it to that point; and if you will accede to this request I will voluntarily send you material which it would give me infinitely more pain to have you use. I do not say I permit you to use it; but I suppose you will do as you like. I am led to write this by the information that you are writing of me as a sort of "circle-squarer" and a "goner."[109]

I think the circle-squaring part, if it will help the circulation of your paper at all, will do so quite as well after my logic is out; and the "goner" part,—well if you think that will help you, you do think so. I request you to postpone such things till early winter.

Your well-wisher
C. S. Peirce

You will remark that this is written in the strictest confidence

Carus to Peirce. MS: RL 77. On OCP Stationery B.

Chicago, July 1st 1894

Charles S. Peirce Esq.

108 "How to Reason," formerly "A Quest for a Method" (neither published). MSS submitted to Ginn & Co. by invitation. See Peirce to Hegeler 11 July 1894 and Peirce to Russell 8 Sept 1894.
109 Here and in the next paragraph, likely for "gorner," as in "Mr. Charles Gorner" of Carus's article, for whom Peirce assumed he was the prototype.

1 of] *inserted above*
11 Upon the] Upon<,> th~~en,~~
16 accede] ac*ede*cede
23 think so] *inserted above*

Milford Pike County, Pa.

My dear sir: Your letter of June 25th of this year is very enigmatic to me. You speak of my hostility towards you and seem to imagine that my little satire "The Circle Squarer" was aimed at you. I assure you that you are mistaken. I cherish not the slightest animosity toward you and on the contrary am very sorry that you broke off our correspondence so suddenly. I fear that your error arises from the fact that you have some hostility towards me; and I was pained to notice symptoms of it,—not because I want your assistance or fear your criticism, but because I can not help liking you personally on account of your spirited temper and ingenious buoyancy.

My last letter had reference to several MSS of yours which we accepted, one for The Monist the others for The Open Court. I returned the MSS to you at Mr. Hegeler's suggestion, so as to leave you at liberty to keep or return them. I told you so in my letter. I have neither received the MSS back, nor did I hear from you again.[110]

In case you publish your books I wish you would not suppress any criticism of my views but write about me exactly as you see fit in the interest of your cause. I shall not take offence at your criticisms; nor did I take offence at former criticisms or yours.

Far from having any intention of injuring you by criticisms or otherwise I shall be pleased to assist you as much as I can. If I can do anything for you in the line of reviews of your new book, please let me know.

As to my story "The Circle-Squarer," it is aimed at real regular Circle-Squarers with whom I had the pleasure of being editorially bored on several occasions, as I might tell you orally should I have the pleasure of meeting you again.

Your remarks on materials which you intend to send me are not quite clearly expressed. I read the passage over several times and must wait till I know what it means.[111]

110 Carus to Peirce 10 June 1893 (though not quite his "last letter"), dealing with Peirce's "Synechism" for *The Monist*, and "Christian Faith" and "Critic of Arguments. That all inferences concerning matters of fact are subject of certain qualifications" for *The Open Court*. Carus enclosed in that letter proofs of the former two articles (though not of "Critic," which he believed "was not sent as a contribution") and confirmed the receipt of the return-proofs from Peirce 6 July 1893 (Carus's actual "last letter") of "Christian Faith," but not of "Synechism." Yet oddly both Carus and Peirce in their remaining July letters here suggest that neither heard from the other in 1893 after Carus to Peirce 27 May.

111 Paragraph beginning "But in several ways . . ." in Peirce's letter. See also Carus's June 28 draft below, paragraph beginning, "As to your remarks . . . ."

22 story "The Circle-Squarer,"] story ~~of~~
"The *S*Circle-Squarer"

I should add that this letter is private, viz., it touches only upon points between you and me. In business matters Mr Hegeler has requested me to submit to him any arrangement I might propose to make with you, for approval.

Hoping to hear from you,

I remain
Yours very truly
P. Carus

Excuse haste

Carus to Peirce. TS: SIU 91.13. Unsent draft. Note along header reads, "Not sent | a similar letter | sent *from Chicago*" (referring to that of 1 July) and *"Copy"* written across first page.

*[*Unsent draft of 1 July letter*]*
La Salle, Illinois, June 28, 1894.

Charles S. Peirce, Esq'r,
Milford, Pike County, Pa.,

My Dear Sir:—

Your letter of June 25th of this year is very enigmatic to me. You speak of my hostility and entertaining rancorous sentiments towards you, and seem to imagine that my little satire "The Circle Squarer" was aimed at you. I assure you that you are mistaken. I cherish not the slightest animosity towards you, and on the contrary am very sorry that you broke off so suddenly our correspondence. I fear that your error arises from the fact that you have some hostility towards me, and I was pained to notice symptoms of it,—not because I want your assistance or fear your criticisms, but because I cannot help having a personal liking for you and because I admire your ingenious buoyancy.

The last letter I wrote you was with reference to several MSS.[112] which we accepted, one for The Monist, the others for The Open Court. I returned the MSS to you at Mr. Hegeler's suggestion, so as to leave you at liberty to keep or return them. You did not return them, nor did you let me have any reply to my letter. In case you publish your books I wish you would not suppress any criticism of my views, but write about me exactly as you see fit in the interest of your cause. You may be assured that I shall not take offense at you.

It is true that I took exception to some of your views in my rejoinder to your reply to my criticism,[113] but if you look at our controversy with any impartiality you

---

112 Carus to Peirce 6 July 1893.
113 "Founder of Tychism," responding to Peirce's "Reply to the Necessitarians" (both July 1893)

---

24 because I admire] *inserted above* 25 letter] *inserted above*

will find that while you impute to me editorial arrogance and other little vices, I do not attack you personally. I only return your very vigorous onslaught, and never failed to show my respect for your unusual abilities and the great merits you have in the cause of the advancement of science.

Far from having any intention of injuring you by criticisms or otherwise, I shall be pleased to assist you as much as I can. If I can do anything for you in the line of reviews I shall be pleased to do so. Only let me know.

As to my story "The Circle Squarer," it is aimed at real, regular circle squarers with whom I had the pleasure of being officially bored in my capacity as editor. We have several of them on hand, and the idea of sketching this type of human beings has been in my mind for several years,—ever since we published Hermann Schubert's "Squaring of the Circle" in the second number of The Monist.[114] I cannot help laughing at the idea of any one, let alone you yourself, imagining that you had been the prototype of Mr. Gorner.

As to your remarks concerning the "materials" which you "will voluntarily send" me, I must confess I do not understand your meaning, even after a repeated perusal of the passage. You say, "I do not say I permit you to use it, but I suppose you will do as you like." Do you expect me to use materials which you would not permit me to use?

I have to add that Mr. Hegeler has requested me to submit to him for his approval all matters of business between you and me.

This letter, of course, is private, being simply in reply to your last letter touching only upon points between you and myself.

Hoping to hear from you again, I remain,

Yours very truly,

Peirce to Carus. MS: SIU 91.13. Note under date reads, "P. O. Date | Jul. 6 | 7 A. M."

Milford Pa 1894 July 4.

My dear Carus:

I am exceedingly glad to get so friendly a letter from you. Get out of your mind that I have or have had any hostility to you. I am a laboratory bred lad, & we don't harbor hostility in laboratories toward those whose experiments seem to point differently from our own. Just the contrary: those are the ones we are after. You are seminary bred, & dont perhaps fully realize how true this is.

I have written many letters to Mr. Hegeler, but, *papa super grammaticam*, he does not deign so much as to acknowledge receipt of them.

My memory of your whole family is more than agreeable.

114 Jan 1891.

I never got any letter from you accepting my MSS. & am much surprized at your telling me you sent one. On the contrary, I was much chagrined that, money having been spent in advertising my articles, you & Mr Hegeler should have been forced (as I supposed) to conclude they were not suitable. I have now written out my whole logical system.[115] It is a bad book, from the point of view of book-making,—jamming & stuffing in as well as I could like an impatient traveller into his trunk, too many things. But it will help young men to learn to think in a very different way from any of the current textbooks, in my opinion.

Whether I could not now, having made this *précis* of my ideas, give you something better than any thing in those MSS (whether I could find or identify them now I dont know) would be a question.

I thank you warmly for your kind offer. I am working 12 and often 15 hours a day on my books. I have a good many things all written. If I could get some cash for some of them, it would be of great consequence to me. My books the publishers promise me great things from. But the most important man has gone to Europe with a sick wife & nothing can be done in his direction till his return, in the autumn.

I dont know what you mean by not "fearing" my criticism, unless it be that you have no fear of not learning anything from it. What we want is to learn, I suppose. I shall not refer to you except as to one point, because that is the only point where such reference seems to me to aid in clearing up the ideas of my readers, or rather students,—for the book is no light reading. The book is confined to logic; and even that is ginned into one volume—[A *gin* is cotton-baling machine.]

very truly
CSPeirce

Peirce to Carus. MS: SIU 91.13. Note under date reads, "P. O. Date | Jul. | 6 | 12 M. | 1894".

Milford Pa 1894 July 6

Dr. Paul Carus

My dear Sir:

I did not in the letter you answered say anything (as you understood I did) with a view to your taking any articles of mine. That had not occurred to me since

115 "How to Reason," though by this point or soon to be retitled "The Art of Reasoning," outlined in Peirce to Russell 8 Sept 1894. Peirce made several attempts at a comprehensive system or treatise on logic, and in 1902 even got so far as to submit an elaborate grant proposal to the Carnegie Institute detailing his logical system in "Memoirs on Minute Logic" (see Ransdell, "MS L75"). By late 1908 he would make his "System of Logic" his "chief work" for the remaining few years of his life (see Peirce to Carus 6 and 7 Jan 1909).

you returned the articles I had sent, without a word. But since you suggest it, I will say that I have a number of sketches of great men, and various things about the History of Science, & a variety of other pieces some of which might be found serviceable for you.

I enclose one of Rienzi as a sample; if you like it, you can take it. I have also sent a long article some 6 000 words on Philosophical Reflections on Table Turning, to be type-written & will send that.[116]

If you could take some things & pay me cash for them, it would be extremely useful; for since I wrote you a week ago I have met with three new thunderbolts, I mean disappointments about money I expected. I never in my life saw anything that approached the present state of the times.

I enclose a note for Mr. Hegeler.[117] Please present my respects to Mrs Hegeler, Mrs Carus and the other members of the families.

yours very truly
C. S. Peirce

Peirce to Carus. MS: SIU 91.13.

Milford, Pa. 1894 July 9

Dr. Paul Carus:
My dear Sir:

I offer to contract to give you a page and a half a week for the *Open Court* for a year, carefully avoiding conflict with Mr. Hegeler's religion, choosing topics of the day in the philosophical & scientific world as much as possible, in other cases dealing chiefly with questions of good & bad reasoning, evidence, etc. and also with points in the history of science & of thought in general, but keeping as much as I can to the concrete, consulting the requirements of the journal as much as possible, & writing in a lively style with a view to enlarging the subscription list.

The condition to be that I am to be allowed 6 000 to 8 000 words in each number of the Monist to further set forth my philosophy; and for pay, say $ 60 a week for the Open Court articles and $ 150 each for the Monist articles.

Yours very Truly
C. S. Peirce

If you prefer a shorter period, it will suit me.

116 "Rienzi, Last of the Tribunes" (R 1318); no MS found for his "Table Turning."
117 Peirce to Hegeler 6 July 1894.

9 three] *above* ~~two~~ 25 journal] *over* ge

Carus to Peirce. TS: RL 77. Stationery headed, "The Monist, | Editorial Department, | La Salle, Ill." LPB: SIU 29.46.369. Bottom-right corner of TS torn off. Missing text supplied by LPB.

La Salle, Illinois, July 10, 1894.

Chas. Peirce, Esq'r,
Milford, Penn.,
My Dear Sir:—

Your letters of July 6th and July 4th, the former including a MS. on Cola Rienzi, were duly received, and I think that your "Philosophical Reflections on Table Turning" would be most in our line. If you send the MS. please send it registered, so as to make sure of its delivery. It is the plan of The Open Court to limit its scope to our special problems and not to give too much space to topics of a purely literary or historical interest, and this is the reason why I expect to be under the obligation of returning your MS. But I will wait and see what Mr. Hegeler will say about it. I read your sketch with great pleasure; it is admirable in it is conciseness.

Yours very truly,

You will confer a favor on me by omitting all remarks on your relation to Mr Hegeler, such as made in your letter of July 4th. I must leave many of your remarks unanswered (so for instance your reflection on the distinction between seminary-bred and laboratory bred). If I do not answer them do not take my silence as an assent. This is especially true concerning your attitude toward Mr Hegeler for I look at the matter in a different light than you do.

With kind regards
P. Carus

If you send the MS on Table turning (I suppose you mean Spiritualistic table turning) send it as it is and I shall have it typewritten here in LaSalle, leaving you the last revision of the copy.

Peirce to Carus. MS: SIU 91.13.

Milford Pa 1894 July 11

Dr Paul Carus
My dear Sir:

---

15 Yours very truly,] *Everything below added in pen.*

24–26 If you … the copy] *This paragraph down left margin.*

I enclose an article[118] for which if you choose to use it I will charge you $ 100 which I beg you to send me.

I would rather if it be question between the two insert four articles concerning my philosophy.

I forgot to mention in my last that I should want to reserve right to reprint my articles. Two of the articles about my philosophy would be taken from my logic & the other two I should expect to use otherwise.

I enclose, especially for Mr Hegeler, some account of what my logic contains.[119]

I think I quoted from the enclosed article in my Evolutionary Love. It is a bad habit of mine to plagiarize myself. But I am not sure & don't believe it will be noticed. At any rate, this is the original & in charging $ 100 I have deducted for that.

Yours Truly
CSPeirce

Peirce to Carus. MS: SIU 91.13.

Milford, Pa. 1894 July 12.

My dear Dr. Carus:

I can have said nothing in my letters taking any "attitude" toward Mr. Hegeler. I never have taken any attitude toward him but that of regard most warm and esteem most high. I believe I said he did not answer my letters; but that is not an attitude but a fact. Whether he gets them I don't know; and, if he does, whether he has anything against me, I cannot guess. But this I have always thought, that if there be any misunderstanding (of which I have never had the slightest intimation) it can only be such as a little conversation would remove.

In my letter to you reopening correspondence,[120] I said nothing about writing for you. I had no such idea in my head. But you expressed a desire to take some of my articles in case I was short, of money, & also informed me that two sent back to me without a word had been so sent *as being accepted*.[121] I thereupon sent you

118 His typewritten "Table Turning" article, mentioned 6 July as forthcoming and referenced in his next letter, paragraph beginning "In my letter to you...."

119 "How to Reason." See Peirce to Hegeler of same date. See also Peirce to Russell 8 Sept 1894 for a detailed outline.

120 25 June 1894.

121 "Christian Faith," or possibly "Critic of Arguments. III.," and "Synechism."

8 Hegeler] H*g*egeler

27 of money,] *inserted above*

two articles; and also a proposition to supply a number.[122] You reject one of them, unless Mr. Hegeler wants to keep it. The other I send is not of my best, & I do not, of course, want you to take it unless you want it. I have a large amount of matter in my portfolio. If you are in want of anything from me, I have of all sorts.

When any money *is* sent, (if any should be sent) it will oblige me if it is sent in a form to be immediately available; for we are in want of food and light. I think it unlikely that we shall pull through. It is now near forty-eight hours since I tasted food and then not much. But do not think I offer myself as object of pity. Whatever suffering there was is done. I have tried hard to do my duty, & I leave to the world a work which amply pays for my keep. I have treated my fellow-man with brotherly love. I am ready to face whatever is in store for me. I feel already far removed from the jealousies and vanities of this world, in regard to which, for that matter, I never did much take them to heart.

If I do pull through I shall feel like a second Lazarus. If I get any money, the first of it has got to pay a certain bill before my wife or I touch food.

I trust your families are all well. It is very beautiful here.

very truly
CSPeirce

Shall I forward the "accepted" articles?

I have no money to Register Letters. You make me grin. Luckily I have few P.O. envelopes.

Peirce to Carus. MS: SIU 91.13.

Milford Pa 1894 July 15

Dr. Paul Carus

My dear Sir:

I desire to say that should my proposal for a general arrangement about my writing for the Open Court & Monist have been accepted (as I suppose by this time it has either been accepted or rejected) I wish to be credited on Mr. Hegeler's private books with the amount from the four Monist articles; for I can scrape along on the $ 60 a week for the Open Court articles

---

122 "Rienzi" (rejected, as Peirce notes here) and "Table Turning," also some "four articles concerning my philosophy" to be taken instead of the former two if needed. See previous letter.

---

5 (if … sent)] *inserted above*
8 and then not much] *inserted above*
8 object] *over* pit
20–21 I have … envelopes.] *This paragraph down left margin.*

Though this has nothing to do with my business with the two journals I send it to you being doubtful whether letters sent to Mr. Hegeler ever reach him.

Yours very truly
CSPeirce

P.S. Having found some .5 cent stamps I will register this, according to your suggestion.

Carus to Peirce. MS: RL 77. Stationery headed, "The Monist, | Editorial Department, | La Salle, Ill." LPB: SIU 29.47.386.

July 16th 1894

Charles S. Peirce Esq.

Dear sir: I handed your two letters[123] to Mr Hegeler and also informed him of the MSS which you offer for publication, but he said he had no answer to make and did not want to have your MSS accepted. When I advised him of the emergency in which you were at present, he handed me for your immediate needs a five dollars-bill which is here enclosed at his request. Under these circumstances I must return your MSS and can do nothing except to hope that you will pull through.

Yours very truly
P. Carus

Peirce to Carus. MS: SIU 91.13.

Milford Pa 1894 July 20.

Dr. Paul Carus

Dear Sir:

Yours of the 16th is received, and it seems plain that Mr. Hegeler is for some reason bent upon being as insulting as he can. This only confirms me in my belief that the difficulty is due to some misapprehension, or to some thoughtless word of mine of which I am quite unaware, and that a little frank conversation would probably set matters right. I am willing for my part to show my good will by contributing occasionally to the Open Court, or, if agreeable to respond to some points in your last year's Monist article[124]; and I should be glad to do so gratuitously, only that I dont think in my circumstances I have a right to be making

123 11 and 12 July.
124 "Founder of Tychism" (July 1893).

11 two] *inserted above*

such presents. Consequently, I shall be obliged to require in the event of anything of mine being printed that I be credited at the rate of $ 25 a thousand words on Mr. Hegeler's books.

One of my letters to Mr. Hegeler called for some action.[125] I should be glad to be notified if he does anything. Kindly hand back to him the enclosed $ 5. My remark about being in need of money had reference to my receiving in cash a part or the whole of any money I might earn. In this connection I may remark that I have never been informed whether I have been credited with the pay for the last article[126] by me printed in the Open Court or whether it was perhaps inadvertently that payment was omitted.

Yours very Truly
CSPeirce

Carus to Peirce. TS: RL 77. Stationery headed, "The Monist, | Editorial Department, | La Salle, Ill." LPB: SIU 29.49.427.

La Salle, Illinois, July 28, 1894.

Chas. S. Peirce, Esq'r,
Milford, Pa.,
Dear Sir:—

Your request concerning payment for an article still due you which appeared in No. 286 of The Open Court under the title "The Marriage of Religion and Science," touches matters of The Open Court which I, as editor, can directly attend to. The article containing 1 095 words, I send you a check for twenty-seven dollars and fifty cents ($ 27.50) in payment for it.

Hoping that you will soon overcome your present difficulties, I am,

Yours very truly,
P. Carus

125 Peirce to Hegeler 6 July 1894, regarding Peirce's library. Footnote corresponding to this point by Diesterweg also indicates such: "refers to taking possession of the library I su/p/pose | c. D."
126 Although his last article was "Christian Faith" (27 July 1893), Peirce is referring to the article before that, "Marriage of Religion and Science" (16 Feb 1893), as Carus also notes at the bottom of Peirce's letter:

> The article referred to was published in No 286 of Vol VII of The Open Court under the title "The Marriage of Religion and Science" and contains according to Dr. Garbutt's ~~exact~~ calculation exactly 1 095 words.

Details remain unclear about the exchange of "Christian Faith" and the other MSS between Carus and Peirce June and July 1893.

Aug 1894–May 1896: For the next two years the correspondence goes silent. By this time Peirce had fallen out of favor and contact with Hegeler due to consistent complications and misunderstandings with agreements, which consequently affected his standing with the OCP, including Carus. Only Russell would remain in touch.

During the two-year silence between Peirce and OCP, Ernst Schröder publishes his 1895 *Algebra und Logik der Relative*, the third volume in his *Algebra der Logik* series. Carus asked Russell ca. Sept 1895 to review the book, but Russell suggested instead that Peirce be invited to do so. Carus refused and Russell consented, although secretly asking for Peirce's help 25 Sept 1895. In reply, Peirce sent Russell 4 Oct several of his papers on logic and remained in touch on related matters.

By 25 Feb 1896, apparently influenced by Schröder's insistence that Peirce was the only one suited to review his *Logik*,[127] Carus asked Russell to invite Peirce to write the review instead. "As matters are," Carus adds, "I cannot ask him to write the review, but if he would send it I would not refuse it.... Mr. Hegeler does not wish me to be in any business connections with him, but that would not exclude the acceptance of a contribution which he might offer at a cash price, the bill to be settled at once ... without further complications."[128]

Per Russell's invitation 22 April 1896, Peirce submitted "The Regenerated Logic" and likely other logic of relatives material to Russell ca. 30 April 1896, which Carus received 8 May and opened correspondence as follows.

1896

Carus to Peirce. TS: RL 77. Stationery headed, "The Monist, | Editorial Department, | La Salle, Ill." LPB: SIU 30.85.100.

May 9, 1896.

Charles S. Peirce, Esq'r,
New York City, N. Y.,

Dear Sir:—

Through Mr. Russell's kindness I received last night an article[129] from your pen on Schröder's New Logic. Mr. Russell informs me that the article was so arranged

127 Brent, 256; also *Peirce Project Newsletter*, 1994.

128 TS: SIU 91.7.

129 "The Regenerated Logic," *Monist* (Oct 1897). Although Peirce submitted the article to Russell ca. 30 April, Russell delayed in forwarding it to Carus so that he could extensively work it over to appear as his own (as Peirce had previously agreed to), despite Carus's request to have it under Peirce's name, and despite his insistent pleadings to Russell for the MS, noting that Peirce was "probably anxiously awaiting his check" (2, 7, and 9 May 1896, LPCs: SIU 30.85–86). This led Peirce to believe that Carus had broken his promise to remit immediately upon receiving the MS, as

as to appear to be written by himself. I would, however, prefer that the article should be published under the signature of its real author, and I know positively that Mr. Hegeler would not be pleased with any pseudo-personality. I have not as yet been able to read the article through, but what I read pleases me sufficiently to accept it on the terms mentioned by Mr. Russell, that is to say, on the payment of $ 210.00, for which amount I enclose a check.

With best wishes for your scientific success and personal welfare, I remain,

Very sincerely yours,
P. Carus

Peirce to Carus. Telegram: SIU 91.14. Form headed, "Night Message. | The Western Union Telegraph Company."

New York May 11th-96

Dr Paul Carus
Open Court Co
Merron Bldg
Chicago

According to promise you should have remitted ten days ago

Peirce
1248am12

Carus to Peirce. TS: SIU 91.14. On OCP Stationery B. Estimated to be an unsent letter replaced by Carus's of 20 May 1896. The fact that the original TS remains in OCP records and is unsigned is also evidence that it was not sent.

[Unsent]
Chicago May 12th, 1896.

Chas. S. Peirce, Esq.,
Stewart House
W. 41st St., Broadway, N. Y.

Dear Sir:—

This morning I received the following telegram: "According to promise you should have remitted ten days ago."

---

he discussed with Russell 25 Feb 1896 and which Russell relayed to Peirce 22 April 1896. Hence Peirce's telegram below.

---

25 Peirce] Pierce *[E]*

(Signed) Peirce.

As I have not the slightest idea what kind of a promise I made I can only conclude that the telegram was mis-directed.

I received a manuscript from your pen through Mr. Russell, on Friday last by the evening express, and Mr. Russell wrote me that you were anxious to receive the honorarium at once; whereupon I made out the check on Saturday, which together with my letter was mailed to you, through Mr. Russell's kindness on Monday, the 11th.

I wish to state here that all our business transactions are on cash terms. If you offer articles, I shall, on the terms mentioned by you and stated in definite figures, accept or reject them. On other terms I cannot enter into business relations.

Hoping that the check has reached you safely, I remain,

Very truly yours,

Peirce to Carus. MS: SIU 91.14.

New York 1896 May 13

Dr. Paul Carus

Dear Sir: Yours of the 9th inst. with enclosures reached me today. After writing the article which went on to you first, it was still my duty to express myself in regard to minuter points contained in Schröder's last volume. I thought perhaps these would be too technical for the Monist, & I proposed printing them in the Bulletin of the Am. Mathematical Society. At the same time, I thought *perhaps* you would think that the dignity of the *Monist* required that the notice should be addressed mainly to students of the particular branch. I therefore sat down to write something which if you want it you can have & in that case I will give still more technically mathematical matter to the Mathematical Society, or if you do not want this new matter, or not all of it, then *that* can go to the Mathematical Society. At the same time, I rewrote certain parts of my article which I feared were too perfunctorily done, being matter I am tired to death of.

I am very glad you have decided to sign my name, first, because I think that course alone befits the dignity of your now leading publication, & then because, being so often referred to by Schröder, it is respectful to him that I combat what I have to combat face to face. While I hold that he has greatly, very greatly, advanced the subject, I nevertheless hold that the problems he has attacked were not those most urgently demanding study, that they have been selected rather by a mathematician than by a logician, and while *very valuable*, leave still more valu-

1 Peirce] Pierce *[E]*

6 honorarium] honararium *[E]*

able results awaiting the man who shall take up the study from a different point of view. It is curious that he has hardly duplicated anything of my extensive unpublished results.

I suggest that the article be printed in small type so as to enable me to say as much as possible; since I have far more to say than I can put into two reviews, and that the very maximum number of words you are willing to allow be mentioned to me with any wishes or suggestions you may have to make & then the copy (after examination by you & Russell) be returned to me to put into final shape.

I suppose you will not object to having a number of new types cut, which will be necessary for the notation. Rather than not have this done, I would pay for it myself.

It is a pleasure to receive a courteous letter from you. There is I think no element in all the misery I have suffered of late years which has cut me more than the consciousness to Mr. Hegeler's bad opinion of me, which is undeserved & is really due to the extreme difference of our educations & experience. No man who has always prospered as he has can possibly have any conception of what it is that misery consists in. I have often suffered of late years from cold & hunger, have gone for days without shelter or a morsel of food; but there is no misery in that. I don't suppose Mr. Hegeler will ever change his opinion about me; to that I cannot look forward. But I am determined that sooner or later I will at least square up with him in a monetary way.

Yours very truly
C. S. Peirce

5 W 50$^{\underline{h}}$ St | New York City.

Carus to Peirce. LPB: SIU 30.86.120.

May 20, 1896.

Charles S. Peirce, Esq'r,
New York City, N. Y.,
Dear Sir:—

Your telegram of May 11th and your letter of May 13 were duly received. As to the telegram I have to say that I made you no promise whatever. I merely said to Mr. Francis C. Russell of Chicago if I received a MS. from you on Schröder's book

6 number] *above* ~~yo~~

I would accept it and pay your price.[130] I ordered a check to be made out on the very date on which I received the MS., and even if I had made a promise you have no reason to complain about my negligence.

As to the offer made in your letter I fear that I cannot accept it. I could not publish it in the present number, and would have to defer the publication to a future number. I mentioned your offer to Mr. Hegeler, and he made the remark that it would probably be enough to publish the article on Schröder which is now in my hands. At any rate I could *not* publish it *in the next number* of The Monist.

With kind regards and best wishes, I remain,

Yours very truly,
P. Carus

Peirce to Carus. MS: SIU 91.14. Undated, in response to Carus of 20 May and responded to by Carus 2 June.

[ca. May 25, 1896]

Dr Paul Carus

Dear Sir:

I did not mean that any engagement in regard to the notice of Schröder had been broken. I may have implied that a *promise* had not been kept to the letter. But a promise is not necessarily a contract, inasmuch as it may be carelessly worded and thus convey hopes that the party making it never had it in his mind to convey. That was the case on this occasion; and I never meant more than that. I ought to have reflected that it could not have been intended as I understood it. But I was sailing so very close to the wind that I was naturally and almost inevitably mistaken. Let this apology for my words of impatience close that matter.

The MS. I sent on first was written in a very bad physical condition. As the precise form it was to take had not been fixed, it necessarily would require much editing. As it is to appear in my name, I wish to do that editing myself. Moreover, I do not like the article very well for several reasons. First, it does not give a sufficient account of Schröder's work nor sufficiently characterize it. Second, it has too much about myself; and is too much taken up with an exposition of my own theories. Third, it does not always do justice to Schröder and contains several mistakes and imperfect statements. I was already engaged to write two notices of Schröder

130 Carus to Russell 25 Feb 1896, as noted in the heading preceding these 1896 letters.

18 broken] *over* p   28 does] *after* ~~has~~

when Russell's letter came.[131] I have since decided to write two more and have spent all the time I could give since to writing about it. I consequently have a large choice of materials. As only one article besides that for the Monist is to be paid for, and that one is short, I am desirous of using my materials first of all so as to give you the best article I can make for your purpose. Accordingly, I am holding back all my other articles, so that they shall not use any matter that might be wanted for the *Monist* notice. But I cannot do so long.

As you are bound to allow me to edit and revise the MS. sent, if you wish to put my signature or initials to it, I suggest that you send it to me at once, and say what treatment of it would be pleasing to you. There is no question of any further payment, unless it should be lengthened, which of course I should not do of my own motion. But for the reasons given, I propose to rewrite it. With my mind actively upon the book as it now is, I could do that in 3 days. I will follow any general wishes you may express in regard to the nature of the notice, as to avoiding excursions, as to addressing myself to those who know more or less about the subject, etc. etc. In the absence of such instructions, my plan would be as follows. The general standpoint of exact logic to be explained. Boole's original system how defective. Question of Quantification of Predicate. Two ways of remedying prime defect of Boole's logic, that adopted by me, and that adopted by Schröder. These methods connected with different views concerning hypotheticals. (Very brief this part). The logic of relatives. Its history. Explanation of the general algebra of relatives and the algebra of dual relatives, which are the two algebras chiefly used by Schröder, both of my invention. Other notations employed by him. His theory of individuals and individual pairs. The general problem of the logic of relatives as conceived by me and as conceived by him. His view equational and aniconic. Why I thinkthat which he aims at is not the principal desideratum. His great skill in treating the problems he aims at fully illustrated. Trusting to hear from you shortly, I remain,

yours very truly,
C. S. Peirce
160 W 87$^{\underline{h}}$ St. New York City.

131 See Peirce to Russell 26 April 1896.

5 the] the *[page break]* the *[E]*
6 any] any~~t~~
18 Question of Quantification of Predicate] *inserted above*
18–19 prime defect of Boole's logic] *above* ~~it~~
25 and aniconic.] *inserted above*
26 that] *over* h

Carus to Peirce. TS: RL 77. Stationery headed, "The Monist, | Editorial Department, | La Salle, Ill." LPB: SIU 30.90.189.

June 2, 1896.

Charles S. Peirce, Esq'r,
160 W. 87th St.,
New York City, N. Y.,

Dear Sir:—

Your letter without date is duly received, but I hardly know what to say in reply. Our next number of The Monist is overcrowded and an extension of your article would on that account not be very desirable. On the other hand I wish that *your article should be exactly as you want it*, and for that reason I leave you at liberty to remodel it and to re-write it just as you see fit. But remember one thing, we must go to press not later than the 10th of June, which leaves a very short margin for making corrections, even if I leave your article for the very last. As to your emendations I give you no instructions, except to do as you see fit to alter without extending the article. Consider, if there is any reason for omitting your article in the present number (by its being belated) I shall be unable to accept another article from your pen for the next number, while otherwise *there is at least a possibility* of doing so. The plan which you submit of writing an article on the subject appeals to me, and I must confess that I wish the present article would discuss the subjects which you mention. I send this letter off in a hurry so that it will reach you in time. Your article is in the hands of the compositors and I directed that the proofs which came to my hands be sent to you. I send them to you unread and uncorrected. They may be full of mistakes. I also enclose the MS. for safety's sake by express; but please return at your very earliest convenience.

With kind regards and best wishes, I remain

Yours very truly,
P. Carus

Please do not add new symbols since there is no time left for having them cut.

Peirce to Carus. MS: SIU 91.14. Not actually sent until 19 June 1896. After writing the letter 6 June, Peirce held it and the enclosed proofs of "Regenerated Logic" for further reflection and, assuming it would not make the 10 June deadline anyway, finally returned the MS to Carus ca. 19 June (see

15–16 to alter … Consider,] *inserted above; before* *I*if
17 (by its being belated)] *inserted above*
20 would discuss] \would/ discuss~~ed~~
24–25 I also … convenience.] *inserted partly above and below*
24–25 safety's sake by express;] safty's sake by xpress *[E]*
29 Please do not … them cut.] *This line added in pen.*

Peirce to Carus ca. 24 June); hence OCP's and Carus's desperate pleas of 19 and 20 June. Carus confirmed its disappointingly late receipt 22 June. The article was ultimately postponed to the October number as a result.

[6 June 1896]

Dr. Paul Carus
Dear Sir

I received the MS. Friday June 5 at 11 1/2 A.M. I might have mailed it back that night but considering it would reach Chicago Sunday morning, I allowed myself till this morning, June 6, for reflection. I would rather it should appear in its original condition than lose the chance of a second article. So I leave you to do what seems best. I have struck out all symbols that could give the compositor trouble. If any are already set up, they may stand. I think I have shortened the article slightly. If you wish to abridge, leave out the scheme of the sciences, and if that is not enough, any one topic, say the last two pages.

Yours very truly
C. S. Peirce

I reenclose the two galleys sent. Also the whole original copy, so that if there is no time for the revision, this may go in & the proofs be corrected.

Carus to Peirce. TS: RL 77. Stationery headed, "The Monist, | Editorial Department, | La Salle, Ill." LPB: SIU 31.1.257.

June 20, 1896.

Charles S. Peirce, Esq'r,
New York City, N. Y.,
Dear Sir:—

I am in great despair at not having received the return-proofs of your article, and as I know that you are very particular I do not dare to insert the article without having received your sanction, especially as there was some confusion in the copy. Several pages were doubled and we arranged them as we thought best.

I would not mind so much if I did not know that you wish to write another article, and therefore I wished to insert the article now on hand in the present number. Moreover, we need the type, and it is very unpleasant to have several galleys of type tied up for several months. I do not know what is the cause of the delay, whether the proofs are lost in the mails, or whether sickness prevents you from attending to the proof reading. Please let me know at once, for we cannot longer wait. We have lost more time than we can conscientiously allow, and we must make up the last forms of The Monist tonight.

The plan of an article such as you sketched in your last letter[132] pleases me greatly. It is really what I wanted you to write. Without making a promise that I shall accept the article, I can only say that I feel confident that such an article would be desirable.

With kind regards and best wishes, I remain,

Yours very truly,
P. Carus

Peirce to Carus. MS: SIU 91.14. Undated, responding to OCP's and Carus's of 19 and 20 June, respectively, and answered by Carus 24 June.

[ca. 22 June 1896]

My dear Sir:

I am both mortified and disappointed to find what a misunderstanding I fell into about the proofsheets. Owing to your mention of June $10^{\underline{h}}$ as the last day, and remembering a former explanation to the same effect, I supposed on receiving the proofsheets *later* than that date, that all possibility of the July insertion was gone. True, the word "forthcoming" was used in the note accompanying them; but nothing more was said of haste; and that word was overweighed in my mind by what I had been informed of the serious necessity of going to press on the $10^{\underline{h}}$. So, as I was for several days so fatigued by other things as almost to have lost the power of speech, which often happens to me, I kept the Proof Sheets till I felt fresh. Now I see I have thus lost the chance which I should have been only too glad to have,—and what is still worse, I have put you to most serious inconvenience for nothing, which I would not have done for the world. I should infinitely have preferred to let the last corrections go unmade.

I enclose herewith a small piece on an interesting bit of scientific history.[133] I have not much hope that you will find it suitable for either of your publications. But as I have just finished it, and do not know what to do with it, except put it in the chest which holds a great many such papers that are never likely to see the light, I will offer it to you. Should you not use it, I should be obliged if you would kindly return the MS.

yours very truly
CSPeirce
160 W $87^{\underline{h}}$ St New York

132 25 May 1896.
133 "Note on the Age of Basil Valentine," rejected by Carus (24 June 1896) but later published in *Science* (12 Aug 1898).

P.S. It would hardly be necessary for me to read the proofs of the enclosed. Only having no books by me, I may have the Emperor Mathias and Rudolph II confounded. In that case, a simple correction is called for.

Carus to Peirce. TS: RL 77. Stationery headed, "The Monist, | Editorial Department, | La Salle, Ill." LPB: SIU 31.1.259.

June 22, 1896.

Charles S. Peirce, Esq'r,
160 West 87th St.,
New York City, N. Y.,
Dear Sir:—

At last your proofs came, but I am greatly vexed that they came too late for insertion in the present number. The last forms left on Saturday and are now on the press. It was quite inconvenient to us to rearrange the contents in The Monist in the very last moment and to replace your article by other material. This is the more disagreeable to us as we have announced your article, in several quarters, and are now obliged to disappoint our readers. Well, it cannot be helped and we must make the best of it. Your article shall appear in the next number, and I hope it will be as effective as we might have expected it if it had appeared this time. As to another article, I repeat that I cannot make any promises. I only say that I like the plan which you sketched out in your last, but one, letter.[134]

With kind regards, I remain,

Yours very truly,
P. Carus

Carus to Peirce. LPB: SIU 31.2.267.

June 24, 1896.

Charles S. Peirce, Esq'r,
160 W. 87th St.,
New York City, N. Y.,
Dear Sir:—

Your MS. and letter were duly received this morning. I am very sorry that the delay was due to a misunderstanding which arose from the very fact of my trying to accommodate you. I kept the forms back as long as possible, but we could wait no longer and you may be sure that there are few for whom we would hold the forms

134 25 May 1896, plan for a second "Logic of Relatives" article.

that long. But never mind, perhaps it is better as it is. Your article will appear in the fall number, which may after all be more widely read than the summer number, and an article on the "New Logic" as outlined by you in a former letter will not suffer should it appear later. On the contrary, if you can elaborate it without being hurried the article would undoubtedly be better.

As to the article on "Basil Valentine" I am sorry that I can do nothing with it, at least not at present. The subject is very curious and interesting. Nevertheless, it does not fall, as you well know, within the lines of either The Open Court or The Monist. I might accept such subjects during a time of dearth, but we are at present, I am sorry to say, extraordinarily overcrowded, and I have several series of articles on hand which I wish to bring out as soon as possible. Moreover, Mr. Hegeler is at present not at home. He has gone East and it is difficult to reach him by letter because he is travelling from place to place and will not be back for quite a while. I do not know how long he will stay.

With kind regards and best wishes for your welfare, I remain,

Yours very truly,
P. Carus

Peirce to Carus. MS: RL 77. Undated, responding to Carus's of 22 and 24 June 1896.

[ca. June 26, 1896]

My dear Sir:

My last letter to you was written before I had received your kind letter. I had only received a letter from the Open Court Co. in Chicago.[135] I do not know, however, that I can add anything except a renewed expression of my extreme regret. I have little hope that you will care for the article on Basil Valentine; but if it can be made available by cutting it down about one half or less, I will do it or you can do so.

You had not given me before any positive hope that you would take a second article from me on the Logic of Relatives. But your last letter makes me think that you might do so. The plan you liked so much would have to be considerably altered on account of what has already been inserted in the first article. I should think the most timely thing now (say for the December number, if the first has to go over till October) would be an account of what the logic of relatives really is, in itself, and also of the two algebras which I invented for the study of it. Were it not

135 Referring to Carus's and Peirce's 22 June 1896 to each other, and OCP to Peirce 19 June 1896.

33 for] *over* to

for Schröder's book I should be content to explain one of these,—that one which he does not prefer, but which is both simplest and most powerful. But in that way, it would be impossible to make Schröder's work comprehensible. In a single article, I could not both explain what Schröder has done and also what I conceive to be the sort of algebraical work which really most illuminates the new logic. (Far less, could I enter upon the philosophical bearings of the doctrine.) But I could do whichever of those two things you would prefer. If you would like the article, I should like to write it as soon as possible, because a company are talking of sending me to Europe to do some chemical work for them; and when I am embarked on that, I shall hardly have time to turn my thoughts to logic. But it is far from certain that the plan will be carried out, though they are very desirous it shall be. If I succeed in making some money, of which there is good prospect, I shall think it my duty (after squaring up with all creditors & putting out some unfinished and finished educational books) to lay out my life so as to get my philosophical ideas fairly developed and set forth. They may be wrong (as you think) but they deserve to be so presented that their value or valuelessness may be intelligently judged. In that contingency, *[ ]*

[...] may have so much of my subsequent results as can be stated in a few pages. I have stated some of the circumstances, in order that in the future my not carrying forward the development may in some measure be understood, on pp. 54 and 55 of the MS. But if this passage is going to call for any reply upon your part, I will not consent to its being printed, and you can simply omit it; for I wish to write about philosophy and not to squabble.

I expect, in a short time, perhaps before the first of April, to be in a situation to commence payments upon my debts.

Yours very faithfully,
C. S. Peirce

Peirce to Carus. MS: SIU 91.14. Undated, referenced in Peirce to Russell 16 July and answered by Carus 24 July.

*[*ca. 15 July 1896*]*

Dr. Paul Carus

My dear Sir:

I have the honor to submit a paper on the Logic of Relatives. It would be impossible for me to write this later. It is excessively long; but it could be cut almost exactly in halves at p 36 of the MS and each half would be in some sense complete by

---

5 algebraical] *inserted above* 24 be] *after* ~~be~~

itself. Each would also contain a positive contribution of undoubted value (if the logic of relatives be of any value.) Namely, the first half gives a graphical method of representing the subject, which throws much light on it, and is remarkably clear considering how far it carries the analysis of the meanings. The second half gives some improved rules for the general algebra, a doctrine of "involution" which has never been published, criticisms of Schröder, and a new demonstrative doctrine concerning infinite multitudes. I am holding back matter suitable for mathematical journals etc. till I know whether you will publish this; and as I shall not later have time to make up those mathematical papers, I should think it a particular kindness if you would let me know as early as quite convenient whether or not you will accept this paper. My material prospects are promising (though not for the *immediate* future), but I fear I shall be unable to work continuously on philosophy for a term of years.

very truly
C. S. Peirce
Care Albert Stickney Esq. | 35 Nassau St, New York City

Carus to Peirce. LPB: SIU 31.5.323.

July 24th 96

Charles S. Peirce Esq.
35 Nassau Street
New York City

My dear sir:

Your MS. was duly received. It came somewhat sooner than I anticipated and is at the same time much longer than I expected. Since you do not state your terms I will make a proposition on which I see my way to accept the MS. I offer 250 Dollars of which I shall pay 100 Dollars within a week or not later than a fortnight after having received a definite reply from you and the remainder not later than the last day of the following calendar month.

It is probable that the article will have to be divided, but I shall endeavor to have offprints made (if possible to be incorporated in the R. of Sc. series) which will contain the whole in pamphlet form and at a cheap price so as to make it generally accessible to students.

Let me know your decision.

With best wishes

Yours very truly
Paul Carus

Carus to Peirce. TS: RL 77. Stationery headed, "The Monist, | Editorial Department, | La Salle, Ill." LPB: SIU 31.8.392.

August 10, 1896.

Chas. S. Peirce, Esq.,
C/O Albert Stickney, Esq.,
35 Nassau St.,
New York City, N. Y.

Dear Sir:—

Your MS. was received some time ago and I replied in a letter of July 24th in which I made a definite offer to pay you for the MS. $ 250. The letter was sent to your last address, "35 Nassau St, New York." Not having heard from you I was at a loss what to think of it. If you wish the MS. returned, please let me know *a reliable and secure address* and I shall send it to you by express. Of course the offer made in my letter of July 24th holds good still, and I should be glad to have a definite reply at your earliest convenience.

Hoping to hear from you soon, I remain, with kind regards,

Yours very truly,
P. Carus

Peirce to Carus. MS: SIU 91.14.

New York 1896 Aug 13

Address: Care Stickney, Spencer, & Ordway | 35 Nassau St. New York

Dr. Paul Carus

My dear Sir:

Your letter of July 24 has never reached me. I only know of it by yours of Aug 10. This is because I have been occupying lodgings temporarily vacant during hot weather. The lodging house keepers treat such temporary lodgers with every species of hostility & confiscate their mails.

Of course, I must accept your offer of $ 250 for the MS; nor do I doubt that the offer is as liberal as you feel justified in making it. Still, I beg to submit the following considerations. It is generally admitted I am clever about logic; and I myself estimate my efficiency in that field as about 3 times what it is in other fields, where nobody grumbles at paying me $ 20 a day when they need my services. The article I sent you is the fruit of very great study & labor. I think it would be quite below the truth to say it cost me 2 years over and above anything spent in work

---

24 July] *after* ~~Aug~~

already published. The pay of $ 250 is therefore at the rate of about 30 cents a day. I cannot repress a feeling of the deepest nausea and disgust, *not by any means with you*, but at the conditions in which the highest & best work of a brain like mine,—which future times will value, I do verily believe—can command less pay than in selling newspapers or blacking boots. Still I must accept, if that is your ultimatum, and will do so without further repining.

I am quite willing to rewrite to any extent; and I have the graphical method all written out in a formal, scientific statement; and will substitute that, if preferred. With unchanged regard, I remain

yours very truly
C. S. Peirce

Peirce to Carus. MS: SIU 91.14.

1896 Aug 14

My dear Dr. Carus:

I had no sooner put my last letter in the letter-box than I regretted having sent such a complaining epistle. Five days of terrible heat put most persons in an irritable condition which they are not aware of, until reflection comes.

I wish to say that I thank you with all my heart for accepting the MS. and paying the very welcome price you do for it, than which nothing could be more liberal. I also think I was not justified without carefully looking at my daybook in saying there was two years' work upon it. I have had the matter very much in my mind for the last 8 years. But perhaps there is not more than one year's solid work over and above what is elsewhere published. I will send you a more formal statement of the graphical method as soon as I can get my box open. I remain, with much gratitude

yours very truly
CSPeirce

Care Messrs Stickney Spencer & Ordway | 35 Nassau Street | New York

Carus to Peirce. TS: RL 77. Stationery headed, "The Monist, | Editorial Department, | La Salle, Ill." LPB: SIU 31.9.415.

August 17, 1896.

---

3 in] *after* ~~at~~

6 , and … repining.] *inserted in-line and between paragraphs*

17 , until reflection comes.] *inserted in-line and between paragraphs*

Mr. Charles S. Peirce,
C/O Messrs. Stickney, Spencer, & Ordway,
35 Nassau Street,
New York City, N.Y.,

My Dear Mr. Peirce:—

Your two letters were duly received—one yesterday, and the other one (dated August 14th) this morning. I feel like entering into an explanation concerning authors' fees, but I suppose it is unnecessary. In the ideal world of my construction I would make it possible that an author receive ten times as much as he does now, but the fact is that the author is not so much paid for the work he has actually spent on an article, but as a rule he only receives what the public interest in his subject may warrant a publisher to offer. Not even the laborer receives full pay for what it costs him to acquire his skill and ability, but only, as it were, the interest on it. One thing I can assure you is, that if The Open Court makes money on your articles we shall give you an extra royalty. However that may be, I am glad that you accept my offer and I am also glad that in your second letter you express your satisfaction in cordial terms. The check is being made out, and I shall send it to you at my earliest convenience.

Hoping that your private affairs are beginning to prosper, and wishing you all success in your scientific aspirations, I remain, with kind regards,

Very faithfully yours,
P. Carus

Peirce to Carus. MS: SIU 91.14.

Stewart's Hotel. 41st St and Broadway | New York. 1896 Aug 31.

Dr. Paul Carus:

My dear Sir:

I enclose here with a paper on logical graphs which I should like to have you examine, thinking you may prefer it to the one you propose to print in your January number.[136] It is in some respects very superior to that. But I fear it is too dry for the *Monist*.

136 "On Logical Graphs," precursor to his expansive work on "Existential Graphs" especially beginning 1902 (large collection in *CP* 4.347–584, and various related material in R 479–514).

13–14 the interest on it] ~~according to~~ the interest *i*on it

I received your check a week ago today, and a letter from Mr. McCormack.[137] But I was busy revising this paper for you & expected each day to get time to finish it & did not do so till today.

I should be pleased if you decided to print this one; but not if it is not really suitable for the *Monist*. I have much more matter on this subject. But what I send is too long as it is. I may mention that in my opinion this system is far superior to the algebra that Schröder likes, and perhaps even to the other algebra for relatives. It is probably more easily used by the non-mathematical. Theoretically I think it very true. In short, I think it has a future & if you were to print this more elaborate paper, it might be good for the reputation of the *Monist* in the end. But you can judge of its suitability. Please return the MS if not used.

yours very truly
CSPeirce

Carus to Peirce. TS: RL 77. Stationery headed, "The Monist, | Editorial Department, | La Salle, Ill." LPB: SIU 31.13.495–496.

September 8, 1896.

Charles S. Peirce, Esq'r,
C/O Messrs. Stickney, Spencer, & Ordway,
35 Nassau St.,
New York City,

My Dear Sir:—

Your letter was duly received and also your reply to Mr. McCormack's inquiry. I am at present very busy and almost overworked. In addition to much work on hand there is the uneasiness of the business world, and the complication of our politics, leaving me little leisure to read articles such as "The Algebra of Logic," which require unusual concentration of thought. From a hasty glance I would say that the article on "Logical Graphs" would be preferable for The Monist as well as for the exposition of your system of logic. I heartily endorse the sentiment in which your article that is appearing now in The Monist has been written.[138] The method of making opinion is such as you describe it, especially in German universities. I intend to bring your article before the public as forcibly as I can, and wish to

137 21 August 1896.
138 "Regenerated Logic."

11 Please … used.] *inserted in-line and below byline*

issue, as soon as possible after its appearance in The Monist, the same article in pamphlet form. If I can conveniently do so I will insert it in the Religion of Science Library, where it would reach another class of readers who might be stirred by your thoughts. Now, since I am unable to read both articles at once, and since, if I were able to read them, I would perhaps not be the right man to judge of the respective merits of either the one or the other for our special purpose, I can do no better than to leave the final decision to you. It seems to me that your article on "Logical Graphs" is more popular, and, therefore, more appropriate as a first introduction to your labors on the Algebra of Logic. Please let me know what you think about it, and I shall do what you deem best.

With kind regards and best wishes, I remain,

Yours very truly,
P. Carus

Peirce to Carus. MS: SIU 91.14.

New York 1896 Sep 11.

My dear Sir:

I have this moment received your very kind letter. Since you have not leisure to decide between the two MSS.,[139] suppose you see what Judge Russell would think, in case he has time to examine them. I should be glad if the decision were the way you incline to make it; but I fear the treatment in the paper on logical graphs is too dry. I wouldn't tell Russell what my preference is, but let us get his independent judgment.

yours very truly
CSPeirce

Dr. Carus
La Salle
Ill.

How would the West like Free Coinage of dollars (not fractions) with cessation of maintenance of parity, contracts & laws already made mentioning "dollars" to be interpreted to mean *gold* dollars, future contracts and laws not specifying gold or silver to be understood to mean *silver* dollars?

Peirce to Carus. MS: RL 77. Undated, answered by Carus 14 Oct 1896.

[ca. 10 October 1896]

139 "Logic of Relatives" and "Logical Graphs."

Dear Sir:

There are a number of persons to whom I should like to have copies of my article *The Regenerated Logic* go. I do not know whether you continue your liberal practice about sending copies or not; but if not I suppose you will send them & charge me the wholesale price. I am sorry you do not send separate articles, on account of the expense of the whole number; although the whole number is exceedingly interesting.

I should like a copy to be sent without delay to M. C Comte d'Aulby 18 W 32nd St New York City; because he sails next week. The other persons to whom I should like the article to go are:[140]—

Dr. W. K. Otis 5 W 50h St. New York
Dr. A. Jacobi 1 W 34h St. New York
Mrs. James M. Barnard Milton, Mass
Geo: S. Morison Esq. 35 Wall Street, New York
Albert Stickney Esq, 35 Nassau Street, New York
Mr. A. D. Risteen, Hartford Steam Boiler Insurance Co., Hartford, Conn.
Prof J. M. Peirce 4 Kirkland Place, Cambridge Mass
H. H. D. Peirce, Esq, American Legation, St. Petersburg, Russia.
E. C. Stedman Esq 16 Broad Street, New York.
Professor E. S. Holden, Lick Observatory, Mt. Hamilton, Cal.
Father Geo: M. Searle, Washington University, Washington D.C.
Prof. Ferdinand Becker, U.S. Geological Survey, Washington D.C.

Carus to Peirce. TS: RL 77. Stationery headed, "The Monist, | Editorial Department, | La Salle, Ill." LPB: SIU 31.21.650.

October 14, 1896.

C. S. Peirce, Esq'r,
Stuart House,
Broadway and 41st St.,
New York City, N. Y.,

Dear Sir:—

Your letter consisting of a list of names was duly received this morning and I shall be glad to attend to it. I wish to say at the same time that the signs in your MS. give more trouble than I anticipated. We have difficulties at the foundry and dif-

140 See Biographical Register.

20 E. S. Holden] E. D. Holden *[E]*
21 M.] *over* S
31 of a list] of list *[E]*

ficulties with the draftsman, and I do not yet know what to do. It seems to be impossible for them to make anything now for the next Monist number. It would take more time than I can afford, but it would be a pity to let the article wait for the next number. Besides, there is so much type tied up that we would be short of type for the present number. If we replace your tailed symbols by Schröder's analogous simpler symbols we can use the type which we got from Germany, but our head compositor tells me that he has only a dozen of them, and has not enough for your article, in which they abound. I do not as yet know how I shall meet the difficulty. I shall write again as soon as I can see some way of arranging it.

With kind regards and best wishes, I remain,

Very truly yours,
P. Carus

Peirce to Carus. MS: SIU 91.14. Stationery headed, "White's Hotel, | Massena, N.Y. | S. S. Danforth, Manager." with sketch of hotel at left.

Oct 21 1896

My dear Dr. Carus:

Your letter reaches me at this place, where I am looking into an industrial work which I am confident is going within a couple of years to pay my debts and to advance philosophy. I shall return to New York tomorrow or next day.

Schröder's tailed character $\curlyvee$ will answer well enough as *a sign of relative addition*. But as for the few formulae like this $\overline{\overline{A \curlyvee B} \curlyvee \overline{C \curlyvee D} \curlyvee E} \curlyvee F$, I think the best way would be to reproduce them entire as if they were drawings. But I supposed that the first thing you would print would be my last MS,[141] of which the figures might be put direct into the engraver's hands. If you want them redrawn, however, for process-reproduction, I can have it done, if desired. But I think the other way, that of letting the engraver draw directly from my figures would be the cheapest.

I would go on there and attend to the matter if thought worth while. I certainly should greatly regret further delay in the publication. When I say I *would* go on, I must not be positive; for I have agreed under certain contingencies to go to England shortly. But I do not think it likely that anything will occur to prevent my going to Chicago.

The enterprise in which I have a share and which promises to be extremely profitable has to float some bonds. No step has been taken & indeed the matter

141 "Regenerated Logic."

20–21 as *a sign of relative addition*] *inserted above*

23 figures] *after* ~~dr~~

has never been divulged, but I have got to state, when the time comes, the results of my observations & that *might* interfere with my movements.

very truly,
C. S. Peirce

Carus to Peirce. TS: RL 77. LPB: SIU 31.22.680. Both corners and a middle portion at top of TS torn off. Missing portions of date and header supplied by LPB. Stationery appears to be headed, "The Monist, | Editorial Department, | La Salle, Ill." At bottom-right corner of recto and verso are two sketches by Peirce of his tailed sign of relative addition. Verso also contains estimated dimensions of capital letter *P* and lowercase letter *g* as points of reference. [142]

October 24, 1896.

Charles S. Peirce, Esq'r,
Broadway and 41st St.,
New York City, N. Y.,

Dear Sir:—

One trouble with Schröder's tailed character is that we have not enough of them, and it is too late to order additional material from Germany. For that reason we must have it cut and I must trouble you to give an exact description to the draftsman or cutter which would be sufficient to ensure against mistakes. I do not know on which points you put the most stress. The formulas with lines overhead can be made by composition, so there is no need of troubling about it. We shall also have to cut the Psi-like plus sign, and the sign "greater than   "[143]. Will you make, on a separate slip for each one, an exact description of the signs, so that there will be no mistaking them?

The signs of the zodiac can be easily had, but I do not see that they are needed in your article. They do not seem to occur in the article, which is now in the hands of the printers. Please let me have your descriptions at your earliest convenience, so that we can go ahead with the composition work. I suggested to our printer to have the formulas photographed in written characters, but he declares that it would look too ugly and interrupt the style of the print.

Very truly yours,
P. Carus

142 Bottom-right corner: recto  verso  (both to scale).

143 Additional space left here during typing presumably for Carus to draw the character he had in mind. The character has either disappeared over time or Carus ultimately did not draw it.

15 trouble] troubled *[E]*

Peirce to Carus. MS: SIU 91.14. Undated, responding to Carus 24 Oct 1896 and answered by him 16 Nov.

[early November 1896]

Dear Dr. Carus

I enclose drawings of the three characters which give you trouble. I have drawn them about the size of double canon. But I do not know the proportions of your heads and tails, nor do I know the thickness of your lines. I have indicated the former by lines drawn horizontally and my sketches will have to be modified to suit the latter. Of course they should all look rather black.

The character ɟ which has given you the most trouble was originally intended to be printed by an ordinary dagger †, as it was in my Studies in Logic.[144] But I always wrote it with the curve; and Schröder has seen my MS. Since he considers the introduction of the curve a great improvement, I give that I always used.

It requires a body about as broad as a capital P, but is a long character. The under side of the horizontal stroke is the height of an ordinary short letter. No kern is required. The line as drawn by me is to be made thicker if required to suit your type. It should be about as thick as a j, that is, as any ordinary letter.

The character ψ is on an *m* body. It is a high character. It should be as thick as a capital; say as I. But the cross line should be not quite so thick, and should end in bulbs which should project a trifle above the upper line of a short letter. The lower edge of the cross-line the same height as the horizontal line of an e, as that letter is cut in modern types.

The character ⤙ ought to have a somewhat Chinese effect. I have drawn the longer line pointed at both ends. But I dont know but it would look better blank at the left. If so cut that edge must not be vertical but slanting ⤙.

The letters Σ and Π which I so often use should be upright sanserifs. These characters are to be had in mathematical printing offices I think.

If the sign of Aries is used, get a curly one and cut off its shank so as to make it look like this ♈

I repeat that I thought you would first print the MS last sent about graphs. I wish you would show these figures to your engraver and see if they will do. If they want to be redrawn by a draughtsman and you want me to have that done, please send on the MS.

Anyway I think you had better send it for the following reason. If promises made to me are kept, I shall in a few days sail for England; and in that case, I should like a copy of that paper to read to the Royal Society or some other society, which will only enhance its value to the *Monist*. If I don't sail (and I dont put

144 1883, specifically "Note B (The Logic of Relatives)," pp. 187–203.

much faith in the positive promises) there is a meeting of the *National Academy of Sciences* in New York in November & I should like to read the paper there, which would also be to the advantage of the Monist.

I hope you wont have a terrible disturbance in Chicago on election day; but it looks from here as if that was what Altgeld counted on.

very truly
C. S. Peirce
Care S. Bierstadt Esq | 1271 Broadway | New York

Carus to Peirce. TS: RL 77. Stationery headed, "The Monist, | Editorial Department, | La Salle, Ill." LPB: SIU 31.29.84–85.

November 16, 1896.

Charles S. Peirce, Esq'r,
Broadway and 41st St.,
New York City, N.Y.,
My Dear Sir:—

Your article is being set up. The characters are in the hands of the type cutter and I hope they will meet your approval. Your draft will be sufficient and there is no need of having them redrawn by a draftsman.

I read your article in the first rough proofs, but they do not as yet contain the formulas. Allow me to call your attention to the very first paragraph of your article in which you attack the present university system.[145] While that paragraph contains a great and important truth, I do not deem it advisable to tell the truth in this connection, introducing a new treatment of a science that has to fight its way to recognition. You antagonise unnecessarily all university professors and they will be prejudiced against you, while otherwise they would read your article with care. I suggest, therefore, in the interest of your cause, if not of yourself, that you alter the first paragraph and remove the objectionable passages. You might retain the warning that unless universities appoint professors who would not be pledged to teach certain doctrines, that must as a foregone conclusion agree with their church dogmas, they will not afford to our young men a sufficient and adequate

145 Paragraph later softened by Carus, as he suggests 30 Nov 1896, to instead address "a few inquirers into logic, sincere and diligent, … " (161).

5 Altgeld] Altgelt *[E] For John Peter Altgeld (1847–1902), Illinois governer 1893–1897*

Dear Dr. Carus

I enclose drawings of the three characters which give you trouble. I have drawn them
about the size of double canon. But I do not
know the proportions of your heads and tails, nor
do I know the thickness of your lines. I have
indicated the former by lines drawn horizontally
and my sketches will have to be modified to
suit the latter. Of course they should all look rather
black. †

The character ⸸ which has given you the
most trouble was originally intended to be
printed by an ordinary dagger †, as it was in
my Studies in Logic. But I always wrote it with
the curve; and Schröder has seen my MS. Since
he considers the introduction of the curve a great
improvement, I give that I always used.

It requires a body about as broad as a capital
P, but is a long character. The under side of
the horizontal stroke is the height of an ordinary
short letter. No kern is required. The line as
drawn by me is to be made thicker if required
to suit your type. It should be about as thick
as a j, that is, as any ordinary letter.

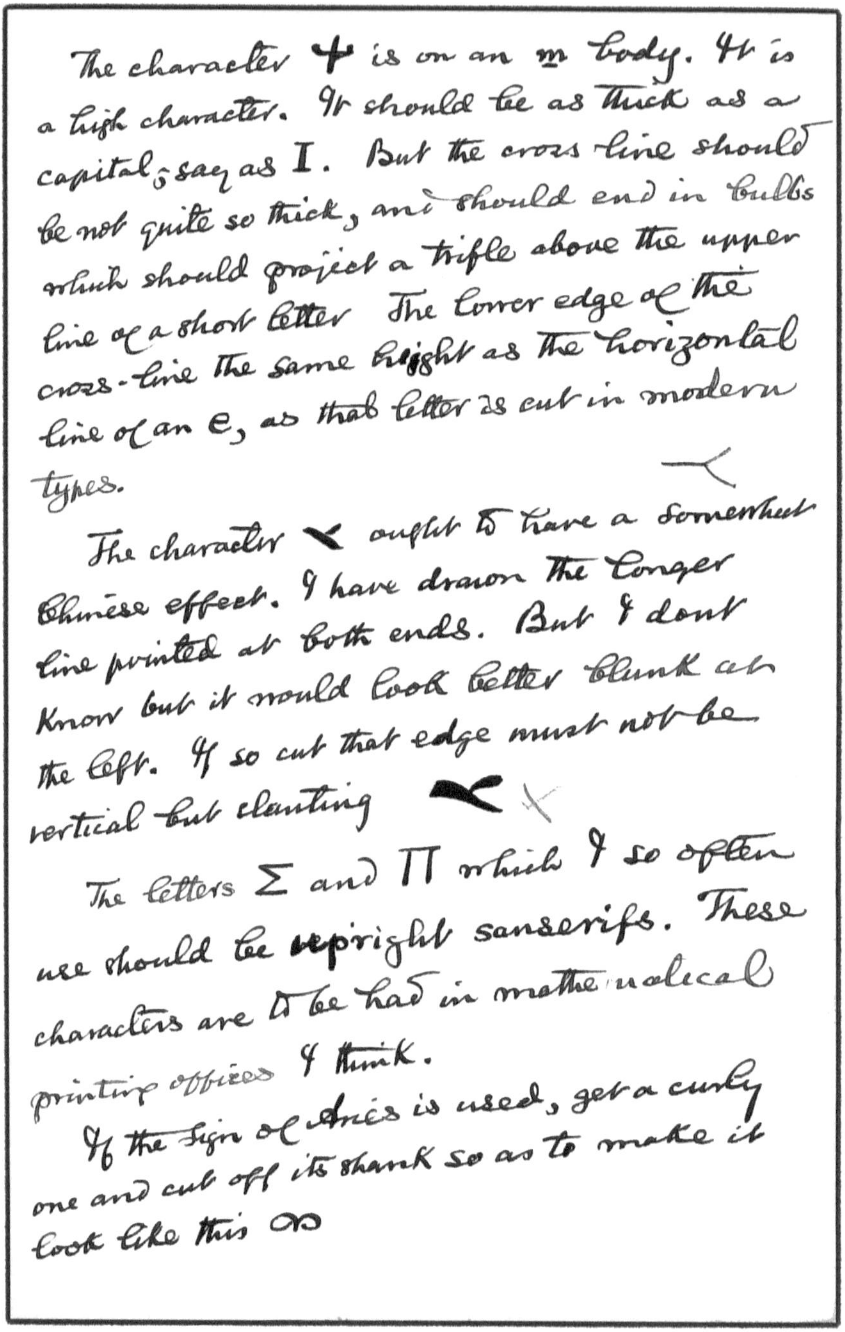
The character ✝ is on an m body. It is
a high character. It should be as thick as a
capital; say as I. But the cross line should
be not quite so thick, and should end in bulbs
which should project a trifle above the upper
line of a short letter The lower edge of the
cross-line the same height as the horizontal
line of an e, as that letter is cut in modern
types.

The character ⤙ ought to have a somewhat
Chinese effect. I have drawn the longer
line pointed at both ends. But I dont
know but it would look better blunt at
the left. If so cut that edge must not be
vertical but slanting ⤙

The letters Σ and Π which I so often
use should be upright sanserifs. These
characters are to be had in mathematical
printing offices I think.

If the sign of Aries is used, get a curly
one and cut off its shank so as to make it
look like this ∞

Figure 3: Peirce to Carus early Nov 1896, first two pages. Explanation of Peirce's logical characters, with illustrations. See also Figure 8, Peirce to McCormack 2 Sept 1896.

education. I would not alter the passage on my own authority, but I only suggest the alteration to you, and you may do as you see fit.

Incidentally I might add that what you say about me with regard to the specialisation of the modern logic, shows that you have misunderstood me as much as you did with regard to the statement that I do not take any stock in your peculiar philosophical doctrines.[146] You misquote me in both cases. I take no stock in your new logic in so far, and in so far only, as you utilise it to prove your favorite theory of tychism. I do not, and I never did, for that reason belittle your labors in the field of pure logic. I believe that your work would be more valuable if you were not hampered by your hobby of tychism, but for that reason I would not say that the whole science of pure logic is erroneous. On the contrary I believe that the logic of relatives is a very important branch of science, and I shall be glad of every advance in that line which you may still accomplish. As to the word specialisation, I would remind you that the work of generalisation, too, is specialised work in so far as the attention is concentrated upon special features only to the exclusion of others. The most comprehensive treatment can at the same time be a specialisation.

But these personal remarks are only incidental, and I do not care specially what you say in your article or anywhere else.

With best wishes for the success of your enterprises which at present seem to promise good results, I remain,

Yours very truly,
P. Carus

146 The quotes were apparently removed by Peirce before publication, as suggested in his response that follows. Based on Carus's remarks here, both quotes were from Carus's "Founder of Tychism" pp. 578–579, the first likely dealing with his paragraph noting that "Every specialist is inclined to look at things through the spectacles of his own speciality" (later modified in Peirce, p. 172, as a general caution to "sedulously avoid the error of regarding [the logic of relatives] as a highly specialised doctrine"). The other was removed without trace, but seems to have referenced Carus's criticism that Peirce's "arguments, to be derived from the logic of relatives, are like promises to pay out of the returns of a gold-mine, just discovered and boomed by the owners. There may be gold in the mine, but I do not as yet take any stock in it" (Carus, p. 579).

3–4 specialisation] ~~special~~ specialisation ~~character~~
13 still] *inserted above*
19 or anywhere else.] *inserted in-line after deleted period*

Peirce to Carus. MS: SIU 91.14.

84 Broad St, N.Y. Thanksgiving 1896

My dear Dr. Carus:

Your letter enclosing the first paragraph of my article only reached me night before last, owing to my having left the Stuart House & my address is now 84 Broad St (Room 6) New York City where I have now an office.

I shall be happy to strike out the paragraph both to please you and because if it makes an unfavorable impression on you it will on others. But I am surprised that you should think for an instant that the average American professor of logic would ever pay the smallest attention to anything I write. He is, as I say in that ¶, usually appointed as being theologically sound.

As to misrepresenting you, it was quite unintentional & I am very glad if you think well of my works in pure logic. The particular phrase you mention was, it is true, applied to my synechism & tychism. But in the same article you spoke with extreme contempt of my purely logical work,[147] and I have heard from various sources of your slight opinion of it. Perhaps, however, Schröder's flattering opinion has caused you to think better of it. At any rate, I am very glad that you have come to do so; and not having any surplus of supporters in that field, I shall naturally desire to modify what I say on that head.

I regret that you should see fit to add that you do not care what I say concerning you.

For my part, I have long been convinced in regard to philosophers in general that their most favorable opinions about one another are as a rule far more nearly just than their least favorable opinions. I have myself the reputation of looking with a cold and sarcastic eye upon new writers on philosophy; but this is largely due to the fact that very many articles of mine have been rejected by editors as containing too much praise, while others have had the praise cut out. But slashing articles almost always please the ordinary editor, so long as they do not affect his advertising business. This has at any rate been my experience with many of them,—I do not say with all. It tends to make a critic take too unfavorable views of original writers,—a tendency against which I find it necessary to be always on my guard.

I do not think that an original American writer on philosophy has a fair chance. Some have received full justice, but others have not, in my opinion. My own writings on pure logic were utterly unnoticed for very many years. My meta-

147 "Founder of Tychism," pp. 178–180.

23 their] the*y*ir

30 critic] *above* ~~writer~~

physics has only an uncertain chance of being ever fairly judged. It depends on whether I can make some money in time.

Yours very truly,
C. S. Peirce

Carus to Peirce. TS: RL 77. Stationery headed, "The Monist, | Editorial Department, | La Salle, Ill." LPB: SIU 31.31.137.

November 30, 1896.

Chas. S. Peirce, Esq'r,
84 Broad St.,
New York City, N. Y.,
My Dear Sir:—

Thanks for your letter. I shall either omit or modify the first paragraph of your article. When I said that I do not care what you say concerning me I meant that I wished you to enjoy the privilege of saying whatever you pleased. I do not wish to influence you to modify expressions which you use in public statements on my account. I have not changed my opinion concerning you through any influence that came from Prof. Schröder, and I believe that I have the same opinion to-day that I had before. If you will read over the passage in which I criticise your logic of relatives you will find that I do so on the presupposition that it is going to prop up your tychism. I have not time now to look up and read it over, but I shall do so when more at leisure.

With kind regards and best wishes, I remain,

Yours very truly,
P. Carus

Peirce to Carus. MS: SIU 91.14.

Confidential

84 Broad St (Room 6) | New York. 1896 Dec 8.

My dear Doctor Carus:

You know some little of the terrible times I have been going through. I wonder that I am alive today. But for three things which gave me courage, the necessity of caring for my wife, the debts I owed especially to Mr. Hegeler, and the desire to give philosophy an impulse which should continue after my death, I should have sunk under it all.

---

1 an uncertain] a$^{\vee}$n$^{\vee}$ ~~chance~~ \uncertain/   33 under] *after* ~~to~~

But now I have a prospect of pulling through; and I feel confident that if I dont starve, or otherwise break down, in a very short time,—a few months, I shall be able, perhaps not to save my wife's life, but to pay my debts and to do something for logic and metaphysics.

I write now to suggest one way in which my debt to Mr. Hegeler might be paid, if he thinks well of it. If he does not, then it will be paid otherwise, if I dont collapse,—and I only fear starvation, which is mighty near sometimes. My health & strength & energy are still great.

In the hopes of accomplishing my purposes, I interested myself in acetylene. I made some inventions & seemed a large interest in the European introduction of a system to generate the gas in dwelling-houses.[148] But it was plain that *that* could not be started until carbide could be made cheaper. I made an invention for that purpose. But after all, it must depend chiefly on getting power very cheaply and advantageously.

In that way I was led to take an interest in a company for developing in this state a little below Montreal, from the water of the St. Lawrence an enormous power,—double what Niagara has yielded. The place is very extraordinary. The fall is less than 50 feet (there will be only 40 directly on the turbines as against 140 at Niagara). But the facility with which it can be taken from the river, and manifold advantages, such as short shafting, cheap tailraces, absence of violent forces on the turbines, ease of loading oceangoing vessels at the mills, proximity to New York City, etc. will make a horse power cost the company $ 2 instead of $ 15 at Niagara. Having rendered some services to the company of a professional nature, I was admitted into a syndicate for placing the bonds to supply all the money needed, and I learn by a cable from London that the bonds have been placed. My profits on this alone, if I am not swindled (and I don't think I shall be,) will be enough for all purposes. But still I do not think it prudent to relax other efforts.

The company has one of the broadest special charters which has been granted by the State of New York for many years. It has positive & pressing offers from very large and unimpeachable companies to take all the power it can possibly develope twice over for $ 15 leaving a clear profit after paying interest on the bonds, wear and tear of machinery, labor, and administration of $ 13. The capital is $ 6 000 000. The very highest engineers in the country are responsible for the

148 See footnoted description in Peirce to Russell 26 April 1896.

11 dwelling-houses] *after* ~~the~~
20 violent] *above* ~~terrific~~
21 turbines,] *before* ~~two trunk lines,~~
21–22 New York City, etc.] New York<,> *etc.*City, etc. ~~giv~~
28 special] *inserted above*
32 and[1]] *above* ~~of~~

calculation that all the work can be done within a year, for under a million, & will develope 200 000 horse power. That would pay 43 percent. A rough calculation made by me in loco of the *minimum* horse power made it 100 000. That would be over 20 percent.

Now I am in a situation to ask to be let in on the ground floor and in the pool. That is, stock to be paid for now at par and privileges of all connected enterprises at par. I propose, or suggest, that if Mr. Hegeler were to advance $ 20 000 to buy such stock, he could keep it until it had paid up all I owe him, and then could give to me so much of the stock as to leave him an annual profit of 15 percent.

If he does *not* do this, I firmly believe that within 18 months what I shall get from the placing of the bonds will enable me to pay him up anyway. Therefore, I only suggest it as being demonstrably a very good investment. And also because I have been cheated so badly before I dont positively calculate on anything not in hand.

If you want further details, I can furnish everything and accompanied by complete documentary proof of every assertion. The sum of $ 20 000 I only mention as being one which would in no very long time cancell my debt. If you would like to go in to this for a few thousand for yourself I should take pleasure in procuring the stock for you.

yours very truly
C. S. Peirce

Time is very limited, for soon they would laugh at me if I wanted stock at par. But I think any time before Jan 1, I could get it.

Peirce to OCP telegram 11 Dec 1896 (missing): Approves the diagrams to be used for "Logic of Relatives."

Carus to Peirce. TS: RL 77. Stationery headed, "The Monist, | Editorial Department, | La Salle, Ill." LPB: SIU 31.36.224.

December 11, 1896.

---

2 pay] *over* be
3 in loco] *above* ~~on the ground~~
3 it] *inserted in-line*
6 to be paid] to *p*be paid
8–9 give … stock] ~~repay~~ \give to/ me so much \of the stock/
12 demonstrably] *after* ~~pro~~
12 investment.] *before* ~~anyw~~
12–14 And … hand.] *inserted in-line and between paragraphs*
18 yourself] *before* ~~& if~~
22–23 Time is … get it] *This paragraph below byline and up right margin.*

Charles S. Peirce, Esq'r,
84 Broad St.,
New York City, N. Y.,
My Dear Sir:—

Your letter of December 8th was duly received, and I spoke with Mr. Hegeler concerning your proposition. He read your letter and told me that he was not inclined to venture into the enterprise. This afternoon we received your telegram which O. K's the diagrams of your article and gives us authority to go ahead with the printing of your article.

Hoping that you will be prosperous in your business undertakings, I remain, with kind regards,

Yours very truly,
P. Carus

Peirce to Carus. MS: SIU 91.14. At bottom of the letter is Carus's text for telegram received by Peirce 29 Dec 1896.

New York, 1896 Dec 21.

My dear Sir:

I started to make a memoir for the National Academy of Sciences out of my article for you on Graphs.[149] I find that what I am writing is much better for you at least for a first article, because it is a simple direct account of the system of graphs which I do not speak of in the body of the paper written for you but only in the final section, and does not begin by first describing the most philosophical method and then proceed to simplify it. It there makes something more definite and striking and will thus leave room for much further development of the applications and uses.

I have got to send my academy memoir in before the first of January. If you will telegraph me your authorization and also the date at which you must have the copy (remembering the numerous figures) I will let the paper written for you go to the Academy and will take a little more time and finish up that I am now writing as a far more striking paper for you. The philosophy of the thing, showing all possible systems of logical graphs might subsequently appear in the Monist

149 From his "On Logical Graphs" sent to Carus 31 Aug 1896. Memoir not finished in time, though a similar paper presented 1906 to the Academy on "Recent Developments of Existential Graphs."

19–20 at least for a first article,] *inserted above*

22 section] *above* ~~chapter~~

27 the] *after* ~~how~~

in a more mature form, if you like. But your telegram will not be understood as committing you to take such subsequent paper.

yours very truly
C. S. Peirce
84 Broad St | New York[150]

Carus to Peirce. TS: RL 77. Stationery headed, "The Monist, | Editorial Department, | La Salle, Ill." LPB: SIU 31.37. 257.

December 24, 1896.

C. S. Peirce, Esq'r,
84 Broad St.,
New York City, N. Y.,

My Dear Sir:—

On my return home after a short absence I found your letter awaiting me, and would say in reply that you better have your Memoir for the National Academy of Sciences made out of your article on Graphs without giving any heed to The Monist, for in the first place I cannot make any promises to publish that or any other article of yours in The Monist, and then should you send me another paper on Graphs, you might have in the meantime changed your ideas or thought of a better plan of presenting the subject. In my opinion it never pays to keep material back for the purpose of using it later on.

With kind regards and best wishes, I remain,

Yours very truly,
P. Carus

With the best wishes of the season.

Carus to Peirce telegram 29 Dec 1896 (missing): "Make memoir for National Academy without minding Monist | Kind regards. | Paul Carus" (Text comes from Carus's transcription at bottom of Peirce's letter of 21 Dec.)

---

150 *Note by Carus at bottom of letter, presumably for his missing telegram to Peirce of 29 Dec:* "Make memoir for National Academy without minding Monist | Kind regards. | Paul Carus"

1 be] *after* ~~commit~~

24 With the … season.] *Added in pen.*

Peirce to Carus. MS: SIU 91.14.

84 Broad St. N.Y. 1896 Dec 29

My dear Dr. Carus:

I receive today your letter and telegram for which receive my thanks. Since you say "I cannot make any promises to publish that or any other article of yours in the Monist," I will at once send you back the article already accepted and paid for.[151] You say "Should you send me another paper on Graphs, you might in the meantime have changed your ideas or thought of a better plan of presenting the subject. In my opinion it never pays to keep material back for the purpose of using it later on." This seems to me to manifest a disappointment or dissatisfaction which rests upon a misconception, as I will proceed to explain.

The logic of relatives is like mathematics, physics, etc. in this, that every discovery opens up further inquiries, so that the more it is investigated the more rapidly new discoveries will be added. Hence, to expect ever to get it into a finished and completed state, would be chimerical. One might as well expect geometry, or the doctrine of elasticity, or astronomy to be completed, instead of advancing faster and faster, as they always have done. As well expect that civilization should be perfected and brought to a final condition. This I remark, not directly with reference to your letter, but because you once objected in print to the Logic of Relatives that it was not completed.[152] I notice it now, because the idea that science should be unchanging may be an element in your sentiments about the present case. I will also remark that in my experience of the logic of relatives, I have always observed a phenomenon which is certainly true in mathematics, in physics, in chemistry, and in other sciences in the investigations of which I have taken part, namely, that a new discovery closely related to an old one generally (if not invariably) enhances the significance and importance of that old one.

151 The "Logical Graphs" sent to Carus 31 Aug, in fact not accepted nor paid for (see Carus to Peirce 18 Jan 1897). Throughout the letter, Peirce refers to this paper as containing his "old," more philosophical graphical system that turned out to be superior to his "new," less philosophical though more reader-friendly system in the version prepared for the Academy.

152 Possibly "Founder of Tychism," p. 579, especially paragraph beginning, "I shall be glad to sit at Mr. Peirce's feet as an attentive student, as soon as he has worked out his logic of relatives, or any other subject."

4 your] you *[E]*
13 opens] *after* ~~b~~
14 rapidly] r*ec*apidly
14 it into] i*n*t into
15–16 geometry,] *after* ~~mathematics, or~~
18 directly] *after* ~~so~~
21 be^2] *after* ~~ha~~
21 sentiments] *after* ~~present~~

The present case is this. Of course, I have far more to say about the logic of relatives,—more positive and important finds,—than I am at all sure of ever being able to get published at all. If I make enough money to pay my own printer, of course I can get them printed; but printing and publication are different things. What I seek to do, therefore, is to select for each public whom I get a chance to address in print such ones of my discoveries as I think will particularly interest them. That there were two principal logical systems of graphs, I used to teach as long ago as my Johns Hopkins lectures.[153] One of these is more philosophical and useful for the purpose of the ultimate analysis of reasoning than the other. That one I have carefully developed in the paper written for you, so as to exhibit its philosophy, and the necessity of every feature in it; and I have made an enumeration of the elementary steps of reasoning as exhibited by that system which could not have been as satisfactorily performed by any other method which has as yet been proposed. I consider that paper as unquestionably the strongest of all my logical papers. In it I mention the other system of graphs and briefly describe it. The new paper which I today send to be printed in the Memoirs of the National Academy of Sciences is a description of that other system of graphs without any attempt to give it a philosophical foundation or development. Of course, I would not have sent any paper to be published in the Memoirs of the Academy unless I felt very sure that its solid value was beyond all question; but when I started to write it, I expected it to be inferior *in every respect* to the paper written for you. But when I came to develope the matter and to write it out, I found that while it was undoubtedly inferior to the other paper in profundity and thoroughness, it was nevertheless very superior to it as something likely to strike the reader, in what might be called brilliancy, and also that that system of graphs has certain remarkable points of superiority over the old one, particularly the extreme simplicity of the conception of it. At one moment, I even felt that it might turn out that this second system was preferable for the ultimate analysis of reasonings, and perhaps even from a philosophical point of view. It is my habit always to call into doubt propositions I have admitted, particularly those I have for a long time admitted; for it is precisely among things that seem unquestionably true, that errors are most likely to lurk. I have since examined these doubts more closely, and after a great deal of conscientious work, I have dismissed them. The first system is as I always

153 Lecturer in Logic at JHU 1879–1884.

9 for the purpose] *above* ~~from the point of view~~
12 exhibited] *after* ~~sh~~
17 of that] of th*e*at
25 that that] that that *[line break]* that *[E]*
25 has] *over* d
32 after] *after* ~~I~~

said the most philosophical, and as I now see more clearly than ever, is the one to employ in fundamental analyses of reasoning to the bottom. But the second system is the more simple,—very slightly so in its proceedures, but quite considerably so in conception, so long as no philosophical conception is attempted.

Such being the state of the case, my judgment was that the new paper would be better suited to the Monist, and the old one to the Memoirs of the National Academy of Sciences. But as you are not prepared to trust implicitly to my judgment in the matter and there is no time to show you the new paper, I shall immediately send it to publication as a Memoir of the Academy (which shows sufficiently what my own opinion of its value is.) Thus, you will remain in possession of the *chef d'oeuvre* of my logical life, and without any ground for dissatisfaction; for its value is certainly enhanced by the new discoveries in the other paper, and you have got it at a price which does not pay me 50 cents a day for the work it has cost me. I trust there is not going to be any doubt about its being printed. I do not understand your expression in your letter to mean that you will hesitate about that. Should you not wish to print it, after expressing your high estimate of the importance of the logic of relatives, I should certainly think it one of the most curious things in the whole history of science, and of course I should never be content till I had bought it back from you.

I am now engaged in writing another paper[154] in which I apply the system of graphs developed in the paper I return to you, in order by its means to set finally at rest one of the most vexed of questions. This it must do to the mind of anybody who applies himself to understand the paper. I intended offering it to you; but as you have much doubt whether you will ever publish another paper by me, and I want a place where there is a strong probability, not only that some paper may sometime be published, but that any given paper, which seems to me to *[*be*]* solid and valuable and has, at any rate, cost me much pains in the sincere desire to adhere to demonstrable certainty, may be published, I seem to be forced to go elsewhere with it, or to put it away, along with many other works of mine which can only see the light if I can earn the money to print them.

With warmest wishes for a happy New Year for you and your family circle in La Salle, I remain

yours very truly,
C. S. Peirce

154 Undetermined.

7 are] *above* ~~do~~
8 the new paper,] the ~~two~~ \new/ paper~~s~~,
10 in possession] *after* ~~with~~
12 discoveries] discover*y*ies
21 set] *over* on
30 can[1]] *above* ~~will~~

The Academy Memoir will not appear for a long time, and then will not be seen by many of the readers of the *Monist*.

1897

Carus to Peirce. TS: RL 77. Stationery headed, "The Monist, | Editorial Department, | La Salle, Ill." LPB: SIU 31.44.388.

January 18, 1897.

Charles S. Peirce, Esq.,
84 Broad St.,
New York City, N. Y.,

My Dear Sir:—

Your long letter was duly received but I do not understand its contents. As far as I know we have paid for the article published and the article in your hands is not paid for.[155]

Hoping that your prospects are good, I remain, with kind regards,

Yours very truly,
P. Carus

Peirce to Carus ca. 25 March 1897 (missing): Returns his "Logical Graphs" on the misunderstanding that it was paid for, and includes an appendix summarizing his receipt of payments.

Carus to Peirce. TS: RL 77. On OCP Stationery C. LPB: SIU 31.68.151–152.

La Salle, Ill., U.S.A. March 29, 1897.

Charles S. Peirce, Esq.,
84 Broad St.,
New York City, N. Y.,

Dear Sir:—

Your letter and MS. came to hand a few minutes ago and I hasten to reply that I am utterly at a loss to make out what you mean. You ascribe to me things which I have never written. You say that I seemed annoyed at your giving the article on

---

155 "Logic of Relatives."

---

1 appear] *over* be   2 of[2]] *over* in

"Logical Graphs" to the National Academy of Sciences. The very opposite is the case. I wrote you that *you had better publish your memoir in the National Academy of Sciences* and *give no heed to The Monist whatever*, suggesting that if you desired to write for The Monist again on the same subject you could present the same ideas in other words, and perhaps even in an improved form. I added that I never keep articles back and I believe that the more one's ideas are made accessible to the public the more fruitful they become. You must have been in a very strange mood when you read my letter of December 24th, to read things into it which it does not contain. I do not know what to think about the return of your MS. on "Logical Graphs", and your appendix of MS. pages 54–57 is based on a misconception of the situation. So far as I know, I have neither accepted nor paid for the article on "Logical Graphs." Is it your idea that the last payment which you received should cover both articles,[156] of which you left me the choice? I am as ignorant about having paid for the article as I am about its acceptance or rejection, and there is no need on your part to refund any money to the Open Court Publishing Co. You say that I want no more articles from you. When did I ever make such a statement? I have never thought of ruling you out of The Monist, and it is quite impossible that I ever said so.

I have instructed our business manager at Chicago to send you copies of the January Monist.

With best wishes for the success of your business plans, I remain,

Yours very truly,
Paul Carus

Peirce to Carus. MS: SIU 91.16.

1897 Apr. 1. 84 Broad St N.Y.

My dear Sir:

It was a very great relief for me to receive this morning your letter. It makes matters plain to me. You do not understand how I could put such an interpretation on your previous letter. You will find the key to it in this underlying misapprehension, that *I understood the last sum you sent me* was for this article on Logical Graphs. Thus, it was not what you said in your letter taken by itself, but taken in connection with that supposed previous acceptance of and payment for that article which was before me for interpretation.

156 Both "Logical Graphs" and "Logic of Relatives."

8 does] *above* ~~did~~

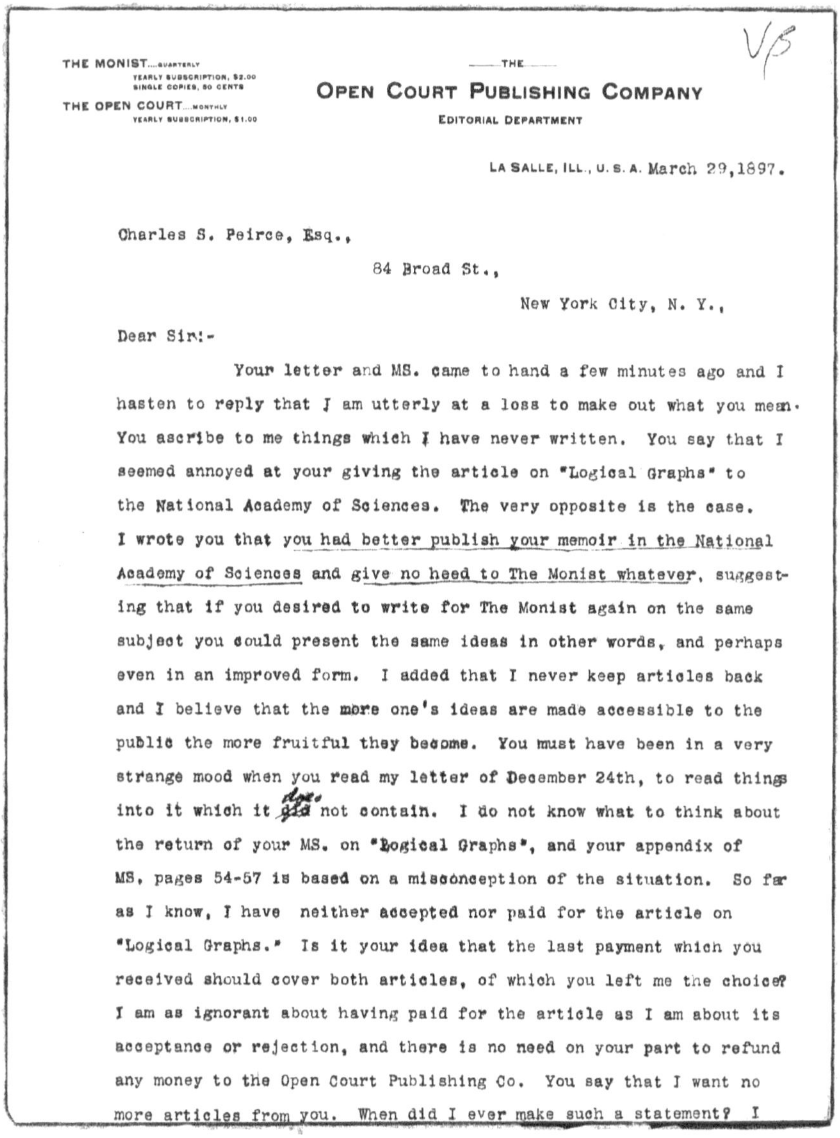

THE MONIST....QUARTERLY
YEARLY SUBSCRIPTION, $2.00
SINGLE COPIES, 50 CENTS
THE OPEN COURT....MONTHLY
YEARLY SUBSCRIPTION, $1.00

THE
OPEN COURT PUBLISHING COMPANY
EDITORIAL DEPARTMENT

LA SALLE, ILL., U. S. A. March 29,1897.

Charles S. Peirce, Esq.,
84 Broad St.,
New York City, N. Y.,

Dear Sir:-

Your letter and MS. came to hand a few minutes ago and I hasten to reply that I am utterly at a loss to make out what you mean. You ascribe to me things which I have never written. You say that I seemed annoyed at your giving the article on "Logical Graphs" to the National Academy of Sciences. The very opposite is the case. I wrote you that you had better publish your memoir in the National Academy of Sciences and give no heed to The Monist whatever, suggesting that if you desired to write for The Monist again on the same subject you could present the same ideas in other words, and perhaps even in an improved form. I added that I never keep articles back and I believe that the more one's ideas are made accessible to the public the more fruitful they become. You must have been in a very strange mood when you read my letter of December 24th, to read things into it which it ~~did~~ does not contain. I do not know what to think about the return of your MS. on "Logical Graphs", and your appendix of MS, pages 54-57 is based on a misconception of the situation. So far as I know, I have neither accepted nor paid for the article on "Logical Graphs." Is it your idea that the last payment which you received should cover both articles, of which you left me the choice? I am as ignorant about having paid for the article as I am about its acceptance or rejection, and there is no need on your part to refund any money to the Open Court Publishing Co. You say that I want no more articles from you. When did I ever make such a statement? I

Figure 4: Carus to Peirce 29 March 1897, on OCP Stationery C, first page. Misunderstanding with Peirce's "On Logical Graphs" MS.

The first thing incumbent upon me, then, is to thank you warmly for that payment (at which I rather grumbled) as so much additional for the long article in the January Monist. I thought it a little below the figures I have often received when supposed to be for so important a step in logic as this article.

Then I desire to explain that I feel badly enough about the light in which I appear to Mr. Hegeler to be extremely sensitive about doing anything that could possibly appear to violate any understanding with the Monist. I have been working with all my might to recover my position with him, & *think* the moment will soon be at hand when I can do so. But if I should be disappointed about that, I shall only struggle the harder.

Next, let me ask you whether you will take the article of Logical Graphs (with suppressions of the 2 or 3 pages based on misapprehension) and if you desire the form changed, so as to be less stiff & technical, to that I have no objection. I could, if desired, add to what is there, another article explaining and illustrating the method of working with these diagrams in the most expeditious and simple way.

In case you do not want any article on logical graphs, I have, among others, three[157] things which might attract you. One[158] is a close examination of Psychological Association & Suggestion especially in their logical aspect. I defend association by Resemblance as highly important, and show the necessity of concluding that much is in consciousness in an obscure way which the man himself is unable to recognize there. So that association is an attraction between ideas which causes them to cluster together & the dimmer ones to become more vivid. The nature of vividness is discussed.

Another paper[159] is designed to show in detail the extraordinary importance in philosophy of the logical conceptions of *one*, *other*, and *third*. It not only shows that they are important but just how each of them enters into philosophical conceptions.

Another paper,[160] which is less original, is an explanation for thinkers who are not very well read mathematicians of the conception of Space in the light of

157 These three articles are not precisely determined, but are estimated as follows based on context and Peirce's descriptions.

158 Possibly a precursor to his work on "Phaneroscopy" especially ca. 1906 (R 298–299, 336–337; partly in *CP* 1.306–311, 4.6–11).

159 Possibly offspring of his 1894 "What Is a Sign?" (*EP* 2:4–10) or precursor to his 1903 "On Phenomenology" (R 304–306; partly in *CP* 5.41–56, 59–65).

160 Possibly "Doctrine of the Census in Geometrical Topics or Topical Geometry" (R 145) or "On Space-Logic" (R 146–147). Other related material on Listing in R 159–163.

1 upon me] upon *t*me   17 three] *above* ~~two~~

the writings of J. B. Listing (But Listing wrote 60 years ago, and to do justice to his ideas they need to be all worked over according to other and more modern studies.) I show that just as Metrical geometry is well known to be nothing but Graphical geometry applied to an individual surface, the "Absolute," so Graphical geometry in its turn is nothing but a special problem in Topical Geometry, or the doctrine of the modes of continuous connection of lines, surfaces, solids, etc. This last article is not actually completely written. But I could soon complete it; and it would be a useful one to anybody who wishes to philosophize about Space.

I am going in a few days to deliver an address at Bryn Mawr.[161] You could have that article if you liked.

I have been and still am surrounded by scoundrels,—a pack of wolves who seeing me weakened attack me on all sides. I had hoped before this to have gained such signal advantage as to discourage them; but it is postponed. What adds to the difficulty of the situation is my wife's health, which has now taken such a definite form that she must shortly undergo the great operation of extirpation of the uterus etc or succumb. If she comes through it successfully, it seems probable her health will not again be critical for many years, & there will be a period of great happiness for us in that. But meantime, it takes up all my time & nervous energy for nights and days so frequently as to cripple and hamper my earning powers terribly.

Consequently should you accept an article from me and find yourself disposed to make me any advance upon it, it would be a great boon. For I only hint at a state of things which if pictured would astonish anybody.

As I appear to be drawing to the end of a long period of misery, I examine myself to see if I have really learned all the lessons of an unusual experience. They have been impressive. The most delightful has been to discover all the virtues of my wife whose life is in such danger at this moment. If I lose her it seems as if every light within my soul must become extinguished. It is a question of days now. An advance for an article would make a great difference in the hopefulness of the case.

yours very faithfully
C. S. Peirce

Dr. Paul Carus

161 "Number: A Study of the Methods," delivered Bryn Mawr College, PA, ca. 5 April 1897.

4 Graphical] *after* ~~the~~
6 modes of] *inserted above*
13 signal] *after* ~~an~~
15 undergo] *after* ~~have~~
27–29 It is … case.] *inserted in-line and above byline*
30 yours very] yours *y*very

Carus to Peirce. TS: RL 77. On OCP Stationery C. LPB: SIU 31.70.193.

La Salle, Ill., U.S.A. April 6, 1897.

Charles S. Peirce, Esq.,
84 Broad St.,
New York City, N. Y.,

My Dear Sir:—

Your long letter of April 1st was duly received, and I hasten to reply to it, although I cannot at present give you any definite answer. I am under the present conditions extremely hampered in my business dealings, and must limit myself in all ventures which imply monetary expense. Mr. Hegeler is not at home now, and while absent he fell ill, and is at present under the treatment of physicians who advise him to drop for some time all business affairs—a condition which has made it impossible for me to communicate with him and have his advice concerning anything relating to the Open Court or Monist. Besides, on account of heavy expenses which I have had of late, I must greatly economise, and unless a change sets in which will render it possible for me to make an explanation to Mr. Hegeler, I cannot venture into anything that would again increase expenses. I write this to you at once lest you may be misguided by expectations that under present circumstances I am not able to fulfil, however gladly I should like to assist you. It is not impossible that I shall have to leave La Salle in order to join Mr. Hegeler, who is planning, at the advice of his physician, a European trip for the sake of relaxation. I shall write to you again to keep you informed as to what I can do. In case I have to leave La Salle for a long time, I would prefer to return you the MS.[162] that is now in my hands because you might see your chance of turning it to account somewhere else. I wish I could meet you personally, and should I go to New York I would not fail to call on you at 84 Broad St.

With kind regards and best wishes, I remain,

Yours very truly,
P. Carus

Carus to Peirce. LPB: SIU 31.72.228. Undated, text found in LPB below letter of same date to another author.

[ca. 13 April 1897]

I am leaving for the East today. Shall call upon you about the 18, 19 or

162 "Logical Graphs."

20th ins—which are the days in which I shall probably be in New York. MS. by express

Your MS will be returned to you and we shall talk matters over. Hoping to see you soon

With kind regards

P. C

Peirce to Carus. MS: SIU 91.16. This and the next undated letter were written two days before Juliette's operation, which, based on combined context of these letters and that of 1 May, occurred on Thursday, 22 April 1897. This was also the last of the days proposed by Carus for a meeting, which, based on Peirce's slow reply and Carus's frustrated note around this same day, never took place.

Please direct reply | *To 108 W 89^h St.*

*[*20 April 1897*]*

My dear Dr. Carus:

I should like to see you very much and if you can mention any time I will meet you at 108 W 89^h St where I have a little flat or at 84 Broad St or anywhere else. I can meet you almost at any definitely set time except Thursday afternoon. But at present I cannot remain all day at my office & don't go there very often. I get my letters from there daily. My wife is at the surgeon's house where a major operation is to be performed this week probably Thursday PM. I should hate to miss seeing you. I am more concerned than you can imagine about Mr. Hegeler.

very truly
C.S. Peirce

Peirce to Carus. MS: SIU 91.16. Stationery headed, "Grand Union Hotel | Opposite Grand Central Station. | Ford & Shaw. | Proprietors."

*[*20 April 1897*]*

My dear Dr. Carus:

I wrote to you today that my wife's operation was to be Thursday afternoon; but it has been changed to Thursday morning. That will only leave me tomorrow forenoon in which I can possibly see you. I will be in my apartment 108 W 89^h St. The cars passing the Railway Station in 42nd St going west go to 89^h St & the Boulevard. You would walk 2 or 3 blocks east if you came to see me in that way. If you take the Broadway cable cars, I am just a little (1/4 block) *west* of the point where

17 at[1]] *inserted above* 15–23 I should … Peirce] *up right margin*

they cross 89$^{\underline{h}}$ St. The elevated (6$^{\underline{h}}$ Avenue or 9$^{\underline{h}}$ Avenue) for Harlem has a convenient station at 93$^{\underline{rd}}$ St. Or, if you will send me word,—probably a note with an immediate delivery stamp would if mailed tonight reach me in time—I would call on you.

At 1 o'clock, I must go to my wife.

very faithfully
C. S. Peirce

Carus to Peirce. MS: SIU 91.16. Unsent draft on torn-out piece of notepad paper.

[Unsent, ca. 20 April 1897]

[recto, scribbled over]

Charles S. Peirce Esq.

Why did you not right at once make an appointment or state where I could meet you? There are so many things which I should have liked to talk over with you, and which are either too trivial to write or can not very well be discussed in letter

[verso]

Dear Professor: It is a great disappointment for me not to be able to trace you. [For when you receive this card, I shall probably be gone and have left New York.] I would have preferred to tell you orally what I have to look over with you.

I saw Mr. Hegeler off for Europe and succeeded in mentioning your offer concerning articles, but I regret to say that I did not find him favorably inclined.

I shall be out of town today and probably tomorrow too. I shall leave Tuesday night or Wednesday morning, but expect to be back during the next week, but then my time will be limited.

I have made the Grand Union Hotel my headquarters for receiving letters but there is a small chance of meeting me there.

---

17 trace] *above* ~~find~~
19 I would … with you.] *inserted in-line and between paragraphs*
21 regret] *after* ~~did not find~~
22 I shall] *this paragraph below* ~~I fear~~
22 tomorrow] tommorrow *[E]*
23 night] *inserted above*
23–24 but … limited.] *after* ~~but it is doubtful whether I shall have time to make an appointment.~~
25–26 I have … there.] I have ~~taken up my abode~~ \made/ the Grand Union Hotel~~,~~ ~~for~~ my headquarters for receiving letters ~~and other and other but th~~ but ~~I won't be th p~~ \there is a/ small chance of meeting me there.

Peirce to Carus. MS: SIU 91.16.

108 W 89$^{\underline{h}}$ St. New York | 1897 May 1

My dear Sir:

I was greatly disappointed at not being able to see you, for various reasons. I should like to know more about Mr. Hegeler, for one thing.

A very grave and difficult operation was performed on my wife a week ago Thursday. I have not yet seen her; but she is doing well, except that her lung-trouble has broken out again since the operation. But I believe she will conquer it.

I have some heavy payments to make & if you will take the article on Graphs & pay for it now, I will take $ 150 for it, which is low considering the great amount of labor bestowed upon it. I will rewrite it in a more literary style if desired. I have been writing in a laboriously plain & unadorned manner because one of the things you said to me when I last saw you was that you would prefer a drier style than that I was then writing. If you think my last two articles too dry (which is my own opinion) I shall be glad to change.

If you don't care for the Graphs, suppose you let me immediately write you an article for $ 125 cash on the extension of the doctrine of probabilities to *relative* logic. This is a very important subject, in my opinion, concerning which I used to lecture in Baltimore. But it is a part of my system of logic on which I have never published anything. I think the value of the thing would be recognized by everybody who has studied the Logic of Relatives, and would draw some others to that study. As the matter is thoroughly digested in my mind, it would not take me long to produce the essay, especially if it were to be written in the manner which is most natural to me, & (I think) more attractive to readers.

I will enclose with this the circular of the Universal Trust Company. I have had some dealings with them & have looked minutely into their methods. It is economically administered & they can profitably perform their promises. You Westerners don't like our Eastern low interest; but it is best not to have all eggs in one basket,—be it even a basket as big as the mighty West. Anyway, this I can recommend as an easy way to make an investment out of which you can draw at any time. Were you to invest anything in it through me, I should get a small commission, which I would divide with you.

I have not heard from Russell for a long time. I trust he is well, prosperous, & happy.

Yours faithfully

---

6 and] *over* o

13 drier] *above* ~~dier~~

17 cash] *over* o

29 be it] be~~ing~~ \it/

C. S. Peirce

Carus to Peirce. TS: RL 77. On OCP Stationery C. LPB: SIU 31.76.320–321.

La Salle, Ill., U.S.A. May 12, 1897.

Charles S. Peirce, Esq'r,
108 W. 89th St.,
New York City, N. Y.,
My Dear Sir:—

On my return home I find your letter and enclosure. I am exceedingly sorry not to be able to accept your proposition concerning your article on "Graphs." I might have accepted it prior to my talk with Mr. Hegeler, but I can do so no longer, and however much I appreciate the value of your thought, I find that the readers who are interested in the Algebra of Logic constitute a very limited number, and the publication of articles of that kind is not looked upon with favor by our readers in general. I am frank in stating the situation, and I think it is always wiser to state the circumstances exactly. At the same time I will say that we must be economical with our means, which is the more necessary as Mr. Hegeler will be absent for several months.

In reply to your remarks on the enclosure of the circular of the Universal Trust Company, I have to say that I have no money to invest and am for this obvious reason not burdened with the cares of looking out for a place for investments.

I stayed only a few hours in Chicago and did not see Mr. Russell, but so far as I know he is very prosperous and doing a booming business.

When I passed through Cambridge, I met, among others, Prof. James, and I was very glad to hear that he appreciates your genius highly and he spoke of an appointment which he expected to procure for you in Cambridge.[163] I suppose you know about his efforts in your behalf and I trust that he will succeed. It would be a great advantage to you in many respects, not only giving you a definite income, but also bringing your name before the University public in a dignified manner.

Hoping that your wife will recover from the effects of the operation, and that her health will be perfectly restored, I remain, with kind regards,

Yours very truly,
P. Carus

---

163 Peirce's Cambridge Conferences Lectures, "Reasoning and the Logic of Things," to be delivered Feb–March 1898 (published in Ketner, 1992).

---

21 stayed] staid *[E]*

May–Nov 1897: Peirce focuses on preparing his 1898 Cambridge lectures in collaboration with William James, and on various other endeavors to alleviate his financial straits.[164] Also plans for another *Monist* article on Schröder dealing with "The Problem of Individuation," opening correspondence with Carus as follows.

Peirce to Carus. MS: SIU 91.16.

108 W 89$^{\underline{h}}$ St. New York, 1897 Nov 28

My dear Sir:

In my remarks upon Schröder's Logic[165] I passed by what he had to say about individuals, partly because the subject is a very broad one, having bearings far beyond the algebra of logic and requiring a broad treatment, and partly because I wished before expressing myself upon the subject to subject my opinions to thorough reëxamination to make sure no blunder had been committed and also to develope them, if possible, from various points of view. During the last year I have spent a large part of my spare time in doing this, and should now like the privilege of saying something concerning the Problem of Individuation. Meantime, Royce and others seem to have been occupied with this same matter; so that perhaps what I have to say may be deemed pertinent to the state of philosophy at this moment.

I undertake to show by logical analysis, supported by mathematical, metaphysical, psychological, and physical considerations that *thisness*, though it is termed "individuality," and "singularity," is nothing but an aspect of duality, so that in so far as any pairing has any truth, in that measure there is thisness; and that thisness arises in no other way. But the question of individuation is not solved until the relation of the individual to the general is made out. I am prepared to generalize our idea of generality so as to throw a new light upon it, and to make its relation to the individual clear.

I believe that if you would let me make you an article for the Monist on this subject, it would, in contrast to my two Schröder articles, which were too severely technical, be found interesting to large classes of readers. It is not in my plan to touch upon the religious bearings of the question; for I not only understand that

164 Peirce to James 30 May 1897: "If I were to have such a charge, I ought to spend every moment from now till October in preparing the first part of my course. Therefore, I beg you to let me know what the prospects are.... We are in truth very poor." (Brent, 261–262)

165 From "Regenerated Logic" and "Logic of Relatives."

12 no] *after* ~~there was~~
19–20 metaphysical,] *after* ~~psych~~
30 bearings] *after* ~~aspects~~

that would not be acceptable, but I also much prefer to keep my speculations free from all such bias. My own religion has nothing to do with metaphysics. But still, perhaps it will not bar out the expression of my line of reasoning, if I mention that it is favorable to the view of immortality which has been expressed by the editors of the Open Court.

While I am writing, I may mention that if promises are kept, as I believe they will be, I shall now in a short time receive some stocks which, if held till the first dividend is declared, will enable me to pay all my debts with interest. Far from being able to do anything for my creditors this year, I have been in the severest destitution, with a wife whose health has threatened to take an alarming course unless she had more than I could afford. It is terribly alarming at this moment, as the cold comes on. But you must not suppose that I am merely trying to make an article as a pot-boiler. It is that I have something which I believe will prove particularly interesting, and which will complete in a more interesting way my comments upon Schröder.

Yours very truly
CSPeirce

Dr. Paul Carus
Editor of the Monist, LaSalle, Ill.

Carus to Peirce. TS: RL 77. On OCP Stationery C. LPB: SIU 32.32.742.

La Salle, Ill., U.S.A. Dec. 3, 1897.

Charles S. Peirce, Esq.,
108 W. 89th St.,
New York City, N. Y.,
My Dear Sir:—

Your letter was duly received two days ago, and I hesitated to answer it because I actually do not know what to say. Your offer to write an article on Schröder's Logic, which you promise should contain new things and should be popular, is very tempting, still I am at present so exceedingly overcrowded that it would not be proper to induce you to write it when I am not able to use it at once. At the same time The Open Court's bank account is pretty low, and I could not accept an article from you unless I had special assistance from Mr. Hegeler. I will speak with him if I can find a good opportunity to approach him on the subject,

2 has nothing] has *l*nothing
7 if] *after* though
27 Your] ~~Although~~ *y*Your

and in case I can procure the article I shall write you again. In the meantime I advise you to have the article accepted in some other quarters.

Hoping that your expectations concerning the stocks of which you speak will not disappoint your hopes, I remain, with best wishes,

Yours very truly,
P. Carus

1898

Jan–Oct 1898: Peirce delivers Cambridge lectures Feb–March. Shortly after, focuses on preparing a "History of Science" book for G. P. Putnam's Sons (never published).[166]

Peirce to OCP 23 Sept 1898 (missing): Asks to be added to the circulation list for *Monist* and *Open Court*, requests a copy of Carus's "Buddhism and Its Christian Critics," and inquires about additional articles for submission. Answered by OCP (Diesterweg) 28 Sept and by Carus below.

Carus to Peirce. TS: RL 77. On OCP Stationery C. LPB: SIU 33.8.631.

La Salle, Ill., U.S.A. | October 13th, 1898.

Mr Charles S. Peirce,
Milford, Pa.
My dear Sir:—

Your letter to The Open Court Publishing Co. was forwarded to me, and I responded to it at once by directing our business manager to send you *Buddhism and its Christian Critics* and *Lao-Tze*, another new publication of mine which I expect will command your interest. I am very glad to have from an authoritative source a review which speaks very favorably about the book, especially as to the interpretation of the Chinese.

I suppose that *The Gospel of Buddha* is in your possession; I send you in this mail a copy of *The Dharma*.

As to your remark concerning contributions, I must confess that we have had very heavy expenses, and that I could not ask Mr. Hegeler for an extra donation; the sacrifice which he makes for *The Open Court* is great enough, and I should not like to increase it even if he were willing to do so. I must also add that although your contributions are valuable, from a scientific standpoint, they are not so from a business standpoint, as the circle of readers is naturally quite limited. My en-

166 Based on Lowell Institute lectures of 1892–1893. Plan for the one-volume edition in R 1290, and publication agreement in Peirce to Putnam 28 Feb 1898, RL 364. See also Brent, 264–265.

deavor at present is to reach a larger circle, and I am at the same time obliged to economise more than formerly.

With kind regards, and hoping that you are well both in health and otherwise, I remain,

Very truly yours,
P. Carus

Peirce to Carus. MS: SIU 91.17. Undated, responding to Carus's 13 Oct and answered by him 31 Oct. In the same SIU folder a postcard from Juliette Peirce to Carus ca. 20 Oct 1896 also indicates that "Since Oct 5th Mr. C. S. Peirce has been ill brain fever."

[ca. October 28, 1898]

Dear Dr. Carus:

I have had brain-fever and now am in a situation difficult in various ways. There are two things I would like to write for the Monist. One is an application of the ideas of Least Squares to Lutoslawski's Plato,[167] with the result of showing his argument is far stronger than he supposes, consisting in fact of *two independent* arguments. I also wish to make a few remarks on the effect of making the "Sophistes and Politicus" so late; and here I wish to suggest some metaphysics of my own.

The other thing I want to say in the Monist is that every system of logical algebra or graphs is subject to a very serious defect hitherto altogether overlooked and seemingly irremediable. This discussion will lead me to a closer examination of one important point made by Schröder.

very truly
C. S. Peirce

I wished much to speak of your grand works Chinese & Buddhist. I want to see the Gospel of Buddha. But what I have written has been a too great effort.

I may say I dont want any price for those papers, which shall be made as simple and readable as possible. I have tried in vain to get leave to notice your works in

167 *The Origin and Growth of Plato's Logic* (1897), reviewed by Paul Shorey in July 1898 number, p. 601. Peirce's proposed article possibly to be used in conjunction with or in follow-up to Shorey's "Mr. Lutoslawski's Plato," which at this time was being prepared for the Jan 1899 *Monist*.

14 Lutoslawski's] Lutoslowski's *[E] after* ~~Lobatchewski~~
25–26 I wished … great effort.] *This paragraph up left margin.*
25 Buddhist] Buddist *[E]*
142.27–143.2 I may … of books.] *This paragraph at top and across header of letter.*

the Nation. They give no reason but they are crowded and must exclude certain classes of books.

Carus to Peirce. TS: RL 77. On OCP Stationery C. LPB: SIU 33.11.753–754.

La Salle, Ill., U.S.A. | October 31st, 1898.

Mr Charles S. Peirce,
Milford, Penna,
My dear Sir:—

Your letter has just come to hand, and I hasten to reply that I am glad to learn of your recovery from brain fever, although you may still be convalescent.

As to the subject matter of your letter, I have at once directed our business manager to send you a copy of the "Gospel of Buddha".

Concerning the *Nation*, I may say that we stopped sending them any books for review, because we saw no reviews; thereupon, they wrote to us, and asked for review copies; I do not know how the matter stands now, but they have in the meantime given us some reviews.

You make two propositions to contribute to *The Monist*; first, on Lutoslawski, and secondly concerning Schröder. As to the former, I must confess that I care very little whether or not the "Sophistes and Politicus" is late or early, and as I am not sufficiently interested in the conclusion, I do not care about arguments either to corroborate or refute his views. I think that upon the whole Professor Shorey is better posted on Plato, for that is his specialty; while Lutoslawski, our learned Polish friend, is very erratic in his philosophy, as well as in his scientific propositions. Note his previous contributions to *The Monist* and his other publications. Personally, of course, he must be a charming man.

As to the other subject, concerning Schröder, I will say that any comment from you pen will be very welcome.

Now, in conclusion, I wish to say that if you write on Lutoslawski I shall publish the article simply as correspondence, and would accept it according to your proposition, without payment; for, in my opinion, it is too unessential, but I would publish it because some one whose opinion I highly appreciate wishes to have his say upon the subject.

As to the article concerning Schröder, I would prefer you to name the price yourself, so that I can accept or reject it, and I shall be much obliged to you if I

---

16 Lutoslawski] Lutosl*o*awski
21 Lutoslawski] Lutosl*o*awski
25 subject] subjecy *[E]*
27 Lutoslawski] Lutoslowski *[E]*

can manage the affair without obliging me to appeal to Mr. Hegeler for a special contribution.

With kind regards, and best wishes for your complete recovery, I remain,

Very truly yours,
P. Carus

Peirce to Carus mid-Nov 1898 (missing): Criticizes Carus's views of the Lord's Prayer, likely as expressed in his 1896 "Buddhism in Its Contrast with Christianity" in *Open Court*, and aspects of his work on *Tao-Teh-King*. Also appears to refer again to Peirce's desire to review Carus's work in *The Nation*.

Carus to Peirce. TS: RL 77. On OCP Stationery C. LPB: SIU 33.16.932.

La Salle, Ill., U.S.A. | November 21st, 1898.

Mr. Charles S. Peirce,
Milford, Penn.

My dear sir:—

Your letter was duly received, and I hasten to say first that the MS. which you intended to send has not arrived; if it is for the next *Monist*, we must have it soon.

Secondly, you can imagine that I have much to say in reply to your criticism of my view of the Lord's Prayer, but I am in a great hurry to-day, as I shall leave on the next train for Chicago, and I must therefore forego the pleasure of answering in detail.

Thirdly, in case you refer to *The Nation*, as I must conclude in thinking of your last letter, I should give this information to our business manager in Chicago.

But now I must repeat that I am in a great hurry, and with kind regards and best wishes, I remain,

Very truly yours,
P. Carus

Carus to Peirce. TS: RL 77. On OCP Stationery C. LPB: SIU 33.16.956.

La Salle, Ill., U.S.A. | November 25th, 1898.

Mr. C. S. Peirce,
Milford, Penn.

1–2 affair … contribution.] aff*i*air without \obliging me to/ appeal~~ing~~ to Mr. Hegeler<.> ˅for a special contribution.˅

16 must] mush *[E]*

My dear Sir:—

Upon my return from Chicago I wish to add to my last letter that the New Testament word βατταλογεῖν to which you refer does not mean "pray *much*", but "talking at random"; the meaning may be considered doubtful, because it is a word that is little used otherwise, and it is almost a ἅπαξ λεγόμενον[168]; but it is apparently onomatopoetic like the German "plappern". Luther properly translates it "plappern"; "to make *vain* repetitions" is a correct translation, but "speaking *much*", or "using *many* words", would not cover the sense of the word.

As to replacing "friendliness" for "benevolence", in Chapter V. of the Tao-Teh-King, I do not yet see your intention. The original Chinese word means benevolence in the sense of humaneness. Etymologically, the word means the inside of a thing, the kernel, especially the heart or inside of man,—man as a humane being, as having a heart and civilised sentiments.

With kind regards and best wishes, I remain,

Very truly yours,
P. Carus

Peirce to Carus ca. 25 Nov 1898 (missing): Discusses various points regarding Chinese history, culture, and literature, and appears to ask for confidentiality in his informal criticisms of the work of others.

Carus to Peirce. TS: RL 77. On OCP Stationery C.

Chicago Dec. 2, 1898.

Chas. S. Peirce, Esq.,
Milford, Pa.

My Dear Sir:

Your letter reached me after a week's absence from home and I hasten to acknowledge its receipt. I will briefly say at once in reply to your questions;

First:—I have never taken any interest in Chinese Chronology and I have no preference for either system. I might look the matter up according to Meyers or

---

168 "(something) said (only) once" (or *hapax legomenon*)

---

3 βατταλογεῖν] *inserted above*
11 Etymologically] Etomologically
12 heart or] *inserted above*

Williams[169] but I suppose the facts and opinions of authorities are also at your disposal and so my opinion would help you very little.

Second:—I believe that there is a great probability that La Couperie's theory[170] is, in the main, right, although the details of his statements do not appeal to me as tenable. Unless the Chinese civilization has gotten an original, independent growth, there is no other people from whom they could have derived their first notions of culture. In addition there are some striking similarities which have convinced me that there must be a connection of some kind, and the idea that some settlers, perhaps exiles, from Mesopotamia, migrated eastward and settled in the fertile valley of the Yanǵ tsé Kiang,[171] appears quite probable.

Third:—I have never myself found any article made of pure antimony in China, but that does not count because I have never been there. I cannot speak for other travelers and I have never given a thought to the subject before.

I do not think that the Chinese have displayed any extraordinary genius in their own literature except that they made some ingenious guesses and inventions which were never actually utilized. They had to be imported to the west to become the great stimulators of civilization. I mention gun-powder and printing, for it seems now more than ever probable that the knowledge of printed books has reached Europe from China.

I found of late a statement of the connection between east and west from a number of unknown travelers whose experiences were not written down, as those of Marco Polo, but at the moment I cannot think of the book in which I read it.

Their accomplishments in astronomy are apparently very moderate and the Jesuits and the Emperor Kong-Hi were by far the superiors of the old Chinese astronomers. The famous historical structures which are now seen at Pekin[172] were made after the Chinese had become familiar with western astronomy through the Jesuits.

In reply to your private remarks I would say, that, as a matter of course, anything you write us about other papers remains confidential.

169 W. F. Meyers (n. d.) and Samuel Wells Williams (1812–1884), American diplomats and sinologists with government posts in China.
170 Albert Étienne Terrien de Lacouperie (1844–1894), French orientalist who wrote extensively on Chinese history, civilization, and linguistics.
171 Chang Jiang, or Yangtze River in China, longest river in Asia.
172 Beijing.

13 before.] *paragraph break inserted*

26 through] *above* ~~from~~

I wish that whenever you send contributions to The Monist you would mark the MS with a price. We have at present ventured into publications which involve a good deal of outlay and I must economize. In addition I hope that you will bear in mind that what we need mainly is popular articles. We frighten the people away with heavy articles even though they be good.

With best wishes for your health and success with your "History of Science"[173] which I expect will be a book of great value, I remain,

Very truly yours,
P. Carus

Peirce to Carus ca. 5 Dec 1898 (missing): Submits "Knotty Points on the Doctrine of Chance" (R 209) for *Monist.*

Carus to Peirce. TS: RL 77. On OCP Stationery C. LPB: SIU 33.19.56.

December 9th, 1898.

Mr. Charles S. Peirce,
Milford, Penn.
My dear Sir:—

I am very sorry that I shall have to return your MS. on "Knotty Points on the Doctrine of Chance", which I find upon my return after an absence of some days. I have talked with Mr. Hegeler on the subject, and he has repeated to me that he wishes us not to swerve too much from the purpose to which our publications are devoted. At the same time, I would suggest to you, whenever you send an article to The Open Court Publishing Co., to mark the price which you put upon it; this will greatly simplify matters, and will serve to prevent complications, in case we accept the article.

With kind regards, and regretting that I cannot publish your article, I remain, dear sir,

Very truly yours,
P. Carus

---

173 See note preceding 1898 letters above for details.

1–2 whenever … price.] wh*at*enever you ~~wish to~~ send \contributions/ to The Monist you would mark \the MS/ with a price.
20 purpose] purpose~~s~~

1899

Peirce to Carus ca. 12 March 1899 (missing): Peirce offers to write an article on Renouvier's 1899 *La Nouvelle Monadologie* for *Monist*. Around this time Peirce was also finishing a related *Nation* review of Latta's *Leibniz: The Monadology* to appear 16 March.

Carus to Peirce. LPB: SIU 33.34.696.

March 14th, 1899.

C. S. Peirce, Esq.,
Milford, Penna.

My dear Sir:--

Your letter is just received, but I regret to say that we are unable to accept your offer. I believe that your impression of the importance of Renouvier's reconstruction of Leibnitz's Monadology is justified; but the ventilation of this problem does not seem urgent enough for us to change our present plans. I wish you could have the article published in some other philosophical magazine, either in *Mind* or in the *Philosophical Review*, where it might do the same good as in *The Monist*.

Hoping that you will not give up the publication of the works which you announced some time ago,[174] I remain, with best wishes,

Very truly yours,
P. Carus

Peirce to Carus. MS: RL 77.

Milford Pa 1899 Aug 17

My dear Dr. Carus:

I find I did an injustice to Kant in one of my Monist papers in which I discussed continuity.[175] I was so much dominated by Cantor's point of view, that I failed to see the true nature of continuity, which is now quite clear to me, and also for the same reason mistook Kant. This was somewhat excusable since Kant himself made the same confusion between his real doctrine and the one I took for his, as can be seen at the bottom of p. 524 of 1st Ed of the Critik.[176] Namely he does not quite say that it is the same to say that Space is infinitely divisible, but he does not draw attention to the important distinction as he would have done, if he had seen

174 Peirce to Carus 1 April 1897.
175 "Law of Mind" (July 1892).
176 *Critique of Pure Reason* (1781).

it. His true doctrine is not that space is divisible without end, but that it *cannot be so* divided as to reach an ultimate part as clearly stated in the last ¶ of p. 169. This definition is really no nearer right than the other, but it is better as looking at the matter from the right point of view and not trying to build up a continuum from points as Cantor does. To the obvious objection that points are ultimate parts of lines, Kant begins to make the right answer, that they are not parts but limits. But that he does not understand this rightly can be seen by p. 209 where he speaks of a change as passing through all the instantaneous intermediate states. He thus looks on the points as existing in the line, while the truth is they do not exist in the continuous line, & if a point is placed on a line it constitutes a discontinuity. As he conceives the matter, it becomes a mere quibble to say that a point is not part of a line. It is not a homogenous part, but it is in the line. The case is an interesting example of Kant's sagacity about points he did not distinctly apprehend.

Very truly
CSPeirce

See Jour. of Spec. Phil. II. 200.

1900

Peirce and Sacksteder mid-late April 1900 (missing): Sacksteder contacts Peirce with some question of "special interest," to which Peirce replies and in some way criticizes Carus, who opens correspondence as follows.

Carus to Peirce. TS: RL 77. On OCP Stationery C. LPB: SIU 34.13.488.

La Salle, Ill., U.S.A. | March 3rd, 1900.

Charles S. Peirce, Esq.,
Milford, Penn.

My dear Sir:—

Our business manager handed me your letter in reply to his question, as one of those which appealed to his as of special interest, and I hasten to say in reply that your criticism is quite just, and if I could I would act upon it. But I cannot do as I please, and must consider circumstances and do the best I can.

I hope that you are doing well, and should be very glad some day to meet you again.

---

16 See Jour. of Spec. Phil. II. 200.] *This line down left margin.*

I sent you some time ago my little book on Kant and Spencer, and I suppose that you noticed that I quoted again as I did on its first publication your article on Kant as being not hard reading.[177] Of course I agree with you perfectly, and your statement being said in criticism of my own views comes out much better in this way than if you had written your remarks in assent.

The difficulty in reading Kant does not consist in his thoughts but originates through an ignorance either of the language or the terms which he uses, or the historical connection in which he appears.

Did you see the amusing little notice in *The Nation* about my Kant and Spencer book? It is written by one who was hit.[178]

I have received a great number of letters on the book, and expect it will contribute its share to overcome the superficiality of philosophic thought of the present day.

With best wishes, I remain,

Very truly yours,
P. Carus

Peirce to Carus ca. 6 March 1900 (missing): Discusses further Carus's *Kant and Spencer*, particularly in defense of Spencer (consistent with Peirce's *Nation* review).

Carus to Peirce. TS: RL 77. On OCP Stationery C. LPB: SIU 34.14.531.

La Salle, Ill., U.S.A. | March 12, 1900.

Charles S. Peirce, Esq.,
Milford, Pa.
My Dear Sir:—

Thanks for your letter. I would forgive Spencer for all his superficiality if he were not so high-handed, but his condemnation of others and his sovereign contempt of criticisms make it necessary to drive the fact home to him. He is only a mortal like ourselves. I did not like to attack him and waited nine years before I brought the pamphlet out in the present form in which it will prove more effective.

177 *Kant and Spencer*, p. 51, quoting Peirce to Carus 6 Sept 1890 (paragraph beginning, "I have heard too much of Kant's being hard reading … "), previously quoted in Carus's "Kant on Evolution," *Monist* (July 1892).
178 8 Feb 1900, p. 109. Accredited to Peirce in *CBPW* P00718, which perhaps Carus suspected.

1 and^2] and and *[E]*

As to your comments to be made in *Science*,[179] I shall be glad to have a copy of the paper, but if you will let me know the number and page, it will be sufficient. I can easily look at it, we have *Science* in file.

Mr. and Mrs. Hegeler are in Europe.

With kind regards and best wishes, I remain,

Yours very truly,
P. Carus

Carus to Peirce. TS: RL 77. On OCP Stationery C. LPB: SIU 34.15.574–575.

La Salle, Ill., U.S.A. | March 17th, 1900.

Charles S. Peirce, Esq.,
Milford, Penn.

My dear Sir:—

Since your last letter was received, I have thought of you and your prior contributions to *The Monist*, and the idea struck me that we might publish the controversy which took place between ourselves, and we could leave it exactly in the shape in which it appeared in *The Monist*.[180] The little incidental tilts of a personal nature will only add zest to the dry reading, which, owing to the subject, could scarcely be elucidated in a more *piquant* style. I would propose to use your picture as it appeared in *Sun and Shade*[181] as a frontispiece, and would publish

179 "Infinitesimals," vol. 11 (16 March 1900).
180 Necessity–Tychism feud April 1892–July 1893.
181 "Charles Sanders Peirce," vol. 4, no. 12, Aug 1892 (photogravure accompanying short biographical sketch):

1 *Science,*] *after* ~~the~~
1 glad] *inserted above*
3 *Science*] *after* ~~the~~
19 *Sun and Shade*] *Sunsh~~ine~~\and Shade/*

the short editorial note which appeared in *The Monist*, Vol. II, page 432, as a kind of introduction.[182]page 432

Of course, you are welcome to add either a prefatory note of your own or a conclusion, which might be as short as my note. I would also suggest, if you see fit, to add incidental remarks which you would like to have added in footnotes, wherever you please. My intention is not to settle the problem of discussion between us, but to offer it to the public in a handy shape, so as to set people to thinking. I would make the word *necessity* prominent in the title. Perhaps The Problem of Necessity: Discussed in a controversy between etc

I have to add that this is simply a proposition of mine on which I wish to sound your intentions and inclinations. I have not as yet asked Mr. Hegeler's consent to it, which of course would be necessary under present conditions. He is now in Europe, and I might formulate the question so as to ask him whether or not he has objections to it. I will make the proposition to him, but of course he has the veto power.

I ought to add first that the book should appear in the Religion of Science Library, which commands quite a good sale, thus making the books published therein popular. There is no business profit in the sale of these cheap books, because we use good paper and the margin of profit is too small to bring any returns. In fact, it scarcely covers the cost of handling the books. Therefore, I must at once state that in proposing this plan of republishing our controversy I cannot pay any honorarium; moreover, the understanding is that The Open Court Publishing Company holds the copyright on these articles, and so I trust that you would scarcely expect a new payment. Further, if you favor the idea, and if Mr. Hegeler does not veto my proposition, I have to inform you that we could not publish it for some time to come, because we have other material on hand; but we might arrange, so far as I can see now, to publish them at the end of this, or the beginning of next, year.

Let me know how you feel about it, and as a matter of course I would not venture into it, and would not even break the idea to Mr. Hegeler, if I knew you would oppose it.

I remain, dear Sir, with kind regards and best wishes for your welfare,

Very truly yours,
P. Carus

---

182 Carus's note "Mr. Charles S. Peirce on Necessity," which followed Peirce's "Doctrine of Necessity Examined" (April 1892).

---

2 page 432] page 442 *[E]*
8–9 Perhaps ... etc.] *inserted in-line*
13 him] *inserted above*

Peirce to Carus 23 March 1900 (missing): Regarding Carus's proposed book, suggests that he be able to revise his necessitarian articles, include other articles of his, and write an introduction to the book himself. Also seems to suggest publishing separately their respective articles.

Carus to Peirce. TS: RL 77. On OCP Stationery C. LPB: SIU 34.16.616.

La Salle, Ill., U.S.A. | March 26th, 1900.

Charles S. Peirce, Esq.,
Milford, Penn.

My dear Sir:—

Your letter of March 23rd just comes to hand, and in reply I wish to say that so far as I am personally concerned I shall comply with all your wishes concerning the publication of your articles in book form. I would be glad to have you write an introduction such as you propose, and also would comply with your desire to insert as many articles as you see fit. The articles are not in plates, and would have to be reset under all circumstances. But I have to state, first, that the funds of The Open Court are limited, and I could not make any new payment as a honorarium for the arrangement of the book or or introduction. Secondly, I myself could not for quite a while undertake to rewrite my answers to you. Thirdly, if you make such considerable changes it might necessitate a rewriting of my answers to you, which I could not do for a long time, although until The Open Court Publishing Company might be ready to republish your articles in book form I might again be more at leisure.

I would suggest to publish your articles alone, without my answers, but I must be pardoned for taking an interest in your views just because I believe they are the most ingenious presentation of the opposite side, and therefore I wish to have my own views represented side by side with yours.

Considering all in all, would you not deem it wise to republish all the other articles of yours, including those on logic from the *Popular Science Monthly*,[183] in a little book by itself, and our controversy in another little book by itself. The

---

183 "Illustrations of the Logic of Science" series 1877–1878, also referenced in the next letter. Carus had encouraged Peirce to re-publish the series throughout their correspondence at least as early as 1893 (see Hegeler to Peirce 27 Feb), and began working with Peirce more seriously to that end 23 Jan 1907 until Peirce's death 1914. The book was never completed. For details, see de Waal, 2014.

---

15 as] *over* or 16 or introduction] *inserted above*

latter should have your introduction, and if you make changes I would of course be entitled on my part to revise my articles, or have a concluding word to say.

I shall be away in the summer, and expect to visit the Paris Exposition. It will be impossible to undertake the publication of your articles before my return, but we might at least come to an understanding concerning the general plan.

I remain, dear Sir, with kind regards and best wishes,

Very truly yours,
P. Carus

Peirce to Carus ca. 1 April 1900 (missing): Replies in a long letter insisting that he be able to update his necessitarian articles and include other articles as well. Suggests again that his articles be published separately from Carus's, or else he might take the work to another publisher. Also criticizes Charles Chase's "A Strange Attack" in the newly released April *Monist*, and points out an apparent flaw in his bracketed formula at p. 464.

Carus to Peirce. TS: RL 77. On OCP Stationery C. LPB: SIU 34.7.680–683.

La Salle, Ill., U.S.A. | April 3rd, 1900.

Charles S. Peirce,
Milford, Pa.
My dear Sir:—

Your long letter was duly received, and having duly perused it I find that you make so many suggestions that in making my reply I fear I shall miss a point; therefore, please take this letter as a preliminary answer.

First, I wish to say that I am desirous of letting you have your full say and bring out your philosophy as clearly and completely as possible. If you wish to publish your articles without my controversy, do so, and you may if you see fit make only such references as you deem necessary. If you wish to publish the essays as a controversy of the discussion of "Tychism and Necessity" do so, and in that case you are welcome to write an answer to my last article;[184] and if conditions require I shall (though not before my return from Europe) be glad to write a rejoinder to your answer. That would, it seems to me, be the best way out of the dilemma, for a revision of all the articles would require much time, and might leave us both dissatisfied.

---

184 "Founder of Tychism" (July 1893).

---

19 having] haveing   28 (though ... Europe)] *inserted above*

As to the little tilts which you began and to which I retorted, I believe they add a great deal of zest to the controversy, and a few words on your part in your prefatory note might set the public right as to the spirit in which they were dealt. They have been frequently commented upon, and I remember especially a French magazine which spoke of it as typically American that the editor did not hesitate to publish attacks on himself and replying to them good-naturedly. I doubt whether all American editors would do the same; you, at least, have an experience of ethical editors who did not. You remember the affair, Abbott *contra* Royce.[185]

Your remarks concerning the three elements of the world, interest me greatly, for I, too, am a Trinitarian, although it may be quite in a different sense. I do not venture to enter into it now, because I would have to write an article on the subject.

Your remarks on page 4 of your letter as to "counting the reply" are not clear to me, and you continue to say on page 5 that my "course in that matter was most unfair and most unjust". This must rest on a misunderstanding, for I was always anxious to let you have your full say, and never restricted you either in using any expression you wanted, or in the size or number of your replies.

Your remarks on page 6, concerning the Lord's Supper are worth quoting, but of course I shall abstain from it; they are good nevertheless.

We returned the MS. of Judge Chase, of Ithaca, Mich., but he was so persistent in clamoring for what he deemed justice to him, that we had to allow him to have his say. I was quite under the impression that this was "cruel" of him, but he would have accused me of unfairness, if I had not published his articles. As to the expression in square brackets at the bottom of page 464 of *The Monist*, I must confess that I did not examine it, and Mr. McCormack, to whom I showed your comment on Judge Chase, declared that he neither read nor examined it. Judge Chase, like so many others, wants to prove a definite soul-conception which smacks of the Brahman Atman, and all men of that type have a twist in their arithmetic and logic; so, it is scarcely worth while to trouble about, or going over their deductions.

As to your postscript, I can only say that if you find a publisher for publishing your articles as you would like to have them published, you are at liberty to repub-

185 1890–1891: Josiah Royce libeled Francis E. Abbot's *The Way Out of Agnosticism* (1890) for philosophical fraud and allegedly plagiarizing Hegel. The debate was examined in several notices in *The Nation* Nov 1891, including Peirce's and James's respective "Abbot against Royce" letters. Also detailed in *W* 8:lvi–lviii, and in Brent, 215–219, who considers this "the nastiest—and the most absurd—philosophical dispute" in the U. S. in the late nineteenth century.

6 good-naturedly] *after* ~~as~~
9 world,] *comma inserted*
16 or number] *inserted above*
21 of] on *[E]*
24 showed] *above* ~~quoted~~
26 smacks] sma\c/ks
28 , or] *inserted in-line and above, respectively*

lish whatever you want from *The Monist*, or to cut up your articles and adapt them to new uses.

As to your logical articles in the *Popular Science Monthly*, containing the doctrine of Pragmatism, I remember that you received Appleton's permission to have them republished by The Open Court Publishing Co. If you wish to make use of it, I shall be glad to receive your propositions.

I conclude by stating again that although as a rule I am free to act as I deem best, I am still subject to the veto of Mr. Hegeler in case he would not approve of my transactions. Thus, all propositions are subject to this veto. Further, I hope you understand that the means at my disposal are now limited, and I could no longer pay the amount which we paid you before. It would simply be ruinous to our business dealings in general. Perhaps it is well that I have to look out for the business part of *The Open Court*. But the point at present is that you understand the situation. I am quite willing to pay for new articles, but although I esteem your contributions highly, and more so than others, I am not able to pay the old prices.

Let it be a general rule, therefore, that whatever claims you may have for articles ought to be stipulated beforehand, on transmitting the MS., and agreed to by me.

I must confess that a general view of your philosophy would be quite welcome to me for *The Monist*, especially if you would condescend to write in a popular style. You may not care for popularity, but if you had seen all the letters we received complaining about your articles on "The Logic of Relatives", you would understand that an editor has also to look out for his readers. Even your stanchest friends wrote me that the article did not do you justice.

I remain, dear Sir, with kind regards and best wishes,

Very truly yours,
P. Carus

Peirce to Carus. MS: SIU 91.19. Answered by OCP 9 Nov 1900.

Milford Pa 1900 Nov 3

My dear Dr. Carus:

I need some types such as you had cut for the characters ⤙ Ψ ʄ of Logic.[186] They are for use in a Philosophical Dictionary going through the Oxford Press.[187]

186 "Logic of Relatives" (Jan 1897).

187 J. M. Baldwin's *Dictionary of Philosophy and Psychology*, for which Peirce wrote extensively 1901–1902 (see *CBPW*).

May I ask you to have a sufficiency of them sent to me? The printer will know how many. Not more than you used, I should think. Or I would take the dies if necessary.

You will ascertain what it is proper should be paid and I will see it is paid. If you will kindly oblige me by doing this as soon as possible, since the press waits, I shall feel it as a particular favor.

very truly
C. S. Peirce

My Express address is Port Jervis, N.Y.

1901

Peirce to Carus ca. 1 Jan 1901 (missing): Proposes to write an article on Ludwig Boltzmann's "Recent Development of Method in Theoretical Physics" in the newly released January *Monist*.

Carus to Peirce. LPB: SIU 34.47.948.

January 16th, 1901.

Charles S. Peirce, Esq.,
Milford, Penns.

Dear Sir:—

Having returned from Europe only a few days ago, I am still overburdened with work, hence my delay in replying to your letter. I should indeed be glad to publish some comments of yours on Boltzmann's article, but I fear that I cannot make any engagement with you for writing such an article as you indicate, because the funds at my disposal are at present greatly reduced, and you are quite an expensive contributor. I wish I could see you once and talk matters over with you. There are so many things that are liable to be misunderstood and which can only be explained orally.

I hope that you and your wife are well, and that your business prospects are flourishing.

I remain, dear Sir, with kind regards,

Very truly yours,
P. Carus

---

2 Not more ... think.] *inserted below*

Jan 1901–Aug 1904: For the next three years, with the exception of his unusual, incomplete correspondence with OCP compositor Fred Sigrist, no letters between Peirce and OCP are found. It is likely that Peirce simply maintained little contact with OCP during what was then a very busy time for him: he translated for the *Annual Report of the Board of Regents of the Smithsonian Institution* (1901), wrote definitions for Baldwin's *Dictionary of Philosophy* (1901–1902), submitted his elaborate grant proposal to the Carnegie Institute for "Memoirs on Minute Logic" (1902),[188] delivered his Harvard lectures on pragmatism[189] and his Lowell Institute lectures on logic (1903),[190] and all along continued reviewing books for *The Nation*.

He was, however, also gearing up for a new series of *Monist* articles on pragmatism to disentangle his own form of it, pragmaticism, from other contemporary variants, such as those of William James, F. C. S. Schiller, and John Dewey.

1904

Peirce to Carus 10 Aug 1904 (missing): Proposes his first pragmatism article, "What Pragmatism Is," to be partly a book review of Herbert Nichols's 1904 *A Treatise on Cosmology*. Also mentions he does not receive *Open Court*.

Carus to Peirce. TS: RL 77. Stationery headed, "The Monist, | The Open Court, | Editorial Department, | La Salle, Ill. | Dr. Paul Carus." LPB: SIU 36.52.574.

August 27, 1904.

C. S. Peirce, Esq.
Milford, Pa.

My dear old Friend:

After a long absence I find your letter of August 10th among a heap of mail which I try to dispose of at my earliest convenience. I am very sorry that your letter had to await my return among the rest. I wish it had been forwarded to me and I could have answered it at once. You must have thought that I purposely refused to answer and that no answer means refusal. I hasten therefore to write if it be not yet too late to *let me have your Ms.* and I shall pay you the honorarium *of $50.00* as

188 Ransdell, "MS L75: Logic, Regarded as Semiotic", and partly in *NEM* 4:13–73
189 *EP* 2:133–241, also in full in Turrisi, 1997.
190 *EP* 2:258–299.

25 purposely] purpose\ly/

stipulated in your letter. I have had my hands full and I shall be glad to have your opinion of Herbert Nichol's "Cosmology," for I must confess that I myself have not yet bothered myself about it. I only wish that Nichols is worthy of your steel, and I scarcely doubt that you will bring out some important and also interesting points in your discussion.

The Ms. will be too late for our present 5th number but I should be very glad to have a good opening article from your trenchant pen for the new volume. As to the *Open Court*, I am sorry to learn that you are no longer receiving it, but I am glad to notice that you missed it. *It was not my intention to cut you off*, which must have been done by our business manager. I remember that he sent out a circular letter to those who received complimentary copies of the *Open Court* and he cut off everyone who did not show that he cared for its continuation.

With kind regards and best wishes, I remain

Yours very truly,
Paul Carus

Peirce to Carus ca. 12 Sept 1904 (missing): Submits "What Pragmatism Is," which wouldn't appear until April 1905.

Carus to Peirce. TS: RL 77. Stationery headed, "The Monist, | The Open Court, | Editorial Department, | La Salle, Ill. | Dr. Paul Carus." LPB: SIU 36.53.631.

September 15, 1904.

C. S. Peirce, Esq.
Milford, Pa.

My dear Sir:

Your Ms. just reaches me on the day on which I intend to leave for St. Louis. I have no time to look at the Ms. but I expect that your plan to have your treatment of pragmatism divided into a book review and an article is proper. You will notice in the same, (viz., the next) number that I did the same with Paul Rée, a philosopher of second degree on whom I published an article and a review of his work.[191] The article is written by one of his believers and I have to let the public

---

191 Hooper, "Paul Rée"; and "Paul Rée: An Obituary."

---

6 5th] *4*5th
7 have] *before* \~~it as~~/
9–10 must have] \must/ ha*s*ve
27 (viz., the next)] ˅(˅viz., the *last*next˅)˅

know that Paul Rée is not up to the mark while for private reasons he deserves all honor as a searcher for, not as a promoter of, philosophical truth.

I shall attend to the business part of our transaction at my earliest convenience and will send you the promised check for $ 50.00 as soon as I can have it made out from Chicago.

I remain with kind regards and best wishes,

Very truly yours,
Paul Carus

Peirce and Carus late Sept 1904: Peirce in a missing letter proposes to write a review (later titled "Substitution in Logic") of Hippolyte Taine's 1870 *De L'Intelligence*. Carus in another missing letter replies accepting the proposal but suggests that Russell sign it in his name.[192] Peirce complies but in turn adds in a letter to Russell 3 Oct 1904 that he work up a version of it in his "own language" and "own point of view," which Carus to Peirce 27 Oct approves and which appeared in *The Monist* April 1905. For a draft of Peirce's original version of the review, see Appendix, "Peirce and Russell."

Peirce to Carus ca. 1 Oct 1904 (missing): Requests a number of books to help with his "What Pragmatism Is" article, including Höffding's *Outlines of Psychology* (1891) and Wilhelm Wundt's *Soul of Man and Animal* (1863) and *System of Philosophy* (1899). Submits a paragraph to be added to "What Pragmatism Is," that beginning "Suffer me to add one more word on this point" (pp. 180–181).

Carus to Peirce. TS: RL 77. On OCP Stationery B. LPB: SIU 36.54.658.

La Salle, Ill., U.S.A. | October 3, 1904.

C. S. Peirce, Esq.
Milford, Pa.

My dear Sir:

In reply to your letter and the addition to your article I will see at once what I can do to forward you the books which you need. As to Höffding's *Psychology*, and Wundt's *The Soul of Man and Animal*, I cannot make a definite promise because I do not have the books handy here nor do I know what you mean by Wundt's "*System*." Is that perhaps the systematised index to his work by some of his disciplines?

192 Apparently because Peirce quotes his own 1868 "Natural Classification of Arguments" in the review. Furthermore, "seeing that Mr. Peirce cannot sign the communication himself," Carus wrote Russell 7 Nov 1904, "and as the editor should not publish it under his own name it seemed to me best to have some one of Mr. Peirce's friends who would unhesitatingly endorse his views, sign the paper" (RL 387).

I have further to state that a paragraph in your addition is misleading. You say that your series of contributions to *The Monist* have been unexpectedly cut short which suggests the idea that our controversy had not been finished and I must state in reply that at the time I offered you to send in your rejoinder, but you refused. The article which I did not encourage you to send had nothing to do with your series of contributions and was offered only after a long pause in our communications.[193] I will be glad to publish your comment as it stands but will have to make an explanatory note so as to avoid misunderstandings. Perhaps you would prefer to change the words so as to render an editorial comment on my part unnecessary. You might very well say that a concluding essay would throw better light on your former contributions but you can*not* say that your series was incomplete because the conclusion was cut short, for it was not done by me. I am not anxious to have controversies but if your note is a challenge or a reproof I shall not shrink from stating my side of the question. I am confident that I always treated you with fairness and if I publish your comment as it stands it would imply a concession on my part that I had not given you a chance to express yourself fully.

Hoping that you will understand the point which I make I remain with kind regards,

Very truly yours,
Paul Carus

Peirce to Carus. MS: RL 77. Undated and unsigned, with MS page missing where indicated in the text. Possibly an unsent draft, but no alternative survives and seems fairly complete. Letter responds to Carus's of 3 Oct 1904 and is answered by him 12 Oct.

193 Carus misinterprets Peirce's comment as referring to a rejoinder to Carus's "Founder of Tychism" that he was not allowed to submit prior to their 1894–1896 gap in correspondence, which reopened May 1896 when Peirce submitted "The Regenerated Logic" on Schröder, the article Carus "did not encourage" him to send. However, Peirce is more likely referring to his "A General Sketch of a Theory of the Universe," the sixth and final article for his 1891–1893 metaphysical series, also listed as such in his outline of "A Quest for a Method" sent to Hegeler 7 March 1893. Peirce similarly reported to Russell 4 Oct 1895 that Hegeler had broken his promise to include this final article, which "would have been the keystone of the whole." This is also consistent with Peirce's revision to the paragraph in his responding letter and in his article as it now stands, beginning, "Had a purposed article concerning the principle of continuity and synthetizing the ideas of the other articles of a series in the early volumes of *The Monist* ever been written …" (p. 180).

9 would] $^\vee$w$^\vee$ould
9 words] *above* ~~note~~
12 , for … me] *inserted above*

[ca. 5 October 1904]

My dear Doctor Carus:

I am shocked to find that anything I sent you for publication could be construed as finding fault with your conduct. There is far too much about myself in my article, anyway. But such matter seemed unavoidable. I think (not having what I sent before me) that if the passage were amended & shortened so as to say, "Had the article about the principle of continuity & synthetizing the whole series, with which the writer had intended to close that series of articles ever been written, it would have more clearly appeared that" etc.

If you will kindly take the trouble to alter it somewhat like this before it goes to press, it will perhaps be the better way. If your amendment don't suit me, it will be open to me to delete the passage. I don't see just now what the public should care about what I meant to say but didn't; but I have forgotten the context. But there is one thing I can say positively which is that my concern is to present certain thoughts about philosophy & not to show that I was formerly right or wrong, still less to occupy readers of the *Monist* with my personality. What I have said about myself I thought far too much, but I did not see any better way of expressing the philosophical ideas I wanted to express. I have on some occasions been found fault with for allowing misstatements about my position to pass unnoticed & I suppose I ought not to do that. I dont doubt that some men bring out their ideas best in controversy; but I am inclined to think that as far as I am concerned *[…]*

[…] myself for the inadvertency, for I really was not conscious, I am sure, of making any complaint. In fact, I remember having thought of saying that there was some misunderstanding in consequence of which the article did not appear, & then thinking that misunderstanding might be understood in the sense of a dispute, and in any case that it smacked of complaint, I changed that; but it seems that I did not sufficiently consider the possible interpretation of what I was writing. I trust now that what I have said on this topic has removed the disagreeable impression which my language made upon your mind & that the matter may be dismissed. I have always felt that the hypothesis of tychism was one which it was, at any rate, of advantage to to philosophy to have brought forward & argued, and though I do not yet feel sure that it is sound, I do not think its unsoundness has ever been made clear. I always thought that it would have been strengthened by presenting the synechism which was to root of tychism in my own mind & which would

---

7 about] *above* ~~with~~

11 your] you *[E]*

20–21 do that. I don't … concerned] *from an aborted draft of page 4, included here in absence of correct page*

20 do] to *[E]*

27 writing] *after* ~~saying~~

31 to philosophy] *inserted above*

have shown the significance & connection of all the different articles of my series. Synechism is a doctrine which may be regarded as beginning in logic by saying that in the nature of things no idea can be exact, so that the principles of excluded middle & of contradiction only refer to an unrealizable state of thought. Then if metaphysical conceptions are only forms of logic in another application, the assumption should be that in *reality* there is no such thing as absolute exactitude. This gives a particular form to the conception of the continuity of space & time.

But my article on continuity in that series[194] I regard as quite the most faulty of the lot. So that after all I could not, at that stage of my studies, have produced an account of synechism which would satisfy me now. It would have appeared to me now as to some extent erroneous & to a much greater extent imperfect.

I am inexpressibly obliged to you for your kind promise to loan me those books. The "System" of Wundt to which I referred is a "System der Philosophie" which has reached a 2$^{\underline{nd}}$ Edition, to which I have seen reference though I never saw the book. Your not knowing it makes me suspect I may have misunderstood; yet I think that cannot be.

Peirce to Carus ca. 10 Oct 1904 (missing): Sends a note (received halfway through Carus's response below) addressing the unclear origins and ownership of the term pragmatism. The note was initially to go unsigned in his article but likely became Peirce's footnote glossing "pragmaticism" at p. 166 (beginning "To show how recent the general use of the word 'pragmatism' is") after Carus directed him 25 Oct to Moore's "Humanism" in the October number addressing the same issue.

Carus to Peirce. LPB: SIU 36.54.692.

La Salle, Ill., October 12, 1904.

My dear Mr. Peirce,

Thanks for your letter which has been duly received and I shall insert the suggestion as you state it. Your letter just comes to hand. Why do you not mention

194 "Law of Mind."

3 principles] principle$^{\vee}$s$^{\vee}$
6 in] *after* ~~rea~~
6 absolute] *after* ~~pre~~
12 inexpressibly] *after* ~~prea~~
14 which … Edition,] *above* ~~I think in 2 vols.~~
15 misunderstood] misunder*t*stood
163.27–164.2 Your letter just comes … wind.] *added in pen*

the names of either periodical or writer? It makes the reading unsatisfactory. The reader has the impression that you are whipping wind.

Very truly yours,
P.C

Peirce to Carus ca. 18 Oct 1904 (missing): Explains further the nature and purpose of the note to be included in his article. Also discusses his political views on the current U.S. president at the time, Theodore Roosevelt, Jr. (1858–1919).

Carus to Peirce. TS: RL 77. Stationery headed, "The Monist, | The Open Court, | Editorial Department, | La Salle, Ill. | Dr. Paul Carus." LPB: SIU 36.54.705.

October 21, 1904.

Mr. C. S. Peirce,
Milford, Pa.
My dear Friend:

Your letter was duly received and I understand now that you intended to have your note go in unsigned. I thought you intended to have it as an addition of your article to go under your signature. With the new information on hand I will reread it and reconsider the advisability of its publication. I prefer of course to let every thing go under your name. I understand also the importance of the note but I must confess that I have not yet studied pragmatism sufficiently in its details to write on the subject. I have preliminarily made up my mind about it and I believe we two would agree pretty well on the subject.

Your reasons for voting for Roosevelt are strange but I think you are right. I think if I were in your place I would vote for Parker.[195] However, I am not in favor of the Democratic party at present. The Democrats have been too vacillating to have my confidence. The silver proposition has spoiled a good deal and if Parker is a good man, his party is too much mixed up with Bryanism.[196]

---

195 Alton Brooks Parker (1852–1926), American judge and U.S. presidential candidate in the 1904 election, which he lost to Roosevelt.

196 Political philosophy associated with William Jennings Bryan (1860–1925), American politician and three-time U.S. presidential candidate (1896, 1900, and 1908) and later U.S. Secretary of State 1913–1915. Bryan was a proponent of the Free Silver Movement that opposed the gold standard, as Carus references. See also footnote in Russell to Peirce 30 Sept 1896.

---

1 writer] *after* ~~auth~~
17 publication] ~~consider~~\public/ation
19 confess] con~~v~~\f/ess
20 preliminarily] *over* temporarily
21 subject.] *paragraph break inserted*
23 am] *inserted above*

With kind regards and best wishes, I remain

Very truly yours,
Paul Carus

Carus to Peirce. TS: RL 77. On OCP Stationery C.

Chicago, Oct. 25, 1904.

Chas. S. Peirce, Esq.,
Milford, Pa.
My dear Mr. Peirce:—

The present number of THE MONIST contains an article by Prof. A. W. Moore of the Chicago University on "Humanism", which is Prof. Schiller's name for "Pragmatism". Prof. Moore mentions your name in it and also Prof. James, referring to the same incidences as you do in your note.[197] This gives you a chance to make your comments under your own name and you may recast your note as you see fit.

With kind regards and best wishes, I remain,

Yours very truly,
Paul Carus

Carus to Peirce. LPB: 36.55.732.

October 27, 1904.

C. S. Peirce, Esq.
Milford, Pa.
My dear Sir:

In reading over your "Substitution in Logic,"[198] I just come to the conclusion that the best is to let it appear as it stands with the exception that you insert the several names which you omitted. I will polish it as it stands and will have it signed by one of your friends. Either let us use a fictitious name or Mr. Russell of Chicago will sign it. It would not do that *I* should sign it, but *he* is an admirer of yours and he will be glad to father this child.

197 pp. 748–749, discussing Peirce's "How to Make Our Ideas Clear," *PSM* (Jan 1878), and James's *Will to Believe* (1897).
198 *Monist* (April 1905). See note preceding Carus to Peirce 3 Oct 1904 for details.

Will you kindly return the Ms. at your earliest convenience.

I remain, with kind regards
Very truly yours,
[Paul Carus]

Carus to Peirce ca. Nov or Dec 1904 (missing): Sends proofs of "What Pragmatism Is" for review.

Peirce to Carus end Dec 1904 (missing): Returns proofs of "What Pragmatism Is" and asks whether or how he will continue to receive proofs of his work.

1905

Jan 1905–Jan 1907: This period of the correspondence is mysteriously incomplete, composed almost exclusively of Carus's LPB copies. Only one letter from Peirce is found (9 Nov 1905) and only one original TS from Carus (8 Sept 1906). This has made it difficult to determine with much certainty the MSS delivered and discussed, especially given the many variations and exchanges Peirce's pragmatism series underwent.[199] Therefore, the articles identified and the gaps bridged are in many cases approximations based on context.

Carus to Peirce. LPB: SIU 36.57.881.

January 9, 1905.

C. S. Peirce, Esq.
Milford, Pa.
Dear Sir:

Your proof and letter reached me during my absence and I wish to say in answer that it is a matter of course that you shall receive proofs. Your article[200] is reserved for the leading essay of the next number of *The Monist*.

With kind regards and best wishes, I remain

Yours very truly,
Paul Carus

---

199 MSS are discussed and included in part in *EP* 2:331–433, and in full in Bisanz, 2009.
200 "What Pragmatism Is."

---

4 [Paul Carus]] *Cut off in LPB. Only* "P" *partially shown.*

Peirce to Carus ca. 10 Feb 1905 (missing): Discusses his second pragmatism article, at this point likely "Consequences of Pragmaticism"[201] to be in two parts, "I. Critical Common-Sensism" and "II. Generality and Vagueness."

Carus to Peirce. LPB: SIU 36.60.993.

February 14, 1905.

Charles S. Peirce, Esq.
Milford, Pa.

My dear Sir:

Your letter just comes to hand and I should like to see and know the price of, your new article on "Pragmatism." I am greatly overcrowded so that I do not know how to insert all the materials but I will try to make room for it in the next number after this. Your reasons seem to me quite plausible and I can understand why you are anxious to have the subject discussed.

Hoping that you are well, I remain with kind regards,

Very truly yours,
Paul Carus

Peirce to Carus ca. 1 April 1905 (missing): Sends "Consequences of Pragmaticism" and, as the third article, likely "Evidences for Pragmaticism."[202] Also possible that "Consequences" in its two parts constituted these second and third articles. Includes a list of names to whom the current *Monist* with his "What Pragmatism Is" should be sent and touches on previous *Monist* and *Open Court* numbers he is missing.

Carus to Peirce. LPB: SIU 36.64.92.

La Salle, Ill., April 3, 1905.

C. S. Peirce,
Milford, Pa.

---

201 R 288–289, in part in Bisanz, 301–306. Reported in *EP* 2:331 and 346 as the proposed second article in Peirce's initial plan, later changed to "Issues of Pragmaticism," the article actually published.

202 No MS under this title; reported in *EP* 2:331 and 539n6 as the proposed third article in Peirce's initial plan.

---

9 see and] *inserted above*

9 of,] *comma inserted in-line*

My dear Sir:—

Your letter came duly to hand, and I wish to say that you forget to mention the honorarium of your articles, but since nothing is said on the subject, I suppose you offer the second and third articles on pragmatism onon the same condition—viz. \$ $50^{\underline{00}}$—and we might as well publish them in book form afterwards. I have frequently been thinking that we might publish our controversy in separate form.[203] In book form it would better remain before the reading public, and so far as I am concerned, I would just let them go as they stand. They touch upon several very important problems and our views form a contrast that ought to be better appreciated and more discussed. There are many who take your view while others would side with me. The little personal tilts only add to the zest of the reading and I don't mind a sharp word in the discussion. What do you think of it?

I will forward your list of names to our business manager, and a copy of the Monist will be sent free of charge provided we have enough. As to the six numbers missing on your file, I will write to our business manager and in case we can supply them he will send them to you. Some copies are sold out, others are held at higher prices because they would break our files and a few can no longer be had except when buying the bound volumes, but I hope the numbers you mention are not among them, and he can send them free of charge. The same is true of Open Court numbers. If you will let me have the list of your numbers that are missing, I will attend to it.

Yours very truly,
Paul Carus

S.K.[204]

Peirce to Carus ca. 15 April 1905 (missing): Discusses a plan to reprint his pragmatism articles in book form later and points out a misprint (undetermined) in his "What Pragmatism Is."

---

203 Necessity–Tychism feud April 1892–July 1893. Proposed book also discussed March and April 1900.

204 Undetermined, OCP staff.

---

4 you offer] *to*you offer ~~to~~
4–5 on pragmatism … \$ $50^{\underline{00}}$—] $^{\vee}$on$^{\vee}$ pragmatism<;> \on the same condition—viz. \$ $50^{\underline{00}}$—/
4 on] *inserted in-line*
7 In book form it] \In book form/ *I*it
11 zest] sest *[E]*
19 , and he … charge.] *inserted in-line and below, after deleted period*

Carus to Peirce. LPB: SIU 36.65.127.

La Salle, Ill., April 17, 1905.

Chas. S. Peirce,
Milford, Pa.
My dear Sir:—

Your letter just comes to hand while I think of going away East for a few weeks. I may stay in Philadelphia and also in New York for a few days. I shall consider all the propositions you make, but having my head full of other things I cannot give you a definite answer at present. I can only say that on the whole I favor your scheme of having the articles reprinted. I shall be in Chicago a day or two during the end of the week, but expect not to leave La Salle before Monday or even Tuesday next.

With kind regards, I remain,

Very truly yours,
Paul Carus

P.S. I shall have the misprint in your article mentioned in the next number of the *Monist.*[205]

My address in New York will | presumably be care of Albert *[?Prat]* Esq. | 231 West 70th Street

---

Peirce to Carus late April and ca. 20 May 1905 (missing): Sends two articles, the first undetermined and the other likely "Issues of Pragmaticism" to replace "Consequences of Pragmaticism." Slated to appear in July number, but delayed by Carus to ensure Peirce could review proofs (see Carus to Peirce 8 July 1905).

Carus to Peirce. LPB: SIU 36.65.150.

La Salle, Ill., May 26, 1905.

Charles S. Peirce,
Milford, Pa.
Dear Sir:—

---

205 No mention found in July 1905 *Monist*.

---

10 a day or two] *inserted above*
11–12 or even Tuesday] *inserted above*
18–19 My address ... Street] *This sentence added in pen.*
26–27 Charles ... Pa.] *Added in pen*

Having returned from the East I found your correspondence and one Ms. The second one of which you wrote in your letter came in today and I hasten to acknowledge its receipt. I shall hand it at once to our compositor, and he will set it at his earliest convenience.

With kind regards and best wishes, I remain,

Very truly yours,
Paul Carus

S.K.
in haste.

Peirce to Carus ca. 5 June 1905 (missing): Requests to have a MS returned, possibly his previous draft of "Consequences" now replaced by "Issues," or the undetermined MS sent late April 1905.

Carus to Peirce. LPB: SIU 36.67.187.

La Salle, Ill., June 8, 1905.

Charles S. Peirce,
Milford, Pa.
My dear Sir:—

Enclosed please find your Ms. which I return upon your request. I have ordered a check to be made out, but for some reason unknown to me it has not yet reached me for signature. I hope you will have it in a few days.

With best wishes, I remain,

Very truly yours,
Paul Carus

S.K.

Peirce to Carus ca. 12 June 1905 (missing): Proposes to write a review of Edward Ross's 1905 *Foundations of Sociology* for *Monist*.

Carus to Peirce. LPB: SIU 36.67.209.

La Salle, Ill., June 19, 1905.

C. S. Peirce,

---

9 *in haste.*] *Added in pen* 17 return] *above* ~~send~~

Milford, Penn.

Dear Sir:—

Your letter reached me during an absence of three days, and I wish you had sent me the manuscript instead of writing about it. We are going to make up *The Monist* at once and if I had the manuscript now it might be inserted in the present number though there may not be time for proof-sheets. Let me have the manuscript on your terms at once, if I cannot crowd it into the present number, which is almost impossible, I will surely insert it in the next.

With kind regards, I remain,

Very truly yours,
Paul Carus

S.K.

Peirce to Carus ca. 28 June 1905 (missing): Submits review of Ross, which would appear July 1906.

Carus to Peirce. LPB: SIU 36.67.230.

La Salle, Ill., July 3, 1905.

Charles S. Peirce,
Milford, Pa.

My dear Sir:—

Please find enclosed a check for $ 15.00 (fifteen dollars), in payment of your review.

I don't think that I can agree with your views, for instance the significance which you attribute to Lester F. Ward. He is a personal friend of mine and I like him very well, but I don't think that his work is epoch making, nor lies the merit of the book in the point which he emphasizes and which you bring out, but I don't care for the review will go under your name.

With kind regards and best wishes, I remain,

Very truly yours,
Paul Carus

S.K.

---

4 about] *above* ~~for~~ 7 into] in$^{\vee}$to$^{\vee}$

Peirce to Carus ca. 5 July 1905 (missing): Informs Carus of a typo in his name on the check sent him and responds to Carus's criticism about his Ross review.

Carus to Peirce. LPB: SIU 36.68.239.

La Salle, Ill., July 8, 1905.

Charles S. Peirce,
Milford, Penn.
Dear Sir:—

Thanks for your letter. I have informed my business manager concerning the mistake in your name.

As to my "criticism" I would say that my comments were not intended as a criticism, but only to state a difference of opinion concerning some passages in your review. I jotted them down rather hurriedly, and so it may be that it is my fault when I speak of Ward's book and you referred to Ross's. I have not seen Ross' book; neither author nor publisher have sent it for review, and your review is the first opinion of the book I have seen. You need not pay any attention to my comments, because I have not formulated them with sufficient exactness.

Your article on Pragmatism[206] will appear in the next *Monist*. I was afraid of having it appear without your reading the proofs.

With kind regards, I remain,

Very truly yours,
Paul Carus

S.K.

Peirce to Carus ca. 1 Oct 1905 (missing): Sends a list of addresses, likely of persons to whom he wants the current *Monist* with his "Issues of Pragmaticism" sent. Submits an article on "The Finite God" (undetermined).[207] Continues discussing a plan to publish in book form his pragmatism articles and now also his "Old Cosmological Articles" from *Monist* Jan 1890–Jan 1893.

---

206 "Issues of Pragmaticism."

207 Perhaps a precursor to his "Neglected Argument for the Reality of God," *Hibbert Journal* (Oct 1908), of which several drafts dated 1905–1908 are extant in R 841–844.

17 on Pragmatism] *inserted above*

Carus to Peirce. LPB: SIU 36.71.383.

La Salle, Ill., October 5, 1905.

Mr. Charles Peirce,
Milford, Penn.

My Dear Sir:

I have forwarded the addresses which you gave me to my business manager in Chicago, and I expect he will attend to it.

As to your article on "The Finite God", I anticipate that you offer it on the same terms as before, viz: $ 50, and I would say that it ought to be a valuable contribution.

Your idea of having your articles on pragmatism appear in book form is a good one, and I do not think that you need look for any other publishing house. I have not seen the articles of 77 to which you refer.[208] I understand that your article on God should be the last of the series, and the book ought to appear soon after its publication.

The October number of the MONIST must have been printed by this time, and I have just finished the makeup of the January number, 1906.

I do not exactly know what you mean by your "Old Cosmological Articles". Do you mean anything that has been published in the MONIST, or did it appear in the MONIST, but as you wrote you are not in a hurry about them, I will not trouble.

Hoping that you are well, I remain, with kind regards,

Yours very truly,
Paul Carus

Peirce to Carus ca. 25 Oct 1905 (missing): Asks to submit two *Monist* articles, one responding to James Peterson's "Some Philosophical Terms" in Oct *Monist*, and the other on Josiah Royce's 1905 *The Relation of the Principles of Logic to the Foundations of Geometry*,[209] likely a version of, if not the very paper, "The Relation of Betweenness and Royce's O-collections" that Peirce would present 14 Nov 1905 to the National Academy of Sciences.

---

208 Undetermined. Might be his 1877–1878 "Illustrations" series in *PSM*, but Carus was very familiar with and admired those papers.
209 Identified in Robinson (OCP) to Peirce 22 Nov 1905.

3 Peirce] Pierce *[E]*
6 forwarded] forwared *[E]*
12–13 I have not … refer.] *inserted above*

Carus to Peirce. LPB: SIU 36.74.437.

La Salle, Ill., Oct. 28, 05.

Charles Peirce,
Milford, Pa.
My Dear Sir:

Send the two pieces at once, and without promising to insert in the next number, I will do my best. Perhaps I can omit one of the other articles. Perhaps you had better not cut down. The trouble of making room for long or short pieces will be practically the same. Send at earliest convenience.

With kindest regards,

Yours very truly,

Peirce to Carus ca. 3 Nov 1905 (missing): Sends "Mr. Peterson's Proposed Discussion" which appeared immediately in Jan 1906, and likely his "The Relation of Betweenness and Royce's O-collections" (or related article), ultimately rejected.

Carus to Peirce. LPB: SIU 36.74.460. Letter rejects Peirce's second article, estimated to be "Relation of Betweenness," yet Carus's postcard the same day, the two letters that follow, and Robinson (OCP) to Peirce 22 Nov 1905 seem all to address it as if accepted.

La Salle, Ill., Nov. 6, 1905.

Charles S. Peirce,
Milford, Pa.
My Dear Sir:

I hasten to acknowledge the receipt of your two articles. The first one has been sent to the compositor, but I fear that I shall not be able to make use of the second one because in our composition room we are in a state of transition and not yet adapted to the difficulty of setting Greek type into the usual English type. We have to saw each line for each insertion of a Greek letter. For this reason I relegate all Greek quotations either to special lines or to foot notes, avoiding the occurrence of the mixing of two types on the same line. The insertion of one letter is as bad as whole words or sentences of foreign type, and it would take days to set your article under present conditions. Probably it would have to be set by hand, but then the difficulty comes in in making the two types match; each type forming a "universe" of its own, and the different "universes" are like oil and water, they do not mix.

---

3 Peirce] Pearce *[E]*
24 composition] compositi*ng*on
29 or sentences] *inserted above*

I will have the business part of our transaction attended to at my earliest convenience.

With kindest regards and best wishes, I remain,

Yours very truly,
Paul Carus

Carus to Peirce. LPB: SIU 36.75.463. LPC of postcard sent 7 Nov 1905. Very faded, so text contains several conjectures.

La Salle, Ill., November 6, 1905.

Charles S. Peirce,
Milford, Pa.

My Dear Sir:

I wish to ask whether the sign resembling a Greek Ψ is the logical sign ♈ invented by you or the Greek letter *[?and that]* whether $\bar{\mu}$ with dash on top can too be replaced *[?by]* some other sign* I do not know how to *[?get that]* combination even in the type foundries.

Can you not invent symbols that *[?are already]* type?

Please answer at once.

Yours very truly
Paul Carus

perhaps μ′ or μ°

Peirce to Carus. MS: SIU 91.21.

Milford Pa | 1905 Nov 9

My dear Dr. Carus

Your letter & postcard reach me this morning. That is as soon as can be expected.

Postmarks letter La Salle $6^{\underline{h}}$ $3^{\underline{30}}$ PM
" " Milford $8^{\underline{h}}$ 1 PM

I only get mail in the morning. Your Postcard LaSalle $7^{\underline{h}}$ $12^{\underline{30}}$ A.M. reached Milford at the same time.

I think if you have such a difficulty with Greek letters you had better substitute some other font, only keeping the two distinct, and *explaining that it is because of a change in your arrangements*. I should be sorry if obelus over the letters had to be changed. Still, an accent will do. The ♈ is my peculiar sign which I call Aries, and

---

26 letter] *inserted above* 33 Still, an accent will do.] *inserted above*

the regular ♈ will do as well, & at a pinch the simple *plus* +. Of course, algebraic work must be set as "*fat*." *That* you had better make up your mind to. Anything ought to be practicable for a type foundry.

P.S. I am writing about Sir Thomas Browne. The *Religio Medici* is an immortal book & would attract you. You might for example pitch into my article which will, I suppose, appear in the *Nation*;[210] since our views are sufficiently different, perhaps, and certainly sufficiently near. As you develope your religious ideas I like them more & more.

CSPeirce

The *Religio Medici* is one of the best edited books in the language. 1st by Wilkin (Bohn's Library) & then by Dr. W. A. Greenhill (Macmillan)[211] Browne's large work Pseudodoxia Epidemica is tiresome to me.

1906

Peirce to Carus ca. 1 Jan 1906 (missing): Sends another pragmatism MS, possibly "The Basis of Pragmaticism (in Phaneroscopy)" (*EP* 2:360, dated ca. Dec 1905), one of six variations of an article by this point intended to be Peirce's third of the series.

Carus to Peirce. LPB: SIU 37.3.640.

La Salle, Ill., January 8, 1906.

Charles S. Peirce,
Milford, Pa.

My Dear Sir:

Thanks for your letter. We shall be glad to publish your article in the July number supposing that we have your MS. in time.

You have not answered Miss L. G. Robinson's question concerning your symbols.[212]

---

210 "Gosse's Sir Thomas Browne" (14 Dec 1905).
211 *The Works of Sir Thomas Browne* (1852) and *Religio Medici, Letter to a Friend* (1881).
212 Robinson (OCP) to Peirce 22 Nov 1905, referring to his "Relations of Betweenness."

3] "Turn over" *written here at bottom of page.*

Hoping that you are doing well, and congratulating you on your success at Harvard,[213] of which I had some information from other quarters, I remain with kind regards,

Yours very truly,
Paul Carus

Carus or general OCP to Peirce ca. Feb 1906 (missing): Sends proofs of Peirce's pragmatism article.

Peirce to Carus mid-March 1906 (missing): Returns proofs of said article with instructions for the proofreader.

Carus to Peirce. LPB: SIU 37.8.866.

La Salle, Ill., March 24, 1906.

Charles S. Peirce,
Milford, Pa.
My Dear Sir:

Your proofs just come to hand, and I hasten to say that we have yesterday made up the OPEN COURT,[214] and not having your proofs your article had to be omitted, but *I will try my best to upset the arrangement and introduce it.* If ever possible it shall be done. If not I suppose you will not worry for it will not lose its intrinsic value by appearing three months later.

I have to thank you, in the name of my proof reader, for the valuable lesson you wrote down, and I called the attention of the proof reader to the money value of your instructions, since the price of your contribution is one and one half pennies per word. We will count the words and estimate your good services accordingly, but I am afraid you will never receive the pay. It is nevertheless fully appreciated and will, with due consideration for your authority, be acted on in this special case.

I need not add that the proof reader is very contrite and overwhelmed by your learned expositions.

With kindest regards and best wishes, I remain,

Yours very truly,
Paul Carus

---

213 1903 Harvard lectures on pragmatism.
214 *Likely for* "the MONIST"

---

20 valuable] valu*e*able 22 is] *above* ~~are~~

Peirce to Carus, "several letters" late March and early April 1906, one on 20 March (all missing): Discusses Carus's position on Christianity and its origin, as well as Hegeler's "immortality idea."

Carus to Peirce. LPC: SIU 91.22.

La Salle, Ill., April 6, 1906.

Mr. Charles S. Peirce,
Milford, Pa.

Dear Sir:

There are several letters of yours which I have on hand for answer but I have not yet found the necessary leisure to write you, and even today I can only write you in acknowledgment of your letters, and to promise that I shall write again as soon as I can. I have accepted several lecture engagements which deprive me of a good deal of time, but I have not forgotten you and you shall hear from me again.

I am pleased to learn of your agreement with my position concerning the significance of Christianity and its origin.

As to Mr. Hegeler's immortality idea referred to in your letter of March 20th., I would say that he appreciates also what you have to say on the subject, and thinks that the author's and scholar's conception of immortality falls in the same line. The spiritual immortality, the preservation of thought and intellect is really more important and grander than the physical or rather physiological immortality.

With kindest regards and best wishes, I remain,

Yours very truly,

Peirce to Carus ca. 25 May 1906 (missing): Submits "Prolegomena to an Apology for Pragmaticism" to replace his previously submitted article as the third in the series, and which would appear Oct 1906.

Carus to Peirce. LPB: SIU 37.14.88.

La Salle, Ill., May 29, 1906.

Chas. S. Peirce
Milford, Pa.

My Dear Sir:

Your article just comes to hand and I hasten to acknowledge its receipt. I have given notice at the business end of the OPEN COURT to make out a check for $ 50, the amount which has been consigned to the payment of your article, and I will see to it that it will be forwarded to you without delay.

The article is a little late for the present number of the MONIST which is practically made up. We are only delayed by the OPEN COURT that is intervening at present. I notice that your article is a little long and in going over its contents I fear a break would not be favorable to the exposition of your thoughts. I will try to keep it together and publish it in one number, but I could scarcely do so in the July number.

Allow me an incidental question. I wonder what you mean by the formula "$1/f^1 + 1/f^2 = 1/f^o$".[215] Do you mean by "f^o" f zero, or is it an abbreviation for ordinary focus. Is the instance a mere abstract logical example or do you mean the focus of a real true lens, in which case it is not clear to me what you mean by 1 divided by the forms of any kind, and how the addition of these relations are to be 1, divided by f^o, is another problem to me. How a third focus comes in is an additional question, and I fear other readers will be as much puzzled as I am. Every line has a definite form—but perhaps you do not mean physical *[?forms]*.

I regret that tomorrow being Decoration Day, it will somewhat delay the making out of the check, but you can count on it *within a few days*, because I have reserved the sum of $ 50 for you in order not to let you wait.

With kindest regards and best wishes, I remain,

Yours very truly,
Paul Carus

I wish you would distinguish your pragmaticism from the pragmaticism that is now current. Otherwise some readers might think you adopt the whole system as your own, when in fact yours is different in spirit.

Peirce to OCP late Aug 1906 (missing): Sends numerous last-minute corrections to his "Prolegomena," frustrating the OCP staff. Carus returns to the situation after a trip and opens correspondence as follows.

---

215 p. 494 of Oct 1906 *Monist*. The formula in print and in Peirce's MS uses subscripts, as follows: $\frac{1}{f_1} + \frac{1}{f_2} = \frac{1}{f_o}$

11–12 of a real true lens] of real true ~~lines~~ \lens/ a *[E]*
12–13 the forms] \the/ f\orms/
15–16 Every … forms.] *inserted in-line and between paragraphs*
15 a] *inserted above*
23–25 I wish … in spirit] *This paragraph down left margin.*

Carus to Peirce. TS: RL 77. Stationery headed, "The Monist | The Open Court | Editorial Department | La Salle, Ill. | Dr. Paul Carus" LPB: 37.18.329–330. Letter actually sent 12 Sept 1906, as Carus explains in the letter that follows.

La Salle, Ill., Sept. 8, 1906.

Chas. S. Peirce,
Milford, Pa.

My Dear Sir:

On my return home from a trip to Lake Michigan I find my office staff in consternation and despair concerning your corrections. The compositors are in rebellion, but I have insisted to let your changes be made as you wanted. I am sorry that there is no more time for reconsideration, nor can I myself give the necessary leisure to reading over and carefully considering your articles and the difficulties which they present. I will this time again follow the advice of our mutual friend, Mr. Holden of West Point to publish your article even though I myself feel rather reluctant in offering it to the public in the shape in which you have written it. I refer here especially to the big insertion which you have made on the bottom of the second galley, and I fear that I shall be again in receipt of a number of letters accusing me of partiality in publishing unintelligible articles. I appreciate your significance in the development of science, and I will publish your article exactly as you want it, but I must confess that if I myself had written it I would first give it a careful revision and render the diction of it more obvious to the reader. In the sentence to which I refer the subject and the predicate are divided by fourteen lines, and I think you yourself would have trouble in keeping it (the subject) all the time in mind, or if you have lost it to find it again. It beats even the worst involved German sentence, and I think Chinese is easy in comparison with it. I will say nothing here of the great expense which your corrections involve. You requested of the compositor to follow your copy exactly, especially in capitalization, italics, spelling and punctuation, and afterwards you made essential changes especially in your capitalization.

The expenses are especially heavy (1) because corrections are charged higher than new material; (2) the running over of lines is as much as setting new type, and you will notice that you are very chary in making paragraphs, and many of your corrections occur either at the beginning or in the middle of the long paragraphs which extend almost over a page. Sometimes the insertion of one little word necessitates the re-setting of a whole paragraph. Of course the compositor must exactly follow the intention of the author, but the author's MS. ought to be

---

11 reconsideration] re*s*consideration
11 myself] *inserted above*
23 (the subject)] *inserted above*
32 chary] char~~r~~y

in final shape and ought not to be altered after the work is done. I write you this so you can be familiar with the difficulties with which I have to cope, and let me in the future have a MS. that would obviate a rebellion in our compositors' room. If I have another experience of the same kind I shall have trouble on my hands, for our head compositor will start in open rebellion and leave me in the lurch. I hope you will understand the situation and not take my comments amiss, for I shall presumably have to refuse the acceptance of another article if I stand before the alternative of discharging my compositors for the sake of publishing your MS., and I am sure if you knew all the difficulties which your MS. has caused here, you would have been courteous enough to avoid them.

Hoping that you are well, and wishing to help you in bringing your thoughts before the public, I remain,

Yours very truly,
Paul Carus

Peirce to Carus ca. 15 Sept 1906 (missing): Suggests further revisions to "Prolegomena."

Carus to Peirce. LPB: SIU 37.18.349–350.

La Salle, Ill., September 17, 1906.

Charles S. Peirce,
Milford, Pa.

My Dear Sir:

Your letter just comes to hand, and I will say in reply that the MONIST has been made up and we can no longer make corrections or alterations. The sentence is grammatically all right, but extremely ponderous, so much so that I myself would rewrite it. I must confess that I would have taken the liberty to alter it but I remembered your very categorical request not to tamper with even the punctuation, spelling, capitalization, etc., of your article, and so I did not know whether you had written the sentence in such a ponderous style on purpose, so I did not venture to make any changes, for I count you among those authors who should come before the public in exactly the form as they want it, even if I do not agree with them in diction, style and thought. Any one who reads your article with care will find out which parts of the sentence belong together, but the average public will

---

2 cope] ~~scope~~
4 hands,] *comma inserted*
8 compositors] compositor\s/

lose patience. I myself deem it worth while to make it easy to the average reader, but I know that you care only for the few select, and your work is exceptional.

As to the date of my letter[216] I have to say in explanation that I doubted for some time whether I should finally let you know the difficulties under which I labored or whether I should let it go, and for that reason I hesitated sending the letter. I left it on my desk over Sunday to reread it in a quiet moment. At any rate I did not want to send it in the rush of business without due reconsideration. Finally I thought it would be better if you would know the exact state of things here, and so I let it go, hence the difference of the dates. The letter was written on the 8th, and it was mailed on the 12th.

You write that you can easily publish your articles in some pragmatist journal, and I wonder whether it would not be advisable for you to have them appear there. I do not know to which journal you refer for pragmatism is at present fashionable in many journals, but it may be that it would be good for the dispersement of your views to have them appear in several different periodicals.

With kindest regards and best wishes, I remain,

Yours very truly,
Paul Carus

1907

Carus to Peirce. TS: RL 77. Stationery headed, "The Monist | The Open Court | Editorial Department | La Salle, Ill. | Dr. Paul Carus" LPB: 37.25.726.

La Salle, Ill., Jan. 23, 1907.

Mr. C. S. Peirce,
Milford, Pa.
My Dear Sir:

I write to inform you that I shall soon go to Europe, and will probably pass through New York in a hurry as here my time will be taken up to the very last moment. I should be glad, however, to meet you before leaving, and will incidentally mention that among your former publications there is a series of essays on Logic, published in the Popular Science Monthly many years ago which I wish you would have republished either by Appleton & Co., or by any one who would bring

216 8 Sept 1906.

14 dispersement] disspersement *[E]* 27 here] *inserted above*

it out.[217] I think I wrote to you on the subject before, and you told me that you had the right to republish it in book form.[218] Is there anything in Europe which I could do for you.

With kindest regards and best wishes, I remain,

Yours very truly,
Paul Carus

Peirce to Carus ca. 27 Jan 1907 (missing): Accepts Carus's offer to publish the book, asks for an advance of $100 to that end, and suggests he might include his 1868–1869 articles in *Journal of Speculative Philosophy* and *Proceedings of the American Academy of Arts and Sciences* (see Bibliography). Possibly mentions his two remaining pragmatism articles to be sent to Carus, footnoted below.

Carus to Peirce. TS: RL 77. Stationery headed, "The Monist | The Open Court | Editorial Department | La Salle, Ill. | Dr. Paul Carus" LPB: 37.26.751–752.

La Salle, Ill., Jan. 30, 1907.

C. S. Peirce,
8 Prescott Hall,
Cambridge, Mass.

My dear Sir:

Your letter just comes to hand, and I have at once taken steps to attend to your wishes by talking matters over concerning both the publication of your book and the advance of $ 100.00. I have strongly recommended the publication of your book which I wish to bring out because I think that your papers on logic in the Popular Science Monthly are the best introduction to the new movement in logic that

---

217 Peirce's 1877–1878 "Illustrations of the Logic of Science" series in *PSM*. Carus had encouraged Peirce to re-publish the series throughout their correspondence at least as early as 1893 (see Hegeler to Peirce 27 Feb), but this letter marks the beginning of Carus's 7-year endeavor (until Peirce's death) to work with Peirce on the book to be published by OCP. In the end, the book was never published. Peirce insisted that the articles not appear in their original form yet he was never satisfied with his extensive revisions. Although the individual articles were included in various posthumous editions of Peirce's writings, they were not treated at book length until a century later in de Waal, 2014 (published by OCP), who chronicles in detail this chaotic episode of Peirce's "Illustrations" and its extensive revisions.

218 See Carus to Peirce 3 Apr 1900; also Peirce to Hegeler 24 Feb 1893, confirming copyright from Appleton.

have been written in the English tongue—perhaps in all tongues—and they are not sufficiently known among our teachers of logic, and have scarcely reached at all broader circles. It is the first I have read of you, and at the time they made an extraordinarily favorable impression on me so as to create a desire to be better acquainted with you and have articles of yours for The Monist. Your articles on Tychism in the Journal of Speculative Philosophy I have not seen, nor those in the Proceedings of the American Academy of Arts and Science, but if they are written as vigorously as those on logic and if they complement the thoughts of your papers on logic, as I suppose they do, they will be a welcome addition to the book. I undertake this publication entirely on my own hook, and have strongly recommended it because I believe that nothing will bring the importance of your best thoughts so well before the public as will these essays, to be printed as a book in handy form. I will promise you for the Open Court Publishing Company a a Royalty of *50% on the net returns* after a deduction of the entire expenses. This as you may know is an arrangement frequently made with authors and it is fair to both publisher and author. As a rule it practically amounts to 10% on the selling price of the book. I would be much obliged to you if you could let me have the copy very soon, for I should like to hand it to the compositor before I leave so as to have this book attended to during my absence. Otherwise it might be delayed until I come back, and if anything should happen to me on the way, no one knows whether the book will be published at all. As to the advance of $ 100 I hope there will be no difficulty and I will send you the check as early as I can conveniently attend to it. (I am just told that Mr Hegeler is opposed to making any advanced payments but I will see what I can do.)

It would be very helpful in our office work if you would send your two remaining articles on pragmatism at your earliest convenience, and at any rate the one that is to appear in summer, if possible, during the coming month for there are always difficulties in MSS. and, especially in my absence, your MS. should not go

---

1 perhaps … are] \perhaps in all tongues—/ and *it*they *is*are
2 have] ha*s*ve
3 they] *inserted above*
4 extraordinarily] extraordinar\il/y
8 as those on logic] *inserted above*
8 complement] comple~~te~~\ment/
12 as will] as w*e*ill as *(omitted second* “as” *mine)*
13–14 a Royalty of] *inserted above*
16 amounts] amount *[E]*
18 soon,] *comma inserted in-line*
23–24 (I am … can do.)] *inserted in-line and between paragraphs (parentheses mine)*
28 and, … absence,] and∨,∨ especially ∨in my absence,∨

to press without being revised by yourself.[219] I myself anticipate much pleasure and profit from your next article because I still stand in need of being shown the significance of your theories for philosophy. Though I think highly of your logical work I have not yet been able to make up my mind to regard your propositions, as made in your Monist articles, as being in final shape.

Your conception of pragmatism that a concept is to be judged *by practical application exclusively* is in my mind a matter of course. If truth means that an idea should tally with experience, it is only expressing the same thought in other words to expect that it shall be proved in its practical application. As to the current pragmatism (as held by others) which slurs theory and proposes to do away with exact thinking wherever an idea would suit our own subjective needs, and if this subjectivism is called "practical application", it is to be a slight improvement only on agnosticism.

---

219 Peirce's plan for the remaining two pragmatism articles changed several times over the next few years and ultimately they never appeared. For the fourth article, to follow "Prolegomena" in the series, Peirce around this time and over the next 18 months or so composed various options, including "Phaneroscopy" (R 298), "The First Part of an Apology for Pragmaticism" (R 296), and later "The Bed-Rock Beneath Pragmaticism" (R 300), all following suit from "Prolegomena" to incorporate his Existential Graphs-based proof of pragmaticism instead of a semiotics-based proof from previous attempts (see *EP* 2:xxviii–xxix; also Robin for R 298 and 300). Although it is unclear what the fifth was to be, by 7 Dec 1907 Peirce to Russell seems to have in mind "The First Part" for the fourth, leaving a fair chance that the fifth might have been "Apology for Pragmatism" (R 297). This would also be in line with Robin's commentary to R 296, and which additionally notes a sixth article that Peirce at some point had in mind: "The fourth article *[*"The First Part"*]* was to begin the apology, the fifth *[*possibly "Apology"*]* to have contained the main argument, and the sixth to have provided the subsidiary arguments and illustrations."

Around the summer of 1907, partly due to illness (reported Peirce to Carus 12 Oct 1908), Peirce took a brief respite from his pragmatism series and turned his attention instead to a separate *Monist* series on "Some Amazing Mazes" dealing with mathematical "curiosities" and other applications of Existential Graphs (Peirce to Carus beginning ca. 28 Sept 1907). By the fall he resumed his pragmatism efforts, and a year later Peirce to Carus 12 Oct 1908 re-proposed the remaining articles, but now in a way more closely resembling his Harvard Philosophy Club lectures on "Logical Methodeutic" and his "letter to the editor" for *The Nation* and *Atlantic Monthly* on Pragmatism (*EP* 2:398–433) from the spring and summer of 1907. Carus shortly after drew Peirce's attention back to the reprinting of his "Illustrations" series, and the pragmatism series was never completed.

---

4 propositions,] *comma inserted in-line*
5 articles,] article∨s,∨
5 shape.] *paragraph break inserted*
7 course. If] course*,*. ~~for~~ *i*If
10 (as held by others)] *inserted above*
11–13 and if … agnosticism.] ∨and∨ if ~~that~~ \this subjectivism/ is called ∨"∨practical application∨",∨ it ~~is simply~~ \is to be/ a slight improvement ∨only∨ on agnosticism.

Hoping to hear from you soon. I remain with kind regards,

Yours very truly,
Paul Carus

Peirce to Carus. Postcard: SIU 91.23. Recto headed, "Postal Card" with portrait at right and American eagle at left. Corresponds with a large manila envelope addressed to Carus the same date in SIU 91.23, presumably enclosing MSS for the "Illustrations" book (to which this card refers) as discussed in Peirce's previous letter.

Boston | Feb 4 | 1-PM | 1907 *[stamped]*

Paul Carus
Open Court Company
LaSalle
Ill.

Let the title of the new volume be

CREAM

Skimmed
By C. S. Peirce

It will be well understood in LaSalle that milk is one thing, cream quite another.

C. S. P.

(Please send $ 50 to pay for typewriting. If there be excess, it will be returned.)

March–July 1907: Peirce prepares a lengthy "letter to the editor" for *The Nation* and *Atlantic Monthly* on pragmatism (rejected, *EP* 2:398–433) and delivers Harvard Philosophy Club lectures on "Logical Methodeutic" 8–13 April (see *W* and *EP* Chronology). Likely begins working on his imminent "Some Amazing Mazes" series as a respite from his pragmatism series, which he was reportedly too ill to complete at this time (see Peirce to Carus 12 Oct 1908).

Peirce to Carus ca. 25 May 1907 (missing): Asks for a copy of Oct 1906 *Monist* with his "Prolegomena" to be sent to Giovanni Vailati, whose "Pragmatism and Mathematical Logic" appeared in the same *Monist* number. Answered by OCP 31 May 1907 (see correspondence).

Carus to Peirce. LPB: SIU 37.34.78.

La Salle, Ill., July 23, 1907.

Mr. Chas. S. Peirce,
Milford, Pa.

Dear Sir:

Having just returned home from Europe, I hasten to ask whether I can count on you for the MS. which we had arranged to publish in the fall number of *The Monist.* I am extremely busy at present, and am in a great hurry to decide concerning the material to be selected, so please let me know whether I shall leave space for you, and if so how much.

Hoping that you are well, I remain, with kindest regards

Yours very truly,
Paul Carus

Peirce to Carus ca. 1 Aug and on 4 Aug 1907 (missing): Two letters, the first not detailed below, the other expressing that Peirce feels rushed by Carus to submit his work. By this point Peirce was likely in the middle of his new "Amazing Mazes" series, hence his delay. Also gives Carus permission to edit his articles going forward to save time if needed.

Carus to Peirce. LPB: SIU 37.35.128.

La Salle, Ill., August 6, 1907.

Chas. S. Peirce,
Milford, Pa.

My Dear Mr. Peirce:

I received a letter from you a few days ago, and now I have another of August 4. I have not yet been able to answer the first, because I am at present too busy, and I will say in reply to the second that *I do not wish you to rush your article* or to be in any hurry sending me your MS. Write your article at your leisure for it will be better if you write them well than if you work at unfavorable moments. Please note that I did not rush you in my first letter. I only wanted to know whether or not I should make room for you in the present number of The Monist. I wanted to accommodate myself to your needs. I did not mean to urge you to be punctual in fulfillment of your promise and I write only to *[?mention]* that I have made up the present number of The Monist and *you need not hurry at all* in sending your MS. I am too busy to write more at present for I have not yet even attended to all my correspondence and the MSS. that have been sent in during my absence.

With best wishes for yourself and your wife, and hoping that she will recover soon, I remain,

---

23 did not … letter.] d~~o~~\id/ not rush you \in my first letter/.

23 wanted to know] *above illegibly faded deletion*

25 did not mean to] d~~o~~\id/ not \mean to/

26 and] *above deleted period*

Yours very truly,
Paul Carus

P.S. It is very flattering to me that you allow me to edit your paper, but I would not accept this privilege for *[?two]* reasons, first I would not dare to meddle with your thoughts because editorial reading would not be as careful as the close study of a student, and I could not give your paper more time at present, and further *for your own sake* I think.

Peirce to Carus ca. 28 Sept 1907 (missing): Sends "The First Curiosity," "Explanation of Curiosity the First," and "A Second Curiosity" (MSS A–C)[220] of his "Amazing Mazes" series. Answered by OCP 3 Oct 1907 (see correspondence).

Peirce to Carus. MS (copy): RL 387. Original MS missing, text from copy made by Russell. Note at top reads, "*(From Peirce's letter to Dr Carus of Oct 18th 1907)* Copied by Russell". Punctuation conventions characteristic of Russell and not likely in Peirce's MS have been silently standardized. (See editorial method in "This Edition" for details.)

[18 October 1907]

As to those curiosities of Mathematics, I have a lot more in my budget, and am thinking of making a volume of them. If I do, of course I must write all that is to go into the volume, although I propose to put the heavier parts into a little smaller type. But there is no reason why those parts should not be altogether omitted from the *Monist* publication. At any rate, if you print my explanation of the first curiosity (which I see no need of your doing)—It is the part of the M.S. marked B.—I would insert it *after* the second curiosity, marked C. I have substantially finished a fifth part, which is chiefly about the secundal system of numerical notation.[221] There are about 70 pages of M.S. I was obliged to give at least a part of this, be-

---

220 Peirce marked his "Amazing Mazes" MSS A to E and often referred to them accordingly, as follows:

MS A: "First Curiosity" (*Monist*, April 1908)
MS B: "Explanation of Curiosity the First" (*Monist*, July 1908)
MS C: "Second Curiosity" (*Monist*, January 1909)
MS D: "Third Curiosity" (n.p., R 199)
MS E: "Fourth Curiosity" (R 200, partly in *CP* 6.318–348 and 4.647–681)

Peirce may have also subdivided some MSS into parts along the way, which he also occasionally references (e.g., in the following letter he mentions a "fifth part" of the series that seems to correlate with MS D, the "Third Curiosity").

221 Likely "Third Curiosity," to be followed by "Fourth Curiosity" as he goes on to describe here.

cause I wish to use that notation in connection with the next following curiosity, which has a philosophical value. This is intended to show what numbers *essentially are*, and why they should be of such vast importance as they are. I show two very curious ways of setting them down in the order of their values, the one *all of the fractions*, the other all of the *fractional values*, without reducing them to common denominators, and without any implication as to the equality of the parts into which the unit is divided; and from either of these arrangements all the relations between and rules of operating with fractions can be deduced. The first method involves no arithmetical operation, though it involves counting either with the children's nonsense, "Eeny, meeny, mony, mi" &c or otherwise. The other method involves addition; but only the addition of whole numbers. Now as this is found in a row of objects whose distances need not be the same, there is no idea of *equality* further than that positions in a row neither of them coming before or after the other are equal in a certain sense.

Once it is established that numbers *per se* express nothing but the relation of before and after along a line,—no matter how irregular the intervals,—

x x xx x xxx xxx

we see that this is the fundamental relation of logic; and therefore it is no wonder that number should be highly important in all difficulties of reasoning.

Logicians are such an indolent minded set of noodles that they have never discovered the high importance for reasoning of the concept of a *collection*. I doubt if there be a dozen logicians in the world who can to day define a collection without a vicious circle. The concept is however essential to the *higher kinds* of necessary inference though it is not indispensable to all inference as the concept of before and after is. Its high importance explains at once the importance of cardinal numbers which Dedekind showed (7 years after I had done the same thing in the Am. Jour. Math.)[222]—not explicitly however,—are quite secondary to ordinal, or rather to *climacal* numbers (κλῖμαξ)

Well, I show that while the addition, multiplication, and involution of multitudes (maninesses of the members of collections) formally agree with those of the climacal operations, yet the idea of them, their definitions, are quite different, and a good deal simpler. It seems odd that when the cardinal numbers are not

222 Peirce's 1881 "On the Logic of Number" and Dedekind's 1888 *Was sind und was sollen die Zahlen?* (trans. Beman, *The Nature and Meaning of Numbers*, OCP, 1901).

4 them] *inserted above*
5 of the] *inserted above (both instances in the sentence)*
5–6 them to common] \them/ to ~~op~~ common
20 an] *inserted above*

pure numbers but mix the essential concept of a number with a different concept, the result should be that the arithmetical operations are more simply defined. But it is explained by two circumstances. The first is that the arithmetical operations are logical operations. One thing that has prevented the exact logicians from seeing *that* has been their regarding logical aggregation as addition. And they have done so because they have neglected the logic of indefinite individuals. Dedekind calls mathematics a branch of logic. I demur because the whole aim of the mathematicians is quite of a contrary kind to that of the logicians. However waiving that point and regarding mathematics as logic, it prevents this striking contrast with the kind of logic that has been cultivated. Namely, that the latter busies itself about *definite generals* (Any man or every man) while mathematics busies itself about *indefinite individuals.* When Euclid has enunciated a proposition which he does in general terms because that is the form in which it is ready for application to particular cases, before he can apply his mathematical methods to proving it, he has to transform it into a statement about indefinite individuals. He begins, "Let A. B. C. be a triangle". He means a single triangle and A. B. C. are not *any* angles of it but are single angles though indefinite. Now the arithmetical operations, addition, multiplication, and involution with their correlatives are logical operations upon indefinite individuals while the logician's *aggregation* is a logical operation upon *definite generals.* This I shall show in my paper.

The other circumstance that goes to complete the explanation is that the idea which is added to the pure idea of number, the idea of a collection, is itself a logical idea of great importance in simplifying logical relations. What wonder is it, then, that it should simplify those logical relations that are involved in logical operations called arithmetical? I can somewhat elaborate this explanation and make it quite as good as could be expected in the present state of the development of logic. Addition is the formation of a collection of the members of two indefinite single collections. In aggregation, $A \mathbin{+\!\!,} A = A$—for the general concept *A* cannot be duplicated. In mathematics however by *A.* is meant *some A.* Now *some A,* taken along with *some A,* makes up *some pair of A.s.* That is addition. Some lion plus some tiger makes some single pair of things composed of a lion and a tiger.

I have much more to say about the logic of mathematics besides what is solved in M.S. – B.[223]

Another rather startling curiosity is this. I have often seen it said, very incautiously, in mathematical books that a divergent series nowise represents the function from which it has been derived. I rather think that no series, by itself, com-

223 "Explanation of Curiosity the First."

2 are] *after* ~~should~~

34 rather] *inserted above*

pletely represents a function, but only one value of it, for each value of the variable. But be that as it may, that a divergent series does *somehow* represent that its function seems to me so simply proved that it is hardly worth talking about. We find in all the books series about tangents which involve Bernoulli's numbers and everybody knows that those series are divergent. "Yes" it may be replied "but they cannot be used after they begin to diverge". Well I propose to show a purely inductive method of employing divergent series by which correct numerical values to four or five places of decimals can be deduced from the most extravagantly divergent series,—divergent from the word "go"! I think this will surprise most people. It is true that special series can be found for which my method fails to give any definite value or gives the value correctly only to one or two significant figures. But that is because being an inductive result. I have not got the entire method. Any rule of approximation would fail in exceptional cases if instead of the entire rule one was merely in the possession of the rule *as it is usually needed.*

I should like to make up a little volume of such stuff. I dont know whether The Open Court Co. has such relations as to be able to sell such a book. In any case its sale must be small. But I think it might sell (enough to pay for the writing of it) in England & France and Italy if not at home. If you Open Court Co. would like to handle it I should feel confidence that I was honorably dealt with that I confess my experience makes me most unpleasantly sceptical about with *some* publishers,—and this *some* is unfortunately quite indefinite. I should not know where else to have implicit confidence.

Meantime I am impatient to get back to pragmatism. The two articles[224] have been most maturely considered and will prove interesting to those who are capable of exact thinking. As for W. James, he is one of my dearest friends, and I should dislike extremely to say *all* I think about his way of handling philosophy. Meantime he and Schiller & the Leonards man whose name just now escapes me, in the first place: have as clear a grasp of the leading ideas of pragmatism and of how they are to be applied *as they have of anything in the world*, and in my opinion are doing splendid work beneficial extremely to a wider public, and in the second place, they sincerely want to learn, and are learning notwithstanding certain difficulties under which they labor. In my *Monist* article my purpose has mainly been to insist upon points that other pragmatists do not see, and thus one does not perceive in what I write the warm sympathy I really have with their work.

I also must write a short primer of logic as I conceive it, for I believe it would be exceedingly useful not only to those who would study it critically but also for

224 By this point likely "The First Part of an Apology for Pragmaticism" and "Apology for Pragmatism" (R 296–297). See Peirce to Russell 7 Dec 1907; also Peirce to Carus 30 Jan 1907.

thousands who would in a measure take my word for the conception of the universe that that would lead up to.

Then on top of that, based on that, I will show what a mighty reality religion is, that its majestic command is such as to make all theological questions seem contemptibly trivial, such as whether Gautama was a little lower or a little higher than Jesus,—though in some respects our traditional relations to the latter make that idea the more profitable while in other respects it is just the other way, and such as whether there is an individual hereafter, *et omne genus*

C. S. P.

Peirce to Carus ca. 23 Nov 1907 (missing): Describes his current financial difficulties and asks for Carus's aid. Also discusses a plan (mentioned in his previous letter) to publish a book of his mathematical "curiosities."

Carus to Peirce. LPB: SIU 37.41.387–388.

La Salle, Ill., December 2, 1907.

Mr. Charles S. Peirce,
Milford, Pa.
My dear Sir:

Your letter just comes to hand, and I will answer at once that I am shocked to hear the news concerning your plight, but I am afraid you bring me into great difficulty, for I have written you before that my means are limited, and I have to be careful not to go beyond them. Mr. Hegeler is away and so I can not approach him, and even if he were here I fear he would turn a deaf ear to your request. Mrs. Carus is absent too and the Open Court bank account is run down. My business manager in Chicago is sick with typhoid fever, and so I can not ask his advice what to do. Perhaps he might know how to help you without involving myself or the Open Court Publishing Company into trouble. You must consider that I have to account for the money I spend, and there is still quite a little sum paid to you for which I have not yet got the returns, and all this comes during the money stringency which, we hope from day to day, will soon pass by.

Accordingly at the present moment I can not do anything but wait for Mr. Hegeler's return. My wife is also absent, and she having the trusteeship of The Open Court Publishing Company would have to be consulted too.

---

5 contemptibly] contemptably *[E]*
24 typhoid] typoid *[E]*
27 paid] pai$^{\vee}$d$^{\vee}$
29 , we … day,] *commas inserted in-line*
29 soon] *inserted above*

I ought to state another thing. Your last article now in our hands[225] which is quite extended is now under preparation. In addition to the honorarium it costs us a goodly sum of money having diagrams drawn and reproduced. Further the typographical difficulties are not without considerable extra expenses. It will fill a great part of the next Monist, and I ought not to accept and pay for more than I can use. Otherwise I would soon see myself involved in the double difficulty of having too much material on hand, and increasing my expenses without being benefitted thereby. While I wish to bring the best of your ideas before the public, and while I am anxious to let the world know what you have done, good and valuable work, I find myself greatly embarrassed by the objections which a great many readers offer to your writings, and I must confess that to some extent their complaints are justified. Though you have really something to say, and although your articles contain valuable propositions, you enwrap them unnecessarily in such difficulties, and sometimes switch off your ideas without explaining why you do so, that you tax the patience even of your most attentive readers. You actually write only for a few men, and even they have to work hard before they can dig out what you are really driving at. That would not be necessary, and it would be to your own interest if you could write in such a way as to keep the leading idea of your thoughts before the reader. That you can write well i.e. intelligibly for common readers is plainly shown in your former articles on logic in the Pop. Sc. M. I wish you would continue in that same forcible style which is at once instructive and to the point. You think that there would be money in a book of yours but I really do not believe it. You limit your public by frightening away your readers with your ponderous expressions of thoughts which could very easily be stated in clear language.

Whether I could at all publish a book of yours depends very much on other conditions. Remember that I had your permission to republish your articles on logic published many years ago in the *Popular Science Monthly*, yet I have not been able to do so, and I think a collection of those articles would take more with the public than your new book.

Hoping that you will pull through and get out of your difficulties, I remain, with kindest regards,

Yours very truly,
Paul Carus

225 "Explanation of Curiosity the First"

1 now in our hands] *inserted above*
3 having] *after* ~~by~~
9 , good and valuable work] *above* ~~and what you can do~~ , *[E]*
11 some] *above* ~~a great~~
19 i.e … readers] *inserted above*
20 in the Pop. Sc. M.] *inserted above*
26 articles] article$^{\vee}$s$^{\vee}$

Peirce to Carus ca. 1 Dec 1907 (missing): Submits "Third Curiosity" and "Fourth Curiosity" of "Amazing Mazes."[226]

Carus to Peirce. LPB: SIU 37.41.400.

La Salle, Ill., December 5, 1907.

Mr. Charles S. Peirce,
Milford, Pa.

Dear Sir:

I am in receipt of a package of MSS. sent by you per Express, and priced at $ 300. You wrote me in your last letter that upon the receipt of your MS. I would be in debt to you, while according to my calculation you still owe me two articles which I have paid for.[227]

Since I do not wish to be in debt for so high a sum as you rate your MS., I prefer to avoid complication and hold the MSS. subject to your order.

Yours very truly,
Paul Carus

1908

Peirce to Carus ca. April 1908 (missing): Likely per Carus's invitation, sends a brief "Letter from Mr. Peirce" endorsing Carus's "Problems of Modern Theology" in the April *Open Court*. Excerpt published May 1908.

OCP to Peirce 19 May 1908 (see correspondence): Sends proofs of "Explanation of Curiosity the First" for review.

Peirce to Carus. MS: SIU 91.24

P.O. Milford Pa 1908 May 26

My dear Dr. Carus:

In the article of which I have just read the proofs, there is a very long footnote that I should prefer to transpose to the end of the article (of course without reset-

---

226 Identified in Peirce to OCP 23 Nov 1907. Never published in *Monist*, possibly due to financial misunderstandings alluded to in Carus's response below.
227 Likely the two pending pragmatism articles to conclude the 1905–1906 series.

ting.) But one of the subjects of this note has been the subject of a great deal of hard work on my part with interesting results. But I have more such unprinted results than I can ever hope to get published. Therefore I have written an addition to that note *hinting at* some of these I have been working out. *If you have the space to insert the enclosed addition* to that note, I should be glad. Otherwise, no matter.[228]

very faithfully
CSPeirce

Peirce to Carus. MS: RL 77.

P.O. Milford Pa 1908 Oct 12

My dear Carus:

I am moved to write to you primarily to ask a question (if you are at liberty to answer it) prompted by the October number of The Monist, which I have just received & have not fully examined. I turned first to Borinski's review of "Der vorchristliche Jesus,"[229] because I have a great admiration for the reasoning of W. B. Smith both in that book and still more in his dealing with MSS. F and G.[230] But what struck me specially was the *life* of that translation (as I take it to be) of parts of Borinski's review. It strikes me as extraordinarily good, although I have not seen the original. For that reason I have curiosity to know who did it. I expected to see a little article of mine in the number, because I had corrected the proofs. But I am much better contented to see the *Vorchristliche Jesus* brought to your readers' attention. I looked over Lovejoy's piece & then turned to "*Esperanto.*"[231] I know very little about this. About all I know is that a friend of mine who has so little faculty of language that he went to the European continent without speaking apparently any language beyond English & German and who had devoted only about 3 weeks, I believe, to studying Esperanto, found himself somehow in some town where there was an Esperanto Congress and attended it; and *though he had never spoken Esperanto before*, found himself at once (so he told me,) perfectly capable of conversing with pleasure and profit with the persons he met. Such being the case, although I do not believe that in the present state of linguistics, it is possible for anybody to construct a really excellent artificial lan-

228 pp. 461–464, addition beginning at p. 463.
229 pp. 487–587.
230 "The Pauline Manuscripts F and G," *American Journal of Theology* (Oct 1903).
231 "The Theory of a Pre-Christian Cult of Jesus," pp. 597–609; and Carus, *Monist* (July 1906).

5 I] *after* ~~and~~ 22 very] *over* a

guage, it seems to me that Esperanto, if it is widely enough understood, and a person of no great faculty or training for learning languages can learn it as well as that in 3 weeks, is at any rate a valuable *stop-gap* for the time being; and if I can get hold of a grammar etc and a text worth reading, I will devote a month to learning it myself. For certainly I don't believe such a person as I have described could in 3 weeks acquire a useful knowledge of English—I mean a knowledge making any real conversation possible.

In 1907 I was far too ill to write anything but such articles as I did write,[232] which I thought quite as valuable and as curious as the articles about magic squares you had been printing. This year I am quite well & have the abstracts of the concluding papers of my series on Pragmatism fully elaborated and one of the articles itself nearly written, treating of logical analysis and definition and its *methodeutic.*[233] But some circumstances have hitherto prevented my completing the article.

Very truly,
C. S. Peirce

Carus to Peirce. LPC: SIU 91.24.

La Salle, Ill., Dec. 9, 1908.

Mr. Charles S. Peirce,
Milford, Pa.
Dear Sir:

I have been away in Europe and have only recently returned. I attended several Congresses that were of interest to me, and interrupted my work which I am vigorously resuming again. I have my hands full, but in thinking over and planning the scope of the work of the Open Court Publishing Company for the coming year, I think again of an old and cherished plan of mine which I had proposed

232 His "Amazing Mazes."
233 By this point the proposed articles are difficult to determine, though they seem to be related to Peirce's Harvard Philosophy Club lectures on "Logical Methodeutic" and his "letter to the editor" for *The Nation* and *Atlantic Monthly* on Pragmatism (*EP* 2:398–433) from the spring and summer of 1907. Compare estimations for Peirce's other plans for these articles in Peirce to Carus 30 Jan 1907 and 26 Aug 1910.

5 such … described] \such/ a person \as I have described/
6 a^2] *after* ~~one~~
13 hitherto prevented] *End of last page, with the note,* "Continued | on page 1." *What follows written up left margin of first page.*
26 mine] my mine *[E]*

to you some time ago,[234] and which consists in the reproduction of your "Illustrations of the Logic of Science" which was published in the Popular Science Monthly, beginning November, 1877, and if I am not mistaken the last article of these five papers end on page 482 of the same volume. The whole amounts to a little over 100 pages in the Popular Science Monthly. If I am not very much mistaken you gave me once the permission, but I hesitated to do so because I was afraid of complications.[235] Now there have elapsed thirty years since then and I think there is not the slightest doubt that you can dispose of it freely. I almost believe that you told me that you and not Appeltons are the owner of the copyright, and that you would have a right to republish them even if the copyright had not expired. Would you accede to my wish to publish it in book form for a honorarium of $ 100, and in addition thereto $ 50 for your trouble in revising these five essays, altering, adding or changing as you may see fit, and the proof reading. My intention is to bring this book of yours broadly before the public because I believe it is of great value. There are a few points where I would take exception to your presentation, for instance what you call "reality" I would call "truth"; but your meaning of the use of the term "reality" is plain enough in the context and so it does not matter. Presumably you would like to change the name of Pius Ninos for a modern pope. Since then we have had two new popes in Rome. It is a pity for the alliteration of Numa Pompilius with Pius Nonos seems to be intentional.[236] Would you prefer to have the diagram on page 292 redrawn in better style, or do you wish it to be as rough as you made it? The other illustrations on page 296 are also typographically objectionable, also the lettering is very carelessly made.[237]

---

234 Carus to Peirce 23 Jan 1907.

235 Carus to Peirce 3 Apr 1900; also Peirce to Hegeler 24 Feb 1893, confirming copyright from Appleton.

236 Numa Pompilius (753–673 BC), second king of Rome, and Pius Nonus (1792–1878), Pope from 1846–1878, mentioned by Peirce in "The Fixation of Belief," *PSM* (Nov 1877), p. 9.

237 "How to Make Our Ideas Clear," *PSM* (Jan 1878):

p. 292

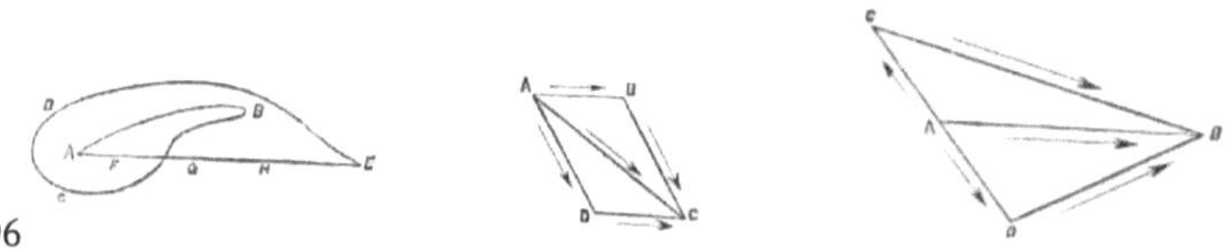

p. 296

---

7 elapsed] elapses *[E]*

19–20 Numa Pompilius with Pius Nonos] Numa Pompelius with Pius Ninos *[E]*

Please let me hear from you, and I will send you at once $ 100, and the remaining $ 50 at the completion of the book.

Some alterations in or a few additions to your science of logic would be welcome in order to justify our taking out a new copyright, and we might put on the title page, “Revised from a Series of Papers published in 1877–78”. The date would insure your priority claims if other people have said the same things at a later date.

Hoping to hear from you again, I remain with kindest regards and best wishes,

Yours very truly,

P. C./N.

1909

Peirce to Carus. MS: SIU 91.25. Two unsent drafts follow.

P.O. Milford, Pa. 1909 Jan 7

My dear Dr. Carus:

I beg your *entire* forgiveness for not replying until this last day of Christmastide for your very kind offer to reprint my papers “Illustrations of the Logic of Science.” For I have been overworked beyond conception. I request you to hold it open for three months or more; for though it is likely that I shall accept it, yet I can only do so when I have done so much of the revision that you can know, and that I can know, what that revision is going to amount to. It is considerably more than 31 years ago since I wrote the two first, constituting my Plea for Pragmatism[238] (or, as I now call it, Pragmaticism,) and as I have been actively engaged ever since in studying the Logic of Science, they naturally require much revision, which must

238 “My Plea for Pragmatism,” to be the new first article of the proposed “Illustrations” book and to comprise the first two “Illustrations” articles in two parts, which Peirce alludes to in the next paragraph and also describes in the second unsent draft of this letter below. At this point the book was to be titled “Essays on the Reasoning of Science,” but would soon cycle through several other titles, including “Studies in the Meaning of Our Thoughts,” “Essays Toward the Interpretation of our Thoughts,” and “Essays on Meaning” (R 618–640). For a detailed discussion and transcriptions of these variant texts, see de Waal, 18–24 and 185–270.

3 additions] addition *[E]*
6 insure] *after* ~~let~~
12 1909] 1908 *[E]*
15 to] *after* ~~of~~
16 For … conception.] *inserted below*
18 that] *inserted above*
20 Pragmatism] Pragmaticism *[E]*

be of my very best quality, both of thought and of expression. I have already of my own motion been considering the doing of this, with a view to preparing a public to take an interest in the Logic book[239] upon which I have been at work for years, and which I cannot hope to make as nearly popular as I should strive to render these "Illustrations etc." If I should be as successful in my revision as I hope to be, I would rather risk publishing on the royalty plan, or some other that would bring *more* than a hundred dollars; for it would involve at least three months of solid work. Very likely more.

My plan is to reprint them with no alteration except mere clerical ones; *but*, firstly, to delete passages which were due to my trying to mollify Youmans's wrath at the engagement which old Mr. Wm. Appleton forced upon him. He, Y, had not an atom of the scientific spirit. He was devoted to Spencerianism just as he might have been to a scheme to get up a revolution in some country; and the idea of allowing an individual to express views in the Pop. Sci. that at all conflicted with Spencer's was disgusting to him. That accounts for various features of the articles; & as I was paid $ 150 an article, I was bound to take great pains with the style and all that. I still think them sound in the main; but I did not then see the whole truth of the matter. Consequently, in revising them, I might enliven them considerably by a little sparring-match with myself on special points. Such passages, to be in brackets, & perhaps with a different font of type. Besides that, secondly, at the end of each, I should *add* a supplement setting forth what I have learned since then. The first and second of the 6 original articles would make one essay in two parts, the third and fourth would make another. The fifth would be one part of an essay in two or three parts. The sixth would be replaced altogether by an account of the Methodeutic of Scientific Research; and I should add two chapters at least in the way of *Illustrations of Science*, setting forth, for example, the nature of the actual reasoning in detail for today's views on the chemical constitution of compounds. This would follow a method essentially different from Mach's treatment

239 To become Peirce's proposed treatise in three volumes on a "System of Logic, Considered as Semiotic," or "Logic as the Theory of Signs in General" as reported to William James on Christmas 1909, addressing the classifications of Signs (Book 1), Critic and different kinds of reasoning (Book 2), and methods of research or "Methodeutic" (Book 3) (letter in *NEM* 3/2:867–875 and in part in *EP* 2:500–502). Also identified in the two unsent drafts that follow.

1 of[1]] *inserted above*
1 already] *after* ~~b~~
2 a[2]] *above* ~~the~~
11 He, Y, had] He$^{\vee}$,$^{\vee}$ ~~was~~\Y, had/
19 Such] *after* ~~To~~
20 secondly,] *inserted above, and under* ~~second~~
21 supplement] *after* ~~sp~~
22 of the 6 original articles] \of the*se*6 *ar*original articles/
27 actual] *inserted above*

of dynamics; for in chemistry there have been too many refutations of theories that always had been bad i.e. unwarranted, from the very start. I shall insert also my classification of the sciences (perhaps the entire scheme, which has never been printed yet,)[240] my criticism of Pearson in the Pop. Sci. for Jan. 1901,[241] and my Evening Post article on the Great Men of Science of the XIXth century.[242] The book will be much stronger & a good deal more popular (if I can succeed in making it so, with hard work,) than the original articles.

Of course, I cannot accept your generous offer of an advance at present.

Accept my most heartfelt wishes for the perfect happiness of your entire circle at LaSalle.

very faithfully
CSPeirce

If I have averaged *four* hours sleep *per diem* for the last month, it can only be because I sometimes fall asleep at my writing-table. My wife needs really *all* my time, her health is so bad.

A hen flew up and snatched off my only pair of spectacles, & it was a good while before I could get a temporary *pis aller*.

My next paper[243] toward my proof of pragmaticism, which I have been holding back for half a dozen years so as to be sure of its maturation can be got ready for the April Monist if you like. It consists in showing what *logical analysis*, or *real definition*, consists in, and how Existential Graphs can be used to aid the operation, and how one is to distinguish an *elementary* from a *composite* idea with illustrations of the working of it.

C.S.P.

If you dont want it, it will be welcomed elsewhere. $ 100 is the price.

240 Many related extant writings, but notably several ca. 1902–1903, including his "An Outline Classification of the Sciences" (*EP* 2:258–262) in the "Syllabus" accompanying his 1903 Lowell Lectures, "The Classification of the Sciences" paper presented to National Academy of Sciences (April), and Chapter II of his proposed "Minute Logic" (R 426–427).

241 "Pearson's *Grammar of Science*."

242 "The Century's Great Men in Science" (12 Jan 1901).

243 Undetermined, though "half a dozen years" ago (specifically "six and a half" " in the unsent draft that follows) would date it around the time of Peirce's 1902–1903 "Minute Logic," from which the mentioned paper may have spawned. Might also be part of or a precursor to his ca. Dec 1909 "Studies of Logical Analysis" and other related material (R 643–650)—see Peirce to Carus 26 Aug 1910, note 1. Described in greater detail in the first unsent draft that follows.

2 i.e. unwarranted,] *inserted above*   21 can] *over* are

Peirce to Carus. MS: RL 77. First unsent draft.

[Unsent draft 1 of 2]
P.O. Milford Pa | 1909 Jan 6

My dear Dr. Carus:

You must pardon my not writing to convey, along with my heartfelt wishes for all blessings on you and yours this year, my warm thanks for your offer to republish my Illustrations of the Logic of Science. Naturally such a proposition requires careful consideration. I shall probably accept it; but of course I cannot accept the generous hundred dollars until I decide and until I can show the text of alterations due to thirty additional years of research put into luminous and convincing form.

I should wish the volume to bring a sale to my chief work, The System of Logic upon which I am now engaged. You know my Pragmatism Articles have still to be completed. The last article, on Existential Graphs, though when I started out to write the series, I hoped to be able to eliminate it, or give but a slight account of it, is an essential step in the reasoning. The next one which gives a theory of Logical Analysis, or Definition, which I have withheld now for about six and a half years in order to make sure of its being well matured, and well fortified by proofs, rests directly on Existential Graphs, and will be acknowledged, I am confident, to be the most *useful* piece of work I have ever done. I have already discovered a small improvement in the theory of my Existential Graphs. Namely, the difference between a proposition which the following exemplifies

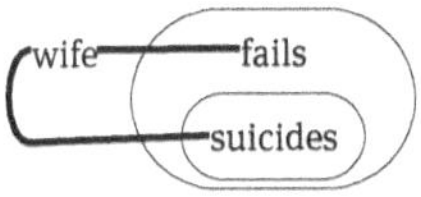

5 writing] *over* con
6 warm] *above* ~~hearty~~
6–7 republish] *after* ~~pub~~
9–10 and until … form.] ~~&~~ \and/ /until I can show the \text of/ alterations *[..]* convincing form.\
13–15 The last … reasoning.] *below* completed*,*. ~~and~~
16–19 which I have withheld … ever done.] ~~based on an application of~~ \which I have withheld *[..]* rests directly on/ Existential Graphs, ~~is~~ \~~will be~~/ ~~no doubt~~ \~~This theory~~ \and/ will be ~~found t~~ acknowledged, ~~I trust to be~~ I am confident, to be/ the most *useful* piece of work I have ever done. ~~I have delayed it, in order that I may be sure it is quite matured before I publish it.~~
19–20 a small … of my] *above* ~~a screw loose in my~~
20–21 Namely, … exemplifies] ~~Namely, one thing which had much to do with my making so much of the different "tinctures" of my Area, was that propositions like~~ \Namely, the difference between ~~such~~ a proposition ~~of~~ a form which I may designate as ~~*B*A, B of~~ which ~~an example is to~~ /~~this is an ex~~ the following exemplifies\ / *[omission of uncancelled* "a form which I may designate as" *mine]*

"There is some married woman who will suicide in case her husband goes bankrupt" and "If every married man goes bankrupt some married woman will suicide," had seemed to me not to be reducible to the difference between "Every catholic adores some woman or other" and "Some woman is adored by all catholics."[244] I discovered this, which is very elusive, and *pinned* it, by means of my method of Logical Analysis, which reposes on and is a consequence of certain consequences of the Existential Graphs. Now Logical Analysis is of course Definition; and this same method applied to Logical Analysis itself,—the definition *of* definition—produces the rule of pragmaticism. The whole thing will be made clear to every mind who has the patience to follow it out. It is the very proof that I intended to bring out when I began this series of papers on Pragmaticism. But on account of my desire of considering every possible objection, I have been led to wander round a good deal & I confess my papers have been pretty badly written. But as it was there was not one of them that did not cost me a full month of solid work, all my impecunious condition permitted me to do for fifty dollars each. But if my original plea for pragmatism is to be reprinted,—and, of course, it will be, in any case, whether I close with your offer or not,—it must be accompanied by as clear and lucid an argument as I can write. I had better not hurry. I am quite sure the thing will have a sale, if merely for the life of the presentation; and I am determined it shall be as clear as possibly I can make it; and I do write clearly when I care not what labor it may cost. But of course I can't accept your generous hundred dollars until I am nearer publication that I am now.

I am sorry that James should have come out so decidedly against *absolutism*, though it is a natural step for a pragmatist to take; but James is much too large a man, and even if he weren't, cares too much for the esteem of the scientific world, not to *perceive* that he is in the wrong as soon as solid scientific argument against his position is adduced,—and to suppose for an instant that, perceiving it, he would not make haste to acknowledging it, would be a foul slander, indeed.

---

244 Also in "Prolegomena," pp. 538–540 (*Monist*, Oct 1906).

---

2 and] *above* ~~could not otherwise be distinguished from~~
3 had seemed … between] *above* ~~, But really the distinction is only a subtiler form of that between~~
5 catholics."] *before abandoned emendation* *,*. \because that transformation ~~would lead to~~ /of\ /
5 is] *after* ~~it~~
11 this] *above* ~~these~~
21–22 But … now.] *inserted in-line and above (new paragraph)* ~~I am s~~
26 as soon as] *above* ~~when~~
27 adduced,] *before* ~~and not to~~

As for Schiller, I cannot be so sure; for I have not the intimate acquaintance with him that I have with James. But I think I shall bring him over too. I shan't ask him to venerate Bradley;[245] for I have no high opinion of him myself. He does very well as logicians go; but he is not the great mind Oxford once thought him. He did not half bring out Schiller's strength as a controversialist; but I guess he will not again be so eager for controversy. In that case, it was really important to show Oxford its mistake & only ridicule would do it. But nothing costs so much to the respect that first rate men have for one as a reputation as a controversialist. I have seen it in a dozen cases. Herbert Spencer was one, though by no means, the most striking.

As for the other parts of any "Illustrations of Science," the controversy about the probability of Induction can be much more tellingly told; and of course must be so. Also the Uniformity of Nature wants enlargement. And I mean to insert more to justify the title, by stating the *whole* argument which has convinced chemists, so far as they are convinced, of the existing theory of chemical constitution. I shall *not*, like Mach, follow the course of *history*, because in the case of chemistry so many theories were adopted & afterwards refuted. I shall only give the argument so far as it appears sound today. I may also give a historical account of Kepler's reasoning about the orbit of Mars; though this has been done (with too much detail for the purpose) by himself.

You thus perceive that there will be much labor to be done before the volume is ready; and it must be made perspicuous & intelligible to ordinary people who are willing to take the trouble to think *when egged on by writing* that provokes thought. I want a sale, so that my System of Logic on which I have spent so many years, may be read sufficiently to do good *mediately* so far as not immediately. For I cannot work that up to such a polish as I mean to do my "Illustrations of the Logic of Science"

With renewed thanks believe me dear Doctor Carus & with best wishes for the year for everybody in the circle at LaSalle & for Russell too.

yours faithfully
CSPeirce

245 Francis Herbert Bradley (1846–1924), British idealist widely known for his *Principles of Logic* (1883) and *Appearance and Reality* (1893).

3 myself.] *after* ~~at presen~~
11 of^2] *over* the
15 follow] *after* ~~tel~~
23 that] *over* as

Peirce to Carus. MS: RL 77. Second unsent draft, and of the three versions, the least considerate of Carus's proposal.

[Unsent draft 2 of 2]

P.O. Milford Pa. 1909 Jan 7

My dear Doctor Carus:

You must pardon my not writing to thank you for your kind offer to reprint my "Illustrations of the Logic of Science" until the last day of christmas, for I have been so driven to death with a multiplicity of matters that I have had little time to consider it. I cannot yet answer very definitely; but can explain my views partially.

But first let me wish you and all the circle at LaSalle all possible happiness & let me thank you for your generous offer which it would not do at all for me to accept at this time, as I shall explain.

I am writing my chief book,—*a System of Logic*, which I hope will be recognized as useful. I have intended to put forth first, however, one volume and maybe two. One at any rate to be a revised, & in large part rewritten "Illustrations of the L. of Sci." The first two articles to appear as two parts of one essay, "A Plea for Pragmaticism." These will be given, with mere clerical corrections, just as they first appeared, but with a correction of one error occupying about 500 words.[246] There will then be a paper about the Doctrine of Chances, taking the same ground in the controversy about inverse probabilities that I originally took, but so stated that readers may better understand the state of the question. There will be as before a discussion of the "Uniformity of Nature"; but this must be much broadened. In place of the Sixth paper there will be a full account of the whole course of scientific research in general; and finally, there will be two or more chapters setting forth the logic *[…]*

Carus to Peirce. LPC: SIU 91.25.

La Salle, Ill., January 11, 1909.

Mr. Chas. S. Peirce,
Milford, Pa.

My dear Sir:

246 The "error" is likely, as Peirce explains in his letter to Carus of 26 Aug 1910, "its *nominalism*, especially as illustrated by what I said about Gray's stanza 'Full many a gem' etc pp 300 et seqq." (of "How to Make Our Ideas Clear," *PSM*, Jan 1878).

5 Doctor] *d*Doctor

15 in] *over* I

Your letter just comes to hand and I will say in reply as follows: I give you all the time you want to revise it but I did not expect you would have to rewrite so much but of course if you have had to adapt yourself to Mr. Youmans I can only wish myself that you would make such changes as you deem necessary, but please do not tamper too much with your own thought of former years. I remember of Goethe that he treated his writings of his younger years with great respect and did not change unnecessarily. Of Cantor I can only say that he changed his article on Pythagoras in the later edition and took out the best, most ingenious and most interesting part of his thought, so I hope you will not touch your own old thoughts but limit your revision to what is foreign to yourself in the articles. I had offered the pecuniary amount for revision on purpose of $ 50 because I did not want you to make supererogatory changes. However, as you state the case it is different.

As to the payment I am now willing to add another $ 100, and so I will pay you for the whole $ 250. I think very much of this article of yours and shall be very glad to have the revised copy within about three months, payment to be made on receipt of copy. I should be very glad to have those papers of yours on logic brought out and have them widely known among all kinds of readers. You have a certain limited number of admirers, most of them are good steady thinkers, but they are very limited indeed. I will mention such men as Francis Russell, of Chicago, and above all Mr. Holden of West Point. Otherwise I am sorry that there is rather a prejudice against your writings, and if I would take articles of yours for the money in it I would not touch any one of your articles. I weight more the opinion according to quality than to numbers and I myself am anxious to have your best thought stated in clearest form presented to the public. I have always been pleased with these papers on logic and hope they will contribute a good deal to clear up the ideas concerning pragmatism and the difference that obtains between pragmatism and pragmaticism.

By the by I have in a little speech at Heidelberg[247] called attention to the big difference that obtains between your pragmaticism and James' pragmatism, and I found a number of men of note curious to get further information on the subject. If I publish your article I shall be able to make the source of this information accessible to them.

247 His remarks are part of the "Continuation of the Discussion on Pragmatism," Third International Congress of Philosophy in Heidelberg (Sept 1908), Elsenhans, 729–740. Carus's "Pragmatism" in the July 1908 *Monist* also addressed the subject in more detail.

12 supererogatory] superrogatory *[E]*
19 Francis] Frances *[E]*
20 Holden] Holm *[E] For Edward Singleton Holden (1846–1914). See Biographical Register.*

You speak of a royalty plan, and I shall be glad to make any modifications in our business conditions which you see fit but I do not think that there is much money for you in a payment in royalties.

I expect that your article on Pragmaticism would be welcome and I would accept also your price, but I am a little afraid of graphs. Our Japanese friend who did your graphs formerly is no longer here and I do not yet know who will be able to do the work. He could do it almost in my presence, while if the work is done in Chicago it has to be attended to through correspondence which will make the details of the work very doubtful and presumably also more expensive.

With kindest regards and best wishes, I remain,

Yours very truly,

Peirce to Carus ca. 30 Jan 1909 (missing): Asks for a statement of payments for his *Monist* and *Open Court* articles to date.

Carus to Peirce. LPC: SIU 91.25.

Feb. 6, 1909.

Mr. Charles S. Peirce,
Milford, Pa.
My dear Sir:

Your letter came to hand a few days ago, and I will say that I have given instructions to my business manager to make a statement of the payments made to you.[248] I remember that when I left for Europe in February, 1907, I sent you a check for $ 100,[249] which I do not remember that you ever acknowledged. I wanted you to write on some popular subject, for instance your criticism of James' pragmatism, but you never followed my wishes and sent me materials which are of a less general or popular character. I expect to hear from my business manager very soon, and as soon as I have the data I shall answer the business part of your letter.

I must confess that I look forward with great satisfaction to the revision of your paper on logic. I believe that the reproduction of this essay would do more good than any new work of yours, and I believe it will help to bring you into prominence better than anything else. It would call attention also to your more recondite work

248 In Appendix, "Peirce and Carus."
249 Carus to Peirce 30 Jan 1907.

9 details] deatils *[E]*   29 help] hep *[E]*

on the Algebra of Logic. I do not want you to make concessions to popularity exactly, but I think that you could in your writings do something to make yourself better understood by the average well educated man, who takes an interest in logical and philosophical questions. As matters stand now you have a limited circle of very good friends while you might have a greater number of students who would read your essays. I am quite anxious to let you understand your own advantage and I should be glad to be of service to you in presenting your ideas in a way which would gain you a larger class of readers. I shall write again as soon as I know more of the payments made to you.

In the meantime, I remain,

Yours very truly,

P. C./N.

Peirce to Carus. MS: RL 77. Incomplete, unsent letter. No draft in its place is known to have been sent.

[Unsent]

Milford, Pa. 1909 Feb 24

My dear Doctor Carus:

Of course, I do not know what it is that prevents you from writing to me, and communicating the result of the information you said you were to obtain from your business manager; but it is a matter of very great importance to me as I shall explain.

In the first place, a philosophic friend of mine who was here and saw how the house is going to ruin for the want of repairs;—and there have been days when we could not pass from one part of the house to the other through the gallery without putting on galoshes;—this friend said to me

Carus to Peirce. LPC: SIU 91.25. Note in pencil at top of letter, "Extracts from | Corr. bet Dr. Paul Carus | to Chas S. P." Likely a label for this letter as one of those Carus quoted in his letter and extract sent to Peirce 8 Dec 1909.

February 26, 1909.

---

18 it is that] *inserted above*
19 you said … obtain] you \said you ~~were to obtain~~/ were to obtain
20–21 of very … explain.] of \very great\~~more~~// importance to me \~~than a little~~ as I shall explain/~~; and I shall not decide upon the publication of my Pop. Sci. articles until matters are squared up; and I will explain to you why I am obliged to be so insistent.~~

Mr. Charles S. Peirce,
Milford, Pa.
My dear Sir:

I have been waiting and waiting for a check for you but for some reason which I do not know my business manager delayed and delayed until at last I received the check this morning, which I am sending to you at once. Excuse this long lapse of time but I really do not know the cause of the delay myself. The statement which they sent me was rather complicated and I have not yet a clear insight into it but I will say at once that it amounts to about $ 1 000 that you drew from the Open Court office, but since I note you are waiting for money I send you at once $ 150 as per enclosed check.

With kindest regards and best wishes, I remain,

Yours very truly,

P. C./N.

Carus to Peirce. LPC: SIU 91.25.

March 9, 1909.

Mr. Charles S. Peirce,
Milford, Pa.
Dear Sir:

You may be interested in reading the few comments which I made on Pragmatism at Heidelberg.[250] It was only a stray comment during a heated debate, and I just received the proof. I refer to you although I do not know how far you would identify yourself with James' Pragmatism. Please return it.

With kindest regards, I remain,

Yours very truly,

P. C./N

Peirce to Carus. MS: SIU 91.25. Likely sent ca. 12 March since in the course of writing Peirce received Carus's letter and proof sheets of 9 March (noted in text).

Milford Pa | 1909 March 9

My dear Doctor Carus:

I have a very heart-felt apology to offer you for not before having thanked you for your kind attention in sending me the check for a hundred & fifty dollars which

250 Elsenhans, 737.

I duly received, and which squares us up. It was not for want of feeling that I omitted to do so but because I was in the midst of a complicated piece of reasoning about the analysis of concepts in which I was so engrossed that every moment of spare time had to be devoted to it for fear I should lose some of the many threads of my problem. Perhaps I will tell you later how it happened that though my method was long ago all perfected,—so far as any scientific method or other result can be said to be perfected,—yet such a fundamental problem should remain unresolved with apparently most redoubtable arguments on both sides.

I might have written a brief note. But I was not satisfied to do that, for in the first place, I had to express my particular obligations to you for explaining at Heidelberg how far my pragmaticism differs from the pragmatism of which James is the apostle. For James is an old and dear friend of mine and for a long time scarce a paper of his has reached me that has not caused me to pass a night in grief that I should have to disagree with him as deeply as I do; and yet loyalty to truth forbids my keeping altogether silent or from minimizing the differences. I therefore feel particularly grateful to you for having taken from my shoulders a great part of the burden of that responsibility.

I am—just as I was beginning to pass to another subject, my dear wife came into my room, and said that if I were writing to you she should like to send her compliments to Mrs. Carus, and to say how warmly grateful she felt for the important aid which your check had given her toward making our roofs tight and so preserving our lives which were really endangered by the condition they were in. I am afraid I am awkward in interpreting what she said, but I can only hope that Mrs. Carus will be able to comprehend what Mrs. Peirce meant, but which I don't seem to have succeeded in expressing.

Now I will return to what I was about to say. The principal reason why I have hesitated to accept your proposition about the volume to contain a revision of all articles I contributed to the Pop. Sci. Monthly (among which I should like to include the one printed in the number for January 1901,[251] since that chiefly relates to my pragmaticism,) has been, 1st, that I continue to be more disposed than ever toward my hypothesis of *Tychism*, and could put more strongly today, & it would find more favour too; while there have been some apparent indications that you are as much opposed to it as ever; and 2nd, although to my mind a well-developed

251 "Pearson's *Grammar of Science*."

1 received, … up.] received*.*, \and which squares us up./
2 a] ~~an~~
8 redoubtable] redoutable *[E]*
12–13 scarce a] *above* ~~no~~
18 as] *after* ~~there~~
20 how] *after* ~~that~~

philosophical theory has nearly as much value if it be false as if it be true. For if it is false, the development of it must throw light upon that. I do not for example hesitate to say that so utterly false a view of logic as that of Hobbes[252] was and still is of great service owing to the thoroughgoing development he gave to it. But I have not observed that you were at all inclined to act upon this view of things & you have seemed to me to regret or dislike the appearance of theories to which you did not assent, all the more if they were worked out thoroughly. Now I certainly expect before very long to be wanting a publisher for my System of Logic, which will be a work addressed to a quite large class of intelligent, or somewhat intellectual men, and is intended to initiate them into my views of reasoning and to put them into condition to examine them critically & to form opinions of their own on the subject.

Now in that work I must certainly introduce and discuss all the fundamental ideas on which the science and scientific philosophy prevalent today reposes. Unquestionably one of the greatest of these ideas (be it true or false,) is that of *evolution*; and according to my logical principles when a given fundamental idea has once been admitted into our hypothesis of the constitution of things, it ought to be carried just as far as it will go, subject to afterward being curtailed as need may be.

My tychism is simply the theory that the variety of the Universe of Existence has been due to evolution; and I suppose that evolution to be, as nearly as may be, of the Darwinian type. Of course, my *absolute chance* may be regarded by anybody so disposed (as a further hypothesis) as *creative activity*. Indeed, that was always my own private opinion; but I did not care to complicate my hypothesis with any other.

Well, my idea has been that you would be averse to publishing any new contributions to my Tychistic theory, while it would probably be for my advantage to find a publisher equally interested in all my writings. However, in view of my wife's health and indeed her *life* being threatened by the state of our roofs etc, and in view of the very important aid that some European friends of hers (who are now in this country but will go away probably never to return considerably before Easter) have offered, namely to double any funds she might have for that purpose before their sailing,—and even if you advanced me the whole $ 250 and they doubled it, she would be about $ 300 short of what she needs; but I think she could get that without embarrassment,—in view of all this, if I could persuade you to advance

252 *De Corpore*, Part I (1655).

6 you[1]] *after* ~~that~~

7 thoroughly.] "no" *inserted above by Carus*

14 scientific philosophy prevalent today] scientific ~~logic~~ philosophy ~~of~~ \prevalent/ today

the whole $ 250 on my solemn promise to devote *all* my energies after drawing the plans for my wife's work (which are nearly complete now) exclusively to those two jobs for you; namely the completion of the revision of the Pop. Sci. articles and the article on my method of analyzing concepts (i.e. of defining them) a theory completely worked out now—then I could close with you with the understanding that your holding these articles should not be a reason for your not accepting *new* articles, being portions of my System of Logic, which would be written with a view to their being widely read, & decidedly more perspicuous than my article in the Hibbert Journal of October 1908.[253] The style in which I have been writing for you has been greatly influenced by a remark you made to me in LaSalle when I was there. It was to the effect that you *preferred* an article which seemed to be written by a man "struggling for expression," rather than one which bespoke a mastery of language. "Lord," thought I, "he is one of those German *savants*, who, as I well know, would be down right ashamed of writing anything that could be understood without much trouble, and to whom awkwardness with the pen is a matter of pride. Nothing is easier than to satisfy him on that score: it will cut my labor in two." Accordingly, I contented myself to sending you first drafts. But it seems that years of experience in editing the *Monist* have brought you to think more highly of style than you formerly did. Very well: I heartily agree with you. Only, you understand, it about doubles,—fully doubles,—the labor of writing. My article in the Hibbert Journal of October 1908 was written four times over; and even so, some have complained of it as being unintelligible. If one has anything like novel forms of thought to communicate, the difficulty of making the reading of them easy is very great. As Byron said "This easy writing makes damned hard reading."

Now as to your having desired something of an amusing nature & my having responded with my Amazing Mazes, you will find if you still have the papers & accompanying letters I sent, that I did not recommend the printing of the explanation of my first curiosity;—at least, not at first. I meant you to go on with the card-tricks, which I thought, & still think, were quite as entertaining as well as intellectually pretty as the Magic Squares which you had been printing. But I will wager that you never took a pack of cards & tried them; and if you didn't, you don't know anything about them. I emphasized the necessity of doing that as strongly as I could. They were written for people who would look out of their eyes, and see

253 "A Neglected Argument for the Reality of God."

12 one which bespoke] ~~b~~ one which ~~seemed~~ bespoke

for themselves & not content themselves with reading about what they *would* see if they did so. [I was at this point in this letter, when I received your proof-sheets.[254] I must beseech you to change one word, at least; you say "Die anderen, *besonders* James, sind ___ Feuilletonisten, nicht wirkliche Philosophen."[255] I don't say it is not true, but do favor me by changing that *besonders* to *abgesehen von* or *ausser*;[256] because I have observed very carefully that *James is more quoted than any other living psychologist*. He is a *foreign member of the Berlin Academy*. He is enormously admired by men of the standing we will say of Hoefding.[257] He is made much of in Italy, in Paris, in Oxford;—and what is to be gained by speaking so contemptuously of him? And indeed I think you might also except Vailati, who is not altogether without strength in severe studies. But I am grateful to you for pointing out that I am not a sharer in the extravagant *expressions* of the other pragmatists. I underscore the word *expressions* because their style of writing prevents me from making sure what they really mean. I don't think their thought is sharply definite enough to be denied.]

I should really like to know what your sentiments are at present in regard to any definite tendency of the universe.

Is there none at all? Then, it is as it were playing an even game, and it is *pure* chance what the future has in store except that whatever is both possible & irreparable may be expected to come about ultimately. No body can play a perfectly even game without ultimate win.

If it has a definite tendency toward becoming what it has not been, we must call that direction improvement & reconcile ourselves to it; and that is evolutionism. Whether or not it be Darwinian or some other type of evolution is a secondary question. What is your view? Can we pull together? Of course I only seek a fair field & no favor.

very faithfully
C. S. Peirce

Peirce to Carus. MS: RL 77. Unsent draft.

[Unsent draft]

---

254 From Carus's letter of 9 March. This letter, therefore, likely sent ca. 12 March.
255 "The others, *particularly* James, are ___ columnists, not real philosophers."
256 "changing that *particularly* to *apart from* or *except*."
257 Harald Høffding (or Hoeffding; 1843–1931), Danish philosopher and theologian.

---

26 favor.] *[?~~I as always~~] inserted above*

Milford Pa | 1909 March 9

My dear Doctor Carus:

I owe you a particular apology for not writing at once to thank you for kindly sending the check of a hundred & fifty dollars.

I did not do it because I wished along with it to say a number of other things; to thank you for your Heidelberg explanation, for one thing. You see that James is a very old friend and such an estimable and delightful man that,—detesting controversy as I do, because it takes up one's thoughts with matters that nobody cares a tuppence about except the immediate parties, unless perhaps some outsiders who think it fun to witness a squabble, because it wastes time that might be employed in advancing our comprehension of logic, even if it does not do worse by tending to give one an irrational leaning toward or away from some opinion, & because it is altogether beneath a philosopher—because of all this, I am embarrassed to point out myself how very contrary to my convictions is much that James says, or *seems to say*,—for his manner of expressing himself is, what no doubt appears to him a broadly generalizing way of thinking but which often seems to me,—rightly or wrongly—to be simply excessively vague. He will say for example that the rule *Nota notae est nota rei ipsius*, (which, by the way, he will be sure to call the *Dictum de omni et de nullo*, a blundering appellation the history of the formation of which & its coming into vogue is, by the bye, very curious, and which ought to be called, as it *invariably* was in the Middle Ages, the *dici de omni* & is quite distinct from the *Nota notae*—in so far as the *Dici de omni* is simply what the Theorists used to call it, a *definition* of "dici de omni," while the *Nota notae* merely notes that whatever is predicated of the predicate genus is predicated of that of which this is predicated,—regarding it simply as a phenomenon.[258]—Well James will call this an instance of the "*axiom of skipped intermediaries*"—see his Psychology about a dozen pages from the end.[259] He refers to what I usually call *transitive* relations,—after DeMorgan;—but there are two objections to that term and possi-

258 In "Methods of Reasoning" (1881), Peirce elaborates: "Several of the rules current in logic are only forms of the statement that Barbara is valid reasoning. Such is the maxim 'Nota notae est nota rei ipsius,' the mark of a mark is a mark of the thing itself, where the mark of a mark must be interpreted as a description applicable to everything to which another description applies. Such too is the dictum de omni, 'whatever is asserted of the whole of a class is asserted of every part thereof.' Another way is to say with DeMorgan that the relation of antecedent to consequent is a transitive one, that is, if *A* is in this relation to *B*, and *B* to *C*, then *A* is in this relation to *C*." (*W* 4:251)

259 *Principles of Psychology*, vol. 2 (1890), p. 646.

8 up] *over* on
12 give] *after* ~~mak~~
19–20 formation … vogue] *above* ~~coming of which into vogue~~
26 his] *over* th

bly *perseritive* relations would be a better term. The idea to be expressed is that a lot of them strung on a string like beads are of the same nature as a single one. Unfortunately, it is difficult to make a word that will fairly express that in one word. He says in that passage "It is the broadest and deepest law of man's thought." Why "man's"? Before the principle of evolution was established, one might think, as Kant did, that it was merely human. Now those of us that believe evolution is the great mode of being of all things, including concepts and forms, there is no longer any reason for inserting "man's." It is so with everything, all causation, all workings of evolution, Time, & Space *[ ]*

Carus to Peirce. LPC: SIU 91.25.

March 15, 1909.

Mr. Chas. S. Peirce,
Milford, Pa.
My dear Sir:

Your letter just comes to hand and I will briefly answer your several points. I am glad to hear what you say concerning pragmatism. I was always astonished why you kept absolutely silent, but I myself waited very long before I criticized that ingenious man, William James. My expression sounds harsh, but it is not meant so. I grant him all the qualities of a poet, but not of a scientist or thinker. I meant what I said. Nevertheless I will try to reach the publishers before they go to press. I have not made the change which you suggest, but I added to the first mention of James the name "*genial*", for this word in German means a good deal and almost more than the English "ingenious". In the second place I omitted his name but I fear the corrections will be too late.

The same may be said concerning the misspelling of your name,[260] but let us hope that it will be in time.

I handed your letter to Mrs. Carus and I will say concerning the money question, that I will have a check made out at once to the amount of $ 250, accepting the conditions under which you desire that sum.

When did you ever see me regret or dislike the appearance of theories to which I did not assent? I published your articles on Tychism, even when they were pointedly formulated against me, and in criticism of my own views. The more you attack

260 Misspelled "Pierce" (see Elsenhans, 737).

1 *perseritive*] *Possibly for "perseverative," unless Peirce is introducing a new term.*

me the more I like it, and I prefer by far to carry on a fight with you than with some incompetent small fellow, who has little or no scientific knowledge or method to back him. Accordingly the subject of Tychism could have been quite welcome, and I would by far have preferred it to the articles you sent me.[261]

By the bye I have noticed in the meantime that your idea of chance is an inheritance from your father, Benjamin Peirce, as appears from the preface in his mathematical book.[262] No wonder that it was deeply seated on your soul.

What you say on page 9 of your letter concerning my preference for articles which seem to be written by a man struggling for expression, is a very strange misconception. It is the very opposite. I like articles which come down to clearness of thought, and I always try to do so. It seems sometimes an inducement to be unclear because on the mediocre mind it makes the impression of profundity, but I appreciate a clear thought, clearly stated to any struggling for expression. What is struggling for expression is still immature. My idea of good style is unadulterated clearness, to speak of the point and state plainly what one has to say. I am only against that literary style which takes delight in flourishes and useless embellishments. A similar error is your idea that I desire you to write something of an amusing nature. It is true I want articles which are intelligible even to those not versed in your specialty, but I do not care for anything amusing.

I prefer not to answer in a letter your question concerning the definite tendency of the universe. Before I answer it I ought to know the specific meaning of the question. In one sense the universe has a definite tendency and this tendency must be read from our knowledge of evolution. This refers to the life in every single spot of the universe, but if you take the universe as a whole, I would say that its tendency does not alter the character of the universe as such. It is in this respect eternally the same. The norm which dominates all the tendencies in the universe has been from the beginning and is and will be to the end. There is no becoming in what religiously speaking is called God. Here I believe lies the main difference between your and my views.

I know that your name is the French Pierre and that its pronunciation has changed frequently. Its first occurrence in English literature is in the title of the oldest English document after the Norman conquest, in the vision of Piers, the ploughman.[263] Perhaps this visionary ploughman was one of your ancestors. Your

261 "Amazing Mazes."

262 Reference unclear, as Peirce explains in his response that follows.

263 Langland, ca. 1370–1390.

30 pronunciation] pronounciation *[E]* 32 Piers] *above* ~~Peirre~~

name occurs in the combination of James Pierre, which has been contracted into Shakespeare.

With kindest regards and best wishes, I remain,

Yours very truly,

Peirce to Carus. MS: SIU 91.25.

1909 Apr 5 | Milford Pa

My dear Doctor Carus:

Your letter and check which I received Saturday afternoon have fired me with the desire to pitch in and make my 'Essays on the Meaning of Thought'[264] as useful to a relatively large circle of readers as I possibly can.

I answer your letter according to the order of the topics you touch.

You don't yet quite understand James. He is *not* a Poet. He is a man of scientific training accustomed to all the usual precautions of science. He early fell under the influence of the strong Mill-ite, Chauncey Wright, which made him for the time being, that is, for many years, an individualistic nominalist. But I have been long endeavoring to turn his thoughts toward the recognition that not only are Laws of Nature and many other Generals Real (i.e. what you have doubtless been accustomed to call "have objective validity") but also that there are *Possibles* which have an "objective,"—or, as I say, a Real,—Possibility. Such for example as *Ellipticity*. This will I think bear fruit in his forthcoming book.[265] No doubt, he is most at home in "society," and consequently avoids what he thinks "pedantry," and is given to ways of expressing himself that I think quite unsuitable for any scientific philosophy. Therefore, he slurs over & contemns many distinctions that the rest of us look upon as important. But you will find that his real meaning, even when mistaken, is eminently scientific, in the sense of being carefully constructed with a view to conforming to experience. He is, as you say, *genial*, in the German sense, though not at all *ingenious* in the modern English sense. He is *genial* in the sense of looking below the surface to underlying human realities. "Ingenious" as

264 One of several titles Peirce proposed for the book. For details, see Peirce to Carus 7 Jan 1909 and de Waal, 18–24.
265 *The Meaning of Truth* (1909).

9 Thought] *after* ~~Concepts~~
10 relatively] *after* ~~som~~
17 many] *inserted above*
17 other] other other *[E]*
20 *Ellipticity*] *e*Ellipticity
28 realities] re*l*alities

used nowadays, implies a sort of mathematical imagination; (Edison and Orville Wright are ingenious) and James is as little mathematical as anybody I ever met.

I must do you justice as to your question "When did you ever see me regret or dislike the appearance of theories to which I did not assent?" I never did find you having any such modification of consciousness. But I did not have *that* in mind, but only a less active interest & eagerness to make public theories one does not believe in. *This*, I think, belongs to all of us! Indeed to be as eager to see what we think deceptive prevail as what we think solid, would not argue much earnestness. I, for example, being very thoroughly convinced by experience how fallible I am, feel much more comfortable if the other side of any question is adequately represented; but I am undoubtedly more in earnest to get a proper hearing for my own views, because to do so is more my proper business than to look out for the interests of the other side. However, what you say about your attitude toward the expression of my Tychism, quite satisfies my mind.

My curiosity is piqued to know what you can possibly allude to in saying that my "idea of chance is an inheritance from" my "father, as appears from the preface *in* his mathematical book." My father wrote 6 mathematical books and 1 non-mathematical one and edited Sir John Herschell's book on Sound.[266] Only two of his books refer to any large subject in their prefaces & they have not the slightest connexion with chance. Therefore, the passage you refer to must be in some other part of the book. But I don't remember anything at all pertinent. My father & I were constantly arguing questions. He never would touch upon anything he thought I could work out without his aid; and I never brought up anything relating to the analysis of concepts as to which I regarded him as the merest child, but a child too old for me to undertake his education. Now my ideas about probabilities were altogether founded on an analysis of the conception as to which I thought, & still think, Laplace was quite beyond his depth. But I knew my father worshipped Laplace and particularly admired his *Théorie des Probabilités*, and therefore I thought it by far the best to devote my time with him to things more profitable to myself & always avoided talking about chance. If you care to know the sort of ideas I *did* inherit from my fatherbeyond the scientific habit of looking

266 Listed in Bibliography.

1–2 (Edison … ingenious)] *inserted above*
4 find] *above* ~~fa~~
5 have] *after* ~~me~~
8 argue] *above* ~~be~~
10 adequately] ad*q*equately
12 to do so] to do *[E]*
14 quite] *after* ~~is~~
17 books] *inserted above*
19 refer] *after* ~~have~~

for the solid facts beneath phrases, and scepticism about popular ideals,—such, for example, at this moment, as "the *right* of suffrage,"—when we are completely swamped by the incapacity of the electorate,—you will find such inherited ideas in his non-mathematical book "Ideality in the Physical Sciences," the preface of which was written by my late brother J.M.P.[267]

I insist you used that phrase "struggling for expression," & it impressed me, just as I said, in connexion with the apparent disdain of the oldtime German professors for any clearness or elegancy of style. But still of course I may have misunderstood exactly how far you meant to go. And meantime, I still guess you have unconsciously shifted your ideals a little in consequence of more experience as an editor. It is not a question of much practical importance as to my *future* style in the *Monist* etc. Of course, there is, I necessarily grant you a sense in which the character of the universe is *one* at all times & in all departments. But I think that one character is the embodiment of ideas,—which is a synonym for "creation"—in the whole course of time. I don't believe anything ever appeared *instantly*,—a sentence I set down knowing that if pressed it resolves itself into an absurdity.

But I only mean that that is vaguely the character of the universe everywhere & always.

And it is a maxim of logic with me that when a given kind of element is once admitted into a hypothesis, we should begin by "going the whole hog," and only put limitations upon it as they appear to be proved.

No there is no implication on my part that there is any "becoming," if you mean subjection to the law of time, in God. In the Hibbert Journal of October 1908, p 98, I said: "The hypothesis [of God], being inevitably *itself* subject to the law of growth, appears in its vagueness to represent God as so, *albeit this is directly contradicted in the hypothesis from its very first phase*. But this apparent attribution of growth to God, since it is ineradicable from the hypothesis, cannot, according to the hypothesis, be flatly false." [That is, *vagueness* of a hypothesis consists in its being in some respects self-contradictory; and we have to look forward to the contradiction being eventually cleared up, or clearing up more and more, at least.] … "A purpose essentially involves growth, and so cannot be attributed to God. Still, it will, according to the hypothesis, be less false to speak of God's purposes, than to speak of him as *purposeless*."[268]

267 James Mills Peirce.
268 "A Neglected Argument for the Reality of God."

7 of] of of *[E]*

I must tell you that the delay in your letter has had the regrettable result of bringing 3 slight interruptions—or possible interruptions,—to my work of getting the revision of the book ready.

For, in the first place, I agreed to review a certain book for the *Nation* in case it should be sent to them. It is the new edition of Roger Bacon's entire extant works.[269]

In the second place, a theatrical fellow came here and offered me $ 100 to "do" Lear in Boston during one week, and I told him that while I considered it so absurd a proposal that I could not believe he really meant it, yet I am in no condition to refuse & if he should *actually send* me $ 100 together with an ample sum for expenses of self and wife, I could do it. I don't in the least believe he will make good; but if he does it will occupy me for a *fortnight*,—a week of preparation all together and a week of performance.

In the third place, I have already nearly got a lot of pamphlets ready to be sent to auction, and they will occupy me 2 or 3 days more.

Please present my most respectful compliments and thanks to Mrs. Carus and believe me

alway very faithfully
CSPeirce

April–Nov 1909: Peirce works over his "Illustrations" extensively and goes through several possible forms and titles that the book would take (see de Waal, 22–25.) By now Peirce has gone well beyond the three months he estimated for the revisions in Jan 1909 and Carus grows impatient, opening correspondence as follows.

Carus to Peirce. LPC: SIU 91.25.

November 26, 1909.

Mr. Charles S. Peirce,
Milford, Pa.
My dear Sir:

I have not yet received your corrected article on Logic although you promised me without fail to send it within two months. I only write to remind you of your promise, and hope you will not forget it. I am quite anxious to keep

269 Steele, *Opera Hactenus Inedita Rogeri Baconi* (1909). No review accredited to Peirce in *CBPW*.

3 revision] *after* ~~see~~

your logical work before the public, although from time to time it seems as if you suspect me of a hostile tendency toward you, but the little differences which obtain between our views, especially concerning the Foundations of mathematics and logic would not prevent me from suppressing or obliterating the great work you have done. Accordingly I wish for the sake of your work, not less than for the sake of the interest I cherish in it, to have your corrections at your earliest convenience.

I have another wish today which I hope you will grant me. Mr. Francis C. Russell informed me of a passage which you had written with reference to a passage of mine condemning a non-Aristotelian logic.[270] Maybe that our views and definitions as to the nature of Aristotelian logic are different. Maybe that I am mistaken, maybe that you are mistaken, but it seems to me that the ventilation of the very idea of non Aristotelian logic would be useful and so I wish to mention it in the forthcoming article of mine,[271] which will discuss Professor Bertrand Russell's views concerning the nature of mathematics. I hope you will have no objection. Mr. Russell has copied for me the passage and I will send it to you together with the text in which I make it as soon as it has been set. I propose neither to refute your proposition nor to make any critical comments on them, but simply state both views, yours and mine.

Hoping to hear from you and if possible to receive an emendated copy of your logical papers, I remain, with kindest regards and best wishes,

Yours very truly,

P. C./N.

Carus to Peirce. LPC: SIU 91.25.

Dec. 8, 1909.

Dear Sir:

---

270 See excerpt in Peirce to Russell ca. early Oct 1895 and in Appendix (original letter not found).
271 "The Nature of Logical and Mathematical Thought," *Monist* (Jan 1910). Peirce quoted at pp. 45 and 158.

---

4 prevent me from suppressing or obliterating] *likely for "provoke me to suppress or obliterate"*

20 emendated] emandated *[E]*

I send you a copy of the most important passages taken from our correspondence.[272]

There was some doubt about certain payments which had been made to you. I thought that you owed an article to The Open Court Publishing Co., and you claimed that the article furnished should be paid separately, and that the $ 100.00 payment I had made to you on a private check the day before I sailed for Europe, had nothing to do with it. On that basis, you claimed that The Open Court was in debt to you. This was discussed about a year ago, when I offered you $ 150.00 for the republication and revision of your logical papers. You wanted $ 250.00 for a revision of these papers, but promised, in addition, an article on your method of analyzing concepts, and I answered you under the date January 11th, 1909, as follows:

"As to the payment I am now willing to add another $ 100.00, and so I will pay you for the whole $ 250.00. I think very much of this article of yours and shall be very glad to have the revised copy within about three months, payment to be made on receipt of copy. I should be very glad to have those papers of yours on logic brought out and have them widely known among all kinds of readers. You have a certain limited number of admirers."

In settlement of your prior claims and giving you the benefit of the doubt, I sent you under date of Feb. 26th, 1909, a check for $ 150.00. I quote from my letter:

"But since I note you are waiting for money, I send you at once $ 150.00 as per enclosed check."

This remittance you receipted under date of March 9th, 1909, as follows:

"Thanking you for your kind attention in sending me the check for a hundred and fifty dollars, which I duly received and which squares us up."

Under the same date you asked for an advance of $ 250.00 as follows:

"In view of all this", (viz, the critical condition of Mrs. Peirce) "if I could persuade you to advance the whole $ 250.00 *on my solemn promise* to devote *all* my energies.....exclusively to those two jobs for you; namely the completion of the revision of the Popular Science articles and the article on my method of analyzing concepts (i.e. of defining them) a theory completely worked out now—then I would

272 See Appendix for enclosure. Variants between quoted portions in this letter, the content in the enclosed "Extracts" document, and Peirce's original letters from which the quotes were taken are minimal and minor—punctuation, capitalization, italicization, and slight variations such as *thanking* for *thank*, *$ 250.00* for *$ 250*, and *could* for *would*—and have not been emended or identified in the text.

27 the] teh *[E]*

close with you with the understanding that your holding these articles should not be a reason for your not accepting new articles, etc."

Not having enough money at this time, I talked the matter over with Mrs. Carus, and we sent you a check for $ 250.00, March 15th, 1909. I quote from my letter which I sent you:

"I handed your letter to Mrs. Carus, and I will say concerning the money question, that I will have a check made out at once to the amount of $ 250.00, accepting the conditions under which you desire that sum."

The check was receipted by you April 5th, 1909, as follows:

"Your letter and check which I received Saturday afternoon have fired me with the desire to pitch in and make my "Essays on Meaning of Thought" as useful to a relatively large circle of readers as I possibly can".......... I had hoped that "having been fired with the desire to pitch in" you would have sufficient interest in your own work to comply with your solemn promise within about three months, but eight months have elapsed and I have received nothing. Please do me the favor of starting this work so that I can publish your papers on logic in an attractive form without any more delay.

With kindest regards and best wishes, I remain

Very truly yours,

Mr. Charles S. Peirce,
Milford, Pa.

Peirce to Carus. MS: SIU 91.25.

Milford Pa | 1909 Dec 14

My dear Doctor Carus

I am suffering from a very bad cold, which partly accounts for my not replying more promptly; but other things besides have stood in my way.

I will mention first that I received a note from you saying that I would find enclosed a proof of the letter from me to Russell that you want to print; *but there was no such enclosure* & none has come since, as I thought it would.[273]

As to the money, it is true that I deposited 1909 March 3 proceeds of your check of $ 150 and somewhat later (before Apr 9) my bank credited me with $ 250 the amount of the check of Mrs Carus (trustee). There were sundry delays at first ow-

273 Proofs sent by OCP 7 Dec 1909 (see correspondence) apparently never arrived.

31 and] *after* ŧ

31 my] *over* the

ing to my having to draw up plans, specifications, and contracts, most distracting to the mind until finally my poor wife though she is suffering agonies from tuberculosis of the lungs & is driven to death with hard work, learned to do without me.

My memory is considerably impaired but my power of logical analysis seems to me to be stronger than ever.

When I came to reflect upon the proposition to write out my method of Logical Analysis & at the same time to revise my 1877 essay on Pragmatism,[274] it seemed to me that the proper way was to join the former to the latter. Whether I did originally or not, certainly I did not *later* appreciate that you cared particularly to have the account of my method of Logical Analysis i.e. of making definitions separate from my Pragmatism essay, this latter being obviously a part of the former.

I have been at work upon that ever since, but my style has become atrocious. I shall not be satisfied until what I write can be followed by any intelligent person *who is willing to take the trouble to think*, and not only *can* be followed but will be read with interest and pleasure. My internal difficulties are great, but I shall be able to overcome them. But I can't say how soon. I cant write much *per diem*. After 3 hours I become fatigued & find it best to stop; and I write very slowly because it is needful to be extremely accurate in this subject.

On receipt of your last letter, I laid aside what I was writing & began to write a separate article on my method of logical analysis. I hardly think I can compress it into one article; but it shall count for one. I expect to have the first part of it ready for your April number of the Monist, but I wont promise absolutely. For my health is bad. Some days I cannot handle a pen at all; and every week I have more than one attack of a nervous nature in which I can move hardly a muscle & am in danger of falling. If I had not had so many that I know how to manage, I should fall often.

When I do send you copy I beg that along with the proofs you send me back by copy; for I am quite sure that I shall so be able to correct the proofs much better and much more easily. It has often been a puzzle to divine what I had written or had intended to write when if I had had my MS. I should have seen at once. You may be sure that I am very grateful for the aid you have given me and that I am determined

---

274 That is, write out one of the pending articles for his 1905–1906 pragmatism series (proposed Peirce to Carus 6 and 7 Jan 1909) while revising his "How to Make Our Ideas Clear" from "Illustrations." By this point the former was likely "Studies of Logical Analysis" (R 643), dated 12–13 Dec 1909 (see Peirce to Carus 26 Aug 1910, note 1).

---

7 revise my 1877] re*f*vise my \1877/

19 aside] *after* ~~as*ide*out~~

24 can] can~~t~~

223.30–224.4 You may be … best.] *inserted in-line, running up right margin of last page and then up left margin of first page*

you shall have your money's worth & I have selfish reasons of the greatest urgency for wishing to complete my task as soon as possible. For I have other things I am most anxious to write; but shan't touch anything else until I finish the work I have engaged to do for you and that in my very best.

very truly
CSPeirce

Carus to Peirce. LPC: SIU 91.25.

December 17, 1909.

Mr. Charles S. Peirce,
Milford, Pa.
My dear Sir:

Your letter just comes to hand, and I shall be very glad to receive your MS. on Method. In case you find a favorable moment of rest you might perhaps gain an impression of your logical papers published in the Popular Science Monthly, and see whether you would not let them stand as they are. They are well written even if they yield in some minor points to the requests of Mr. Youmans. If after a perusal you are satisfied with them, I might publish them as they stand or perhaps I might add in an appendix notes of yours in which you state either a change of view or a dissatisfaction with the form in which they have appeared.

I enclose the proof of your passage quoted from your letter to Mr. Russell.[275] It is preceded by a quotation from my comment on the *Primer of Philosophy* to which you possibly referred (p. 109). It is not followed by a discussion of it but I let the two views stand side by side. The rest of the article is attributed to the discussion of other topics, especially Bertrand Russell's views on Metageometry and Mr. Francis C. Russell's mistake concerning the right angle.

With best wishes of the season, I remain,

Yours very truly,

P. C./N.

---

275 ca. early Oct 1895; see also Appendix.

---

16 Youmans] Youman *[E]*
21 on] *i*on
22 (p. 109)] (p. 199) *[E]*
24 Metageometry] Metsgeometry *[E]*
25 Mr. Francis C. Russell's] Mr. Charles S. Russell's *[E]*

Peirce to Carus ca. 25 Dec 1909 (missing): Sends edited proofs of the excerpt from his letter to be included in Carus's "Nature of Logical and Mathematical Thought." (See Peirce's edits in Appendix.)

Carus to Peirce. LPC: SIU 91.25.

December 27, 1909.

Mr. Charles S. Peirce,
Milford, Pa.
My dear Sir:

I had given up hearing from you, but your letter and enclosure came at the last moment when we intended to send away the forms of the *Monist*. Now I could not without making up all the pages after page 45 insert your entire addition, but since I believe that you care very much for having it in the present number, I made the alterations which you requested me to make, and we are setting your addition to insert it on the last page. I will cancel a few words, but I fear that you have written your addition rather hastily, and if I were sure that you would be pleased with my decision I would omit it, but according to your letter you seem to care very much for having it inserted, and so I will let it go leaving it to you to give further explanations if you see fit.

With kindest regards and best wishes of the season, I remain,

Yours very truly,

P. C./N.

1910

Carus to Peirce. TS: RL 77. On OCP Stationery D. LPC: SIU 91.26.

La Salle, Ills. April 30, 1910.

Mr. Chas. S. Peirce,
Milford, Pa.
Dear Sir:

You have promised again and again to send me the revised copy of your logical papers. I wish you would bear your promise in mind and send them at your earliest convenience. I am anxious to publish them, not only because I think they are good, but also to keep before the public that kind of pragmaticism which I deem to be sound. You wanted to make alterations and perhaps write a few words of introduction for the new edition.

Perhaps it would not be advisable to change too much because the papers are good as they stand, and the little shortcomings which they may have according to your present conception, or which may have slipped in through the influence of Mr. Youmans, are perhaps not very important. Please let me hear from you, and if possible send me the corrected logical papers.

You ought to do so not only to make good your promise but in your own interest as an author and a thinker.

With kindest regards and best wishes, I remain,

Yours very truly,
Paul Carus

Carus to Peirce. RL 77. On OCP Stationery D. LPC: SIU 91.26.

La Salle, Ills. August 1, 1910.

Mr. Chas. S. Peirce,
Milford, Pa.
My dear Mr. Peirce:

I have been waiting in vain for a review of your papers "Illustrations of the Logic of Science". You had even forgotten that I had paid you for them, and I am afraid you have again forgotten that after some reluctance you had acknowledged your promise. It was for sundry considerations including the right of re-publishing your Logical papers, revising and proof reading them, writing a new preface, if you were so minded, etc., that I paid you in April 1909 the sum of $ 250. You have not yet revised your articles and I think that I might publish your articles as they stand. They may contain some objectionable points, which you do not mean to endorse but they are good *as they stand* and contain many valuable ideas. Perhaps it would even be sufficient if you would point out to me the passages which you would have altered and where as you stated in a former letter you yielded to Mr. Youmans's wishes,[276] but at any rate I should like to have your acknowledgment under your own signature that the Open Court Publishing Company has acquired the right of republishing these essays, in which statement you have also to insert that (as you

276 Peirce to Carus 7 Jan 1909.

1 to^{1}] *over* it
6 Youmans] Youman$^{\vee}$s$^{\vee}$
15 S] *typed over* P
28 Youmans's] Youman's *[E]*

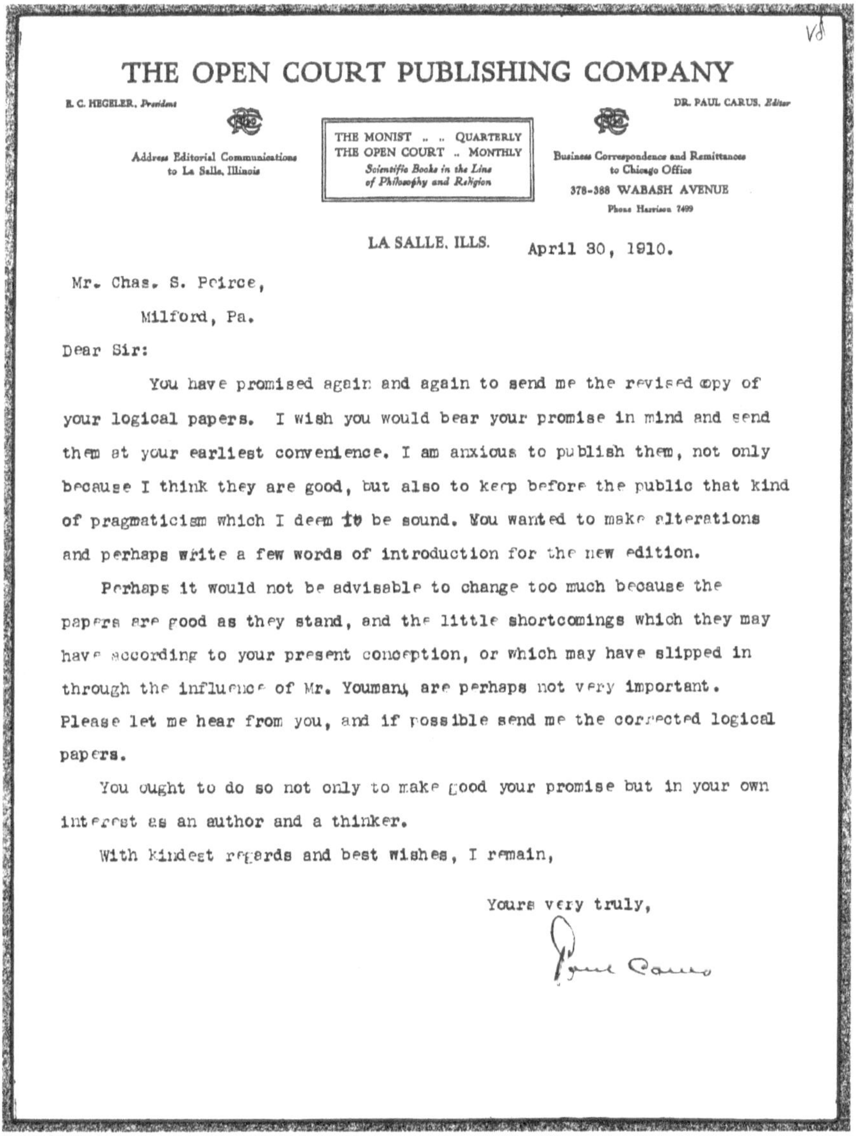

THE OPEN COURT PUBLISHING COMPANY

E. C. HEGELER, President
DR. PAUL CARUS, Editor

Address Editorial Communications to La Salle, Illinois

THE MONIST .. .. QUARTERLY
THE OPEN COURT .. MONTHLY
Scientific Books in the Line of Philosophy and Religion

Business Correspondence and Remittances to Chicago Office
378-388 WABASH AVENUE
Phone Harrison 7499

LA SALLE, ILLS. April 30, 1910.

Mr. Chas. S. Peirce,

Milford, Pa.

Dear Sir:

You have promised again and again to send me the revised copy of your logical papers. I wish you would bear your promise in mind and send them at your earliest convenience. I am anxious to publish them, not only because I think they are good, but also to keep before the public that kind of pragmaticism which I deem to be sound. You wanted to make alterations and perhaps write a few words of introduction for the new edition.

Perhaps it would not be advisable to change too much because the papers are good as they stand, and the little shortcomings which they may have according to your present conception, or which may have slipped in through the influence of Mr. Youmans, are perhaps not very important. Please let me hear from you, and if possible send me the corrected logical papers.

You ought to do so not only to make good your promise but in your own interest as an author and a thinker.

With kindest regards and best wishes, I remain,

Yours very truly,

Paul Carus

Figure 5: Carus to Peirce 30 April 1910, on OCP Stationery D. Eager request for Peirce's much-delayed "Illustrations" revisions.

informed me in a former letter) *you*, not Appletons, hold the copyright and are the sole owner of these said essays. *I wish to be on the* safe side in case Appleton or their successors or heirs would later on trouble me, and claim an infringement on their rights. Yet I may add that the copyright has either or will soon run out for the papers were published in 1878. I do not wish to seem to appropriate the article without a just title. I could prove my claims by submitting to any judge our correspondence, but the reference to these papers are scattered through several letters and I would prefer to have the statement made in *one* document; either in your own handwriting or under your signature. I enclose a form[277] which I wish you would sign and return.

Do you feel like writing a new preface to these papers? Perhaps when you see the proofs you would get into the spirit of saying something. It is not impossible that the papers would find the more recognition, the less attention is called to the fact that they were, first published more than thirty years ago.

With best wishes for your good health and hoping to hear from you, I remain,

Yours very truly,
Paul Carus

How would it do if you would mention the fact that you first excogitated these thoughts and propounded them more than thirty years ago and otherwise no mention of their former appearance were made?

Carus to Peirce. LPC: SIU 91.26. Previous, unsent draft.

*[*Unsent draft of 1 Aug 1910 letter*]*

July 27, 1910.

Mr. Chas. S. Peirce,
Milford, Pa.
My dear Mr. Peirce:

I have been waiting in vain for a revision of your papers "Illustrations of the Logic of Science". You had even forgotten that I had paid you for them $ 250, and I am afraid you have again forgotten that after some reluctance you had acknowledged your promise. In the meantime I think I might publish your articles as

277 See Appendix.

2 sole] sol~~t~~e
7–8 the reference … letters and] *inserted above*
18–20 How would … were made?] *This paragraph added in pen.*
19 more than] *inserted above*

they stand. They may contain some objectionable points, which you do not mean to endorse but they are good as they stand and contain many valuable things. Perhaps it would even be sufficient to point out to me the passages which you would have altered, but at any rate I should like to have your acknowledgment in your own handwriting that the Open Court Publishing Company has acquired the right of republishing these essays, in which statement you have also to insert that you hold the copyright and are the sole owner of these said essays. I wish to be on the safe side in case Appleton or their successors or heirs would later on trouble me, and claim an infringement on their rights. I could prove their claims by submitting to any judge our correspondence, but I would prefer to have the whole statement made in one letter from you, either in your own handwriting or under your signature.

Hoping to hear from you, I remain with kindest regards,

Yours very truly,

Peirce to Carus. MS: RL 77. Brief letter enclosing his long and much-delayed response below to Carus's repeated requests for Peirce's "Illustrations" revisions. Also encloses Peirce's own statement of copyright, included in the Appendix.

Milford Pa. | 1910 Aug 26

My dear Doctor Carus:

When I came to look at what you wanted me to sign, I found it did not quite agree with my understanding. I have therefore written out my own, which I trust will satisfy you. If not, I beg you will send it back to me saying how you wish it changed.

I send you to day a long letter that would have gone two days ago, but I was not equal to going out to the road & waiting for the mail. I can't stand still long yet, though I am, I think, getting health & strength

Ever faithfully
CSPeirce

Peirce to Carus. MS: RL 77. TS1 (copy for Peirce): RL 77. TS2 and TS3 (copies for Carus and Russell): SIU 91.27. MS enclosed with letter above. Also in full in de Waal, 271–290. Long-awaited response to Carus's requests for "Illustrations" revisions. Explains the difficulties Peirce has had in revising the work and agrees to simply add a preface to the original articles stating the alterations he would have made, which are described here. Letter written in at least three sittings, benchmarked in the letter at two points, "Aug 23" and later simply "Aug", and finally sent 26 Aug 1910 with the

1 which] *above* ~~but~~
22 send] *after* ~~let me~~
24 would] *over* s

preceding letter. Two unsent drafts also follow, one written before this (19 July) and the other during the course of it (ca. 20 Aug).

Three TS copies of this letter were later made by Russell, one for Peirce, Carus, and himself, presumably in preparation for it or parts of it to be included in the book as the preface. The first four pages of this letter, roughly the first three paragraphs, are missing from Peirce's original MS in RL 77, so the text comes from TS1, the copy provided to Peirce.

[26 August 1910]

My Dear Doctor Carus:

Ever since I was paid that money by you and Mrs. Carus, I have been engaged with all my energy, allowing only for such as I had to expend upon my wife's health and upon getting this house habitable and in salable condition, in trying to write an article or articles[278] for you upon the second grade of clearness, i.e., that which results from analytic definition and upon corrections to the errors and other faults of the articles of mine that appeared in the Pop. Sci. Monthly in 1877 and 1878 to which I should be glad if you would add a reprint of the article of Jan. 1901[279] which requires no correction.

I have written a great deal but am satisfied with but the smaller part of it, and the result is that I worked, until,—what with worries, too,—I was in a state of downright nervous exhaustion from which I have now been for a good number of months recovering, but owing to the decay of my powers and being so sick of writing about the same things, I cannot even yet write but a few equivalents of pages of this letter very slowly and for not over three hours and with many days when I can do nothing. Since I got your letter I have—often trying in vain to accelerate my speed,—the slowness of which is largely due to the labor of writing with the extreme precision required, gradually been forced to the conclusion that since you are very reasonably impatient, my best course is simply to write a preface[280] in which I state in general terms how what I then say ought to be altered, and I will here (if my strength holds out) try to indicate the points I should make.

In regard to the first essay consisting of the first two articles, the principal positive error is its *nominalism*, especially as illustrated by what I said about Gray's

278 "Studies of Logical Analysis" and others on "Definition" in R 643–650, ca. 12 Dec 1909 to 5 Aug 1910 (also identified in de Waal at this point in the letter).

279 "Pearson's *Grammar of Science*"

280 For a draft of Peirce's proposed preface, see Appendix, "Peirce and Carus."

14 to] *inserted above*

21 yet write] write yet

29 essay] *E*essay

29–30 the principal positive] *Beyond this point the remainder of the text and variants come from Peirce's original MS in RL 77.*

stanza “Full many a gem” etc pp 300 et seqq.[281] I must show that the *will be*’s, the actually *is*’s and the *have been*’s are not the sum of the reals. They only cover actuality. There are besides *would be*’s and *can be*’s that are real. The distinction is that the *actual* is subject both to the principles of contradiction and of excluded middle; and in *one* way so are the *would be*’s and *can be*’s. In *that* way a *would be* is but the negation of a *can be* and conversely. But in another way, a *would be* is not subject to the principle of excluded middle; both *would be X* and *would be not X* may be false. And in this latter way a *can be* may be defined as that which is not subject to the principle of contradiction. On the contrary if of anything it is *only* true that it *can be X* it *can be not X* as well.

It can certainly be proved very clearly that the Universe does contain both *would be*’s and *can be*’s.

Then in regard to the second article, I ought to say that my three grades of clearness are *not*, as I seemed then to think, such that either the 1st or the 2nd are superseded by the third although we may say that they are acquired,—*mostly*,—in the order of those numbers. I ought to describe, if only in a paragraph, how to train oneself and one’s children in the first grade of clearness, so that, for example, one will recognize a millimetre length when one meets with it; and so with colors. I have done a great deal of work in training myself to this kind of clearness. It would if put together amount to two or three years of industry; and I should recommend systematic exercises of the sort to everybody. Useful as that is, however, I don’t hesitate to say that the second grade of clearness is far more important, & all my writings of late years illustrate that. Still, I continue to admit that the 3rd grade is the most important of all & a good example of it is Wm. James who is so phenomenally weak in the second grade, yet ever so high above most men in the third. But there is no reason why all three should not be symmetrically developed.

The bulk of these Pop. Sci. articles, after the first two, are occupied with a criticism of the underlying principles of Laplace’s *Théorie analytique des probabilités* and Mill’s *System of Logic*,—two works of a high order which have had and still have a great a deplorable influence.

Before the 3rd Article on Probability I should like to insert a short and easy account of my Existential Graphs; because when that system is well in hand, it

281 “How to Make Our Ideas Clear,” *PSM* (1878), pp. 300–302, beginning “But I may be asked ...”

1 many] man *[E]*
4 contradiction] con-*[line break]* ~~tradistincti~~ tradiction
6 a *would be*] \a/ *would be~~s~~*
9 of^2] *inserted above*
11 proved] *after* ~~shown~~
25 ever so] *above* ~~decidedly~~
26 why] wh*y*ich

becomes so much easier to show great faults of Laplace and Mill; and that shorter account, *I could now easily write.*

It would also be well to show how all numbers involve *essentially* nothing but ideas of *succession*. Then I should like to point out how utterly Laplace fails to define what he means by *probabilité*, his account of it resting upon what he calls the *également possibles* which I maintain has none but the vaguest meaning. I ought on my side to define *probability*. For that purpose, I should have to begin by distinguishing three ways—three quite different *directions* so to speak, as different as the *X*, *Y*, *Z* of a system of orthogonal coördinates—in which cognitions can fall short of absolute certainty,—or rather of *mathematical* certainty, which is not absolute, because blunders may have been committed in reaching it.

The names which I would propose for general adoption for the three different kinds of *acceptability* of propositions are

plausibility
verisimilitude[282]
probability

The last alone seems to be capable of a certain degree of exactitude of measurement.

By plausibility, I mean the degree to which a theory ought to recommend itself to our belief independently of any kind of evidence other than our instinct urging us to regard it favorably. All the other races of animals certainly have such instincts: why refuse them to mankind? Have not all men some notions of right and wrong as well as purely theoretical instincts. For example, if any man finds that an object of no great size in his chamber behaves in any surprising manner, he wonders what makes it do so; and his instinct suggests that the cause, most plausibly, is also in his chamber or in the neighbourhood.

It is true that the alchemists used to think it might be some configuration of the planets; but in my opinion this was due to a special derangement of natural instinct. Physicists certainly today continue largely to be influenced by such plausibilities in selecting which of several hypotheses they will first put to the test.

By *verisimilitude* I mean that kind of recommendation of a proposition which consists in evidence which is insufficient because there is not enough of it, but which will amount to proof if that evidence which is not yet examined continues to be of the same complexion as that already examined, or if the evidence not at

282 Left marginal note pointing here reads: "I am not satisfied with this. I would rather suggest the phrase 'So it would appear.' "

4 ideas] *after* ~~th~~
8 ways] *before* ~~in~~
20 instinct] *after* ~~sen~~
24 of] *after* ~~which~~

hand and that never will be complete should be like that which is at hand. All determinations of probability ultimately rest on such verisimilitudes. I mean that if we throw a die 216 times in order to ascertain whether the probability of its turning up a six at any one throw differs decidedly from 1/6 or not, our conclusion is an affair not of *probability* as Laplace would have it, by assuming that the antecedent probabilities of the different values of the probability are equal, but is a verisimilitude or as we say a "*likelihood.*" That Laplace is wrong can be demonstrated since his theory leads to contradictory results. But perhaps the easiest way to show it is wrong is to point out that there is no more reason for assuming that all the values of the probability are equally probable than for assuming that all the values of the *odds* are so; or that all the values of the logarithms of the odds are so, since this is our instinctive way of judging of probabilities; as is shown by our "balancing the probabilities."

Having thus defined *plausibility* and *verisimilitude*, I come to define probability. None of the books contain a definition of mathematical probability (which is what I mean by "probability," however measured) which will hold water. For the sake of simplicity, I will define it in a particular example. If, then, I say that the probability that if a certain die be thrown in the usual way it will turn up a number divisible by 3 (i.e. either 3 or 6) is 1/3, what do I mean? I mean, of course, to state that that die has a certain habit or disposition of behaviour, in its present state of wear. It is a *would-be* and does not consist in actualities or single events in any multitude finite or infinite. Nevertheless a habit does consist in what *would* happen under certain circumstances if it should remain unchanged throughout an endless series of actual occurrences. I must therefore define that habit of the die in question which we express by saying that there is a probability of 1/3 (or odds of 1 to 2) that if it be thrown it will turn up a number divisible by 3 by saying how it *would* behave if, while remaining with its shape etc. just as they are now, it *were to be* thrown an endless succession of times. Now it is very true that it is quite impossible that it should be thrown an infinite succession of times. But this is no objection to my supposing it, since that impossibility is merely a physical, or, if you please, a metaphysical one, and is not due to any logical impossibility to the occurrence in a finite time of an endless succession of events each occupying a finite time. For when Achilles overtook the tortoise he had to go through such an endless series (endless *in the series*, but not endless *in time*) and actually did so; Or if he didn't another did.

---

1 and] *over* cre
1–2 determinations] *after* ~~pr~~
3 we] we we *[E]*
3 216] *above* ~~3600~~
27 with] *over* in

Very well, I will further suppose that tallies are kept during the throwings, one tally of the throws turning up 6 or 3, and the other tally of the throws turning up 1, 2, 4, 5; and further I will suppose that after each throw the number that the latter tally has reached shall be divided by the number that the former tally has reached. And I will use the expression that this quotient changes its value at every new throw, instead of saying that a new quotient differs from the last. When the quotient changes from being greater than 2 to being less than or greater than 2 to being either just 2 or on the opposite side of 2 to what it was before, as for example if it passes from being 21:10 to being 21:11 or from 21:11 to 22:11 or from 25:12 to 25:13 etc. I shall say it *touches* 2 (meaning strictly that it either comes to 2 or passes across 2). Then after the first throw it will be either 0 or ∞ and there it may remain for any number of throws. But after it has once moved away it never will return to either of these values, but after it has finally recovered from the effects of the first throws it will oscillate in a very irregular way, and soon it will "touch" (or pass over) some other values *for the last time* although nobody can *know* that it is to prove to have been for the last time; and then values still nearer to 2 will be touched or traversed for the last time. And in its endless series there will be no value except only 2 that it would not touch or traverse for the last time *excepting* only the value 2. And this "would be" is what constitutes the habit which we state in saying that the odds against its turning up a number divisible by 3 are 2:1, or that the probability of its turning up a 6 or a 3 is $1/3$.

Aug 23

I could not write before. Now there are two points to be particularly noticed about the definition of probability. The first is that the probability of an event may be an *irrational* number. In that case *every* value of the quotient of one tally divided by the other will get finally left (so that the quotient never again touches or traverses it). Some will be left because they are too small and others because they are too large. But any one you please will get left for one or the other of these two reasons. This must be taken into account in drawing up the definition.

---

2 turning] *after* ~~giving~~
7 being less than] being ~~less~~ \~~not greater~~ /~~less\ / than 2 (the ratio changing say from 21:10 to 21:11) or changes the other way~~ *[(first revision)* being ~~less~~ \not greater/ than 2 *[. .]* other way *(second revision)* being ~~less~~ \~~not greater~~ /less\ / than 2 *[. .]* other way *(third revision shown in letter)]*
8 of] *after* ~~from~~
10 comes to 2] *above* ~~touches~~
17 there] *after* ~~it will~~
18 would] *above* ~~will~~
19 And] *after* ~~If however the proba~~

The other point is that our throws and tallies will give the probability in question only on condition that the fact that at certain throws (determined by their ordinal places in the series) had turned up, such and such of them numbers divisible by 3, and the rest of them numbers not of that kind, will make absolutely no difference in the probability that any given throw (determined as above) will turn up a number divisible by three. This statement, of course, refers to what would happen if a series of throws long enough to cover the highest *in order* of throwing of the determinate throws were to be repeated *endlessly*.

In short that the numbers turned up are each absolutely independent of what has been or will be turned up on other occasions. This comes to saying that the tallies need not be of consecutive throws but may be every truth or any other endless series so long as whether a given throw be counted on the tally or not is nowise influenced by whether that throw turns up a number divisible by three or another kind of number. The definition must be made explicit upon this point.

[This latter condition will make quite a puzzle for those who deny absolute chance; yet experience proves (*far* more certainly than it proves determinism), that this condition can be fulfilled *in all cases*.]

The definition of probability resulting in this way.

Now from the above definition of probability together with the ordinal definitions of

1. The whole numbers, as expressed in the secundal system in which

10	means	2
11	"	3
100	"	4
101	"	5
110	"	6
111	"	7
1000	"	8
	etc.	

But of course this description is not a definition.

2. Of addition
3. Of multiplication
4. Of involution
5 Of ratio

---

3 numbers] *after* ~~d~~
7 *in order* of throwing] *inserted above*
10 saying] say ~~th~~ing
11 need] *after* ~~can~~
12 whether] *after* ~~no~~
30 But] By *[E]*

6. Of the rational positive values as those which result from putting

$$N : 1 \equiv N$$

whatever whole ordinal number (including 0) N may be. And such ratios I call the ratios of the zero grade

And from running through any complete series of all the ratios of any given *rational grade* from any ratio $N : 1$ to $N + 1 : 1$ inclusive and putting between every successive pair of ratios of grade m, (which two we may denote by $p : q$ and $r : s$) a new ratio

$$(p + r) : (q + s)$$

and all the ratios of grade m together with all so inserted constitute the ratios of grade $m + 1$.

And all the ratios of all finite grades are all the positive rational values.

7. And if $(p : q)(r : s)$ are two ratios of the same grade, whether successive or not, then if they have different values and $p : q$ is less than $r : s$ then $p : q < (p + r) : (q + s) < r : s$. But if $p : q \nless r : s$ then $p : q \nless (p + r) : (q + s) \nless r : s$. And all ratios *of real grades* are real values and two real values only differ by what other real values one higher (or lower) than while the other is not. From these definitions I will deduce every valid rule of the doctrine of chances.

I will then show that Laplace's definition of probability, $1^{\underline{st}}$, has no meaning at all unless it be the expression of a subjective state of mind, and $2^{\underline{nd}}$, that necessary, and quite easily perceived, consequences which can be drawn from his own explanations and expositions of it are revolting to common sense.

---

3 ordinal] *inserted above*
3–4 And such … grade] *inserted in-line and between paragraphs*
5–6 ratios … *grade*] ~~ordinals~~ \ratios/ of any given *rational*/ *grade* ~~and~~
7 and] *after* ~~e~~
12 positive] *inserted above*
13 two ratios of the same] two ~~successive~~ ratios of ~~any~~ \the/ same
14 and] *above* ~~$p + r$~~
15 $(q + s)$] $(q : s)$ *(here and in the next equation) [TS1, TS2, TS3]*
16 *of real grades*] *inserted above*
16 two] *after* ~~rea~~
17 while] *after* ~~and~~
20–21 that … perceived,] that ~~the~~ necessary$^\vee$,$^\vee$ \and quite easily perceived,/

In order to get this matter straightened out, I think it would be well to change the place of the Sixth Paper and place it directly after the third.[283] Then I would append a *Correction* in which I should state that the division of the elementary kinds of reasoning into three heads was made by me in my first lectures[284] and was published in 1869 in Harris's Journal of Speculative Philosophy.[285] I still consider that it had a sound basis. Only in almost everything I printed before this century (?) I more or less mixed up Hypothesis and Induction. *Hypothesis* is, however, the most expressive term as well as the best supported by the historical usage of logicians,[286] generally, for what I had in mind (rather confused by my unsound reflexions upon it.)

The general body of logicians had also at all times come very near recognizing the trichotomy. They only failed to do so by having so narrow and formalistic a conception of inference (as necessarily having formulated judgments for its premisses) that they did not recognize Hypothesis (or, as I now term it, *retroduction*[287]) as an *inference*. The thing was always recognized & all logicians *almost* (perhaps all, though the stoics and others denied the validity of induction) distinguished *induction* from necessary reasoning.

When one contemplates a surprizing or otherwise perplexing state of things (often so perplexing that he cannot definitely state what the perplexing character is) he may formulate it into a judgment or many apparently connected judgments, he will often finally strike out a hypothesis, or problematical judgment, as a mere possibility, from which he either fully perceives or more or less suspects that the perplexing phenomenon would-be a necessary or quite *probable* consequence.

That is a *retroduction*. Now three lines of reasoning are open to him. He may proceed by mathematical or syllogistic reasoning at once to demonstrate that consequence. That of course will be deduction.

Or he may proceed still further to study the phenomenon in order to find other features that the hypothesis will *explain* (i.c. in the English sense of explain, to

---

283 Meaning, to move "Deduction, Induction, and Hypothesis" between "The Doctrine of Chances" and "The Probability of Induction," as later recommended in *CP* 8:227n.
284 His 1865 Harvard Lectures on "The Logic of Science" (*W* 1:162–357).
285 "Grounds of Validity of the Laws of Logic," although Peirce may have had in mind all three articles in the series 1868–1869.
286 See Peirce's extensive footnote in "Some Consequences" (*W* 2:218f) on its historical usage.
287 Later "*abduction*."

---

12 having] *above* ~~giving~~
13 formulated] *after* ~~ad~~
20 judgment] judment *[E]*

deduce the facts from the hypothesis as its necessary or *probable* consequences.) That will be to continue reasoning *retroductively*, i.e. by hypothesis.

Or, what is usually the best way, he may turn to the consideration of the hypothesis, study it thoroughly and deduce miscellaneous observable consequences, and *then* return to the phenomena to find how nearly these agree with the actual facts. This is not essentially different from *induction*. Only it is (most usually) an induction from instances which are not discrete and numerable. I now call it Qualitative Induction. It is this which I used to confound with the second line of procedure, or at least, not to distinguish it sharply.

A good account of Quantitative Induction is given in my paper in Studies in Logic by Members of the J. H. Univ.[288] and its two rules are there well developed. But what I there call Hypothesis is so far from being that that it is rather Quantitative than Qualitative Induction. At any rate, it is treated *mostly* as Quantitative. Hypothesis proper is in that paper only touched upon in the last section.

There is a third kind of Induction. In order to show this, it is requisite to define Induction.

Now the essential character of induction is that it infers a *would-be* from *actual singulars*. These singulars must, in general, be finite in multitude and then, as I show in my J. H. paper, the inductive conclusion can be (usually) but *indefinite* and can *never* be *certain*. Even if the series of singular instances, as in the case of Achilles and the Tortoise be known to hold *without end*, a special postulate is requisite to make the conclusion hold. [It is a distinguished merit of the strictly ordinal view of fractions that it brings this point out in perfect clearness, in conjunction with F. Klein's doctrine of Measure (in one of the early volumes of the Math. Annalen).[289] For there may be a discontinuity at that very point, so that, e.g.

0.111111̣ (continued without end,) does not amount to 1/9, just as if 1/9 and all following numbers were shoved along, which would not derange the arithmetic at all if the definitions are strictly ordinal.]

But in ordinary cases an induction would become both *precise* and *certain*,— though even then it would not be apodictic certainty,—if the instances were of de-

288 "A Theory of Probable Inference," pp. 126–181.
289 "On the So-Called Non-Euclidian Geometry," *Mathematische Annalen* (1871).

3 usually] *inserted above*
5 find how] *see*find ~~what~~ how
6–7 (most usually)] *inserted above*
9 procedure] proce*e*dure
11 there] *over* w
21 known] *after* ~~app~~
23 it] *after* ~~Klein's~~
29 are] *after* ~~were~~
30 an] *after* ~~we should~~

numeral (or simply endless) multitude. Therefore, defining induction as the sort of inference which produces verisimilitude or likelihood, that is which regards an endless series of *actualities* as conclusive evidence of a *would be* since it is the best evidence possible when we are not behind the scenes.

From this point of view any plausible proposition that is supported by instances in every respect is justifiable so long as one keeps on the alert for the first exception. Of course, such an induction has the very minimum of likelihood yet it has *some*; and we very often find ourselves driven to accept it. The world has always turned on its axis so far as we know about once every 24 hours & therefore we presume (vaguely) that it always will continue to do so. In every case that has been sufficiently inquired into, every human being has been born of a woman not a maiden. So almost everybody feels sure it always will be found so. People are far more confident of it than they have any right to be. All former generations of men have died off. Therefore, people say, they always will. In one sense I suppose this is certain. But that they always *would* even if there were no accidents, seems to me as weak an inference as any that I would not positively condemn as utterly worthless. I call this kind of thing *crude induction.*

I must confess that although my explanation of the validity of induction seems to me to be far superior to any other, I am not altogether satisfied with it, or rather with its results. Quantitative Induction depends upon the possibility of making a truly representative sample. That is to say, the examples composing it must be chosen as possessing the conditional character, which is easy enough, but also so that the choice of them shall not be influenced, one way or the other, by whether or not they possess the consequent character. They must be such on the whole in that course of experience to which the induction is to be applied. That, to be sure, cannot but be the case, should the entire class sampled be alike in respect to the consequent character. But the further this ideal state of things is from being realized the more extremely difficult it becomes to get a truly representative sample, and the result, after every precaution has been taken, is that we cannot expect any great precision in inductive conclusions when the class is anywhere near being equally divided between individuals that do and that do not possess the consequent character. However this is not owing to any *falsity* in

---

1 defining induction] ~~defining~~ \defining/ ~~*verisimilitude*, or *likelihood*, as the sort of approach to perfect certainty that <an> \a sound/ induction possesses,~~ induction
3 of *actualities*] *inserted above*
7 has] *above* ~~is~~
10 presume] *above* ~~suppose~~
10 In] *after* ~~Every~~
15 seems] *after* ~~I~~
18 explanation] *after* ~~account~~
20 or … results.] *.*, \or rather with its results./
23 also so] also *as*so ~~not possessing~~
24 character.] character*,*. ~~and~~

my theory, but to the essential imperfection of induction itself when applied to these cases.

As for the validity of the hypothesis, the retroduction, there seems at first to be no room at all for the question of what supports it, since from an actual fact it only infers a *may-be*,—*may-be* and *may-be-not*. But there is a decided leaning to the affirmative side and the frequency with which that turns out to be an actual fact is to me quite the most surprizing of all the wonders of the universe. I hope there are few men who are so often deceived as I. They surely cannot be many. Yet I could tell you of conjectures pronounced by me with a confidence that I could not comprehend, and that were verified amazingly. I simply should not dare to tell them; I feel my credit would not support such tales: I should be a fool to expect it. So I will not enter upon that. But I will point to some of the great steps of science. Faraday[290] was decidedly a cautious reasoner. What made him so strangely confident that electricity in all the variety of its workings, never acts at a distance? What an unaccountable thing was Dalton's atomic theory and its ready reception. What in the world should prevent,—even if there were atoms,—197 atoms of one substance from joining 719 of another, to make a chemical compound, both numbers being prime? That question prevented me, as a boy, from believing in atoms; and I was only convinced when new facts were discovered. How was it that Galileo, after glancing at one pretty evidently absurd hypothesis, at once struck upon the idea that the speed of a falling body increases uniformly with the time? I am ashamed at being obliged to confess that this volume contains a very false and foolish remark about Kepler.[291] When I wrote it I had never studied the original book as I have since. It is now my deliberate opinion that it is the most marvellous piece of inductive reasoning I have been able to find. J. S. Mill says it was not reasoning at all, but only description.[292] But that only shows that he had not the slightest idea

290 Michael Faraday (1791–1867), English scientist and magnetist, pioneered the concept of the electromagnetic field in physics.

291 In "Fixation of Belief," beginning "Kepler undertook to draw a curve through the places of Mars ..." (*EP* 1:110–111).

292 In *System of Logic*, Book III, Ch. II, Sect. 3, pp. 356–357, beginning "The error in question is that of confounding a mere description of a set of observed phenomena, with an induction from them...."

7 to me] *inserted above*
7 are] a *[E]*
14 in all ... workings,] *inserted above*
17 chemical] *inserted above*
21 speed] *after* ~~vel~~
23 as] *inserted above*
25 J. S. Mill says] $^{\vee}$J. S.$^{\vee}$ Mill *do*says

of the problem. There may be nineteen hypotheses in the book, though I cannot now say how I made out that number. But there are certainly nothing like that number to explain any one of the numerous circumstances of the motion of Mars which Keppler succeeded in reducing to order. The facts he had before him were simply observations of the place of Mars in the heavens at various dates. His problem, as Mill seems to suppose was not simply to draw a line on the celestial globe, through all these points, and to see that that line was an ellipse with the sun at one of the foci. For, to begin with, Mill seems to forget that Kepler told where Mars was at every time *in space* with its three dimensions, and not merely, what the *projection of its path* upon the celestial sphere was.

Aug

A pretty good example of a retroduction is the case of stopping an unknown person in the streets of a large city, asking some bit of information and crediting it. What makes this a more characteristic case is that one will have selected the man to interrogate, and will have made the selection almost without being conscious of having done so.

The justification of retroduction is that one *must* trust one's instincts through life or be content with a passivity that cannot content him.

All positive knowledge must be reached if at all by an operation that begins with a conjecture. I do not mean that one reflects upon this at the outset, I mean that afterward when one subjects one's behaviour for the day or for any other marked period to self-criticism this is how one ought to vindicate good retroductions.

As for deduction, Kant gave the principle of that. In the premisses the conclusion is, in substance, asserted.

The practical lesson that experience with retroductions ought to teach men,—especially *male* men,—is that there is much to be learned from one's instincts, and that consequently pains should be taken to let them develope in directions in which they prove to be sound, and to correct them in directions in which they have proved treacherous.

1 book,] book*.*, ~~But there~~
2 I] *after* ~~that~~
5 in] *after* ~~relatively to different stars at many di~~
15 made] *after* ~~don~~
16 of] *inserted above*
28 taken] take *[E]*

The universe which man inhabits, the Real Universe, is the Universe of his conclusions, such as they *would be*, if properly corrected and if duly enlarged by new experiences. This is the true Kantian doctrine.

Often we ought to put much faith,—or, at the very least, high hope,—in retroductions wholly unsupported by inductions—i.e. by positive evidence, (*positive* meaning pragmatistic, but with emphasis on the restraining tendency of experience). If one feels a strong inward tendency to hope for a future life of activity, work, further enlightenment, it is bad logic not to yield to it, while remembering that it is but a justifiable hope, and while on the watch to note whether it is really doing us good or not. Nothing has been a greater curse to a large part of the human race than religions that have been outlived. At the same time that they may be keeping the ruder part of the community from crime and ruin, they may be undermining the characters of a higher class, who ought to remember that they are of the stuff that dreams are made of.[293] That they have been worked over with a view to squeezing money out of the laity to support a despicable clergy. Fortunately no protestant congregation is yet rich enough to answer to this description. Whether or not the mormons, the "christian scientists,", the Romanists, the spiritualists do or do not, it is not my business to say. They are all loaded with gold, or keenly in quest of it. Religions are wholesome while they are *poor and persecuted.*

I want to put into the volume a sufficient refutation of Mill's system of Logic. To begin with it falls into the fatal error of Nominalism. Now how hard it is to form an exact and true definition of Nominalism, is shown by my own treatment of Gray's stanza ten years after I had declared myself a Realist (in the sense in which we call most of the scholastics realists,—or say "a scholastic realist.")

Nominalism might from the etymology of the name be supposed to consist in holding "would-be's" to be of the general nature of names.

But it ought not to remain long in doubt that a *would-be* is something that can be appropriately clothed in words or thoughts, that it is, or would be, *applicable* to singular actuals, and that it has a meaning, that is, is capable of definition, or logical analysis. Now these are the essential characteristics of a particular kind of *sign.*

A sign in general is 1st something Real that is 2nd applicable to an object different from itself and already known to the person to whom it is a sign, and 3rd, is capable of interpretation in the mind of that person, so that it will (or would if ac-

[293] *The Tempest*, Act 4, Scene 1.

1 Real Universe] *r*Real Universe
13 of^{2}] *inserted above*
17 "christian scientists,"] ∨"∨christian <">scientists,"
30 the essential] *inserted above*
31 1st something … applicable] ∨1st∨ something \Real/ that is \$2nd$/applicable
33 in] *above* ~~to~~

cepted as veracious sign) have some effect upon him of a kind that was calculated or fit to have.

Signs in general I divide in ten ways.[294] One of these divisions is according to the substance, or mode of being, of the sign itself. For a sign may be, 1st an existent thing or actual occurrence, 2nd it may be a would-be, or habit, 3rd, it may be a mere *can-be*.

Then there are 3 divisions that relate to the Object. One according to the form under which the Sign presents its Object. This is of course the object *as the sign represents it*, i.e. the Immediate Object. There are two divisions that concern the Real Object. One according to the Mode of Being of it, according to which the Sign will be Abstract, Concrete, or the Sign of a Habit, or Would be. The other according to the relation, or connexion between the Sign and the Real Object, making the Sign either an Icon, an Index, or a Symbol.

There are then 6 divisions of signs that refer to the Interpretant. Now we can regard the Interpretant in 3 ways, or rather there are 3 distinct things which may properly be regarded as the Interpretant. For any thing that the sign, *as such*, effects may be considered as the Interpretant. And this may be 1st something merely subjective, the vague determination of consciousness effected by the sign, 2nd the *actual event* that some signs by virtue of really acting as such bring about. For instance, let the sign be a military word of command. Then the instant action of the whole rank of men addressed will be the Dynamical Interpretant, as I call it. The third sense in which we may properly speak of the Interpretant is that in which I speak of the Final Interpretant meaning that Habit in the production of which the function of the Sign, as such, is exhausted.

I only make use of one division according to the 1st of the above, the Initial Interpretant, as I call it. This relates to the number of independent respects in which the sign is indefinite, that is respects which the sign points out but leaves indefinite. For example, the main difference between the verbs "is beneficial," "benefits," "gives," and "sells" is of this nature.

I recognize 2 distinctions referring to the Dynamical, or Actual, Interpretant. First as to its Nature as Feeling, Action, or Affecting habit, and then as to the manner in which the sign makes its appeal, as Imperative, or calling for Assent, or submitting to consideration. Then I take account of 3 divisions referring to the Final

294 For more details, see especially "Nomenclature and Divisions of Triadic Relations" (*EP* 2:289–299) and Peirce to Lady Welby and William James 1906–1909 (*EP* 2:477–502).

28 main] *over* d
32 as] *after* ~~as merely exciting an Idea, or~~
33 referring] *after* ~~of~~

Interpretant. 1st according to the nature of its Purpose, 2nd according to the nature of the influence the sign is intended to exert, and 3rd according to the nature of the assurance the sign affords to an interpreter that has rightly apprehended it. The last 3 divisions are especially logical in the ordinary idea attached to that word.

Of course there are any number of other distinctions which a student of signs may occasionally have to attend to; but I think that these ten are as many as an ordinary Intellect can carry ready for.

Peirce to Carus. MS and TS (copy): RL 77. Ten-page incomplete draft of the letter sent 26 Aug 1910.

*[*Unsent draft 1 of 2, of 26 Aug 1910 letter*]*

1910 July 19

My dear Doctor Carus:

I now will begin the explanation of the corrections I deem indispensable in the reprint of my Pop. Sci. papers.

The first two[295] constitute one Essay upon what I call *Pragmatism* or the doctrine of *pragmatic clearness* which consists in a clear conception of that *Habit of Conduct* in which any given concept would work out its actualization.

I decide upon more mature reflexion that it will be best to reprint this, just as it originally appeared with the following 4 exceptions

1st, Minor changes to render the English etc better.

2nd, The Articles of 1877 Nov. and 1878 Jan. to have a common title, say "Pragmatic Clearness of Thought."

And be divided into two parts. Namely, the two articles, each with its original title.

3rd, With a prefatial Note setting forth what I conceive to be the main error of the Essay, as it originally appeared. And this I will here state, writing only on one side of the paper, so that, if I should suddenly die in one of the fits to which I am almost daily subject, what I write would be set up & printed.

The error of the Essay lies in its *Nominalism*. Now I should have been not a little surprised and shocked, had I been told I was a nominalist; and the same is true of many other students of philosophy. Few discern just what it is that consti-

295 "Fixation of Belief" and "How to Make Our Ideas Clear," also the "1877 Nov. and 1878 Jan." articles referenced in the next paragraph.

12 indispensable] indispensible *[E]*
16 in] *after* ~~to~~
17 decide] *after* ~~com~~
18 4] *over* 3
22 divided] divide *[E]*
24 a] *after* ~~an~~

tutes "nominalism." Ockham, for instance, would have defined it as the doctrine that "universale est intentio animae nata praedicari de pluris,"[296] I suppose. If he had said that to me, I should have told him that though I could not admit that that defined nominalism, yet what was probably meant, though not distinctly said, is *a* nominalistic opinion. But to make sure I should like to know what he meant by an "*intentio animae*," or *animi*, or *mentis*.[297] If he said it meant *ipse actus intelligendi*,[298] then I should say, *that* is clearly nominalistic; and if he would omit some of the limitations and simply say "nominalismus (a word he never heard of) est opinio quod omne universale sit *actus* natus praedicari de pluribus,"[299] *that* I would admit to be a correct definition of nominalism. Perhaps it strikes you (or more likely some less mature student) as a bit audacious for me to undertake to instruct Ockham as to what nominalism is. But in reply to that I should simply remark that if every philosopher were able to state accurately just what characterize his own way of thinking, the principal difficulty of the study of philosophy would be overcome. Besides, Ockham perfectly recognizes that his own doctrine is only one among several varieties of the "sententia de vocibus,"[300] or nominalism. I found my opinion as to what is the essence of nominalism upon deep meditation upon all that has come down to us of the views of the different parties in the great controversy of the XIIth century, without neglecting what had previously been written both among the ancient Greeks and among Mohammedans & Christians (though I admit I don't read Arabic and have no first hand knowledge of Arabian philosophy.) And I have added to those meditations others on modern philosophies, the result of all which has been to impress me with the supreme importance of the questions that vexed the XIIth century,—not such psychological distinctions as that between *nominalism* and *conceptualism*, which though truly very important absolutely, is *relatively* the merest trifle.

The greatest and most decisive of all questions of metaphysics, is, according to the way of thinking which has gradually ripened in me since 1873 (which is

296 "a premise of the mind is universal, designed by being predicated upon many things."
297 "*premise [*or *idea] of the mind*" or of the "*intellect*" or of one's "*reason*."
298 "*the very product [*or the product itself*] of comprehension*."
299 "Nominalism is the opinion that every universal is a *product* created by being predicated on many things."
300 "thought by words" (i.e., expression of opinion).

2 "universale] *after* ~~a universal~~
3 told] *after* ~~said~~
9 *actus*] *before deleted* ,"
10–11 (or … student)] *inserted above*
13 philosopher] phi-*[line break]*~~lop~~ losopher
17 found] *above* ~~form~~
23 has] *after* ~~is to~~

the date of my formulating the opinion expressed in the two articles that are the two parts of that Essay on Pragmatistic Clearness), is the question Whether generals are *real*; which question I would like to phrase as Whether anything *Indeterminate* is Real (since we define an Individual, meaning what is the contrary of a General, as that which is *omnimodo determinatum*), and which question I would supplement by asking Whether anything Indefinite is Real. The difference I make between the Indeterminate and the Indefinite is the difference between *Anything* and *Something*,—between *Irgendein* and *Etwas*, *Ce que ce soit* and *Quelque chose*, *quodlibet* and *quoddam*, ὅ τι and τὶ.[301] Only all the first ones of these pairs of pronouns, by *emphasizing* the Indeterminacy, convert it into a Positively *determinate* indetermination, in place of the native indeterminacy that I mean.

I think of the distinction from yet another point of view. Pragmatism (pragmaticism) might be defined as that mode of thinking that never results in a concept that is equivalent to a noun substantive, since all that it thinks is an assertion, or a qualified assertion such as "Suppose that (so and so,)," or else it is equivalent to an expression of the speaker's will not *asserting* it so much as *exhibiting* it. Now these propositions *de inesse*, affirmations of actuality or of existence, and their denials belong to the very same mode. To deny the actuality of one predicate is to affirm the actuality of another. Such an assertion represents a state of things as being both determinate and definite. It acknowledges both the principle of contradiction and that of excluded middle.

But the assertion of a *Universal Law*, does not affirm any predicate as actual, it only says what *would be*, under a specified condition. It does not acknowledge the Principle of Excluded Middle. For it does not assert even so much as the possibility of the condition. It does not give assurance that it is not absurd; and the reduction *ad absurdum* runs thus:

If A then B

But If A then not B

Ergo in any case A *would not* be.

---

301 German, French, Latin, and Greek variations of indefinite and definite qualifiers relating to *any/anything/whatever* vs. *some/something*.

---

1 my] *after* ~~the~~
9–10 Only all … pronouns,] Only all ~~three~~ \the first ones of the these pairs of/ pronouns, these] the these *[E]*
15–16 equivalent to] *inserted above*
18 predicate] *above* ~~thing~~
22 predicate] *after* ~~thin~~
24 Excluded] *e*Excluded
24 it does] *above* ~~the cond~~
25 does] *after* ~~wil~~

To deny an actuality is to affirm an actuality. But to deny a *would be* is not affirming any would be whatever, it is only affirming a *might be*, or *can be*, which is a very *[…]*

Peirce to Carus. MS: RL 77. Second unsent draft. The "two of several longer letters" Peirce intended to enclose possibly refer to the draft of 19 July and what to that point was written of the letter (perhaps preceding the "Aug 23" cutoff) that he decided after all to send and "inflict upon" Carus 26 Aug 1910.

*[*Unsent draft 2 of 2, of 26 Aug 1910 letter*]*
*[*ca. 20 Aug 1910*]*

My Dear Doctor Carus

I enclose *two* of several longer letters I have written to you but decided not to inflict upon you. The first of these I send is short comparatively & I hope you will read it as it explains my attitude & what I have been doing.

I am *greatly* in need of a long rest. But if you want to make haste with the publication of my Pop. Sci. Monthly papers, I want (as the best I can do) to append to them an Explanation of how I now view the same subject, this explanation to be about the length of the dozen sheet letter I now send. And if you like, please send me back those dozen sheets and I will lick them into shape and send to you to be copied and printed, I to correct the proofs & the pages of my copy to be sent with the proofs so that I may do so. For sometimes your printers have sent matter from which I could not, until long afterwards, guess what it was that I had written. So I want my copy to accompany each proof.

most truly yours
C.S.P.

Of course I will sign that paper[302] & forward it. I can't at the moment lay hand on it.

Carus to Peirce. TS: RL 77. On OCP Stationery D. LPC: SIU 91.26.

La Salle. Ills. August 29, 1910.

Mr. Charles S. Peirce,
Milford, Pa.
My dear Sir:

302 Copyright permission form enclosed in Carus to Peirce 1 Aug 1910. Included in Appendix.

20 sometimes] *after* ~~often~~ 22 my] *after* ~~copy~~

Your letter and enclosure came duly to hand this morning and I hasten to acknowledge their receipt and express to you my best thanks. I have not yet been able to read your enclosure amounting to 52 pages, but I expect to do so at my earliest convenience. In the meantime I will thank you for your kind words, written in behalf of Mr. Hegeler.[303] You shall hear from me again as soon as I have read the entire MS. on your logical ideas. I may not be able to publish your logical papers in the near future, I only wish to be ready to do so whenever the time may arrive. I have in preparation an English edition of Couturat's Logic[304] and when I see the attempts of others to deal with the logical problem I wish you not to be forgotten, hence my anxiety to be ready to bring you before the public as soon as the time comes. At present our funds are low and I have to economize.

With kindest regards and best wishes, I remain,

Yours very truly,
Paul Carus

Carus and Russell 1–3 Sept 1910 (LPCs and MS: SIU 91.7): Carus and Russell correspond regarding Peirce's long instructive letter and agree to have Russell edit the "Illustrations" papers accordingly, without waiting for any more input from Peirce for fear "that Peirce may die at any time."

Carus to Russell 7 Nov 1910 (LPC: SIU 91.17): Not having heard from Russell in a long time, asks eagerly of the status of the work on the "Illustrations" revisions.

Carus to Peirce. LPC: SIU 91.26.

November 22, 1910.

Mr. Charles S. Peirce,
Milford, Pa.
My dear Sir:

I am rather disappointed that nothing is doing concerning the republication of your logical papers. I handed it over to Mr. Russell, and he is anxious to do the work. In my opinion he is well prepared for it for he has read every line you have written, and is most enthusiastic admirer of yours. He is not only attractive, but also so far as I can judge intelligent and well posted. He will accomplish the work by adding footnotes and references, to the text and when he is done the re-

303 Reference unclear.
304 *L'algeèbre de la Logique* (1905), trans. OCP, *The Algebra of Logic* (1914).

11 At present ... economize.] *inserted in-line and between paragraphs*

sult of his labors shall be submitted to you for your approval. I believe that he is better fitted than even you yourself to revise the work for the simple reason that he will stick to the plan more as an editor than you would do. You seem to be anxious to introduce new material which would be foreign to the subject of these logical papers. Your existential graphs should not be introduced here nor your theory of probability. They are a subject by themselves and would only burden this splendidly popular exposition of logical papers. Accordingly I advise you to leave out what is not absolutely necessary for the present purpose.

I hope to be able to have this book of yours edited by Russell ready by the end of this year, and would like to be able to present you with a copy during the spring of 1911.

With kindest regards and best wishes, I remain,

Yours very truly,

Russell to Carus 22 Nov 1910 (MS: SIU 91.7): Explains that he has been preparing footnotes throughout Peirce's papers and attempting to collaborate with Peirce to that end but with no reply. Feels that Peirce's suggested alterations and additions, especially dealing with his Existential Graphs, are troublesome and out of scope for the book. Believes he and Carus will have "a time" in getting together Peirce's contributions and suggests that they "could make up a better book without him and still not underrepresent nor mis-represent him."

Carus to Russell 28 Nov 1910 (LPC: SIU 91.7): Agrees with Russell to move forward with the work with little revision and without overly involving Peirce. Reiterates his eagerness to publish this soon for "Peirce is old and might break down" and Carus will shortly leave for Europe.

Russell to Peirce postcard ca. 24 Dec 1910 (missing): Explains that Carus plans to avoid substantial alterations and annotations in the "Illustrations" book and to note only Peirce's dissents as described in Peirce to Carus 26 Aug 1910.

Peirce to Carus. MS: SIU 91.26.

Dont read this to F. C. R.

Milford Pa | 1910 Dec 24

My dear Doctor Carus:

A postal card from Russell informs me that you do not want the book (i.e. my papers in the Pop. Sci. of 1877–8) added to or subtracted from or altered over very much. Only to have *noted* my *present* dissents.

I feel that I owe you my thanks for this decision. I hope no notes will be added, in any form, by anybody else; for I should regard it as very unfair to exclude my own explanations and print in the volume any other person's. Your decision gives me the advantage of getting out another volume, perhaps next year.

But I want to tell you, lest you think I have not done as much about the volume as I might, the following circumstances, which I only mention in order to prevent your getting that idea.

My wife, to whom I am so deeply attached, is suffering the last tortures of pulmonary tuberculosis, with all the perturbation that that implies; and we are far too poor to have any kind of servant. The house is large. The carpenters, masons, plumbers, that we paid our last dollars to, all signed contracts without the smallest intention of adhering to them, and wheedled out of my wife, into whose hands I had put the money (for in her condition to attempt to control her would mean killing her,) nearly all the contracts called for & then kept up a pretence that they were coming to finish things, and never did and never did come and so the house is in about as bad a condition as it was at first. (In some particulars worse) The early frosts prevented my efforts to make things a little better. You will see,—or would, if you had had any such experience,—the hard labor this throws on me. Now add that I am in a critical condition of health myself & am on most days so inert, that it's a tremendous effort to do what would otherwise be a trifle, & you will *begin* to understand why I have been unable to get much done on this book, though I have written a good deal which your kind decision will enable me to publish separately, after a reasonable interval. You desire that my "present dissents" should be "noted." I shall take the liberty of understanding that to allow me to state whether each "dissent" is based on demonstration to be given later, or whether it is an inference of an inductive or retroductive kind, and about what degree of probative force I attach to it.

Our friend F. C. R. has I suppose seen his best days. I wrote a few words of thanks to him for his attempt to say what in the *Monist* of July 1908 I was driving at in the last part of my article therein.[305] I did it simply because he so hopelessly missed the point as to convince me that it would be futile to attempt to set him right, & I wanted to spare the poor fellow's feelings. His maunderings about the doctrine of parallels (which might better be called the principle of similar triangles) *fixed* my estimate, and I hope he wont do much in the volume to be printed. But, above all, I don't want to hurt his feelings; and I think you had better not show him this letter.

305 Russell's "Hints for the Elucidation of Mr. Peirce's Logical Work," pp. 406–415, preceding Peirce's "Explanation of Curiosity the First" in the same number.

2 prevent] *after* ~~avoid that idea~~
6 poor] *above* ~~che~~
7 to] to~~o~~
9 to[1]] *after* ~~the~~
11 and never did come] *inserted above*
12 (In some particulars worse)] *inserted above*
22 of[1]] *after* ~~from~~
25 I] what I *[E]*
31 I[2]] *after* ~~if~~

very truly
C S Peirce

Carus to Russell 30 Dec 1910 (LPC: SIU 91.7): Relays the general sentiments of Peirce's confidential letter and suggests that since Peirce does not wish to have Russell's footnotes that Russell instead collect the notes and publish them as a comment in *The Monist.*

Jan 1911–Aug 1913: Peirce loses contact with OCP and re-engages in a commitment from 1909 to submit an essay for a volume in honor of Victoria Lady Welby, Peirce's semiotic confidant and correspondent 1903–1908. Prepares several drafts of "A Sketch of Logical Critics" to that end ca. spring and summer 1911, though never published (R 673–677, one version in *EP* 2:451–462). In Nov 1911 presents papers on "A Method of Computation" and "The Reasons of Reasoning" to National Academy of Sciences, his last public address. Likely continues to work on his "System of Logic" (see footnote in Peirce to Carus draft 6 Jan 1909) though greatly impeded by growing health challenges.

Meanwhile, Carus and Russell continue collaborating and preparing for the "Illustrations" book, especially during spring 1911 (LPCs and MSS: SIU 91.7). By summer 1913, Carus had written a proposed preface to the work and opens correspondence with Peirce.

1913

Carus to Peirce. LPC: SIU 91.28.

August 23, 1913.

Mr. Charles S. Peirce,
Milford, Pa.
My dear Sir:

I have gone over your *Logical Papers* and have written a preface of which I enclose a copy.[306] I believe you will be quite pleased at having these lectures on Logic published. You will remember that I paid you for the republication, further that you have assured me that you possess the right to republish it. Moreover the copyright of the Popular Science Monthly either has expired or will expire very soon. In case your copyright should be doubtful in any way, I would rather wait until the copyright expires.

You wish to have it revised because you think that you could improve several passages, but I believe that the article is very good as it stands. Moreover I

---

306 See Appendix.

have submitted this series of articles to one of the most critical modern logicians, Mr. Philip E. B. Jourdain.[307] He has returned the copy to me, and approves its publication as it stands.

I have not heard from you for a long time and hope that you are in good health. Hoping to hear from you, and expecting that the preface which I enclose will be satisfactory, I remain, with kindest regards,

Yours very truly,

PC/N

Peirce to Carus. MS: SIU 91.28.

(P.O.) Milford, Pa. | 1913 Aug 28

My dear Dr. Carus:

I was extremely glad to hear from you. Through a great part of the year 1911 I was puzzling how to get those Pop. Sci. papers into any decent shape; and I finally decided that the best way would be to let them go with minor corrections and to add a new essay giving an outline of my more mature studies;[308] especially as I can now convince any reasonable man that every class of scientific men are now accepting (apparently) inferences that are absolutely unsound while rejecting others that ought to be admitted. There are so few men who are willing, even if they are able, to devote the requisite long years for a correct analysis of these questions, that it seems to me a duty to print my reasonings on the subject which can be made very lucid without losing accuracy.

On Dec 13, 1911, I was engaged in writing such an additional paper to be sent to you, when an important new idea occurred to me. I jumped up to think it over while walking up & down my study. I began, as frequently, with a sort of frisk for exercise. There was on the floor a wrapper of glazed paper. I trod on this. It slipped on the waxed floor, which was as smooth as that of the Louvre. I was thrown on my back violently & broke some ribs, the broken ends sticking into the pleura and almost stopping my breath. I could neither get up nor call out. At last my wife

307 Philip Edward Bertrand Jourdain (1879–1919), British logician and London editor for *The Monist*.

308 Possibly related to his "A Sketch of Logical Critics" for Lady Welby, which discusses logical critics as "the theory of the kinds and degrees of assurance that can be afforded by the different ways of reasoning." Also addresses faults in his 1877–1878 "Fixation of Belief" and "How to Make Our Ideas Clear."

1 logicians] logisticians *[E]*

17 inferences that are] infe-*[line break]* $^{\vee}$rences$^{\vee}$ \that are/

18 men] *over* p

came, and telephoned for one of the country doctors. If he had confined himself to dressing my back I should have been all right in a month or so, But after making a characteristically slouchy diagnosis, he insisted upon putting me upon a regime the consequence of which is that I cannot yet put in 1/4 of a day's work. In the long interval, though I could do no close thinking, I did much rumination to advantage. I have for the past twelve-month been trying my very best to get some good work done & I *think* I am on the way to accomplish something.

It is needless to say that we are on the verge of indigence, of course unable to keep a servant, my wife struggling against effects of overwork, and all the old friends dead, Stickney, Garrison, Alex Agassiz, and several others. As I am right in the midst of writing something that looks as if it might be worthy of your acceptance, I will postpone for a few days longer replying further to your letter. It may be better that I should renew my understandings with several people.

I confess I should be glad if you would decide to include some of my earlier papers & my review of Pearson's Book[309] which appeared in The Pop. Sci. for Jan 1901.

I have a lot of notes on my old paper on Logical Extension & Comprehension[310] which I will send you as they are pasted into stiff covers and can be easily looked over.

I trust all your family & connections are well.

Believe me
very faithfully
CSPeirce

Carus to Peirce. TS: RL 77. On OCP Stationery D. LPC: SIU 91.28.

La Salle, Ills. September 10, 1913.

Mr. Charles S. Peirce,
Milford, Pa.
My dear Sir:

Your letter of August 28th reached me on my return from Benton Harbor, and I will say in reply that I am on the verge of leaving for Europe. I have not yet reserved berths, but I expect to sail on the S. S. "Berlin" of the North German Lloyd, leaving September 20. My European address in October shall be Deutsche Bank, Berlin W. 8, Germany.

I am very sorry to learn of your misfortune, and I hope that by this time you have overcome the worst of your troubles.

309 "Pearson's *Grammar of Science*."
310 "Logical Comprehension and Extension," *American Academy of Arts and Sciences* (1868)

In case you should work out a resume of your theories which will whet the appetite of readers to study your system more carefully in your former publications, I shall be glad to incorporate it as an appendix to your logical papers. Perhaps I might also incorporate as you suggest some other articles mentioned by you, as for instance your criticism of Karl Pearson. I do not expect that the Open Court Publishing Company will make any money from your papers, but my intention is to keep your thought before the public. I have always liked your logical papers, and though some of your more ponderous articles may contain more original ideas, I have always been impressed with the ingenious and pleasant treatment of obstruse topics which will make logic more interesting than logical text books, and will at any rate remain an introduction to your more ponderous works. I propose to use it mainly to send it out broadcast to lovers of logic to make a propaganda for a sound study of logic. Books on logic are not much bought.

I also intend to republish your criticism of my position together with my answer.[311] You remember they appeared in the first and second volume of THE MONIST, and at the time I acquired the right of republication, but all our funds are taken up now, and I have so many other books in publication, for instance Morgan's "Paradoxes",[312] that I can not think of tackling the work of republication now. When the time comes I shall write to you again on the subject.

With kindest regards and best wishes, I remain,

Yours very truly,
Paul Carus

PC/N

April 1914: Peirce dies of cancer 19 April at Arisbe. Carus invites Russell 22 April to write Peirce's obituary. Russell accepts and sends 24 April his "In Memoriam Charles S. Peirce," which appeared in the July *Monist* (also included in Appendix, "Peirce and Russell").

Carus never published Peirce's "Illustrations."

---

311 Necessity–Tychism feud April 1892–July 1893. Proposed book also discussed March–April 1900 and April 1905.
312 *A Budget of Paradoxes* (1915).

2 system] sustem *[E]*　　5 Pearson] Pierson *[E]*

2.2 Peirce and Hegeler

The bulk of the correspondence between Peirce and Edward Hegeler, OCP founder and owner, lasted only about eighteen months between January 1893 and the summer of 1894 and centered around the failed attempt to publish Peirce's "A Quest for a Method" logic book and his arithmetic textbook.[1] Hegeler took a special interest in Peirce's logical and mathematical work in an effort to further "the scientific missionary work of the Open Court Publ. Co. in a general way," "to do a service to science," and to engage Peirce more with OCP, whether "as its missionary or opponent."[2] But Hegeler soon lost patience with Peirce's inability to comply with agreements and deadlines, and after acquiring only an overdue and apparently unsatisfactory draft of Peirce's Arithmetic sometime mid-1894, broke off contact with Peirce, never to work with him personally again. Their unpleasant separation deeply hurt Peirce and for years, bordering obsession, Peirce tried fruitlessly to regain Hegeler's acceptance. In one of many similar lamentations, Peirce noted sorrowfully to Carus 13 May 1896: "There is I think no element in all the misery I have suffered of late years which has cut me more than the consciousness to Mr. Hegeler's bad opinion of me, which is undeserved & is really due to the extreme difference of our educations & experience." The remainder of Peirce's dealings with OCP would be under the auspices of Russell and later Carus.

1890

Peirce to Hegeler. MS: SIU 91.9

Milford Pa 1890 Sep 7

My dear Sir:

Dr. Carus writes to me that you would be glad to hear what I have to say to your article in No 157 of the Open Court.[3] I write in order to respond courteously to that invitation, & not at all because I have any burning desire to dispute the question, for it is one of those to make up one's mind about clearly.

The Open Court is I am quite sure doing good work. What you call "conciliating religion with science" appears to me, from your article, to be just taking the side of science in advance. As a scientific man, I cannot object to that; and I dont

---

1 See "Peirce and the Open Court" for details. See also Peirce to Hegeler 7 March and 22 May 1893 for preliminary outlines.

2 Hegeler to Peirce 16 March 1893.

3 "Religion and Science" (28 Aug 1890); see Carus to Peirce 3 Sept 1890.

doubt you will have considerable success in persuading religious people to adopt what you call "rational religion," which stands ready to bow itself out whenever science is tired of it. For I believe the Churches are beginning to be disposed to make great concessions,—perhaps *too* great.

You say religious ideas are our deepest convictions. Not yours, I fancy, or you would not be so complacent about them. I should say religion was what a man takes to be his dealings with higher powers. If he really thinks he has such dealings, he will naturally think human science should not be permitted to insult his divine friend. But you do not seem to be convinced you have such mystical relations with Divinity. What you call your religion would appear to be too rational to deserve that name. It is, perhaps, something more elevating and every way more satisfactory; but it does not seem to me to be what people mean & what I mean by Religion. I do not call Matthew Arnold's creed[4] religion, but a sentimental philosophy, the reminiscence of a religion; and if you think that a power enforces a line of conduct by punishing every individual transgressor, this is a bit of religion which I should think could hardly stand alone.

Thus I have little to object to in what the Open Court is doing but I do not call it reconciling Religion,—or anything,—with science. The religious man thinks his God tells him certain things. Science says these things are not true. Shall he go and ask his God if he is not mistaken? He can conclude his religion is false & when he is ready to do so, the priest can come just in time to say that God never dreamt of saying any such thing.

Yours very Truly
C. S. Peirce

Mr. Edward C. Hegeler—

1891

Hegeler to Peirce. MS: RL 77. LPC: SIU 91.9.

Lasalle Jan 21st 1891

C. S. Peirce Esqr
Milford Pa

4 Considered God an abstract influence for good, and religion "morality touched with emotion" (Super, 6:176).

10 would appear to be] *above* ~~is~~
19 certain] *t*certain
30 Milford Pa] *In LPC, below here is written* "near Port Jervis | Erie R.R."

My dear Sir,

Your esteemed favor of the 7^{h} of September 1890 reached me in the regular course of mails in Colorado where a daughter of mine was then sojourning on account of her health. I was in hope soon fully to reply to your letter, but my daughter's health and worries of other kind have kept me from doing this.

To day I can also do nothing but thank you for your letter, in which, though it has not altered my opinions, I take great interest. I am at the point of departing for Las Vegas—Hot Springs New Mexico, where my daughter now is. Perhaps I can answer you from there.

I saw with pleasure your contribution in the Monist.[5]

I am dear Sir
Sincerely yours
Edward Hegeler

1892

Hegeler to Peirce. MS: SIU 91.12. Undated fragment bound third in sequence with the letters Peirce to Hegeler 24 Feb 1893 and Hegeler to Peirce 24 Aug 1893; however, as with other bound letter sets, the thread may not necessarily be related. Furthermore, the final line of this MS is found in a note by Hegeler in *The Open Court*, 11 Aug 1892, p. 3349, thus dating this MS from possibly around that time.

[early August 1892]

Professor C. S. Peirce
with warm regards from

E. C Hegeler

The soul is not the activity of the nervous system; but the form of its activity

Soul is Form.

The Soul is Form.

---

5 "The Architecture of Theories."

6 in] *inserted above*
21 Peirce] Pierce *[E]*

24 The soul … activity] The soul is not the activity ^of the nervous system;^ *fr*but the form of ~~the~~ \its/ activity ~~of the nervous system;~~

The soul is form[6]

We ARE the memories and ideas of which in the language of the day we erroneously say that have them.

1893

Hegeler to Peirce. MS: RL 77. LPC: SIU 91.12.

Lasalle Jan 23^{d} 1893

C S. Peirce Esq
New York.

Dear Sir,

Your article[7] for the Monist came to hand. It is intended to publish the same in the July numbers. I enclose my check for \$ 200$^{\underline{00}}$ on account as desired.

I think it would be a help to our work if we became nearer acquainted and invite you to visit me here at Lasalle. I expect to be home for some time from now.

Shall be glad to hear favorably from you.

With best regards from Dr Carus

Yours truly
Edward Hegeler

Peirce to Hegeler. Telegram: SIU 91.12. The text for this and all other telegrams in the Peirce–Hegeler correspondence comes from OCP transcriptions in company records.

Telegram | New York Jan. 25 · 93

E. C. Hegeler
La Salle Ill

Arrive Chicago—ten—Saturday forenoon, hope to find afternoon train to La Salle[8]

sig. C. S. Peirce

---

6 Possibly an allusion to Edmund Spenser's "An Hymn in Honour of Beauty," *Fowre Hymnes* (1596): "For of the soul the body form doth take: | For soul is form, and doth the body make."

7 "a Necessitarians," submitted to Carus 20 Jan 1893 (first of two letters that date).

8 For Peirce's meeting with Hegeler late Jan–early Feb 1893.

257.25–258.1 Soul is ... form] *All three "soul is form" lines lightly struck out, but retained here.*

7 Peirce] Pierce *[E]*

Hegeler to Peirce. Telegram: SIU 91.12. Message is recorded below Peirce's Jan 25 message on the same sheet.

821 Sixth Avenue New York *[ca. 25 January 1893]*

C. S. Peirce

Glad of your coming Expect you by Rock island railroad leaving Chicago at twelve fifty

Edward C. Hegeler

14 words

Between 28 Jan and 6 Feb 1893: Peirce meets with Hegeler and likely Carus at Hegeler's home in La Salle to discuss larger publication projects, including Peirce's "A Quest for a Method" (or "Search for a Method")[9] and several math books. Hegeler seems also to have asked Peirce to begin investigating into printers to contract with OCP on the books, especially his arithmetic book. These matters dominate the remainder of the correspondence between Peirce and Hegeler.

7 Feb 1893: Returning from La Salle, Peirce meets with Russell at Richelieu in Chicago[10]—details undetermined, though an unsatisfactory meeting for Peirce due to Russell's "not showing me the side of him that I wanted to see" (Peirce to Hegeler 11 Feb 1893).

Peirce to Hegeler. Telegram: SIU 91.12.

New York | Feb 11. 93

E. C. Hegeler

Wilson telegraphs cannot possibly undertake that printing. Shall I try to find another printer?

C. S. Peirce

Hegeler to Peirce. Telegram: SIU 91.12. On verso of telegram Peirce to Hegeler same date.

821 Sixth Avenue | New York *[February 11, 1893]*

Chas S Peirce

Please look for other good printer and report

Edward C Hegeler

8 words

---

9 (R 592–594). For details, see Peirce to Hegeler 24 Feb and 7 Mar 1893. Also de Waal, 15–17, and Brent, 221–224.

10 Carus to Russell telegram 6 Feb 1893: "Peirce will arrive tonight about seven invite you to lunch tomorrow Richlieu" (RL 387).

Peirce to Hegeler. Telegram: SIU 91.12.

N.Y. Feb 11 *[1893]*

Dr E. C. Hegeler

J. J Little & Co. 10 10 Astor Place N.Y excellent firm can do job would doubtless wish binding and all

sig.
C. S. Peirce

Hegeler to Peirce. Telegram: SIU 91.12.

821 Sixth Avenue | New York City | Febr 11th 3$^{\underline{00}}$ p.m. *[1893]*

C. S Peirce

Please have Little mail proposal to Open Court Publishing Company Chicago

Edward C Hegeler

11 words

Peirce to Hegeler. MS: SIU 91.12. Stationery headed, "Buckingham Hotel, | Fifth Avenue and 50th Street. | Wetherbee & Fuller, | Proprietors."

New York, Feb 11, 1893

My dear Mr. Hegeler:

I arrived last Monday without incident in Chicago, and put up at the Richelieu, where I sat up until past two working over improvements in my reply to the Necessitarians, and then got up at 5 in the morning to continue. I hoped to finish the job; but I did not, and am glad; for, properly done, it will take a week. At noon, I succeeded in finding Judge Russell. I spent several hours with him; but it was not satisfactory, and I think the fault was a want of tact on his part in not showing me the side of him that I wanted to see. But please do not tell him I said the interview was not satisfactory, because it may have been so to him. At 5:30 Tuesday afternoon I took the limited express and arrived in Niagara Falls in the morning. It was bitter, bitter cold; and the wind was blowing down the river so strongly that from the bridge one was so battered with frozen spray that hardly anything could be made out. But I went down by the side of the Horseshoe Fall and being there out of the wind had a very good sight. I did not go down on the ice on account of the wind. I went, of course, to the Hotel Kaltenbach and found it all you had said. I saw my favorite Berncastler Doctor on the Wine List; & they

---

4 10 Astor] *over* sout   19 until] *above* ~~pas~~

said there was a man in Rochester who imported it. Had it been August instead of February, I might have been tempted to try it. I left Niagara in the afternoon and arrived in New York at 7:30. My wife did not welcome me very warmly. With French women *C'est toujours l'inattendu qui arrive.*[11] Yesterday, Friday, I went to see Mr. Willie Appleton and also Mr. Vail the head of the Editorial Department of the Book Company, or school-book publisher's trust. I had most satisfactory interviews with both; and I found Appleton had already talked to Vail about my arithmetics. Vail said I must write *two* arithmetics and have the MS of both ready in June, and then they could be published in the September of the *following* year. He said that ordinarily the profits were small for the first year after publication. I had also a long discussion with Mr. Vail of an Elementary Geometry,[12] which he wants me to try, though he distrusts his own judgment as to whether it would prove profitable. The truth is nobody has ever succeeded in writing a really good geometry. The difficulties are almost insuperable. Whoever first conquers them, will have a great prize in his book.

I wrote to the printers in Cambridge that I would be there Monday morning. I shall go on Sunday night.

I do not know what to do with my wife. It would not do at all to let her go to the farm at this season. Cambridge or Boston would be bad, too. Besides, my brother is an old bachelor and would be thrown off his balance entirely at the notion of having a lady in his house. I should hardly venture to beg a night's lodging of him for myself. "He wants to have things just so," and he does have them just so, with a precision quite wonderful.

I found a large bundle of books from the Open Court Publishing Company. But several of those on the list in the last *Monist* were absent. I am surprized to find you do not publish *Ribot's* Diseases of Memory. I give a list of those *not* sent; though probably some of them are intentional omissions.

---

11 "*It's always the unexpected that happens.*"

12 To become his 1894 "The New Elements of Geometry" (R 94; *NEM* 2:233–474), prepared by invitation of Ginn & Co. ca. March 1894 (see Peirce to Hegeler 11 July 1894 and Peirce to Russell 8 Sept 1894). However, as Eisele argues, at this point the MS in question might be his "Elements of Mathematics" (see *NEM* 2:xiv–xxvii), which Peirce to Russell ca. 10 March 1896 also identifies as his "Geometry."

---

19 or Boston] *inserted above*

Gould. Modern Frankenstein
Dreams, Sleep, and Consciousness
Palmer. Artificial Persons.
Romanes. Darwinism Illustrated.
Trumbull. Making Bread Dear
Turner. Only Good Thing
The Vedantasara.[13]
The Vicharsagar
The Panchadasi
Vicharmala

I have an extra copy of the Lost Manuscript[14]; but that is none too many. I could place more than one where it would further the cause.

To return to Mrs. Peirce. I think I shall be forced to send her to Lakewood, a resort in New Jersey where several of her lady friends are at this moment. Later, when the weather is a little milder, I should like to take her to La Salle and leave her there, if the printer will give me the few days' vacation.

Good bye, my dear friend, if you will permit me to call you so. I have brought away from your household the priceless treasure of a sweet memory. Please do present my deep regards to Mrs. Hegeler, Mrs. Carus, Mrs. Bucherer, Miss Annie, Miss Zuleika, Miss Olga, Julius, Carus, and Carus *jeune*,—& Mr. Bucherer.[15]

very faithfully
C. S. Peirce

Peirce to Hegeler. MS: SIU 91.12. Stationery headed, "Century Club | 7 West Forty-Third St."

New York 1893 Feb 15

My dear Mr. Hegeler

On learning that Wilson could not undertake the book I called on De Vinne who is the printer of the Century Company. I found he was too much occupied also; but he recommended me to J J Little & Co, whom I already knew of as good printers, & I think they did the binding of the Century Dictionary. I find that if we put 400 words on a page, two volumes of 400 pages each will not include all my philosophical papers, but will include the principal ones. If they are found to pay for themselves, a third volume can be issued later to embrace articles not yet

13 These four works (of six) acquired by the OCP from the *Vedanta Series* spring 1890, discussed in detail in Gunlogsen, "The Philosophy of the Vedanta," *Open Court* (6 March 1890). See also *Vedanta Series* in the Bibliography of this edition.
14 Freytag, 1892.
15 See all under "Hegeler" and "Carus" in Biographical Register.

written—and also some which, though lively in style, are not of sufficient importance to be included in the first collection. I selected a handsome page and handsome paper and a suitable plain binding. I said nothing about the price; but if my selections are too expensive, the cost may be considerably reduced, by printing more closely. I shall be perfectly contented with any form of page; but I picked out one which was handsome and which I thought would bear a price which might make the thing remunerative with a small sale. But I do not pretend to be a judge of those matters.

In order to enable me to draw immediately upon the check you handed me, I took the liberty of having it *certified* at the bank in Chicago; for the New York banks, though they do not charge expenses of collection like to find any excuse to avoid immediate drafts. It had the desired effect; and I am sure you will not take offence. I shall leave here tomorrow morning. I can, if necessary, come back to New York; but the proof sheets can be sent to me in Milford. Kind regards to all the family.

very truly
C. S. Peirce

Hegeler to Peirce. Telegram: SIU 91.12.

821 Sixth Avenue | Newyork | 2/15 · 93

Professor C S Peirce
Printers DeVinne New York also Cushing Boston recommended to us

Edward C Hegeler

Hegeler to Peirce. MS: RL 77. LPC: SIU 91.12.

Lasalle Febr 20th 1893

Professor C S Peirce
Milford Pa

My dear Professor,

Your kind favors of the 11th and 15th inst. have both reached me in due course of mails.

From J.J. Little & Co I have not heard yet, but it may be that a proposal from them has arrived in the Chicago office, as Mr Sacksteder the office manager was here from Thursday to Saturday. My instructions to him when he left here Satur-

---

14–15 but … C. S. Peirce] *up right margin*
21 Cushing] *before* ~~& Co.~~
30 J.J.] *over* M
32 from] *over* si

day morning were to open the proposal so that he could see from it exactly what you wanted, so that he might report to me what the same work of as good quality would cost in Chicago—and then send all papers to me for a decision. Mr. Sacksteder is to examine too, if the Appletons did renew the copyright on those articles which appeared in the Popular Science Monthly. My impression is that you were not quite certain if you had reserved for yourself the copyright on them outside of their publication in that magazine.

In case you have made or expect to make an arrangement with the Appletons for the publication of the Arithmetics, I believe it will be your interest to ask them if they wish to publish the books on logic. They might otherwise take offense at this publication by us, what might prejudice them in regard to the publication of the Arithmetics.

I was glad to see from your first letter[16] that you found Mrs Peirce in as good health as you could expect. That you send or did send her to a resort in the country among friends seems to me to be the right thing. Have you received the message we sent after you to the Erie station? it came from here probably?[17]

The certifying of the checks is quite satisfactory to me. I should have ordered my Chicago Bank to hand you New York Exchange to your order to New York.

My family and myself remember your visit to us with great pleasure and we all send our regards to you and Mrs Peirce. Hoping to hear from you soon I remain

Sincerely yours
Edward Hegeler

I told Mr Sacksteder to send you the other books we published and also copies of the Disease of Memory and Dis. of the Will.[18] The first part of the Disease of Memory is philosophically the most important.

16 11 Feb 1893.
17 Erie railroad station of Port Jervis, NY, Peirce's frequent mailing address while at Milford, PA. Possibly referring to telegram 15 Feb 1893, forwarded.
18 Ribot, 1882 and 1884.

1 from it] *inserted above*
5 were] *after* ~~y~~
11 this] *over* our
14 send or did send] sen*t*d \or did send/
15–16 Have you … probably?] *inserted in-line*
20 soon] *over* aga
23–24 I told … Will.] *down left margin*
24–25 The first … important.] *down left margin of previous MS page*

Peirce to Hegeler. MS: SIU 91.12

Port Jervis, N. Y. 1893 Feb 24

My dear Mr. Hegeler:

This has been the hardest winter here for many years. I find my house quite uninhabitable for the moment, the steam boiler having burst, with other damage. The roads are in bad condition owing to drifts. My address continues to be Milford, Pa.

Yours of the 20th reached me last night. It must be one of my letters failed to reach you; for I told you I had seen Appleton and have come to a thorough and satisfactory understanding with them (not going into details.) Having said what I could to make them not wish to publish my collected papers, it would now be awkward to go back and ask them to do so. Still, I think if I should they would comply. I had it deeply at heart to have the book appear in time for the Chicago Exposition[19]; and Little had promised it should. I have the copy of the first volume ready, all but one paper. At your request, I told Little to write to the Open Court Co. in Chicago. You telegraph me about De Vinne and Cushing. As I wrote to you, I saw De Vinne first of all in New York. He cannot undertake it. If it is to be printed in Boston, I will pull up stakes & go there. *[?Or if]* the printing is to be done in Chicago will go immediately there. I was in hopes we should have been well on with the printing by this time.

I have mailed you a photogravure of myself. I will mail another to Professor Schröder.

My wife's health troubles me very much. Not that there is now any active tuberculosis, or very little; but this climate especially in the months of March and April frightens me. Then she suffers & her health suffers from the mental effects of her surroundings in this country. If I can sell my place I will send her somewhere in Europe and join her later. But she shall not go until you have all seen her, unless there is some very strong reason against it.

19 *Chicago World's Fair: Columbian Exposition*, to be held 1 May–30 October, 1893, celebrating the 400th anniversary the previous year of Christopher Columbus's discovery of the New World. Neither this arithmetic book nor his "Quest for a Method" (see Peirce to Hegeler 7 March and 6 May 1893) were completed for the event.

13 had] *above* ~~was~~
18–19 *[?Or if]* ... there.] *inserted up left margin; tear in MS through first two words*
26–28 If I can ... against it.] *inserted between paragraphs*

I wish you would let me know if you did not receive my letter describing my talks with the Appletons and the American Book Company. They were satisfactory inasmuch as they showed me those people think the enterprize an important one.

I STILL DO NOT RECEIVE THE OPEN COURT!

With warm regards to all the family,

Yours very faithfully
C. S. Peirce

Hegeler to Peirce. MS: RL 77. LPC: SIU 91.12.

Lasalle Febr 24th 1893

Professor Charles S. Peirce
Milford
Pa
near Port Jervis—(Erie R.R.)

My dear Professor,

We have now proposals for the printing etc. of the books from New York and Chicago—but also the information through Mr Sacksteder that a copyright was for 28 years—so that Appleton's copyright is not expired on the Pop. Sc. Monthly articles; and this was the consent on a written agreement made at the time of the publication of the articles seems to be a necessity now.

Hoping yourself and Mrs Peirce are well

I remain
Sincerely yours
Edward Hegeler

Hegeler to Peirce. MS: RL 77. LPC: SIU 91.12.

Lasalle Febr 27$^{\underline{\text{th}}}$ 1893

Prof C S Peirce
Milford Pa.

My dear Professor,

Your letter of the 24th inst reached me yesterday morning. In my letter of the 20th inst. I opened with acknowledging the receipt of your letters of the 11th and 15th inst. I wrote you again on the 24th inst. In your letter of the 11th you say on business matters: "Yesterday, Friday, I went to see Mr. Willie Appleton

18 a] *inserted above*
25 27$^{\underline{\text{th}}}$] 2*6*$\underline{7}^{\text{th}}$
32 on business matters:] *inserted above*

and also Mr. Vail the head of the Editorial Department of the Book Company, or school-book publishers Trust. I had most satisfactory interviews with both; and I found Appleton had already talked to Vail about my arithmetics. Vail said I must write *two* arithmetics and have the MS of both ready in June, and then they could be published in the September of the *following* year. He said that ordinarily the profits were small for the first year after publication. I had also a long discussion with Mr. Vail of an Elementary Geometry, which he wants me to try, though he distrusts his own judgment as to whether it would prove profitable. The truth is nobody has ever succeeded in writing a really good geometry. The difficulties are almost insuperable. Whoever first conquers them, will have a great prize in his book.

I wrote to the printers in Cambridge that I would be there Monday morning. I shall go on Sunday night."

You had apparently no copy of your letter of the eleventh when you wrote your letter of the 24th inst. You will see the former refers almost altogether to your proposed arithmetics.

I request that you write me a memorandum of all our arrangements to avoid misunderstandings.[20] Concerning the books, Dr Carus made the remark yesterday that he thought the logical papers would best appear for themselves in one volume.[21]

I must say that my impression was (the memorandum you gave me is not before me. it was taken to Chicago by Mr. Sacksteder we think) that the papers to be published were principally logical papers. I never closely examined the list.

For the publication on my part of the papers that appeared in the Pop Sc. M. the permission of the firm Appleton & Co in writing is necessary as I wrote you already on the 24th.

---

20 See Peirce to Hegeler 7 March 1893 for his revised memorandum. The original, likely written for the late Jan–early Feb 1893 meeting in Chicago, was lost by OCP, as Hegeler confesses in the following paragraph. It also included a proposed outline of his "Quest for a Method," shown in the Appendix, "Peirce and Hegeler," late Jan–early Feb 1893.

21 Likely his 1877–1878 "Illustrations of the Logic of Science" series in *PSM*, referenced further in the letter. Carus encouraged Peirce to re-publish the series throughout his correspondence with him and began working with him more seriously to that end 23 Jan 1907 until Peirce's death 1914. The book was never completed, though it did appear in book form, with Peirce's extensive revisions, in de Waal, 2014.

---

6 for] *inserted above*
12 morning] *over* d
21 was] *is*<u>was</u> \~~and is~~\and// *(omitted* "and" *mine)*

Your photogravure has arrived. It is a fine picture. Accept my thanks for the same.

I read with sympathy your case for Mrs Peirce's health. Is she at Lakewood now as thought of by you in your former letters?

The Chicago prices for getting up the book are lower than Little's and as good work as theirs is likely to be expected; but I am undecided yet.

The cost of the whole could be reduced by having say only five hundred copies printed and bound at present—preserving the electrotypers.

If we limit ourselves to the logical papers and print them in one volume a larger saving will result, and also they can be completed sooner.

I understand now from Dr Carus that among the philosophical papers *[it]* had been thought of re-publishing the papers we printed in the Monist[22] and must say that I do not like to see them published by us except accompanied with our remarks as to how far we agree or disagree with them as in general way it is understood that we endorse the views of the books we publish. Hoping to hear from you soon I remain

Sincerely Yours
Edward Hegeler

Peirce to Hegeler. MS: SIU 91.12. Stationery headed, "Fowler House | J. E. Wickham, Prop."

Please address me | at Port Jervis, N.Y. | till March 3
Port Jervis, N. Y., Feb 28 1893

My dear Mr. Hegeler

Yours of the $24^{\underline{h}}$ reached me last night. I have W. W. Appleton's positive promise; so there can be no doubt of the written permission, for which I have just written.

I found my fire had been allowed to go out and the steam heater boiler burst & a lot of water pipes. The house is not yet inhabitable, and there will be a bill which will greatly increase my difficulty in keeping my wife alive. The house is so pretty and characteristic, I hate to part with it, and should dread the mental effect of doing so on my wife, if it were not a simple question of life and death.

With kind regards to all,

---

22 His five *Monist* metaphysical series articles, Jan 1891–Jan 1893.

11–12 that among ... thought] that \among the philosophical papers/ ~~you~~ \~~we~~/ had /been\ thought
12 Monist] *before* ~~with them~~
14 as to how far we] as to ~~not~~ how *we*far ~~dis~~ we
28 alive] *after* ~~p~~

yours very truly
C. S. Peirce

Will you kindly remit check for $ 250

Hegeler to Peirce. Telegram: SIU 91.12. Message recorded on an aborted second page (beginning "The printing it seems to me ...") of Hegeler to Peirce letter the same date (which follows). Message also duplicated on the verso of Hegeler to Peirce telegram 2 Feb 1893.

[March 2, 1893]
Telegram / Professor C. S. Peirce | Port Jervis—New York on Erie R.R.

Permission to copy for other articles apparently necessary have written also ordered mailing check what money will you need for visit with your wife of her home in France

Edward C Hegeler

29 words

Peirce to Hegeler. Telegram: SIU 91.12.

Port Jervis | Mar. 2nd *[1893]*

E. C. Hegeler

Telegram received, warm thanks Carry wife to-morrow to New York doctor

C. S. Peirce

Peirce to Hegeler. MS: SIU 91.12.

Port Jervis N.Y. 1893 March 2

My dear Mr. Hegeler:

Your telegram reached me today. Many warm thanks for you practical sympathy. Before answering, the question now coming up as one of acting I must have a serious talk with the doctor. For my wife's condition alarms me, and I must take care not to do the wrong thing. My wife is so much younger than I am, that she is not only a wife, but like a child to me. Nobody who hasn't some experience of it can know how dependent this makes my life upon hers. I will go to New York with her tomorrow & return if I can Saturday.

Yours faithfully
C. S. Peirce

---

9 to] *over* for

Hegeler to Peirce. MS: RL 77. LPC: SIU 91.12.

Lasalle March 2^{d} 1893

Professor C S. Peirce
Port Jervis
N.Y.

My dear Professor,

I received your letter of the 28th ult this morning—enclosing a receipt for \$ $250^{\underline{00}}$.

I have just ordered the First Nat Bank of Chicago to remit to your N.Y. Exchange to that amount to Port Jervis Pa payable to your order.

I see you have written to Appleton for their written permission.

I learn this morning that of the other articles on the list the following have appeared[23]

17. Theory of Probable Inference in Studies in Logic
18. 19. 20. 21 & 22 (?) On the Logic of Relatives in Memoirs of American Academy of Arts and Sciences
Economy of Research—Smithsonian Institute.

Their written permission to republish will also have to be obtained—

The printing it seems to me now will be done best in Chicago or possibly—if after a little while—here

Your telegram of this morning came about noon

Considering the whole situation from your verbal communications and letters I answer

"Permission to copy for other articles apparently necessary have written also ordered mailing check what money will you need for visit with your wife of her home in France"[24]

Sincerely yours
Edward Hegeler

Peirce to Hegeler. Telegram: SIU 91.12.

3/5 · 93
Telegram from New York.

23 From Peirce's original memorandum delivered in late Jan–early Feb 1893 meeting in Chicago. See Appendix, "Peirce and Hegeler."

24 Telegram sent to Peirce earlier that day.

8 just] *inserted above*

E. C. Hegeler

Will try and see book through press Return Monday Port Jervis.

C. S. Peirce

Peirce to Hegeler. MS: SIU 91.12.

Port Jervis N.Y. 1893 March 7

My dear Mr. Hegeler:

Let me begin by acknowledging receipt of bill of exchange for $ 250. It is made to the order of C. L. Peirce[25] instead of C. S. Peirce; but I do not think that will matter. The letter was addressed to Port Jervis, Pa. instead of Port Jervis, N.Y.; but the Post Office officials discovered where it was meant to go. This shows the advantage of such a name as Port Jervis. Now, there are 3 or 4 *Milfords* in Pennsylvania alone, and numbers in other states.

I carried my wife down to New York on Friday. It turned out that her physician had gone on to the Inauguration with the 7th Regiment, of which he is surgeon. I therefore waited over till Monday; but on his return, he had so many urgent cases, that he did not get round to us. Under the circumstances, we had to come to a conclusion without his advice. Mrs. Peirce was most anxious that nothing should delay the publication of my philosophical papers; and besides, it is possible we might sell the place for $ 12 000 or $ 15 000; and if that were once off our hands, we should be free to make arrangements in another country. We therefore decided that we would try to send her away to the Adirondacks or somewhere at present, and see if we cannot manage to sell the place. I desire, once more, to express my very warm thanks for your very kind suggestion that money might be found to enable me to take her abroad. It would have required an additional $ 500.

You ask me to write "a memorandum of all our arrangements to avoid misunderstanding." Here it is.[26]

25 In his *Nation* review of the National Academy of Sciences meeting of 19 April 1902 in Washington, Peirce also mentions "three papers by Mr. C. L. Peirce" (*CN* 3:74). Whether this was a coincidental recurrence of the typo or the middle initial has a back story is unclear. (Peirce also went by Charles "Santiago" Peirce. See note to Carus's "Publisher's Preface" of "Illustrations" in Appendix for details.)

26 This and all other Hegeler quotations in this letter come from Hegeler to Peirce 27 Feb 1893. A summary of the memorandum made by Diesterweg is found in SIU 91.12 (not transcribed).

23 for] *over* that

I have sold you my books, and received $ 2 000 for the same. I am to send you a list of them; but they are not to be delivered at present.*[27] I retain the Insurance in my name, since there would be difficulty in getting a change made. If I sell the place, I shall offer to buy the books back. Besides this, it is understood that you will advance me $ 250 a month for seven months, at least, to be repaid out of the profits of the Arithmetic books which I am writing.[28] By saying "at least," I mean that there is no express agreement in regard to any longer time, although if the Arithmetics are then in the hands of the publishers, I am led to hope that you may then be willing to help me a little further, if I am in need of it. It is understood that the money so advanced will be repaid from the first five years profits on the books.

As I understand it, this is all the positive arrangement there is between us. You were to publish my philosophical papers, to which I had assented with the understanding that 1 500 copies were to be printed in time for the Columbian exposition and a considerable number was to be given away. You now say that your impression was the papers were principally logical papers. I gave you a list of those I had in mind; but when I got back I formed a new list, slightly differing from that, by including two or three things I had forgotten and omitting some of the less important ones. In a broad sense, all are Logical, except the Monist papers; but these are the inevitable sequel of the logical doctrine, and form in my mind an inseparable whole with that. I had intended the first volume should contain papers on the first principles of Logic and Philosophy, including the Monist Papers, as follows:[29]

I.	Classification of Arguments.	Logic.	No copyright
II.	Categories.[30]	Logic	" "
III.	Logical Extension & Comprehension	Logic	" "
IV.	Questions Concerning Certain Faculties	{Erkennt-nisslehre}[31]	W.T. Harris's copyright (?)
V.	Consequences of	Erkennt-	W.T.Harris's

27 Peirce's own footnote down left margin: "*I will send the list as soon as I can get it ~~made~~ \corrected/."

28 For an outline of his arithmetic book, see Peirce to Hegeler 22 May 1893. Compare also Peirce's original outline in Appendix, "Peirce and Hegeler."

29 See Bibliography for publication details of all. Some clarifying notes have been added below.

30 " New List of Categories."

31 "epistemology."

8 Arithmetics] *above* ~~books~~

17 two] *after* ~~some~~

18 ones] onces *[E]*

19–20 inseparable] insepable *[E]*

	Incapacities	nisslehre	copyright (?)
VI.	Validity of the Laws of Logic	ditto.	ditto.
VII.	Berkeley and British Logicians [32]	Logic	Lloyd Bryce's copyright (To whom I have written for permission.)
VIII.	Fixation of belief.	Philosophy	Appleton's copyright
IX.	How to Make Ideas Clear	"	"
X.	Doctrine of Chances	Logic	"
XI.	Probability of Induction	"	
XII.	Order of Nature	Philosophy	"
XIII.	Deduction Induction Hypothesis	Logic	"
XIV.	Theory of Probable Inference	"	C. S. P's copyright
XV.	Architecture of Theories	Philosophy	Hegeler's copyright
XVI.	Necessity Examined	Logic	" "
XVII:	Law of Mind	Psychology & Logic	" "
XVIII.	Man's Glassy Essence	Philosophy	" "
XIX.	Evolutionary Love	"	" "
XX.	General Sketch of a Theory of the Universe[33]	"	" "

In the second volume I proposed to put all my papers on the logic of relatives, which would make 200 pages, all my writings on probabilities (which is a branch of logic) and on methods of research, and, if there was room, some papers on experimental psychology. At the end, I proposed to have a copious index. These

32 Review of *The Works of George Berkeley* and Harvard "Lectures on British Logicians."
33 MS not found, but identified in *W* 8:xciv–xcv as the proposed sixth and final article of his *Monist* metaphysical series, rejected by Hegeler, as reported in Peirce to Russell 4 Oct 1895: "Had Hegeler kept his positive promise to allow me six articles in the Monist … the 6th article … would have been the keystone of the whole." See also the postscript in Peirce to Hegeler 27 July 1893 addressing the broken promise. MS possibly related to his "Six Lectures of Hints toward a Theory of the Universe" (R 972), though *W* 8 estimates this to be a preliminary plan for his entire *Monist* series (*W* 8:xxxv).

papers would be annotated and woven together. The whole to be entitled "A Quest for a Method."[34] All those papers the most attractive to the general reader, and some involving great study, would be left over for a possible third volume.

You say, "I do not like to see them [my contributions to the *Monist*] published by us, except accompanied by our remarks as to how far we agree or disagree with them." To that I could not assent; but should have no objection to a general disclaimer of responsibility for the doctrines on the part of the Open Court Publishing Co. appearing on the title page.*[35] Upon that I would remark in the preface that my principles would not permit me to publish any contribution to any branch of science or philosophy that should bear the certificate or endorsement of any authority; that my articles appeal to reason alone, and that I do not recommend anybody to accept anything laid down further than the arguments given may be found convincing. It appears to me that this should be satisfactory to both parties.

But you further say: "In a general way, it is understood that we endorse the views of the books we publish." I am very sorry, and think it greatly weakens your cause. A scientific Academy, of course, approves of all it publishes. I hope every respectable publisher does the same. But these people draw a broad line between approving a publication and accepting all its doctrines. Not to make this distinction, that is, not to approve publications whose doctrines you reject, is to sink to a mere sectarian level. I adhere to the Episcopal church, because, though somewhat sectarian, it recognizes the duty of rising above this position, and does so as far as it can. But never would I consent that anything of mine, on Logic, Mathematics, Physics, or any other branch of Science, should appear under the auspices of the church. Nor do I see how with your views you can wish that your purely sectarian publishing concern should take up things like mine which repose on antagonistic ground. My book is to be entitled *A Quest*. Now the very idea of a quest implies that what is said is not in harmony with any fixed creed, like yours. Worse yet, it is A quest for *a Method*, and what that method is to result in cannot be predetermined. The presumption must be that it will be hostile to any creed for-

34 a.k.a. "Search for a Method" (R 592–594), never published. Served as a precursor to his also unpublished "How to Reason" (see Peirce to Hegeler 11 July 1894), posthumously a.k.a. "Grand Logic" (R 397–434).

35 Peirce's own footnote down left margin: "*You might add a reference to any other place where my doctrines would be refuted." Peirce provides such a disclaimer in the unsent draft that follows this letter.

1 to] *over* wou
14 In a] *over* It is
24 your[2]] *over* the
26 *A Quest*] *A after* ~~a~~

mulated in advance. Consequently, if your purpose is only to support books that are pledged or wedded to a given doctrine, everything that I ever could under any circumstances be brought to write must be profoundly and fundamentally antagonistic to your position. I am engaged in promoting free inquiry; while you are engaged in supporting a platform.

[Pasted here in the letter and transcribed below is a clipping from Peirce's "The Marriage of Religion and Science," The Open Court, 1893 (7:359–60). The italics, bolded insertion, and deletion in the last two lines are Peirce's.]

> What is science? The dictionary will say that it is systematised knowledge. Dictionary definitions, however, are too apt to repose upon derivations; which is as much as to say that they neglect too much the later steps in the evolution of meanings. Mere knowledge, though it be systematised, may be a dead memory; while by science we all habitually mean a living and growing body of truth. We might even say that knowledge is not necessary to science. The astronomical researches of Ptolemy, though they are in great measure false, must be acknowledged by every modern mathematician who reads them to be truly and genuinely scientific. That which constitutes science, then, is not so much correct conclusions, as it is a correct method. But the method of science is itself a scientific result. It did not spring out of the brain of a beginner: it was a historic attainment and a scientific achievement. So that not even this method ought to be regarded as essential to the beginnings of science. That which is essential, however, is the scientific spirit, which is determined *not to rest satisfied with* ˅***our own***˅ *existing opinions*, but to press on to the real truth of nature. ~~To science once enthroned in this sense, among any people, science in every other sense is heir apparent.~~

As long as I am animated by that spirit, and as long as your deepest wish is to make certain opinions prevail, any agreement between us can, I fear, be but accidental and temporary.

Mathematics in former days was a body of theorems. Now, we no longer value theorems as pearls of great price; we search for methods. To *search* for a method implies that not even existing *methods* are satisfactory. And such is the spirit of modern science, generally. I do not mean to uncritically reject the results of previous scientific work. We neither uncritically reject nor uncritically accept. But we do not accept anything unqualifiedly nor as a finality nor as the exact truth nor as the last word. We are engaged in *research*. Research implies criticism, criticism implies doubt, doubt implies denial of absolute apodicticity.

This is a doctrine of *logic*. This is the doctrine which you and Carus do not like in my Monist papers. So you see it is precisely on logic that we split.

This being so, I think you had better give me leave to use my *Monist* papers, and let me find another publisher, when I can. When one comes to the foundations of logic, one is on *ethical* ground; and when people differ on fundamental questions of ethics, compromises are out of place.

Those books which I value the highest are books with which I do not agree. For why? Science is inquiry, and equally opposed to agnosticism, which declares certain things cannot be found out, and to dogmatism, which declares that certain things are settled once for all. Consequently, I am always anxious to learn what I do not yet know; and value books in proportion as they may modify my existing opinions, which I hope may be modified as quickly as possible. That would be the sense of "science" which I would attach to it in the phrase a Religion of Science.

I admit there may be, and probably are, things which will never get found out, and others that are already settled once for all. But I do not think that any of these things can be *specified*.

Could you join me in these views, we should be on common ground. But, then, you would not be so disinclined to the opinions I have set forth in the *Monist*; nor, even if you were, would you think them for that the less recommendable. But as long as you have your fixed creed, you must think that I am pushing free inquiry too far; and I dont see how you can consistently wish to publish my logical researches.

Or rather, let me be quite frank. I see the thing clearly. You simply wish to do so as an act of friendship to me, although the things the books contain are not the things you wish to be serviceable in getting conveyed to our public. That is why I emphasize as strongly as I can do, the difference between your proselytizing spirit and my inquiring spirit. One of these days, some large publishing house, animated by an ulterior motive, will choose to shoulder my philosophical book. I can wait for that. It is not what you are interested in; and it is too bad that you should throw away money upon it. It would, perhaps, be ungracious for me to say you *shan't* have it; but I think you had better send me your kind permission to reprint my *Monist* papers, & also return to me the Appleton's note. In any case, the one thing which tempted me to let you do it, when I thought you were not much inclined to it, namely, the prospect of seeing the book in the Exposition, is now, I take it, out of the question.

---

3 my] *above* ~~the~~

12–13 That would … of Science] of "science" *and* it in the phrase *inserted above*

31 my] *above* ~~the~~

As for your Chicago printer, perhaps he does not know that there will be 200 pages like the paper on the logic of relatives that I now send. Let him take a look at p. 31. J. J. Little on seeing this (or rather his *locum tenens*) said “Oh, that is the very sort of thing we are specially prepared for.”

Please pardon my prolixity, & understand that part of my *scolding* is because it is office of a friend to scold & part because I dont want you to discover later that we are not so close to same ground as you had thought. With my best compliments to Mrs. Hegeler & the family, I remain

very faithfully & gratefully,
C. S. Peirce

Peirce to Hegeler. MS: RL 77. Undated, estimated to be part of a draft of Peirce’s 7 March 1893 letter to Hegeler and related to a disclaimer Peirce mentions in the paragraph of that letter beginning, “You say ‘I do not like to see them …’ ”

[Unsent partial draft, 7 March 1893]

[…] I should make this comment. “In reference to the Open Court Publishing Co’s disclaimer, Mr. Peirce wishes to say that he would be unwilling to publish anything in Mathematics, Logic, Physics, or any branch of science or philosophy which should bear any endorsement. He does not guarantee the correctness of his own doctrines, nor recommend their acceptance, except so far as the reasoning may prove convincing.” So far, I do not see that there ought to be any difficulty. But here comes the obstacle. You say: “In a general way, it is understood that we endorse the views of the books we publish.” *[ ]*

Hegeler to Peirce. MS: RL 77. LPC: SIU 91.12.

Lasalle March 16th 93

Professor C. S. Peirce
Port Jervis N.Y.

My dear Professor,

Your favor of the 7th inst has duly reached me. I am glad to see from it, that Mrs Peirce’s health is better, than I thought by your former letters.

My state of health does not permit me to enter into any long explanations or a controversy by letter. I will state today only that we are ready to commence *at*

6 office] *after* ~~the fr~~
18 which should] *above* ~~with any~~
29 than] th*at*an
30–31 long explanations or a] *inserted above*

Chicago the printing of the two volumes (and are promised dispatch)—as soon as all the copy right of those papers which you want to go in are obtained by you, and also the percentages to be paid to you are arranged in definite words.

When I went into the verbal arrangement with you to publish the books the leading thought was that I thereby helped the scientific missionary work of the Open Court Publ. Co. in a general way and thereby attracting attention to the same, as also to do a service to science, and also to interest you in our work—as its missionary or opponent.

Wherein we disagree with the contents of the books, we can speak in our reviews of the same in the Monist and Open Court and the advertisements of the same.

If the books are published by us, you thereby give a certain credit to our work in a general way—and that should be done quite voluntarily only.

So the written permission from Appleton's is at your disposal as also a written permission by me on the same terms as Appleton's to have the Monist articles republished in book form—due acknowledgement, I understand, to be given at the head of each article.

I remain, dear Sir
Sincerely yours
Edward Hegeler

Peirce to Hegeler. MS: SIU 91.12

Milford Pa 1893 April 1

My dear Mr. Hegeler:

My arithmetics progress, though pretty slowly. What I do is, I am sure, a great advance over any other books of the sort now extant, both in pedagogical method and in the methods of computing. Nor will there be their only points of superiority.

---

2 of those ... go in] *inserted above*
2–3 , and also ... words.] *inserted in-line and below*
5 helped the scientific] *d*helped the ~~gen~~ scientific
6 in a general way and thereby] a *and* thereby *inserted above*
7 as[1]] *after* ~~al~~
7–8 , and also ... opponent.] *inserted in-line and below*
7 in] *over* per
10 the[3]] *inserted above*
14–15 a written permission by me] \a written ~~by~~/ permission \by me/
26 there] *on* ~~be~~

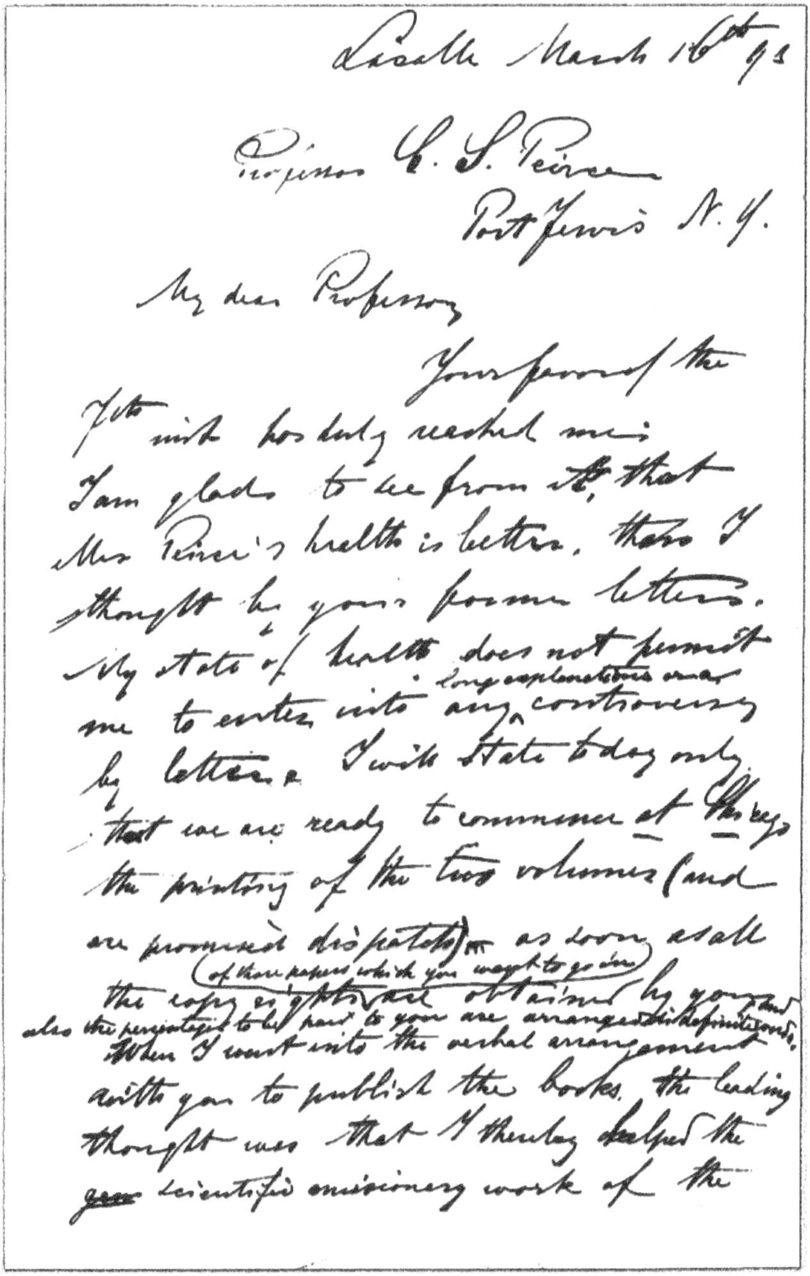
LaSalle March 16th 93

Professor C. S. Peirce
Port Jervis N. Y.

My dear Professor

Your favor of the 7th inst. has duly reached me; I am glad to see from it, that Mrs Peirce's health is better, than I thought by your former letters. My state of health does not permit me to enter into any long explanations or controversy by letters. I will state today only that we are ready to commence at [illegible] the printing of the two volumes (and are promised dispatch) as soon as all the copy (of those papers which you want to go in) is obtained by you and also the percentage to be paid to you are arranged definitely.
When I went into the verbal arrangement with you to publish the books, the leading thought was that I thereby helped the scientific missionary work of the

Figure 6: Hegeler to Peirce 16 March 1893, first page. The "misdirected letter" acknowledged by Peirce 9 April 1893 and which resulted in a miscommunication between the two regarding the fate of Peirce's "Quest for a Method." See letters that follow beginning 6 May 1893 for details.

I am well advanced in my revised catalogue of the books.[36] Suppose I were to succeed in selling my place, or were otherwise to find myself in a situation to be able to do so, how would you feel about allowing me to repay the $ 2 000 with interest?

I do not think I have much prospect of selling the place this year, nor hardly of renting it. The reason is that I am too much behind-hand with all my preparations,—a circumstance partly due to the impossibility of getting workmen here, and partly to my wife's poor health. Nor do I see much chance of getting the house into a proper state to receive visitors.

After the delightful visit I had in La Salle, I was most desirous that my wife should know you all, and that you should see her. Indeed, I shall do all in my power to bring about that result. But, alas, I am much hampered. My own inadequacy is one of the things I have always had to contend with. Much has to be sacrificed because my brain is not large enough to do more than just a little. If I can make these arithmetics in the time proposed, and keep my philosophy-mill going enough not to let the machinery rust, I fear that will be all I can accomplish this summer.

The recognition of the social element in philosophy,—the exhibition of it in psychology, in logic, and in metaphysics,—the tracing out its connections and the conditions of its development,—in short, the reconciliation of the I and the IT through the THOU,—that will be the direction in which I shall be of service to mankind, if I prove of any service; and that I have kept steadily in view for the more than 30 years I have been working in philosophy. There is where you and I come close together.

Please remit $ 250.

With warm regards to all your charming family
C. S. Peirce

Hegeler to Peirce. MS: RL 77. LPC: SIU 91.12.

---

36 Meaning, of his personal library sold to Hegeler in faith (without actually sending the books, apparently part of their verbal agreement), the catalogue to be sent to Hegeler as agreed upon in the memorandum summarized in Peirce to Hegeler 7 March 1893. Despite his promises to comply with Hegeler's repeated requests for it in the months to come, Peirce seems never to have sent the complete catalogue, only a preliminary list ca. 20 Aug 1893. (Related MS material survives in R 1554–1559.)

---

5 much] *after* ~~any~~

LaSalle April 1st 1893

Professor C S. Peirce
Port Jervis—Erie RR.
New York

My dear Professor,

I enclose check to your order for \$ 250$^{\underline{00}}$. Please acknowledge receipt
I wrote you last on the 16th ult.
Hoping that you are well yourself and that Mrs Peirce is in fair health
I remain

Yours truly
Edward Hegeler

Hegeler to Peirce. Telegram: SIU 91.12.

Telegram LaSalle 4 /3. 93

Prof. C. S. Peirce
Milford Pa.
near Port Jervis

Letter of April 1 containing draft was addressed to Port Jervis.

s.g. E. C. Hegeler

Hegeler to Peirce. LPC: SIU 91.12.

La Salle Ill April 3 93

Professor C. S. Peirce
Milford Pa.

Dear Sir:

Since you have given me no new address my letters to you of March 16, April 1$^{\underline{st}}$ (containing draft for \$ 250$^{\underline{00}}$) and of this morning were all directed to Port Jervis N.Y. My letter of March 16 was returned *to me*.

It will be necessary for you to instruct the Postmaster at Port Jervis to forward these letters to you.

Respectfully
Edward C. Hegeler
per C Diesterweg

---

7 on] *after* ~~it~~

8 Hoping ... health] is *and both* that's *inserted above*

27 My] *above* ~~The~~

Hegeler to Peirce. MS: RL 77. LPC: SIU 91.12.

LaSalle April 3^{d} 1893

Professor C. S. Peirce
Port Jervis
on Erie RR
New York State

My dear Professor,

I wrote you April 1st enclosing check for \$ 250$^{\underline{00}}$. This morning my letter to you of the 16th ult is returned to me. For explanation I enclose the envelopes. I also send the letter.

As the letter of April 1st may have been misdirected in a similar way, I cause to be written and telegraphed to the Postmaster at Port Jervis N. Jersey to forward same to Port Jervis, New York, in such case.

Your friendly letter of the same date—April 1st—also came this morning—also yesterday your compliments through my son-in-law—Mr. Bucherer

I write in haste this morning, and await to hear from you in answer to my letter of the 16th ult—more especially in regard to the publication of the books on logic and philosophy—

With my regards to Mrs Peirce I remain

Sincerely yours
Edward Hegeler

PS. I find there is *no* Port Jervis New Jersey.

Peirce to Hegeler. MS: SIU 91.12.

Milford Pa. | 1893 Apr 5

My dear Mr. Hegeler

Your enclosing draught comes to hand today.

Yours very Truly
C S Peirce

Yours of March 16$^{\underline{h}}$ did not reach me.

---

12 and telegraphed] *inserted above*
17 more] *inserted above*
22 PS. I … New Jersey] *This line in pencil up left margin.*

Peirce to Hegeler. MS: SIU 91.12. Stationery headed, "Express Address. | Port Jervis, N.Y. | Wells, Fargo Exp. Co. | C. S. Peirce."Arisbe" | P.O. Milford, PA."

1893 April 7

My dear Mr. Hegeler:

I hasten to say that my address is Milford, Pa. till further notice

In haste
C. S. Peirce

Peirce to Hegeler. MS: SIU 91.12. Stationery headed, "Express Address. | Port Jervis, N.Y. | Wells, Fargo Exp. Co. | C. S. Peirce."Arisbe" | P.O. Milford, PA."

1893 April 9

My dear Mr. Hegeler:

Yours of the 16$^{\underline{h}}$ March is to hand. I am very glad indeed to get it. I had given up entirely all hopes of such a solution of the matter & had consequently begun a new revision or rather rewriting of the whole series of essays. As so much time has now been lost, it will probably make little difference if I take a week or two now to complete the new arrangement of the text.

I am sending Dr. Carus a Shelly's Poems (of course, *not* the copy I sold you) and with it I will put some of my papers. Your Chicago printer will have no difficulty with the logic of relatives, will he?

very faithfully
C. S. Peirce

Address as above till further notice.

Hegeler to Peirce. MS: RL 77. LPC: SIU 91.12.

Lasalle May 1st 1893.

My dear Professor,

I enclose check for $ 250$^{\underline{00}}$. Please ackn receipt.

I received your favor of the 9th ult.—the Chicago printer I hold to be competent.

Expecting yourself to be well and hoping Mrs Peirce to be in fair health I remain,

Yours truly
Edward Hegeler

---

17 am] *over* sen

28 hoping] *over* wishing

Peirce to Hegeler. MS: SIU 91.12.

Milford Pa 1893 May 6

My dear Mr. Hegeler:

I write to thank you for your draft for $250 dated May 1 which I received yesterday.

I must also explain why I have not yet forwarded my copy of "The Quest of a Method." You must know that at the time when in consequence of the non-receipt of your misdirected letter[37] I had concluded you had decided not to publish the book, I gave up the idea of getting it ready for the "Exposition,"[38] and commenced tearing the copy to pieces, with a view of transferring passages from one essay to another so as to bring together all that related to one question, correcting imperfect presentations supplying omissions, correcting blunders, and making material additions. I had done a great deal of this work & of a relatively fine quality, when I got word you were ready to print. Then I thought it would be an easy thing to get the copy back nearly into its original condition. But I found myself unwilling to do that & was tempted to try to make one little change and then another, so as to utilize the work of revision that I had done. Now I feel that I cannot make the thing satisfactory, without a great deal of work. Besides, it would be absurd to print the Monist articles as they are; for without those which are to come later no reader can understand what it is that I am really driving at. It would certainly be necessary to add several chapters on that account, and these ought to be written with the greatest deliberation. Now I can only state the case to you. I have agreed to get the articles together for printing. That I can do in a day or two, and will do so if you desire it. But I have come to think that the book without any material

37 Referring to Hegeler's letter of 16 March—in response to Peirce's long letter of 7 March providing a memorandum per Hegeler's request and outlining Peirce's "Quest for Method"—which was returned to Hegeler and later re-sent to Peirce. Peirce replied to the finally received letter 9 April, explaining, as he does here in detail, that he had assumed Hegeler was not interested and begun reworking the "Quest" extensively. Hegeler responded 15 May (below) and, contrary to his of 16 March and the "verbal agreement" there referenced, declined Peirce's "Quest" proposal, whether partly or not due to this miscommunication, for which he blamed Peirce and even enclosed "a special sheet" explaining why. Peirce later wrote in defense 19 May and again 20 May that the discrepant mail resulted mostly from clerical error at the post office. But the damage had already been done. Whatever the reasons for it, this small mis-step possibly played a role in the failed publication of Peirce's "Quest."

38 *Chicago World's Fair* of 1893. More information footnoted in Peirce to Hegeler 24 Feb 1893.

9 gave up] *after* ~~commen~~

11–12 correcting imperfect presentations] ~~to~~, correcting imperfect ~~passages~~ presentations

addition to its bulk will be infinitely cleaner, more forcible, and more worthy of you & of me if I lay aside the idea of printing it at once & complete the revision I had commenced & which will take a long time, at least six months, perhaps nine. Then I think the book will be one that posterity will hold in remembrance. What shall I do?

During the last month, this work has taken too much of my time from my arithmetic. If I could get some kind of contract to do work for the *Open Court*, I might be able to see my way toward getting that done. My wife, although her lungs cause me less & less uneasiness, is certainly far from well. It would do us both great mental good to go to France & once there our expenses would be about what they are here. But I couldn't go until I had made some disposition of my house & got my arithmetic into such a state that I could be out of reach of personal interviews in New York. There are many men whom I want greatly to talk to before I publish my book; yet at present I seem to be tied down here.

My wife and I want very much,—more than I can tell you,—to have the pleasure of a visit from you and Mrs. Hegeler; and now is the finest season. I wish you would grant me this favor. If you would like to bring Carus, I should enjoy it. Our house is large enough to accommodate a larger number of you, but at present we haven't got many large beds. It takes just as long to come over the Erie road from Chicago to Port Jervis as over the Central to New York and owing to the trains not being crowded I should imagine it was a very pleasant way to do the journey. I shall go that way if I go to Chicago again.

yours very truly
C. S. Peirce

Hegeler to Peirce. MS: RL 77. LPC: SIU 91.12.

Lasalle May 15th 1893

Professor C. S. Peirce
Milford Pa
My dear Professor,

I received your favor of the 6th inst in due course of mails.

It seems that you are under the impression that I or my secretary had misdirected a letter to whereby you had received it much later. We hold this to be a mistake on your part. My secretary has written details on a special sheet which I enclose.

20 to Port Jervis] *inserted above*

32 later] later$^{\vee}$.$^{\vee}$ ~~than~~

The cordial invitation from Mrs. Peirce and yourself to Mrs. Hegeler and myself to visit you at your home we cannot follow for the present. Please express our thanks to Mrs. Peirce.

I have to give a negative answer to what you say about a sojourn in France and for the present also to what you say about the rewriting of your papers for publication in book form.

I think you should concentrate yourself totally on the completion of the arithmetic. I request for your report of the status of this matter. It is a large sum of money that I am to invest therein.

You ask about getting some kind of a contract to do work for the Open Court. To clear this matter and all our business affairs, we have next to make a full settlement for the work already done by you.[39]

You are informed that a special line of work is laid out for this publication. If any work falls in that line and is suitable, I with the assistance of Dr Carus have to *[*?resend for report*]* my opinions. A contract in advance cannot be made.

I have to close for today. As to the publication of the papers in book form I confirm the release given you in a former letter.[40]

I remain
Sincerely yours
Edward Hegeler

Peirce to Hegeler. MS: SIU 91.12.

Milford Pa 1893 May 18

My dear Mr. Hegeler

I am deeply pained by what seems to me the tone of Carus's refusal of my invitation as if it were pitched to just escape the charge of incivility.[41] And you do not yet answer at all. Apart from the humiliation which this puts upon me, it makes me much more miserable for other reasons. I conceived a deep veneration and affection for you; and left your house in the greatest comfort of mind to have met such a man, and one who in several ways had shown signs of a real sympathy

39 Final page of letter missing, so remaining text comes from LPC.

40 Hegeler to Peirce 16 March 1893, though Peirce was unsure, as he notes frustratingly in his 19 May response (paragraph beginning "You close by saying ...").

41 Carus to Peirce 9 May 1893, declining the invitation to visit Peirce.

15 *[*?resend for report*]*] ~~*[*?render*]* the independent judge~~ *[*?resend for*]*/ *[*?report*]*

with my deepest and most passionate aspirations.[42] I had already seen some signs that you were a little narrower than the sort of man I had known best; but that was not surprizing in a man whose life had been a contest with men. But I received a surprizing evidence of it, afterward; and wrote you the earnest letter I did.[43] Since then, I hear so little from you, it cuts me badly. You must know that though the small difference which exists between us seems to have removed your interest in me, it has not at all affected my feeling for you. Though I am a thinker, I am a man of passionate human feelings. I lament that you should overestimate, as you appear to me to do, the importance of any particular opinions, as compared with that spirit which is sure in the end to make opinions right, namely the spirit of moderation combined with earnest seeking. But though I lament that, it does not alter at all my feeling of need to be in hearty accord with you and to help you in the glorious enterprise into which you have thrown yourself. The coolness which seems to have sprung up is to me inexplicable. I have been strongly tempted to send you on my books (which would easily sell for all you have advanced) and to refuse all further advances, because I thought my having accepted them might have made you doubt the sincerity of my sentiments toward you personally. I may yet decide to do so; though I know you are a generous—& that's a poor word—a man who knows just what money is worth & what it is not worth. You are not so sentimental as I am; but I am too much so. I will do anything in conscience to further the things you have at heart. I will go about the country & lecture to increase the circulation of the *Open Court*. I will do anything to show the sincerity of the feelings I profess toward you. But I beg for some slight reciprocation.

C. S. Peirce

I enclose one[44] of the papers I have ready for the *Open Court*, not to have you publish it but in hopes it may lead you to take a less unfavorable view of my opinions.

---

42 Referring to their late Jan–early Feb 1893 meeting.

43 Perhaps referring to Hegeler to Peirce 16 March 1893 and then Peirce to Hegeler 6 May 1893.

44 "Critic of Arguments. That all inferences concerning matters of fact are subject of certain qualifications," R 590 (never published). That Peirce truly did not intend to publish this piece, as he claims here, is unclear. See Peirce to Carus 10 May 1893, where he seems to propose this and "Critic of Arguments. III." as subsequent numbers in the "Critic of Arguments" series in *The Open Court*.

---

1 seen] *after* ~~had~~
5 from] *above* ~~of~~
5 cuts] *after* ~~makes~~
7 it] *after* ~~that~~
9 importance of any] *inserted above*
15 (which] *over* & to
22 the[3]] *above* ~~my~~
22–26 show the sincerity ... opinions.] *up left margin and across header on first page*
25 one] *over* a p

Peirce to Hegeler. MS: SIU 91.12.

Milford Pa 1893 May 19

My dear Mr. Hegeler:

Yours of the 15$^{\underline{h}}$ reached me last night; but I was suffering from a little bilious attack and not able to answer it.

I still think the letter in which you informed me you would go on with the publication of the book was misdirected[45]; for my direction of Feb 28 to send letters to Port Jervis till March 3, coupled with the contents of the letter, in which I stated that I was then to go to my house, was, I think, clear enough. Your secretary says I should have notified the Port Jervis Postmaster. To plead contributory negligence ought to be taken as admitting one's own negligence. As a fact, I dont remember whether that was done or not. As he truly says, it *ought* to have been done. As well as I remember, and most likely I have forgotten facts in my favor, this is what happened. I went down to New York with my wife, leaving my things or some of them at the hotel & requesting them to keep my letters. On my return, I went to the hotel, asked for letters, asked them to notify Post Master to send mail to Milford as usual, and went home. I will remark that Port Jervis is only a village; I am well known there; and the Post Master has standing orders to forward any mail for me. But on this occasion, knowing my house had not been occupied by me, he perhaps thought he would send the letter back. In short, like a country postmaster, he perhaps supplemented his formal orders with his private knowledge, without taking pains to have the latter accurate. Now, if you think there was contributory negligence on my part, all right; I admit there *may* have been, though it is not *proved.* It's a mere guess. If the whole question was whether you had not injured me, that would be a good answer. But that is not the question at all. The question is, why, when I had said my copy was all ready, I did *not send it* on promptly when informed the printer was ready for it. To this I reply that it was because, being left a long time without hearing anything, I concluded you weren't going to publish, and commenced picking to pieces & partly rewriting and greatly rearranging the copy; so that now I no longer feel satisfied to print just as it is. This is my *defence.* This defence you will not admit, because you say *[I]* did not give you my address

45 Hegeler to Peirce 16 March 1893.

4 suffering] *over* d
9 to go] *above* ~~going~~
12 truly] *inserted above*
13 and most … favor,] *inserted above*
14 to] ~~to~~ *(cancallation reverted) [E]*
19 had] *above* ~~was~~
19–20 perhaps] *inserted above*
21 perhaps] *inserted above*
23–24 there *may* … guess.] *above* ~~it.~~
30 no] *d*n̲o

correctly. I maintain my directions were explicit. You say "You should have notified the Port Jervis Postmaster." This I admit; but you have not proved I did not notify him; and even if there were negligence here, it could have done no harm without negligence likewise on the part of you or your secretary. Hence, my defence holds good, and ought to be admitted by you.

I cant tell you how sorry I am you will not come to visit me. I have many things to talk about with you. But, alas, you seem, for some reason I cannot penetrate, to have lost all interest in me. But you could, at least, if you came, get an idea whether your money was safe or not.

I am sure I did not ask you about my sojourn in France. I should not think of such a thing, without first freeing myself from my obligations to you. You mistook some general remarks addressed to a supposed friend, for business;—another illustration of your change of heart.

Your admonition that I should concentrate myself totally on the arithmetic would be almost mandatory, even if I did not think so myself, as I do. On this, I have three remarks to make.

1st. My philosophical book, when done as I want to see it done, and as it would take me a year to do it, will be certainly the greatest philosophical work for half a century, unless Herbert Spencer be considered as a satisfactory exponent of evolutionism. If Spencer has really the place his admirers give him, then nothing can compare with him. Putting him aside, no very great book has been published since Mill's logic,[46]—not even any book of great pretensions. Before that, there was nothing later than Hegel's Logic,[47] about 1816, if I remember rightly. Taking my book at *my* valuation of it, there has been nothing equal to it since Hegel. But taking it at any other common sense appraisal, it is a book expounding a view of logic admitted to have great weight on all hands, a theory reaching from the widest general philosophy of the subject down to minute details, set forth with full historical learning, and with a good English style,—and this theory of logic is shown to lead irresistibly to a philosophy of as great novelty as anything so

46 *A System of Logic* (1843).
47 *The Science of Logic* (1812–1816).

3 even if there were] *if*even if there ~~was~~ \were/
4 likewise] *above* ~~also~~
14 admonition] *after* ~~opinion~~
16 three] *above* ~~two~~
23 Logic] Logic~~k~~
25 at] a*n*t
25 appraisal] *after* ~~p~~
26 admitted] admitt*ing*ed

fundamental can have,—which if true is incontestably of the highest importance. This philosophical consequence of the logical theory is followed out into all departments of philosophy and is shown to revolutionize them everywhere—with results most favorable to natural and mathematical science and equally favorable to the natural sentiments of man and especially to religion. Finally, this theory is put to the task of *predicting phenomena* and is rigidly compared with *observation*. When we come to that part, I think there will be some surprise.

Now even supposing my theory to be all a mistake, yet granting my competence and mental power, it must be admitted, I think, to be as important a book as Mill's logic, or at least more important than any book since then.

The object of all this self-laudation is to show you that it is well worth while to give a year and perfect this book, since you are to have the publication of it, if we come to terms. I have by this time pretty much decided I do not want the articles reprinted without revision. There would be too much repetition, and too little arrangement. Therefore, I will *just put that whole matter entirely aside, while I write the arithmetic.*

2nd. The rule to concentrate myself totally on the arithmetic must necessarily be a little relaxed in one direction. Writing an arithmetic is an irksome task; but to impart to it novelty requires invention to be continually on the stretch. The *irksomeness* and the *invention* both render it very fatiguing work. Consequently, there remain some hours each day, which must be filled with some kind of intellectual activity to take the taste of the arithmetic out. I quite agree the "*great book*" must not be touched; for that would become too absorbing. But something must be done.

Now I *must* be on the look out for something to do when the advances on the arithmetic cease.

You decline to contract with me to write for the *Open Court*. The effect of that is, as you must clearly see yourself, that it will not be in my power to write for the Open Court. Nobody nowadays ever writes for any journal or magazine except under a contract,—I mean no professional writers do so. Of course, amateurs do all sorts of silly things. All the writers for New York papers, for instance, are under contract. Of course, the editor is not bound to *print*, but he is bound to *pay*.

It is quite clear to me that for some reason utterly incomprehensible to me, your tone to me is entirely changed. You do not want me to write for the *Open Court*.

1 incontestably] incontestibly *[E]*
2 all] *above* ~~every~~
8 my theory] *above* ~~this~~
19 impart] *above* ~~make it~~
19 invention] *a*invention

Very well; then I must form other relations. To do so, I shall have to do some preliminary writing; and this can well occupy some of the time that is left on my hands each day after I have done as much work on the Arithmetic as it is wise to do.

3rd. I have another scheme in view. You are so little friendly to me, if not positively the reverse, in your *tone*,—for there has been no occasion to see it in acts,—that I really hesitate to broach my scheme, particularly as it would interfere somewhat with the arithmetic. However, here it is. I have been thinking of advertizing a summer school of philosophy here. Courses of twenty-five lectures. Tickets for each course $ 5. I would begin with a course on Reasoning and another on The English Dictionary each three times a week. Lecture one hour. Another hour would be consumed in talk. The *preparation* would occupy no useful time. I should give the money from my lectures to transportation of the people to and from Milford etc. and other conveniences. They would board in Milford at $ 10 a week. I would get some of the soundest minds in the country to help, and they would stay in my house, one at a time, or for a day or two, two gentlemen since that entails no expense to speak of, and the effect of the thing would be to create and diffuse an interest in my philosophy. I should only be too glad if I could get you to lecture on religion; but I feel that would be to surrender your present air and and attitude. So I fear there would be little hope of that. I think this scheme would greatly further what I have at heart. It would fill many people with that broad scientific spirit of which I am an apostle, it would make them feel how in religious matters the points of agreement are great and important, the points of difference small & trivial. It would improve their reasoning powers, and enable them to judge better, especially about abstract and philosophical questions. It would imbue their minds with conceptions which would aid them to see the truth of the fruits of my studies. I am sorry you do not care for any of these things enough for me to expect any countenance from you. If you would only come into it and come and deliver 25 lectures on your religion, you would help the purposes of the *Open Court*. Of course, I should take every occasion to talk up the *Open Court*, though it seems to me very likely that by that time it will be settled down into something almost like hostility to me; and I shall soon have a weekly journal of my own, or one that I shall be able to

1 other] *inserted in-line*
6 it] *over* a
7 my scheme] *above* ~~it~~
16 a[2]] *above* ~~e~~
16 gentlemen] *inserted in-line*
19 and attitude] *inserted above*
20 fear] *after* ~~feel~~
20 think] *after* ~~feel~~
29 should] *above* ~~shall~~
31 into] *above* ~~in~~

put my energies into. But of course, not before the arithmetic is done. But I shall always praise the *Open Court* no matter how it may run me down.

To return to the school. It would serve to make intellectual people acquainted with one of the most ravishing pieces of country that are still left just as it was turned out of God's studio, without a railway, without a single factory, for fifty miles. But with a road like Macadam and cooks like Vatel. To look out my window is to praise God.

This would again tend to fill the valley with the kind of people I want. It would also make my place known. I have been informed that if this were done and the school seemed to have life I might hope that a certain gentleman would make an endowment for it.[48] I am far from confident of it, however. But the thing certainly would enhance the value of my real estate.

If I do not do it, it will be solely because I feel myself bound to you not to permit anything to interfere with the arithmetic.

These are the three remarks I had to make on that head.

You desire a report on the state of the arithmetic, very properly. I cannot well send on the copy now because it is all confusion. You could not make it out. But I will make up a description of it in its present state. There are several hundred pages of MS. (I tell you this in advance of the details which I will try to send you Monday.) The plan of the whole is laid out and details as to the methods of treating various subjects. Some of these notes might be sent to you; but you will please return them. A good many things are all done just as they will appear; but no chapter, I guess is complete. A good many examples have been made and materials collected for others, and a general scheme of examples is ready. In short, the work will probably be ready during the summer. But I admit *much time has been lost* in one way and another.

---

48 Investor unknown, possibly one of Peirce's affluent acquaintances through his well-to-do Milford neighbors and close friends James W. and Mary E. Pinchot. As Brent notes (186–187 and 237), Peirce frequently interacted with and carried correspondence with high-profile visitors to the Pinchot estate, such as Mrs. Oliver H. P. Belmont, the Vanderbilt and Harriman families, and New York publisher George A. Plimpton, whom Peirce approached for money and employment in 1894.

1–2 But I … down.] *inserted in-line and between paragraphs*
3 would] *over* will
4 it was] *above* ~~they were~~
5 fifty] *above* ~~twenty~~
8 would] *above* ~~will~~
8 would] *above* ~~will~~
16 cannot] *after* ~~will~~
18 several] *after* ~~about~~

You say you want to make a full settlement for work already done by me. As far as I know, there is nothing due, except for reading the proofs of Mach. For that, I said I would take whatever Dr. Carus saw fit to give. The work has amounted to considerable all together but I shall be compelled to take whatever you give, no matter how liberal you may choose to be; if it errs on the *other* side, it wont cut me half so badly as your coldness cuts me. My furnace freezing involved terrible expense; repairs to the house have been too urgent to be put off; Mrs. Peirce's health has necessitated large disbursements; one monstrous bill came to my knowledge only this month and had to be met immediately; and consequently, whatever may be allowed for the Mach revision would be most welcome in the form of cash, instead of being charged to my credit. The same may be said of the pay of any articles you may accept for the *Open Court*. I sent you two articles about religion,[49] which were the fruit of long reflection, and which I hoped would be sympathetic to you. I was more than disappointed, I was wounded, when in answer to the first,—after inquiring whether I meant to attack the management of the *Open Court*, for which I have only had laudations, except in some personal letters on minor points, he then merely added you wanted to know what *honorarium* I expected,[50] as if it went against the grain for you to take the article, for you to take the article and you did not wish to pay what was clearly stated at the outset as the price I should expect for all my articles. My second religious article was also considered unsuitable for the *Open Court*. You are certainly not bound to accept religious articles from me. You can just return the two. But if you take them, you should know yourself, that maturely considered articles upon difficult questions cannot be written for less than the price I got at the outset. I also sent you personally a logical article which I hoped you might be interested in reading. I marked it as "not for publication" because I doubted if you wanted anything more from me for the *Open Court*, and from this last letter, I rather think you would like to be released from any understanding that you were to publish logical articles, even if such an understanding exists, and you would very likely think not. So you can just send that back too.

You close by saying that you confirm the release given in another letter about the republication of my articles. You express yourself with a sententious brevity,

49 "What Is Christian Faith," *Open Court* (27 July 1893), and "Immortality in the Light of Synechism," *EP* 2:1–3 (never published by OCP).
50 Carus to Peirce 4 May 1893.

3 The work] *above* ~~It~~
15 for] *inserted above*
18 for you to take the article] \for you/ to take ~~it~~ \the article/
19 was] *after* ~~it~~
19 as the price] *above* ~~that~~
24 logical] logical~~l~~
29 and] *above* ~~which~~
31 You] *over* S
31 brevity] *after* ~~pr~~

which leaves one to guess what you mean, even if it does not imply that you dont care what interpretation is put on it. I shall endeavor to interpret the remark in a way honorable to you. Sometimes I have heard a street boy extend to another a permission to kiss an unrespected part of his person. When you tell me that I can go and get another publisher, I will not suppose that you mean simply to insult me, or that you really want to change your mind. I desire to have you publish the book when it is ready, though I propose to lay it aside for the present. I hope I shall have it ready next year, and then, unless I should be compelled to make a different arrangement, I propose to send you the copy. I trust you will publish it, and pay me a royalty. I think that royalty should be high, because my whole life goes into the book, and because the sale of such a book is not very sensitive to differences of price. But I shall not quarrel about the amount of the royalty. That there should be a royalty you have already admitted.

I spoke above of "Lectures on the English Dictionary." These would consist of the history of words,—not their prehistoric history,—but their history in the writings of philosophers,—showing how their present shades of meaning have grown up, thus giving lessons 1$^{\underline{st}}$ in English, 2$^{\underline{nd}}$ in Philosophy, 3$^{\underline{rd}}$ in the history of thought. These things are substantially written on the interleaved pages of my Century Dictionary. Easily enough for more than 25 lectures.

The warmth of my friendship has not changed one bit; and no friend can say that my fidelity ever wavered or my steadfastness changed without an imperative reason and a frank explanation.

yours as ever
C. S. Peirce

Kind regards to all the family.

Peirce to Hegeler. Postcard: SIU 91.12. Recto headed, "Postal Card One Cent | United States of America" with portrait at right.

[recto]

Port Jervis, N.Y.| May 19 | 6 PM | 93 *[stamped]*

E. C. Hegeler esq^{re}
La Salle
Illinois

[verso]

---

10 I] *over* T
11 very] *inserted above*
12 there] *over* y
21 wavered] *over* wanted
25 to] too *[E]*

1893 May 19

Dear Sir:

I send you a volume of my papers in quarto with two or three others.[51] These will give you some idea of what my work has been, though there are many others, & in directions not here represented.

C. S. Peirce

Peirce to Hegeler. MS: SIU 91.12. Stationery headed, "Express Address. | Port Jervis, N.Y. | Wells, Fargo Exp. Co. | C. S. Peirce."Arisbe" | P.O. Milford, PA."

1893 May 20

Dear Mr. Hegeler:

Since writing you yesterday I have looked up the envelope of the letter which did not reach me on time.[52] It is directed in your secretary's hand to me at Port Jervis, N.Y.

The La Salle Post Mark is indistinct. The date looks like this

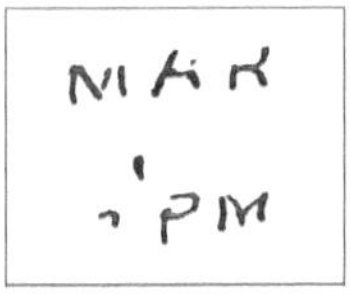

Figure 7: Peirce's rendering of illegible postmark.

It was received in Port Jervis March 18. $7^{\underline{30}}$ P.M. Before that I had sent special word to the Port Jervis Post Master that I thought a letter must be there for me. I also called in Person Monday May 13 and desired letters forwarded to Milford. But this letter instead of being sent to *Milford* was sent to Middletown, N.Y. an obvious mistake of the P.O. clerk. It was then sent back to La Salle.

I consider that my direction about my address was specific & if it had been followed the letter would not have gone astray.

51 Another portion of his "Quest for a Method" MSS, in addition to those sent to Carus late April 1893, perhaps for Hegeler's reconsideration.
52 Hegeler to Peirce 16 March 1893.

16 for] fore

yours very faithfully
C. S. Peirce

P.S. Whether or not there is any use in writing on the envelope "On Erie RR near Port Jervis" or not, I don't know enough about the business of carrying mails to know. Port Jervis *is* on the Erie RR (Milford is *not*) and I suppose the letter ought to go that route without doubt.

Peirce to Hegeler. MS: SIU 91.12. Stationery headed, "Express Address. | Port Jervis, N.Y. | Wells, Fargo Exp. Co. | C. S. Peirce."Arisbe" | P.O. Milford, PA."

1893 May 21

Dear Mr. Hegeler:

The Mach work seems to be finished, and I shall be very glad to receive a check for it. I agreed to leave the amount to Dr. Carus. I prefer to abide by that. In case he should insist upon my naming the amount I enclose my figure in a sealed envelope,[53] with the request that you will burn it if Dr. Carus names the amount, and *in no case* to tell him I have enclosed this unless he positively refuses to name a figure. I enclose it simply to avoid loss of time.

My Report on the Arithmetic will go forward today. The copy is not so much in arrears as my expressions may have led you to fear.

Yours very Truly
C. S. Peirce

Peirce to Hegeler. MS: SIU 91.12. MS Copy: RL 77, headed "Copy".

Milford, Pa. 1893 May 22

E. C. Hegeler, Esq
La Salle, Ill.

Dear Sir:

The following MS. is in my hands for my Arithmetics:

Copy for Primary Arithmetic.

About 50 pages, with very numerous rough sketches

---

53 Enclosed note reads: "$ 150 if I am to be free to review the book at once $ 250 if it is desired that I should refrain from doing so for a year." (SIU 91.12) Hegeler remitted the $ 150 on 25 May 1893.

---

12 that] *over* this 26 my] *over* th

for illustrations, equivalent to about	5 000 words
Copy for Advanced Arithmetic	
About 20 pages, type-written, 60 lines, or 1 000 words to page, say	20 000 words
About 50 pages written, about 120 words to page, say	6 000 "
Total copy	about 31 000

Preparation for copy		
Detailed notes for Primary Arithmetic	50	pages, about
Examples mostly statistical for Advanced A	50	" "
Calculations for physical examples	30	" "

In addition to the above, there is a quantity of MS, which, though it will not serve for copy for the arithmetic, was prepared to guide me in writing it, and out of which one or two articles can be made. Namely, there are

Notes on previous arithmetics. About 40 textbooks now in the schools are carefully noticed. (I much regret not having German, French, and Italian textbooks). About 20 older books are noticed. In all about 150 pages MS.

Notes on apparatus to be used in teaching arithmetic, on the psychology of the subject, on my method of teaching each operation, various attempts at laying out the work so as to get the whole subject in moderate compass. Memoranda of matters to be introduced, etc. About 100 pages MS.

In all, there are about 500 pages MS.[54]

The work is in rather a backward condition; but by no means desperately so.

Yours very Truly
C. S. Peirce

54 In *NEM* 1:xxxii (also Martin, 324), Eisele notes that on a separate sheet is another breakdown as shown below (MS unidentified):

20	Pages typewritten copy for Advanced Arithmetic
50	Pages written copy for Advanced Arithmetic
50	Pages examples for Advanced Arithmetic
30	Pages calculations for examples for Advanced Arithmetic
50	Pages for Primary Arithmetic
50	Pages detailed notes for same
150	Pages notes on Existing arithmetics
100	Pages Remarks on methods, apparatus, Plan of the Work
500	pages of MS.

13 textbooks] texts-books *[E] (for consistency with next occurrence)*

14–15 textbooks] *above* ~~books~~

Hegeler to Peirce. MS: RL 77. LPC: SIU 91.12.

La Salle Ill. May 25/93

Professor C. S. Peirce
Milford Pa.

Dear Sir

Enclosed please find check to your order for $ 150$^{\underline{00}}$ in payment of your work on the translation of Mach.

Yours truly
Edward C. Hegeler
per C Diesterweg

Peirce to Hegeler. MS: SIU 91.12. Stationery headed, "Express Address. | Port Jervis, N.Y. | Wells, Fargo Exp. Co. | C. S. Peirce."Arisbe" | P.O. Milford, PA."

1893 May 27

E. C. Hegeler Esq

Dear Sir:

Please remit $ 250 advance upon my arithmetics as per agreement. The books are now advancing rapidly and successfully, and will, I think, be very popular.

Yours very truly
C. S. Peirce

Peirce to Hegeler. MS: SIU 91.12. Stationery headed, "Express Address. | Port Jervis, N.Y. | Wells, Fargo Exp. Co. | C. S. Peirce."Arisbe" | P.O. Milford, PA."

1893 May 30

E. C. Hegeler Esq

Dear Sir

I have to acknowledge receipt of yours of 25$^{\underline{h}}$ inst enclosing draught for $ 150 in payment of work on Mach.

Yours very truly
CSPeirce

---

6 to] *after* ~~for~~

6 of] *after* ~~for~~

Hegeler to Peirce. MS: RL 77. LPC: SIU 91.12.

La Salle Ill. June 1st 93

Professor C. S. Peirce
Milford Pa

Dear Sir

Enclosed please find N.Y. exchange to your order for \$ 250^{00} being the fifth advance payment on a/c of Arithmetic, for which please acknowledge receipt.

Respectfully yours
Edward C. Hegeler
per C. Diesterweg

Peirce to Hegeler. MS: SIU 91.12. Stationery headed, "Express Address. | Port Jervis, N.Y. | Wells, Fargo Exp. Co. | C. S. Peirce."Arisbe" | P.O. Milford, PA."

1893 June 5

My dear Mr. Hegeler:

I am weekly getting letters plainly addressed to other persons, and hence conclude that I fail to receive many letters addressed to me. Accordingly, when a correspondent surprizes me by his silence, as you now do, I always write and ask him what the matter is.

—I ask you that now.

—I also ask, whether you have changed your mind, and are sorry you ever took me up; in which case, I will do my utmost to set you free at the earliest possible moment, without pecuniary loss to you, whatever it may cost me.

—I also ask whether the Open Court desires further articles from me in the Critic of Arguments series, as Dr. Carus told me you did, when I was in La Salle.

—I also ask whether you are going to keep the articles I sent on religion.[55] If not, please return them.

—I also ask whether you sent me the \$ 150 for Mach in consequence of opening my envelope, or whether this amount was fixed without opening that envelope. This concerns me particularly.

55 "Christian Faith" and "Synechism."

17 now] *inserted above*

Except the letter transmitting the money for the Mach work, I have received no letter from you since the one in which you asked for a report on the state of the arithmetic.[56] I trust the \$ 250 will come to hand soon.

My wife continues in very poor health; although I have little anxiety about her lungs, now.

With warm regards to Mrs. Hegeler and all the rest of your charming family, I remain, Dear Mr. Hegeler,

Your devoted
C. S. Peirce

The early volumes of the Open Court, containing contributions from many different thinkers of different stripes, of special views, are very delightful and very valuable.

Hegeler to Peirce. MS: RL 77. LPC: SIU 91.12.

La Salle Ill. June 7/93

Professor C. S. Peirce
Milford Pa
Dear Sir

A bank draft for \$ 250$^{\underline{00}}$ was mailed to you on *June 1$^{\underline{st}}$* addressed like this envelope viz: Milford, Pennsylvania near Port Jervis on E.R.R.[57]

Respectfully
Edward C. Hegeler
per C. Diesterweg

Peirce to Hegeler. MS: SIU 91.12. Stationery headed, "Express Address. | Port Jervis, N.Y. | Wells, Fargo Exp. Co. | C. S. Peirce."Arisbe" | P.O. Milford, PA."

1893 June 7

E. C. Hegeler Esq
La Salle, Ill.
Dear Sir:

56 15 May 1893.
57 Erie railroad line.

3 I trust … soon.] *inserted in-line and between paragraphs*

18–19 like this envelope viz:] *inserted above*
19 Pennsylvania] Pensylvania *[E]*

Yours enclosing the 5th advance of $ 250 each on the Arithmetics came to hand. There remain two more similar advances to be made, according to our agreement.

I have to be thinking what I am to do after that, until the money from the Arithmetics comes in. I will make the following proposition:—

You to add 4 pages a week to the Open Court, devoting 2 of these to a "Scientific Department, edited by C. S. Peirce."

I to get the most eminent men of my acquaintance, not now writing for the Open Court, to fill half of this space, or one page, I paying them. The rest I write myself, on Logic, Philosophy, History of Science, and what I please, only avoiding as far as possible religious questions, at least, so far as there is the slightest danger of conflict with you & Carus. Now, let me tell you, that while I may find fault with my friends in private (and I never find fault in private with people I do not reckon among my friends) nobody, I believe, ever accused me of not being thoroughly loyal to my friends. Carus seems to be filled with animosity against me, for some unknown reason; but I have no feeling against him, but on the contrary, think he is a good man & wish to forward his work. My tychism, or doctrine of chance, may seem to him and to you as opposed to your theories; but I do not think so. It is not opposed to physical science; why should it be opposed to a philosophy based on physical science?

The remaining two pages I fill with advertisements, of the most select kind, which would be a real attraction. For these I should set low prices.

You turn over the proceeds of the advertisements to pay me.

I guarantee you against loss. That is, the increase in the subscription list to cover the extra printing.

The contract to be for three years, but to fall to the ground if I dont make it pay for you the first year.

I ask you to say that you will come into this, provided I can satisfy you that I can really accomplish my part. When you have given that promise, I will show you what my expectations are founded on in regard to writers, to advertisements, and to subscriptions & you can pass your final judgment. Please reply.

yours faithfully
C. S. Peirce

P.S. I forgot to say I should want to have the copyright on my department.

---

16 work] *over* cr

22 the[1]] *over* p

Hegeler to Peirce. MS: RL 77. LPC: SIU 91.12.

La Salle Ill. 6/9 · 93

Professor C. S. Peirce
Milford Pa.

Dear Sir

Your letter of June 7$^{\underline{th}}$ 93 came to hand.

With reference to what you say of adding 4 pages to the Open Court etc. I will now say that I have to decline the proposition. As to the articles for publication referred to in your letter of 5$^{\underline{th}}$ inst. Dr. Carus will write you.[58]

Respectfully
Edward C. Hegeler
per C. Diesterweg

Peirce to Hegeler. MS: SIU 91.12. Stationery headed, "The Century | 7 West Forty-Third St."

1893 June 16

Dear Sir:

The Elementary Arithmetic, the more important of the two, is now substantially complete. There remain some 1500 examples to be added, which I wish to have written, in accordance with my detailed directions by a practical teacher. This will take a person about a month & I shall have to pay about $ 100 for it.

The Complete Arithmetic is well under way.

Yours truly
C. S. Peirce

Hegeler to Peirce. MS: RL 77. LPC: SIU 91.12.

La Salle Ill. June 22/93

Professor C. S. Peirce
Milford Pa.

Dear Sir

You say in your letter of April 1$^{\underline{st}}$ 93 "I am well advanced in my revised catalogue of the Books." — I expected to be in possession of the catalogue before this. Please forward it.

---

58 Carus to Peirce 10 June 1893.

---

8 for publication] *inserted above*

Respectfully
Edward C. Hegeler
per C. Diesterweg

Hegeler to Peirce. MS: RL 77. LPC: SIU 91.12.

La Salle Ill. 7/4. 1893

Professor C. S. Peirce
Milford Pa.
Dear Sir

I have ordered Bank draft to be sent you at Milford Pa. for $ 250$^{\underline{00}}$ for which please advise receipt.

Respectfully
Edward C. Hegeler
per C. Diesterweg

Hegeler to Peirce. MS: RL 77. LPC: SIU 91.12.

La Salle Ill. July 13. 93

Professor C. S. Peirce
Milford Pa.
Dear Sir

Referring to my letter of June 22. 93 I will say that I am still without the catalogue of your Library which you was to send me; neither have I received from you acknowledgement of receipt for the $ 250$^{\underline{00}}$ remitted to you in the beginning of the month through the First National Bank of Chicago.

Respectfully
Edward C. Hegeler
per C. Diesterweg

Peirce to Hegeler. MS: SIU 91.12.

326 W 77$^{\underline{h}}$ St. New York | 1893 July 27.

E. C. Hegeler, Esq.
LaSalle, Ill.
Dear Sir:

20 you was to send] you was to sent
(stylistic "was" *left unemended) [E]*

You will please send the next installment of advance on the arithmetics to the above address. I hope to be able to get the copy into the hands of the publisher within the next two months. This result certainly would have been reached before this if it had not been for your total abandonment of all courtesy in your treatment of me, leaving letters of the most vital consequence to me utterly unnoticed, etc.

I call upon you to remember that it was a point in the agreement between us that you should not lay hold of my library before my death; and I request you to make a memorandum to that effect upon the paper I left in your hands. I had a chance to make a disposition of the books at a price which would have reimbursed you even aside from the arithmetic; but you did not see fit to notice my letter on the subject.[59]

As for the catalogue, I regret that it has been impossible for me to complete it up to date. But I will do so as soon as I possibly can. I am compelled to remain in New York in order to be in a condition to support my family in the coming months.

Yours Truly
C. S. Peirce

P.S. I consider that a positive wrong has been done to me in not allowing me the six articles promised last summer in the Monist on my philosophy. In sending the rejoinder[60] to Carus, I distinctly said I did not consider that one of them; and therefore it could have been rejected, if another had been considered excessive. Besides that, I think it unjust that when my preliminary plan proved insufficient, I should have been left with an important philosophy half developed, because you would not give a *little more* space than was promised,—instead of breaking the positive promise made to me.

Another positive promise in reference to the Open Court has been broken by you.

It is difficult for me to believe that you are a man to break an explicit promise when the fact is pointed out to you.

59 Peirce to Hegeler 18 May 1893.
60 "Reply to the Necessitarians" (July 1893).

7 before] *l*before
18 six] five *[E] Peirce's sixth and final* Monist *metaphysical series article was to be a "General Sketch of a Theory of the Universe," the final article listed in his outline of "A Quest for a Method" sent to Hegeler 7 March 1893.*
19 rejoinder] *above* ~~answer~~

Hegeler to Peirce. MS: RL 77. LPC: SIU 91.12.

La Salle Ill. July 29. 1893

Professor C. S. Peirce
326 W 77th St.New York
Dear Sir

The last installment of advance on the arithmetics will be remitted to you upon the receipt by me of the catalogue of your Library due me.

Respectfully
Edward C. Hegeler
per C Diesterweg

Peirce to Hegeler. MS: SIU 91.12.

326 W 77^{h} St.New York | 1893 Aug 4

E. C. Hegeler Esq
Dear Sir:

I have your promise to send the advance of $ 250 as soon as I send a catalogue. I already had your promise to send the same unconditionally, Aug 1.

You on your side have my promise to send the catalogue as soon as possible.

Your disappointing me does not facilitate my sending you the catalogue: it hinders me from sending such a one as I should wish.

It will also tend to delay the preparation of the arithmetic; perhaps, by causing my death, stop it altogether.

I mention these things not with the slightest idea of influencing you to adhere to your engagements, but simply because I think it my duty to afford you some information concerning the catalogue and the arithmetics.

yours truly
C. S. Peirce

Peirce to Hegeler ca. 20 Aug 1893 (missing): Sends a preliminary list of books in his personal library with intent to send a complete catalogue later (though does not seem ever to do so).

---

15 of $ 250] *inserted above*
16 Aug 1.] *Possibly for* "June 1", *the last time Hegeler promised an advance.*
18 Your] *over* Our
20 tend to] *inserted above*
20 perhaps] perhas *[E]*

Hegeler to Peirce. MS: RL 77. LPC: SIU 91.12.

La Salle Ill. Aug. 24. 93

Professor C. S. Peirce
Milford Pa

Dear Sir

A List of Books in your Library was received, with a promise that it will be replaced by your writing a more complete and bibliographically perfect one when you may have means and leisure to make one.

I have to day directed the First Nat. Bank of Chicago to remit to you the sum of $ $250^{\underline{00}}$

This sum completes a loan to you of $ $1\,750^{\underline{00}}$ made in 7 equal instalments of $ $250^{\underline{00}}$ each, to be paid to me with interest @ 6% from future profits of your Arithmetic Book. You will please acknowledge Receipt in full in conformity with this.

Respectfully
Edward Hegeler

Peirce to Hegeler. MS: SIU 91.12. Stationery headed, "Express Address. | Port Jervis, N.Y. | Wells, Fargo Exp. Co. | C. S. Peirce."Arisbe" | P.O. Milford, PA." Note at top-right corner reads, "No Answer".

1893 Dec 31.

Mr. E. C. Hegeler

Sir:

Several circumstances impossible to foresee prevented the completion of my arithmetics within the time I had set. Among these I may mention the bursting of my boiler and blocking of the roads with snow afterward, rendering my house uninhabitable for a long time, the breaking of a bank in which my wife was a depositor, the breaking of another which indirectly affected me. Other calamities there were, still worse, which I shall not further mention.

Notwithstanding all this, the books are now nearly finished; but I am counselled not in the present state of business to carry them to the publisher. As soon as the proper time comes, they will be ready.

I will thank you to send me a copy of any agreement I have made with you, together with a statement of any verbal contract which you may claim that I have made, in order that I may ascertain precisely what my obligations to you are.

30–31 As soon … ready.] *inserted in-line and between paragraphs*

34 ascertain] *after* ~~see~~

The matter of the catalogue of books has received my attention, and some work has been done upon it. But I have been obliged to give most of my industry to avoiding immediate starvation; and besides, some questions have arisen in regard to the proper way of making the catalogue which I have a special reason for wishing to answer right, and they have not been sufficiently considered. The list I sent you[61] seems to be sufficient to insure you from any material loss. I have already asked you to name a price at which you would sell the books.

Yours Truly
C. S. Peirce

1894

Peirce to Hegeler. MS: SIU 91.13.

1894 April 2

E. C. Hegeler
Dear Sir:

I have an opportunity of exchanging the whole, or nearly the whole, of my *old arithmetics*, at their full value, for other works of equal value and of more use to me now that I have completed my study of the history of arithmetic. Unless I hear from you within a reasonable time, I shall consummate this arrangement which will not affect the market value of my library but will greatly enhance its interest to 99 persons out of 100. The books I shall get in exchange are also for the most part old books bearing on the history of Science & of the human mind.

C. S. Peirce
"Arisbe" | Milford | Pa.

Peirce to Hegeler. MS: SIU 91.13. Stationery headed, "Express Address. | Port Jervis, N.Y. | Wells, Fargo Exp. Co. | C. S. Peirce."Arisbe" | P.O. Milford, PA."

1894 July 6.

E. C. Hegeler Esq
Dear Sir:

I am going to send Dr. Carus some articles I have on hand which may possibly be found suitable for the Open Court and Monist.[62] Unless it turns out there are

61 ca. 20 Aug 1893.
62 See Peirce to Carus 6 July 1894.

enough of these to pay my expenses for two months or more, and unless you see fit to pay me for them, I think it my duty to say that the books you paid me for should be promptly removed; for I cannot support the anxiety occasioned by the dread that they may be attached for payment of bills.

Take them away I beg, unless you see fit to help pull me through. I have seen enough of the publishers to be convinced that if they see evidences of my being in extreme need, they will attempt to drive such a bargain with me that I (who shall care very little when my wife has succumbed to privations, and my logic is in press) shall not accept.

I have some patent rights that I could do something with even in these times. But I have no time for anything except getting my books ready for the press.

I cannot take a step about the arithmetics until autumn, when the publisher returns from Europe. But meantime there will be another book which may enable me to pay you some interest, at any rate. However, I think it altogether doubtful whether (although they tell me it will be worth $ 2 000 a year to me) I shall receive anything from this this year. I am assured I shall be able at once to draw money from the arithmetic. But this may not prove true.

Yours very Truly
C. S. Peirce

Peirce to Hegeler. MS: SIU 91.13.

Milford, Pa. 1894 July 11.

Mr. E. C. Hegeler

Dear Sir:

The New York partner of Ginn & Co. came to me and said his firm had been recommended to get me to write a logic for them;[63] and strongly urged me to do so. I said, "Am I to understand your firm promises to take the book?" He said, "Why, not precisely. We must see the MS, but it is a mere question of size of book, cost, & so forth." "Very well," I said, "then on the understanding that you want to see the MS for that purpose only, and not to give it to a reader, I will write it." I did so, & sent it to Ginn & Co, who promptly placed it in the hands of a reader. The reader recommended its publication, and they at once decided not to publish it.

---

63 "How to Reason: A Critick of Arguments," a logic book that evolved from Peirce's "Quest for a Method." Referred to posthumously as Peirce's "Grand Logic" (see R 397–434 and *CP* 8:278). For a discussion of this work in relation to Peirce's other OCP endeavors, see de Waal, 16–18.

---

1 or] *after* ~~and~~ 5 help] *after* ~~p~~

Then they write me and say "We strongly advise you to have the book published, and do not hesitate in the least to predict for it a cordial welcome," with other gammon like this.

I do not wish to bother with the work further until my Geometry and Arithmetic are off my hands. I should then like to revise it, and write a popular abridgment of it for wide circulation.

I may mention that the book as written states, or partly states (I forgot precisely) my view of personal identity, which closely resembles if it is not the same as yours. This can easily be struck out. The closer ideas come to yours the less you seemed to like them. Some of my friends, students of philosophy, expressed surprize that my Monist articles should stop just as I was evidently coming out on your ground. I told them you didn't want me to trespass on your property.

This logic MS will make a book the size of Mach's Mechanics. Several professors of the highest universities we have have promised to use such a book by me. There will be a steady & increasing sale; and the abridgment will go off "like hot cakes."

I do not mean, at present, to offer the book for publication by the Open Court Company. Though if P.C.[64] will consent *not* to write a preface to it, that might on consideration be judged expedient by one side & the other. But for the sake of keeping along while I get my other books & finally that out, I would like to assign all future profits from any and every work on logic by me, as security on any sum I can borrow on such security. I send some description of the book.

I cannot but think that if I were to see you we might arrive at some understanding useful for your purposes & for mine. But of course I can do nothing as long as you ignore all my letters.

Yours very truly
CSPeirce

Peirce to Hegeler. MS: SIU 91.13. Stationery headed, "Arisbe, | Milford, Pa." Date uncertain. The type of stationery and the context suggest April or May 1893, but there is no clear place for it in that period nor elsewhere in the correspondence. Note on MS dates it around late 1894, during Peirce's tandem work on his arithmetic, geometry, and "How to Reason" books, but it likewise does not fit neatly here either (e.g., "the sum I name" he mentions is undetermined).

64 Paul Carus.

8–9 if it is not the same as] *inserted above*
15–16 "like hot cakes."] *after* ~~beautifu~~
25 all] *over* ev

[late 1894]

My dear Mr. Hegeler

In looking over the MS matter I have in hand, I find I can do what is here proposed without such a very large expenditure of time now that it will prevent my books from being ready before the publishers are ready for them. The sum I name is just what is necessary to enable me to get along with a very moderate degree of comfort. If you wish to reduce the amount, situated as I am toward you, I should feel disposed to make any concession I could. But it is difficult to write one's best when not comfortable

Yours very Truly
C. S. Peirce

From this point, due to the consistent complications and misunderstandings with agreements, Peirce officially fell out of favor and contact with Hegeler, ending completely their correspondence and postponing all other OCP contact for the next two years—only Russell remained in touch during the lapse.

The remainder of Peirce's dealings with OCP would be under the auspices of Russell and later Carus.

2.3 Peirce and McCormack

Thomas J. McCormack, assistant editor of *The Open Court* and OCP translator (German and French), collaborated with Peirce mostly between the summer of 1892 and spring of 1893 on the first OCP translation of Ernst Mach's second German edition of *Science of Mechanics* (1889) that Peirce had been contracted to help with and in which he played a more substantial role than anticipated.[1] McCormack also accepted and edited several of Peirce's articles for *The Open Court*, and briefly in the second half of 1896 worked with Peirce to prepare the type for his complex "The Logic of Relatives" article for *The Monist*. Peirce's custom logical symbols and graphs in the article posed typesetting challenges for OCP, and the instructions and illustrations he sent to McCormack (and Carus) provide valuable clarifications of Peirce's nuanced formulae that appear in those and many other of his logical writings. Although little more is known of McCormack's relationship with Peirce, he appears to have maintained good standing with Peirce and OCP, considered by OCP compositor Fred Sigrist to be the "best man and intellectual backbone of the enterprise."[2]

1892

McCormack to Peirce. LPC: SIU 91.12.

La Salle, Ill., June 28, 1892.

Mr. Charles S. Peirce,
12 West 39th Street,
New York.
Dear Sir:

By the accompanying mail we send you a copy of Prof. Mach's work on Mechanics in the International Scientific Library.[3] Dr. Carus has informed us of your consenting to revise the translation of this work, the first proofs of which will reach you during the present week.

So far as is possible we shall endeavor to send you the work in completed sections; and we expect to make ready about twelve galleys each week. A memo-

1 His most significant contribution was Chapter III, Section III, Sub-section 8 of the edition on the mechanical units in use at the time in the U.S. and Great Britain, included in Appendix, "Peirce and McCormack."

2 Sigrist to Peirce 26 June 1903.

3 1889 edition of *The Science of Mechanics* (*Die Mechanik In Ihrer Entwickelung*).

randum will accompany the proof sheets, calling attention to any points of technical difficulty which may occur; questions of literary form and expression will, as regards alteration, be left wholely to your judgement: so that any points of this character, with respect to which we are in doubt, will not be noted on the proof sheet or in the memoranda.

The proofs will be re-read for typographical errors after their receipt from you, and subsequently page-proofs will be sent, to be used for reference.

Any suggestions you may wish to make, that will render the work less onerous to you, for example as regards the time and mode of sending the proofs, etc., etc., will be gladly carried into effect by us. We feel greatly gratified, , indeed, that you have consented to undertake this work, and we shall do all in our power to spare you unnecessary labor.

Very truly yours,
Thomas J. McCormack
for The Open Court Pub. Co.

McCormack to Peirce. LPC: SIU 91.11.

La Salle, Ill., July 2, 1892.

Mr. Charles S. Peirce,
12 West 39th Street,
New York.

Dear Sir:

Enclosed you will find. 1) Type written copies of a rough draft of the prefaces to Prof. Mach's work.[4] 2) The proofs of the introduction and of the first chapter as far as the principle of the composition of forces.

The following are the points I desire to call to your attention.

(1) The phrase "magnitudes of equal weight" is a literal translation of the Greek original and of all foreign renderings that I have seen. I suppose there is no objection to this in point of idiom.

(2) The German phrase "eindeutig bestimmt", used frequently by Prof. Mach throughout the work, I have translated by "positively and univocally determined". I have tried many other words, among them "uniquely", "unequivocally", etc., etc., but none seems to express so fully as the phrase mentioned the meaning of

4 These preface drafts are not found, but later drafts in May 1893 are included in Appendix, "Peirce and McCormack."

10 , indeed,] *commas inserted in-line* 24 composition] compostion *[E]*

"being determined in a single, univocal sense, to the exclusion of any and every other sense."

When you return the proofs will you kindly address them to me personally.

—Respectfully Yours
Thomas J. McCormack

Peirce to McCormack. MS: RL 77. Only first page survives.

New York 12 W 39$^{\underline{h}}$ St. 1892 July 5[5]

Mr. Thomas J. McCormack
La Salle Ill
My dear Sir:

Yours of the 2$^{\underline{nd}}$ inst with enclosure just comes to hand. I wish you to keep all I send you relating to Dr Mach's book to be shown to Dr. Carus as evidence of how much work there has been. Please keep the galleys, if convenient. Please send me finally corrected page proofs.

Dr. Carus said I was to be final arbiter on questions of literary form. Now let us not treat Dr Mach's book as if it were a Bible; but just find out what he means to say & express that. The Germans are so habituated to bad taste, and especially to the violation of every principle of style, that the vilest writing no longer jars upon them or even produces an agreeable titillation. This won't produce the same effect on English ears. To explain what I mean I shall be forced at first to be a little verbose. I take up your translation of the preface

5 The following fragment in RL 77 is the final page of an undetermined letter from Peirce, estimated to have been written around the same day (5 July 1892) to a correspondent at Funk & Wagnalls, the publisher with whom Peirce had been working to prepare definitions for their *Standard Dictionary of the English Language* (1894). For more on Peirce and Funk, see *W* 8:xxvii and related correspondence in RL 153.

[…] Carina. There is no such constellation as Parvo. And "Serpens caput" seems to be here first made a distinct constellation. I would spell *Chamæleon* not *Chameleon*. The definition of *constellation* is very loose and inaccurate.

I have already answered your questions about the table. It will be accurate and serviceable as it is. There are no such disks in the market.

Yours very truly
CSPeirce

14 page] *after* ~~sh~~
18 the^2] *after* ~~they~~
19 titillation] titilation *[E]*
21 your] *over* tr

"The present treatise" instead of simply "This treatise" *[…]*

McCormack to Peirce. TS: RL 77. On OCP Stationery B. LPC: SIU 91.11.

La Salle, Ill., July 14, 1892.

Mr. Charles S. Peirce,
12 West 39th Street,
New York City.

Dear Sir:

Enclosed you will find the proofs of the section "The Composition of Forces" of Mach's book. We have been much delayed during the last week by other matters, and are unable to send more.

The corrected proofs of the preceding sections were duly received. The corrections will be made and the proofs of the *first two galleys* sent to you again. (These proofs are enclosed.) There are one or two points to which I should like to call your attention once more in these proofs. Among them are

1. The phrase "eindeutig bestimmt". This phrase occurs repeatedly throughout the book, and as there is no objection to its being translated by "uniquely determined", I have thought it best to introduce that phrase in the first instance together with the other phrases by which you explain its meaning. I shall send you a proof of the page on which this is done.

2. The phrase "vorbilden und nachbilden". These two words are also repeatedly used by Prof. Mach: they are catch-words with him. When used alone "nachbilden" may be translated by "represented" etc. But when used with "vorbilden", as in the sentence, "Die Thatsachen der Natur nachzubilden und vorzubilden", the words are used in contrast. They are so often used in this manner that I have tried to find two English words exactly corresponding to them, and hit upon the words "*re*figure" and "*pre*figure", as in the sentence "to refigure and prefigure facts in the mind". Once the meaning of these words is established as the exact equivalent of "vorbilden und nachbilden", there is no need of a paraphrase on every occasion the latter words appear; while nouns and compound phrases are readily formed from them. There are subsequent chapters in the book where these difficulties especially occur. In the "Introduction" where you replaced "to refigure in thought" by "to represent", the word "nachbilden" is used alone; my only reason

11 sections] *over* d
12–13 (These proofs are enclosed.)] *inserted above*
23 Thatsachen der Natur] *t*Thatsachen der *n*Natur
31–32 "to refigure … represent"] "to refigure in ~~the mind~~ \thought/ by "to represent$^{\vee}$"$^{\vee}$ ~~in the mind"~~

for introducing the word "refigure" there was to prepare the way for it in subsequent chapters. As it is, it has been left out. When the term is used again you will be better able to judge whether I am right or wrong in this.

I perfectly agree with you as to the objects of translation. If in Mach's case I have been too literal it has been because, apart from his peculiar expressions and phraseology, his expositions are not much different from those of others and, by not being literal, would be shorn of their supposed originality.

I am not at all sensitive to corrections of my work and shall always be glad to reset whole galleys if necessary. For the last, I thank you.

Yours Very Truly
Thomas J. McCormack

McCormack to Peirce. TS: RL 77. On OCP Stationery B. LPC: SIU 91.11.

La Salle, Ill., July 20, 1892.

Mr. Charles S. Peirce,
12 West 39th Street,
New York City.

Dear Sir:

Your MS[6] for the next Monist received today. It will be set up as soon as possible and the proofs sent you. Dr. Carus left for a short trip to Europe on Saturday last. He mentioned to me before leaving that you intended to contribute to the Open Court a series of articles to be entitled "The Critic of Arguments". I do not know whether he again communicated with you upon the subject, but the present is a very good time for their publication. A note announcing their appearance might be published in The Open Court a few weeks previously. I simply suggest this, fearing that Dr. Carus might not have written to you in regard to the articles.

Enclosed you will find the first few made-up pages of Mach, with the phrase "uniquely determined" introduced on page 22 as I proposed in my last letter to you. Will you kindly mark it "O.K.", if it has been correctly done.

The proofs of the next section of Mach will be sent to you tomorrow.

Very Truly Yours
Thomas J. McCormack

---

6 "Man's Glassy Essence," *Monist* (October 1892), submitted to Carus 15 July 1893.

---

1 there] $^{\vee}$t$^{\vee}$here
2 As … out.] *inserted above*
27 on page 22] *inserted above*
28 kindly] *l*ki*k*ndly

McCormack to Peirce. TS: RL 77. LPC: SIU 91.11.

La Salle, Ill., July 22, 1892.

Mr. Charles S. Peirce,
12 West 39th Street,
New York City.

Dear Sir:

We received the last proofs of Mach from you just as we were about to send those of the following section, which are enclosed[7] I had intended to make a foot-note to the title "Principle of Virtual Displacements", to be presented to you when all the proofs of this section were ready. As it is, I have introduced yours without any modification except the omission of the date of John Bernoulli's communication to Varignon, which Mach mentions on a subsequent page.[8] I thought it best to leave the main title as it stands, namely "Virtual Displacements". If the term "virtual velocity" is universally used in English, nearly every one who uses it admits that it is inappropriate: and I notice that Minchin calls it "the principle of virtual work".[9]

I also intended to ask you about the term "Archimedean or potential pulley", and have inserted the foot-note you suggest saying that the term is not in use in English.[10]

You remark in one place "that the line of action, not merely the direction is required to determine the equilibrium". Mach uses the word "Richtung", I think, to mean line of action, and when he means direction simply, he uses "Sinn" or *sense*.

On page 61, line eight from bottom, of the text, in the present proofs, should not $2n''$ be n'', and in the first formula on page 62, $n''p''$ read $2n''p''$?

Also, is the phrase "mixed equilibrium" used in English?

Very Truly Yours,
Thomas J. McCormack

---

7 "Retrospect of the Development of Statics" (Ch. 1, Sec. V).
8 For that footnote, see Mach Ch. 1, Sec. IV ("The Principle of Virtual Velocities"), first paragraph. Note begins, "Termed in English ..." MS in SIU 92.1. See also p. 56 for what Mach "mentions on a subsequent page" about the 1717 letter between Swiss and French mathematicians Johann Bernoulli (1667–1748) and Pierre Varignon (1654–1722).
9 George Minchin, *A Treatise on Statics* (1877).
10 Bottom of p. 50.

---

8 section, which are enclosed] section~~s.~~$^{\vee}$,$^{\vee}$ \which are enclosed./
10 As it is,] *comma inserted inline*
22 simply,] *comma inserted inline*
24 formula] formulae

Peirce to McCormack. MS: RL 77. Only first two pages survive. Based on McCormack's responding letter, the missing content apparently announces Peirce's forthcoming series of *Open Court* articles, "Pythagorics," "Dmesis," and two "Critic of Arguments" pieces.

[25 July 1892]

My dear Sir:

I admit that the phrase Virtual Velocities is not quite appropriate to describe the principle known by that name. But we have not the whole of Bernoulli's letter to Varignon. Probably if we had we should see the matter in another light. In fact, Varignon probably did not take in the whole of Bernoulli's meaning, for the term *virtual velocity* sounds precisely as if he had enunciated D'Alembert's principle,[11] or something approaching that. Anyway, we can see exactly how he *might* have understood the matter so as to make the phrase appropriate. But the phrase *Virtual Displacements* is utterly inexcusable. *Virtual* does not mean *möglich*,[12] as Mach says, and never did—*Virtualis* is a term introduced by Duns Scotus.[13] It means that which does not actually exist, but yet produces effects as if it existed. If any one chose to call the principle that of *possible displacements*, I could not quarrel with them,—but *virtual displacements* won't do. *Virtual velocities* implies the state of rest is the resultant of velocities which practically exist as components of the whole effect. In the next place, in English *usage* is supreme; and usage does not permit calling the principle *Virtual Displacements*—that is a thoroughly vicious corruption. Minchin is no authority,—and his *virtual work* has been influenced by the German. Dont use the German phrase in the *title* of the section. It wont be understood.

Richtung means *direction* and <u>not</u> *line*. *Sinn* has a distinct meaning, also called *direction* in English, which is defective in confounding these ideas.

In § 4 ¶ 11, I cannot admit that *discerned* is a proper translation of *erschant*.[14] *[…]*

---

11 Jean Le Rond d'Alembert (1717–1783), French mathematician and philosopher. "D'Alembert's Principle" applies Newton's second law of motion to both dynamic and static objects, first introduced in his 1743 *Traité de dynamique*. (See "D'Alembert's Principle" in *Encyclopedia Britannica* for details.)

12 "*possible*" or "*potential*," as Peirce implies in the next sentences.

13 e.g., "*differentia virtualis*" and "*distinctio virtualis*" in his *Oratio* I 2.402. For further discussion, see Vos, "Conceptual Devices" in *The Philosophy of Duns Scotus* (2006).

14 For "*erscheint*" ("appears" or "seems").

---

8 Probably] *over* B

9 probably] p*l*<u>r</u>obably

18 the[1]] *after* ~~du~~

22–23 Dont … understood.] *inserted in-line and between paragraphs*

McCormack to Peirce. TS: RL 77. LPC: SIU 91.11.

La Salle, Ill., July 27, 1892.

Mr. Charles S. Peirce,
Milford, Pike Co.,
Pa.

Dear Sir:

Your letter of July 25th stating that you will commence the articles for The Open Court, received. When you send the first article will you also kindly write & send a short note to be published in The Open Court a week or so in advance of the first article, stating what the articles will treat of, their plan, etc.?

We have ordered a copy of "Fundamental Problems"[15] to be sent to you.

Enclosed you will find, on a separate sheet of paper, a translation of the *title* of Mach's work. Do you think it will do? We must decide upon one now, as the page-headings are to be set up in advance.

The page-corrections you suggest have been noticed.

Very truly yours,
Thomas J. McCormack

Peirce to McCormack. Postcard: SIU 91.11. Recto headed, "United States | Postal Card | One Cent" with center portrait.

[recto]

[?Milford] | PA | *[Aug | 1]* | 1892 *[stamped]*

T. J. McCormack Esq
Open Court Co.
La Salle
Ill.

[verso]

Milford Pa | 1892 July 31.

Dear Sir

The title page seems to me all right. Will send article for Open Court Tuesday

CSPeirce

15 Carus, 1889.

8–9 & send] *inserted above*

10 first article,] \first/ article~~s~~,

Peirce to McCormack (or general OCP) 5 Aug 1892 (missing): Returns proofs of "Man's Glassy Essence" sent him by Carus late July.

Peirce to McCormack. MS: SIU 91.11. Stationery headed, "Arisbe, | Milford, Pa."

1892 Aug 6

My dear Mr. McCormack

I enclose my first paper for the *Open Court.*[16] Will forward others weekly. I yesterday returned proofs of my "Glassy Essence," but I have my doubts of the fidelity of him to whom the attaching of the postage stamps was entrusted. If you don't get them please send others, as the corrections were important.

very truly
C. S. Peirce

My mail facilities are not facile either for sending or receiving.

McCormack to Peirce. TS: RL 77. On OCP Stationery B. LPC: SIU 91.11.

La Salle, Ill., August 13, 1892.

Mr. Charles S. Peirce,
Milford, Pike Co.,
Pa.

Dear Sir:

The proof-sheets of the Monist article safely arrived. We shall send you the corrected page-proofs today, which you may keep.

The work on Mach has been delayed by circumstances over which we have had no control; but it will soon be commenced again.

Your article for The Open Court, and letter dated Aug. 6th, arrived late last night, Aug. 12th. As you feared, the proper postage-stamps were not attached. The article is superb—in lightness, humor, point, and allusion. The series, in my judgment, will be of *very* great interest to our readers. The length of the article is also about right.

16 "Pythagorics" (8 Sept 1892).

23 6th] 16th *[E]*

We shall wait until we have two or three of the series on hand, before publishing.

Very truly yours,
Thomas J. McCormack

McCormack to Peirce. MS: RL 77. On OCP Stationery B. LPC: SIU 91.11.

La Salle, Aug 29th/92

Dear Mr. Peirce

Your second article for the Open Court, with the enclosed letter,[17] reached us on Saturday. Unfortunately, I could not find Mr. Hegeler during the whole day, and was obliged to defer sending you the check you requested, until to-day. I hope you have not been inconvenienced by the delay. The check, which you will find enclosed, is for two hundred & fifty dollars. The payment is on account.

You will receive the proofs of your second article, this week. The first will be published in No. 263 of The Open Court.

The New York affair which you mention in your letter is very unfortunate. I sincerely hope you will be able to recover something from them.

Very Truly Yours
Thomas J. McCormack

Peirce to McCormack. MS: SIU 91.11.

Milford Pa 1892 Sep 1

T. J. MacCormack Esq
My dear Sir:

I am much obliged to you for the check which was very considerate. I do not understand, I confess, how it is the mails are so slow. Yours mailed Aug 29 · 7 PM only reaches me Sep 1 at 5 PM. I wonder if the mail in the opposite direction is equally snail-like. It was so long since I sent you my second article[18] that I began to think you might be ill or gone away, & therefore mailed my last to E. C. Hegeler *or* Acting Editor Open Court.[19]

---

17 "The Critic of Arguments. I. Exact Thinking" (22 Sept 1892), submitted to Carus 25 Aug 1892.
18 "Critic of Arguments. I."
19 "Dmesis," in Peirce to Carus 31 Aug 1892.

---

21 T. J. MacCormack] *Mr.*T. J. Ma*C*c Cormack

Yours very truly
C. S. Peirce

My next will be on *relations* and the following one on *libraries*.[20]

McCormack to Peirce. LPC: SIU 91.11.

La Salle, Ill., Sept. 5, 1892.

Mr. Charles S. Peirce,
Milford,
Pa.

Dear Sir:

Your third article[21] for The Open Court reached us this morning; also your letter of September 1st, acknowledging the receipt of the check. The first letter was stamped at Port Jervis on September 1st, and the second at the same place on September 3rd: both arrived here on the same day. The mails seem indeed to be very slow.

Your first article[22] appeared in The Open Court of this week; the others will appear in immediate succession, except where, for editorial reasons, it is absolutely necessary to make an intermission.

Very truly yours,
Thomas J. McCormack

McCormack to Peirce. TS: RL 77. On OCP Stationery B. LPC (p. 1 only): SIU 91.11.

La Salle, Ill., Sept. 20, 1892.

Mr. Charles S. Peirce,
Milford,
Pike Co., Pa.

Dear Sir:

It is our custom to electrotype all articles of permanent value appearing in The Open Court, with the view of possible future publication. This will be the case with your articles. To avoid the trouble of corrections in the plates, we send you by the same mail with this letter proofs of the two articles which have already appeared,

20 "Critic of Arguments. II. The Reader is Introduced to Relatives"; the second unidentified.
21 "Dmesis."
22 "Pythagorics."

with the request to alter, if you think it advisable, the opening sentence of the article "Pythagorics", and to make such other suggestions as you deem fit. The titles, or rather, the headings of the plates corresponding to left-hand pages will read "The Critic of Arguments.", and the headings of the right-hand pages will read "Pythagorics", "Exact Thinking", etc. Each article will begin a new page. If your articles on this subject should, therefore, ever be sufficiently numerous, we shall be glad to publish them in pamphlet or book form without any difficulty. In such a series articles like "Dmesis" would probably not be included.

The proofs of "Dmesis" were sent from La Salle on Sept. 12,[23] and are not yet back. Your last article "The reader is introduced to relatives" will have to be divided; owing to the size of The Open Court it is difficult for us to handle articles of more than three pages in length.

"The Pythagorean Brotherhood"[24] seems to have caught the fancy of a number of our readers. I believe there have been several inquiries concerning where and what it is.

We shall have some galleys of Mach ready very soon.

Very truly yours,
Thomas J. McCormack

The proofs of "Dmesis" just arrived.

Peirce to McCormack. MS: SIU 91.11.

Milford Pa | 1892 Sep 24

T. J. MacCormack
Dear Sir:

The proper thing to do with "Pythagorics" is to delete the first paragraph & begin with Lowell's lines. The acknowledgement will appear in the Preface. Of course I mean there shall be a book of the "Critic of Arguments." Pythagorics, Dmesis, etc will appear in appendix or in a second volume—I like *Dmesis* particularly myself. An author naturally values what gives him the most trouble; and very much and very disagreeable pains have gone into that.

Yours very truly
CSPeirce

23 OCP (W.J.G.) to Peirce 12 Sept 1892.
24 A "secret society of scientific students" discussed in "Pythagorics."

1 opening] *over* p

19 The proofs of "Dmesis" just arrived.] *This line added in pen.*

McCormack to Peirce ca. 24 Sept 1892 (missing): Sends Mach proofs for Peirce's "immediate examination," as Peirce notes in his response.

Peirce to McCormack. Postcard: SIU 91.11. Recto headed, "Postal Card One Cent | United States of America" with portrait at right.

[recto]

Milford | PA | Sep | 26 *[stamped]*

Thomas J. MacCormack Esq
Open Court Publishing Co.
La Salle
Ill.

[verso]

I am very sorry I am unable to respond to your sudden call for an immediate examination of the Mach proofs. I had no reason to expect them, & am crowded with work. They require much time. I will take them up on Monday afternoon.

C. S. Peirce

McCormack to Peirce. TS: RL 77. On OCP Stationery B. LPC: SIU 91.11.

La Salle, Ill., Oct. 3, 1892.

Mr. Charles S. Peirce,
Milford, Pike Co.,
Pennsylvania.

Dear Sir:

At page 115, line 2 from top of the original text of Mach, the phrase "optische Schlierenmethode"[25] occurs. I do not know what the method referred to is, at least under that name. Will you kindly inform me what it is. Although in the galley which you corrected the last part of this sentence was omitted, you did not insert the missing phrase or complete the sentence.

25 "Optical Schlieren Method," invented 1864 by German physicist August Toepler (1836–1912), method for using collimated light to study the flow of fluids and to make determinations about supersonic motion. (See Degen, "History of the schlieren method" in *Schlieren Optics*, 2012.)

I notice that the name "Perier"[26] is written without the accent in the English cyclopaedia, and in the Encyclopaedia Britannica.

Will you please tell me the distinction between "fluid" and "liquid". I notice that in some cases you let "a fluid" stand, while in others, and apparently not different ones, you changed it.

Very truly yours,
Thomas J. McCormack

Peirce to McCormack. MS: SIU 91.11. Undated. This and the next letter estimated to be in follow-up to the set of proofs received ca. 24 Sept 1892 (missing) or around his letter 27 April 1893, or sometime in between. McCormack's "remark" that Peirce references is unidentified.

[between October 1892 and April 1893]

Dear Mr. McCormack

A remark of yours suggests you dont intend to send me the rest of the proofs; but I think you should do so.

C.S.P.

Peirce to McCormack. MS: SIU 91.11. Undated.

[between October 1892 and April 1893]

Dear Sir:

Please send page-proofs of Mach. It is awful stuff.

C.S.P.

Mr. MacCormack

Peirce to McCormack. MS: SIU 91.11. Answered by Carus 19 Oct 1892.

Milford Pa. Oct 17 *[1892]*

Dear Mr. McCormack

A fortnight ago I sent on an article for the Jan Monist *Registered* and with *Hasty Delivery* stamp.[27] I have received no notification of its receipt and know not what to make of it. I propose to call the page of Open Court 900 words; but first page 700.

26 Florin Périer (1605–1672), French chemist and brother-in-law of Pascal.
27 "Evolutionary Love," in Peirce to OCP 4 Oct 1892.

2 Britannica] Brittannica *[E]*

Is this correct? If so my four pieces will come to over the $250 last sent, and the Monist article would entitle me, if accepted, to further payment. The terms for the Mach work have never been fixed.

I trust my article is not lost; for it is very near my best.

Yours very truly
C. S. Peirce

I have a lot more written for the Critic of Arguments; but it is not arranged. Will forward shortly.

1893

Peirce to McCormack ca. Jan–Feb 1893 (missing): Submits for the Mach book a lengthy statement on measurements, later included in the edition as Chapter III, Section III, Sub-section 8. A transcription of Peirce's MS included in Appendix, "Peirce and McCormack."

McCormack to Peirce. MS: SIU 91.12. On OCP Stationery B.

La Salle, March 11th, 1893

Mr. Charles S. Peirce
Milford
Pa.

Dear Sir

Enclosed you will find the proofs of the paragraphs you wrote for Mach's chapter on measurements. The matter will be electrotyped as it is; and if there be any alterations, they must occupy the same space.

Clarke's statement of the length of the earth's meridian quadrant is in metres (Ency. Brit Vol VII, p 607), and not in inches (see Galley 71, A).[28] If you want the statement in inches to stand, please mark it.

To-morrow, we shall begin sending you the rest of the Mach proofs. The book has been greatly delayed by our having an insufficient quantity of type for all our

28 Alexander Ross Clarke (1828–1914), British geodesist. Peirce refers to him at the end of the second paragraph of his MS (see Appendix).

20 and if there be any] *above* ~~but I should like you to see the proofs <again> before printing.~~
22 statement] statement~~s~~
23 Brit] *before* ~~) p~~
23 Galley] *before deleted*

work, & we should be greatly obliged to you if you would read the proofs soon after receiving them.

Very Truly Yours
T. J McCormack

McCormack to Peirce. LPC: SIU 91.12.

La Salle, March 29, 3.

Chas. S. Peirce, Esq'r,
Milford, Pa.,
Dear Sir:—

Enclosed you will find the proofs of the two next sections of Mach. Will you kindly read them at your earliest convenience?

If it will save you any time, I might mention that you need not compare the translation with the original, except where it is unclear. I shall look out for the correspondence in all matters of detail.

Yours very truly,
Thomas J. McCormack

McCormack to Peirce. TS: RL 77. On OCP Stationery B. LPC: SIU 91.12.

La Salle, April 24, 1893.

Charles S. Peirce, Esq'r,
Milford, Pa.,
Dear Sir:—

Enclosed you will find pages 387 and 388 of Mach. I have transferred the footnote which you inserted at the place marked on 387 to its present place on 388.[29] Is there any objection to this, or to leaving the sentence in which the error occurs stand?

The proofs of your next Monist article[30] will reach you about the first of next week.

Yours very truly,
T. J. McCormack

29 Note beginning "The normal at any point of a surface …", glossing the "principal radii of curvature" mentioned on that page.
30 "Reply to the Necessitarians" (July 1893).

Peirce to McCormack. MS: SIU 91.12. Stationery headed, "Express Address. | Port Jervis, N.Y. | Wells, Fargo Exp. Co. | C. S. Peirce."Arisbe" | P.O. Milford, PA."

1893 Apr 27

Dear Mr. M^c^Cormack:

The foot-note is of little importance. Let it be inserted where you like, or not at all.

But what *is* of importance is that we should not do so *treacherous* an act as to notice Mach's geometrical slip and not correct it. It is not corrected in the text you send me. But I return the proof with the necessary alteration. The point is that r and r' are *not* the radii of curvature of the lines of curvature as stated. There can, of course, not be two opinions about this. Mach simply committed a momentary blunder.

very truly
C. S. Peirce

McCormack to Peirce. TS: RL 77. On OCP Stationery B. LPC: SIU 91.12.

La Salle, May 10, 1893.

Chas. S. Peirce, Esq'r,
Milford, Pa.,
Dear Sir:—

Enclosed you will find a draught of a Translator's Preface.[31] Will you kindly return it as soon as you can?

We shall send you to-morrow the next section of Mach; and the rest of the book will be sent within a week.

Very truly yours,
T. J M^c^Cormack

McCormack to Peirce. TS: RL 77. On OCP Stationery B. LPC: SIU 91.12.

Chicago, May 23, 1893.

Chas. S. Peirce, Esq'r,
Milford, Pa.,
Dear Sir:—

31 See Appendix, "Peirce and McCormack," for Peirce's edited version of this preface and the TS version published.

Enclosed you will find a literal translation of Prof. Mach's Preface[32] to the translation of his book. At Prof. Mach's suggestion I should also have inserted your name where the thanks to the Translator are mentioned. But knowing your opinion of the work I have thought it best to leave this to you to decide. Will you kindly return it as soon as possible?

Very truly yours,
T. J. McCormack

1894–1896: Two-year lapse in correspondence between Peirce and OCP due to a falling out with Hegeler. For details, see note in Hegeler and in Carus correspondences around this time.

1896

McCormack to Peirce. TS: RL 77. Stationery headed, "The Monist, | Editorial Department, | La Salle, Ill." LPB: SIU 31.10.437.

August 21, 1896.

Charles S. Peirce, Esq'r,
C/O Messrs. Stickney, Spencer & Ordway,
35 Nassau St.,
New York City, N.Y.,

Dear Sir:—

We intend to have your article "The Logic of Relatives" set up immediately after the appearance of the October Monist, and desire to make the necessary preparation now for putting it in type. Will you kindly make a full list of the symbols for which new type will be necessary? Perhaps dies of some of the type have been cast for former essays of yours, and may be obtained by us from publishing houses in the East. Or, you might send us copies of those articles, and we could then have the characters photographed and reproduced by the electrotyping process. This would be better. We may mention that we have acquired from Teubner of Leipsic a dozen or so of all the characters used in Schröder's book.

If you find it necessary to draw up a written list, please make it as accurate as possible, as it will have to be redrawn, then photographed, and electrotyped, in which last condition we can reproduce the characters *ad libitum*. Also note whether the characters in question are ever likely to occur as superior or inferior

---

32 See Appendix, "Peirce and McCormack."

figures, that is to say, whether their "body" is to be always the same as that of the type of the running text.

Very truly yours,
The Monist
sig T. J. McC

Peirce to McCormack. MS: SIU 91.14.

Stewart House, Broadway and 41st St.New York City | 1896 Sep. 2.

Mr. T. J. MacCormack:

Dear Sir:

I have postponed answering your letter, not by neglect, but because I wanted before answering it to have sent to Dr. Carus another writing of the explanation of graphs. I did not suppose it would involve so long a delay. I have now sent that new MS.[33] If he adopts it, the figures there are drawn carefully enough to serve as sketches for the draughtsman especially if he reads the paper. But if my previously sent MS is used, it must be sent back or copies of the figures sent me; so that I can redraw them.

The most difficult character I use is $\overline{A\,+}\,B$, $\overline{\overline{\overline{A\,+}\,B\,+}\,\overline{C\,+}\,D\,+}\,E$. I use as you will see several sizes in each formula and if it were cut as a type, there would have to be three or four sizes, and the justification, or whatever they call it, would be troublesome. I only use a few such formulae; and do not propose the character as one to be permanently used. It is only a stepping-stone to others. If cut as a type, it is little but a + with the body so cut that a brass can be made to join it. Or it might be cut thus

[8 attempts in-line above (6 of which deleted) and 3 attempts in this space. See Figure 8.]

so that a brass will join it. It will have to be cut of four heights, at least, so that complicated formulae may be written.

$$\overline{\overline{A\,+}\,B\,+}\,\overline{\overline{B\,+}\,C\,+}\,\overline{A\,+}\,C$$

Perhaps it would be better to photograph such formulae entire.

33 "On Logical Graphs," Peirce to Carus 31 Aug 1896.

14 especially … paper] *inserted above*
17 use[1]] use~~d~~
26 brass] *after* ~~p~~

The sign of logical addition, much used by me, is my modification of Jevons's ·|· which is an oldfashioned sign of division (a sign by the way never used to any extent I believe) modified so as to be more easily written & more closely resemble +. This I make thus:

[6 attempts in this space. See Figure 8.]

I dont want it to look like a Greek Ψ. Hence I rather prefer a form like the last but one, or like the one below.

ψ For logical multiplication I use a not too light dot above the line.

Schröder's sign of relative addition ⨨ is stiff. I *write* a ϯ b. Still ϯ would not be much better than Schröder's. Besides, you have it. *So that will do.*

My sign of the copula has always been printed —<. But I always *write* ⤙, and I should like to have this cursive form cut by an intelligent, artistic, typecutter's draughtsman, and used.

I make great use of Σ and Π. They can be got in the German offices. For such as I like frequently occur in the *Mathematische Annalen*; but these continental types wont justify will they? They should be *upright, all of one thickness, and devoid of the little finishing lines* (whose name I forgot.)

Σ Π

My rule in all formulae is, Capital Letters to be Roman, *l.c* letters Italics.

I use l.c. letters as subscript, that is, below the line; rarely any others.

I sometimes like to use a pair of heavy parentheses.

My figures of infinity are zodiacal signs ♉ can be bought. It ought to look as much like a zero with a line over it as possible Ʊ Ʊ. But existing forms will do. But my Aries is peculiar: ♈. It is in American types usually: ♈. This is because these figures are no longer so constantly used cursively as they used to be. They are beautifully easy characters to write

♈ ♉ ♊ ♋ ♌ ♍ ♎ ♏ ♐ ♑ ♒ ♓

My Aries is cut particularly short, so that while preserving its resemblance to a ram's horns, it shall look like ∞. ♈ If the draughtsman understands this, he can cut the figure. Should ♉ be cut, it should preserve the resemblance to a bull's face, and yet recall $\overline{0}$. It wants a short-horn bull, not this ♉ but this: Ʊ. These two occur above the line, sometimes.

---

1 modification] *after* ~~modo~~

9 Schröder's] *after* ~~Jevons~~

11 always[1]] alwas *[E]*

24 types] *over* p

25 no longer] *after* ~~dat~~

31 Ʊ.] *after one deleted attempt at this*

Yours very truly
CSPeirce

Peirce to McCormack. MS: RL 77. First page of an aborted draft of 2 Sept 1896 letter.

[Unsent draft]

Stewarts Hotel, Broadway & 41st St.New York City | 1896 Sep. 2.

Mr. T. J. MacCormack

Dear Sir:

I have postponed answering yours of the 21st ult.,—without meaning, however, to put it off so long,—because I was anxious Dr. Carus should have before him a substitute which he may desire to use instead of what I have said about logical graphs in the MS forwarded. This I have now sent to him.

The most difficult characters to reproduce are those which I use for the copula as in this formula A ⤙ B ⤙ C ⤙ D ⤙ E ⤙ F. The cost of justification, etc. in those cases, together with the facts that the uprights would have to be of several sizes, and that the number of such formulae are few, and the sign is not intended to remain permanently in use, makes me think the cheapest way would be to have each whole formula drawn as a cut.

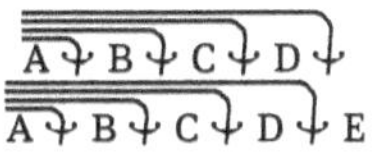

These should be sufficient hints to an intelligent draughtsman. If my *last* sent MS. on graphs is used, the figures are well enough to serve the purposes of the draughtsman, especially if he reads the article. I can have them drawn under my eye, if desired. If the former MS is used I must see the figures & redraw them. The character ⨍ used by Schröder suits me well enough. The character ⤙ I have always printed on this stiff form. I *write* the former ⨍ with a scorpion-tail curve, and the latter ⤙, and I think it would sub-*[…]*

17 each whole formula] ~~the~~ \each/ whole ~~thing~~ \formula/
21 well] well~~s~~
23 If the … them.] *inserted above*
26 ⤙] *after 4 attempts of the same*

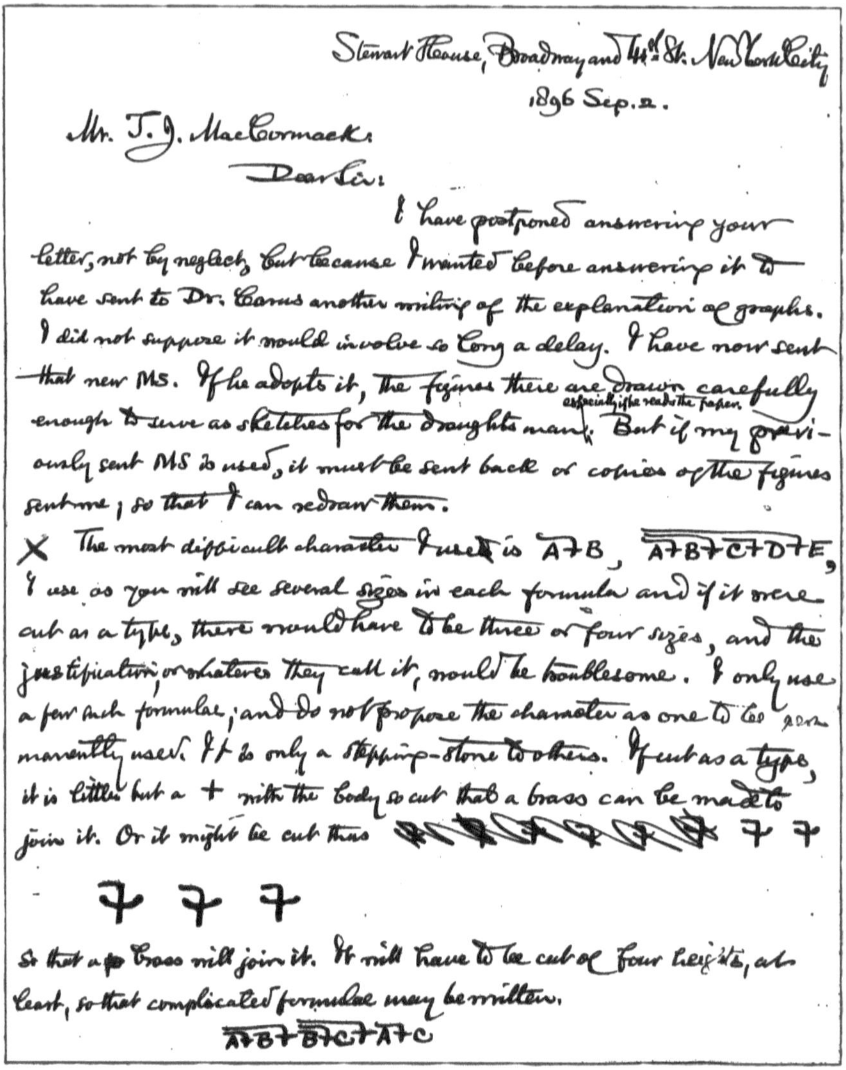

Stewart House, Broadway and 41st St. New York City
1896 Sep. 2.

Mr. T. J. MacCormack:

Dear Sir:

I have postponed answering your letter, not by neglect, but because I wanted before answering it to have sent to Dr. Carus another writing of the explanation of graphs. I did not suppose it would involve so long a delay. I have now sent that new MS. If he adopts it, the figures there are drawn carefully enough to serve as sketches for the draughtsman, especially if he reads the paper. But if my previously sent MS is used, it must be sent back or copies of the figures sent me; so that I can redraw them.

X The most difficult character I use is $\overline{A ⅂ B}$, $\overline{\overline{\overline{A ⅂ B} ⅂ C} ⅂ D} ⅂ E$, I use as you will see several sizes in each formula and if it were cut as a type, there would have to be three or four sizes, and the justification, or whatever they call it, would be troublesome. I only use a few such formulae, and do not propose the character as one to be permanently used. It is only a stepping-stone to others. If cut as a type, it is little but a + with the body so cut that a brass can be made to join it. Or it might be cut thus ⅂ ⅂

⅂ ⅂ ⅂

So that a brass will join it. It will have to be cut of four heights, at least, so that complicated formulae may be written.

$\overline{A ⅂ B} ⅂ \overline{B ⅂ C} ⅂ A ⅂ C$

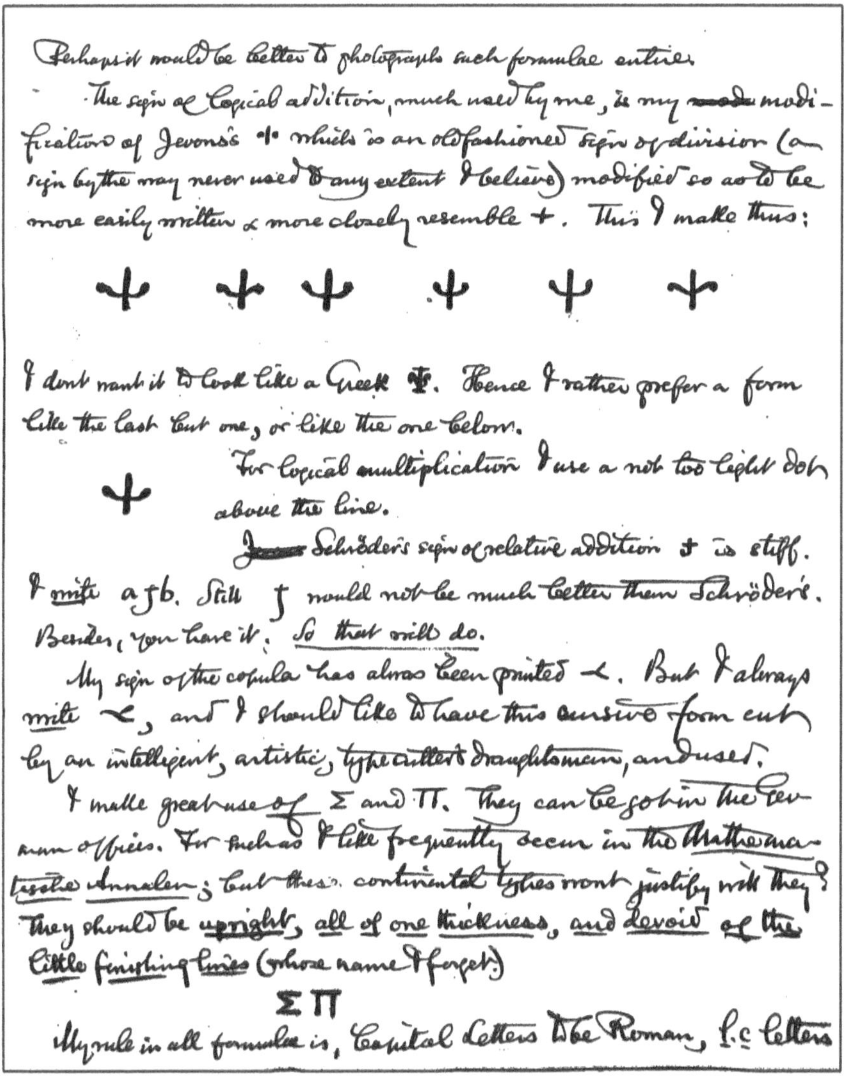

Perhaps it would be better to photograph such formulae entire.

The sign of logical addition, much used by me, is my ~~mode~~ modification of Jevons's ·|· which is an oldfashioned sign of division (a sign by the way never used to any extent I believe) modified so as to be more easily written & more closely resemble +. This I make thus:

I dont want it to look like a Greek Ψ. Hence I rather prefer a form like the last but one, or like the one below.

For logical multiplication I use a not too light dot above the line.

~~I~~ Schröder's sign of relative addition ‡ is stiff. I write a ɟ b. Still ɟ would not be much better than Schröder's. Besides, you have it; So that will do.

My sign of the copula has always been printed ≺. But I always write ≺, and I should like to have this cursive form cut by an intelligent, artistic, typecutter's draughtsman, and used.

I make great use of Σ and Π. They can be got in the German offices. For such as I like frequently occur in the Mathematische Annalen; but these continental types wont justify will they? They should be upright, all of one thickness, and devoid of the little finishing lines (whose name I forget)

Σ Π

My rule in all formulae is, Capital letters to be Roman, l.c. letters

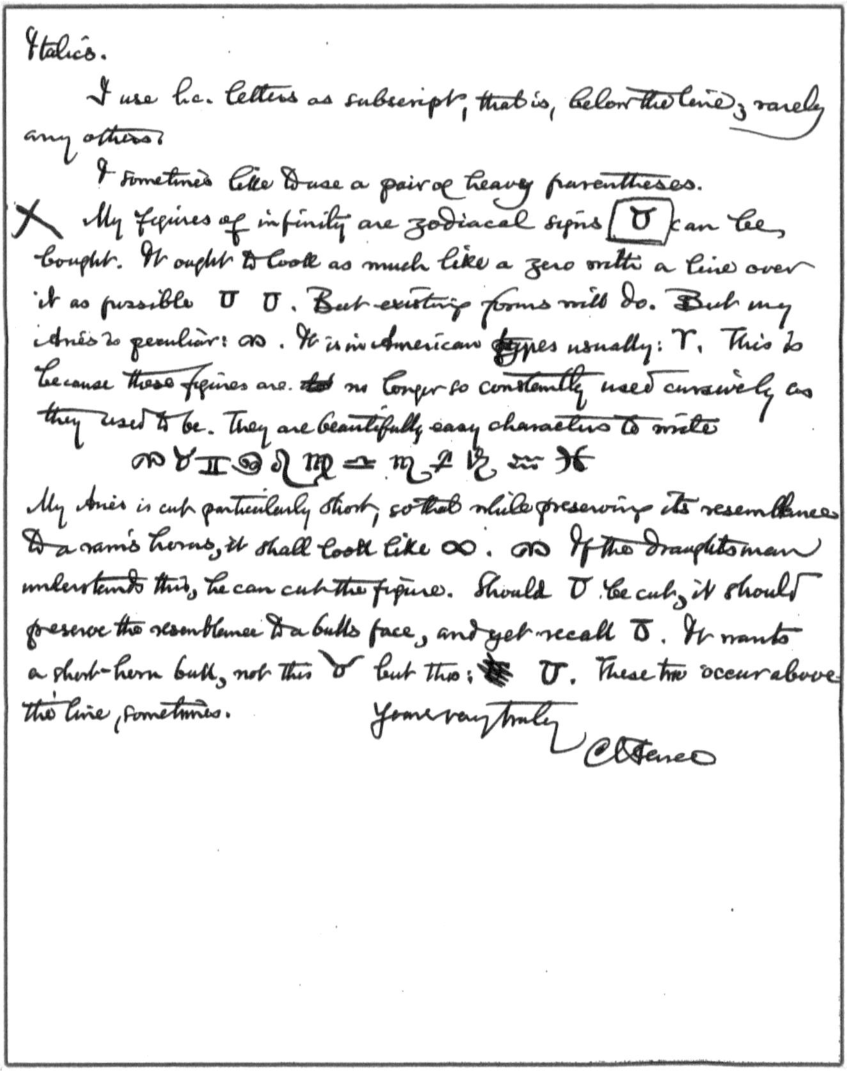

Italics.

I use l.c. letters as subscript, that is, below the line; rarely any others.

I sometimes like to use a pair of heavy parentheses.

X My figures of infinity are zodiacal signs ♉ can be bought. It ought to look as much like a zero with a line over it as possible ♉ ♉. But existing forms will do. But my Aries is peculiar: ♈. It is in American types usually: ♈. This is because these figures are no longer so constantly used cursively as they used to be. They are beautifully easy characters to write

♈ ♉ ♊ ♋ ♌ ♍ ♎ ♏ ♐ ♑ ♒ ♓

My Aries is cut particularly short, so that while preserving its resemblance to a ram's horns, it shall look like ∞. ♈ If the draughtsman understands this, he can cut the figure. Should ♉ be cut, it should preserve the resemblance to a bull's face, and yet recall ♉. It wants a short-horn bull, not this ♉ but this: ♉. These two occur above the line, sometimes.

Yours very truly

C. S. Peirce

Figure 8: Peirce to McCormack 2 Sept 1896, 3 pages. Explanation of Peirce's logical characters, with illustrations.

McCormack to Peirce. LPB: SIU 31.33.171.

December 5, 1896.

Charles S. Peirce Esq'r,
84 Broad Street
New York City, N. Y.,
Dear Sir:—

We send with this mail a number of galleys of your article.[34] By a delay in Chicago we have been unable to insert the figures, but shall send you later a duplicate set of galleys with the missing matter inserted. For the present, will you correct the text thoroughly, and send it back as soon as possible? Then when you receive the second set of galleys you will only have to approve the figures.

Very truly yours,
THE OPEN COURT PUBLISHING CO.
sig T.McC

Peirce to OCP ca. 5 Dec 1896 (missing): Sends revised diagrams to be used for "Logic of Relatives."

McCormack to Peirce. TS: RL 77. Stationery headed, "The Monist, | Editorial Department, | La Salle, Ill." LPB: SIU 31.35.205. The final in-line insertion and the signature are in McCormack's hand. This letter therefore included in this correspondence.

December 9, 1896.

Charles S. Peirce, Esq'r,
84 Broad St.,
New York City, N. Y.,
Dear Sir:—

Galleys 2 to 6 inclusive of your article were received to-day. The new drawings of the graphs which you sent differ from those in the copy. If in any way possible, please approve the drawings which you find in the last batch of galleys sent to you. It takes several days to send copy to Chicago and have it returned with the cuts. All the drawings sent to you had to be made in Chicago by a draftsman, photographed, and electrotyped at considerable expense. The Monist is quite late already, and

34 "Logic of Relatives," *Monist* (Jan 1897).

22 84] 64 *[E]*

even if we were inclined to make the changes we should not have time. This is really important.

Yours very truly,
[signed] The Monist

Peirce to OCP telegram 11 Dec 1896 (missing): Approves the diagrams to be used for "Logic of Relatives." Answered by Carus same day.

1–2 This is really important.] *inserted in-line*

2.4 Peirce and Other OCP Staff

2.4.1 Part I: Miscellaneous

Part I of this section contains letters exchanged between Peirce and various OCP staff members with whom he did not hold regular correspondence, including Carus's editorial assistant and OCP translator Lydia Robinson in La Salle, Hegeler's secretary Charles Diesterweg in Chicago, OCP clerk and sales manager Martin Sacksteder in Chicago, and several others, some not yet identified. These letters deal with editorial, financial, and administrative matters that provide supplementary context to Peirce's dealings with OCP and to the correspondence as a whole.

1890

Sacksteder to Peirce. LPC: SIU 91.9.

July 15 0

Chas S. Peirce Esq
Milford, Pa.
Dear Sir: The enclosed letter[1] was written before Dr. Carus had obtained your address from Mr. Russell.

We take pleasure in placing your work upon our complimentary (free) list.

Respectfully
The Open Court Pub Co
S

1891

Peirce to OCP. MS: SIU 91.10. Note in red pencil at bottom of message from Carus to OCP, "Make out check & send it for signature at once!"

Milford Pa 1891 Nov 8

The Monist
To C. S. Peirce Dr

1 Carus to Peirce 2 July 1890.

For article entitled "Necessity Examined" 6 400 words $160

Rec'd payment
CSPeirce

Sacksteder to Peirce. TS: SIU 91.10. On OCP Stationery B. In response to Peirce to Carus 5 Nov 1891, regarding his "Art of Reasoning" advertisement.

Chicago, Nov. 15th. 189*[1]*

C. S. Peirce Esq.
Milford, Pa.

Dear Sir: In reply to your request for terms of advertising in The Open Court, we shall be pleased to publish your announcements at the rate of 15 cents per agate line each insertion, allowing a discount of 15% on six months, and 25% on a year's advertising.

As a rule we refuse all advertising matter for The Open Court, but are glad to make an exception in your favor; thereby giving unusual prominence to your card.

Respectfully yours,
The Open Court Pub. Co.
S.

1892

Sacksteder to Peirce. MS: RL 77. On OCP Stationery B. LPC: SIU 91.11.

Chicago, June 10, 1892

Chas S. Peirce Esq
12—W. 39$^{\underline{th}}$ st
New York City

Dear Sir: Enclosed we hand you,—with compliments of Dr Carus,—check of 200^{\underline{00}}$ for your article contributed to the July *Monist* entitled *"The Law of Mind"*

Respectfully Yours
The Open Court Pub. Co
S.

Sacksteder to Peirce. MS: RL 77. On OCP Stationery B. LPC: SIU 91.11.

Chicago, June 16, 1892

Chas S Peirce Esq
12. W. $39^{\underline{th}}$ st New York City
Dear Sir: Several days since we sent you check of $\$200^{\underline{00}}$ on a/c of contribution to the *Monist* for July –

Enclosed we hand you with compliments of Dr Carus a check of $\$100^{\underline{00}}$ on account of the article you are now engaged on.[2]

Respectfully Yours
The Open Court Pub Co.
S.

Sacksteder to Peirce. TS: SIU 91.12. On OCP Stationery B.

Chicago, July 21 189*[2]*

Mr. Peirce.

Your order for a copy of Fundamental Problems to Mr. Blood[3] is at hand. Shall it be charged to him or is it complimentary?

Respectfully yours
M A Sacksteder

Peirce to Sacksteder. Postcard: SIU 91.11. Recto headed, "Postal Card One Cent | United States of America" with portrait at right.

[recto]

Port Jervis | Aug 17 | 12M | 92 *[stamped]*

Mr. M. A. Sacksteder
P. O. Drawer F.,
Chicago, Ill.

[verso]

October Monist. p10. $10^{\underline{h}}$ line from the bottom for "it seems to take," read: it takes

2 "Man's Glassy Essence"; advance per Peirce's request to Carus 9 June 1892.
3 Copy of Carus, 1889; likely to Benjamin Paul Blood (1832–1919), American philosopher and poet.

27 October … takes] *inserted above message*

If I wrote Delbœuf's name Debœuf, it was the effect of an inveterate habit of learning letters out of words. The name is rightly spelled in the proof.[4]

CSPeirce

Peirce to OCP. Postcard: SIU 91.11. Recto headed, "United States | Postal Card | One Cent" with center portrait.

[recto]

Milford | PA | Aug 31 *[stamped]*

Open Court Publishing Co
La Salle
Ill.

[verso]

1892 Aug 30

Did the copy of my first article on the Critic of Arguments, "Exact Thinking" reach you? I have not heard. Sent a week ago.

C. S. Peirce

Garbutt to Peirce. MS: RL 77. On OCP Stationery B.

La Salle, Sep 12 1892

Chas. S. Peirce Esq.
Milford, Pa.

Dear Sir:

The proofs of your article "The Critic of Argument," bearing the post-mark of Port Jervis Sep. 8 did not arrive till this morning, after *The Open Court* was partly made up, and for that reason it does not appear in this week's number. We enclose you herewith the proofs of your next article "Dmesis".

Yours very truly
The Open Court Pub. Co.
W. J. G.

Peirce to OCP. MS: SIU 91.11. Recto headed, "United States | Postal Card | One Cent" with center portrait. Hole punched through part of second digit of day. Repeated note in red pencil from OCP across recto and down right margin of verso: "Please attend to this."

4 "Man's Glassy Essence."

[recto]

Port Jervis | Sep 2*[8]* | 12M | 92 *[stamped]*

Open Court Publishing Co
La Salle
Ill.

[verso]

Please send 100 copies Open Court with Pythagorics and 50 copies of future numbers with my article. I shall use these to the advantage of publishers.

C. S. Peirce

Peirce to OCP. Postcard: SIU 91.11. Recto headed, "Postal Card One Cent | United States of America".

[recto]

Port Jervis | Oct 4 | 6 PM | 92 *[stamped]*

Open Court Publishing Co
La Salle
Ill.

[verso]

My article[5] for January Monist will be forwarded tomorrow. It has delayed one set Mach proofs.

C. S. Peirce

Peirce to OCP. Telegram: SIU 91.11. Form headed, "The Western Union Telegraph Company." Note on verso, "35¢ repeating | to Dr Carus | La Salle | Oct 19/92" Answered by Carus same day via postcard and letter.

Oct 19 1892 | New York

Open Court Publishing Co.
175 La Salle st Chicago Ill

Was my article for January Monist forwarded fortnight ago, received telegraph reply Buckingham Hotel new york

C S Peirce

---

5 "Evolutionary Love."

---

7 100] *above* ~~50~~

Garbutt to Peirce. MS: RL 77. On OCP Stationery B.

La Salle, Nov 1 1892

C. S. Peirce Esq
Milford, Pa.
Dear Sir:

We send you herewith proof on your article "Evolutionary Love," and also page 57 of the copy as we think there is an omission

Yours very truly
The Open Court Pub. Co.
W. J. G.

1893

OCP to Peirce. TS: RL 77. On OCP Stationery B, with PS added in pen by Carus. LPC: SIU 91.12.

Chicago, February 15, 1893.

Mr. Charles S. Peirce,
New York City, N. Y.,
Dear Sir:—

All the type which we have is now set up and we are hardly able to proceed with our work. We should be very glad if you could send the "Mach" proofs as soon as your convenience permits, that the type may be set up in pages and disposed of. We desire very much to get the book in press sometime during this spring, but as our supply of type is limited we can only work at the book when there is a lull in our other work.

Yours very truly,
The Open Court Pub. Co

My dear sir: The Mach proofs are at present more urgent than anything else.

With kind regards
Yours very truly
P Carus

Peirce to OCP. MS: SIU 91.12. Stationery headed, "Express Address. | Port Jervis, N.Y. | Wells, Fargo Exp. Co. | C. S. Peirce."Arisbe" | P.O. Milford, PA."

1893 March 30

Open Court Publishing Co.

175 La Salle Street
Chicago

Gentlemen:

The 6th Volume of the Open Court has duly reached me and I will thank you to express my particular thanks for it to Mr. Hegeler & Dr. Carus. I dont want to miss a number; and like to keep it bound.

Might I ask you to do me the favor to send me 6 copies each of the Monist for

Jan 1891 Vol I. No 2
Apr 1892 " II 3
Oct 1892 III 1

and 3 copies of the same for

Jan 1893 Vol III No 2.

I try to give away as few as I can of my papers. If the Open Court Publishing Co. thinks it would serve its purpose to have more given away, I could easily give away more, without giving them to people who would not read them.

Yours very Truly
C. S. Peirce

Sacksteder to Peirce. TS: RL 77. On OCP Stationery B.

Chicago, April 7th. 1893

Prof. C. S. Peirce
Milford, Pa.

Dear Sir: In reply to your favor of March 30th., we sent some days ago the desired numbers of the Monist, and would say with reference to your giving away copies of these, that we should be pleased to send you as many copies as you can use to advantage. If you would kindly supply us with the names and addresses of the persons you give them to, we might afterwards circularize them, providing it is not troubling you too much; or better still, if you would send us the names in advance, we could send the copies from here at our expense and mark your articles "with compliments of the author".

Respectfully yours,
The Open Court Pub. Co.
S.

---

10 1] *over* 4

1894–1896: Two-year lapse in correspondence between Peirce and OCP due to a falling out with Hegeler. For details, see note in Hegeler and in Carus correspondences around this time.

1896

Sacksteder to Peirce. Postcard: RL 77. Recto headed, "Postal Card—One Cent | United States of America" with portrait at right.

[recto]

La Salle, Ill. | June 11 | 7 PM | 96 *[stamped]*

Chas. S. Peirce Esq
160 W. 87th Street
New York City

[verso]

La Salle | Ills. | June 11/96

Dear Sir

We have sent you proofs of your forthcoming article[6] to-day; please return them direct to La Salle.

We hope the paragraphs follow in proper order.

Some of the pages of copy were not numbered, so to avoid an error we have sent copy with the proofs, at the same time we ask you to return it to us with the galleys.

Yours truly
The Open Court Pub Co.
S.

OCP (likely Sacksteder) to Peirce. TS: RL 77. Stationery headed, "The Monist, | Editorial Department, | La Salle, Ill." LPB: SIU 31.1.252.

June 19, 1896.

Chas. S. Peirce, Esq'r,
160 West 87th St.,
New York City, N. Y.,

Dear Sir:—

6 "The Regenerated Logic," *Monist* (Oct 1896).

We have been waiting very patiently for the return of the proofs of your article on "Regenerated Logic" which was to have gone into the present number. The proofs were sent from here a week ago yesterday, that is, on the 11th of June, and should have reached you on Saturday morning, June 13th. As it is, the Monist is a little late, and we cannot defer any longer making up the last Long Primer form which must be sent to Chicago to-morrow. If your proofs come in the morning, we may perhaps be able to put the article in this number, otherwise it cannot appear until October.

Very truly yours,
The Open Court Pub Co

Peirce to OCP. Postcard: SIU 91.14. Recto headed, "Postal Card—One Cent | United States of America" with portrait at right. Answered by Carus 10 Aug.

[recto]

New York | Aug 7 | 1896 *[stamped]*

Open Court Company
La Salle
Illinois

[verso]

I some time ago sent a MS. of continuation of notice of Schröder's Logik to Dr. Carus. Was it received?[7]

C. S. Peirce
Care Albert Stickney Esq | 35 Nassau St | New York

Peirce to OCP ca. 5 Dec 1896 (missing): Sends revised diagrams to be used for "Logic of Relatives."

Peirce to OCP. Postcard: SIU 91.14. Recto headed, "Postal Card—One Cent | United States of America" with portrait at right.

[recto]

New York N.Y. State | Dec 8 | 1896 *[stamped]*

Open Court Publishing Co.
La Salle

7 Peirce to Carus 15 July 1896.

Ill.

[verso]

The list of numbers given in my new foot-note (which, if decidedly inconvenient to insert, might be omitted altogether, though it has some value) is wrong and the true series is both complicated and uninteresting. So this row of numbers can be omitted. It really runs 1-2-12-21-3-13-31-23-32-123-132-312-213-231-321-4-14-41-24-42-124-142-412-etc.[8]

C. S. Peirce

The number of terms, if N is the highest number is

N+N(N−1)+N(N−1)(N−2)+N(N−1)(N−2)(N−3)+etc.

Peirce to OCP telegram 11 Dec 1896 (missing): Approves the diagrams to be used for "Logic of Relatives." Answered by Carus same day.

1897

Sacksteder to Peirce. MS: RL 77. On OCP Stationery B.

Chicago Oct 28/97

Prof Chas S Peirce
84 Broad st
New York City

Dear Sir:

Kindly advise B. F. Stevens 4 Trafalgar Square Charing Cross London Engl, a dealer in books,[9] of the pamphlets and articles published by you and where they may be had—omitting those which have appeared in the *Open Court* & *The Monist*—as we have supplied these for a customer of Mr Stevens who wishes also to secure all of the other contributions from your pen. If you should be able to supply these pamphlets, it would save time if you sent them at once with your bill as the order is final and for every thing you have done.

---

8 "Logic of Relatives," p. 208.

9 Benjamin Franklin Stevens (1833–1902), later partnered with Henry J. Brown ca. 1899 to form B. F. Stevens & Brown, Library and Fine Arts Agents of London.

If you should prepare a list of your works for this purpose it would be agreeable to the undersigned to have a copy of it—as it is likely to be asked for again and we would be able to give definite information without troubling you

Respectfully
M A Sacksteder

1898

Peirce to OCP 23 Sept 1898 (missing): Asks to be added to the circulation list for *Monist* and *Open Court*, requests a copy of Carus's "Buddhism and Its Christian Critics," and inquires about additional articles for submission. Also answered by Carus 13 Oct 1898.

Diesterweg to Peirce. On OCP Stationery B.

Chicago Sept. 28th, 1898

Mr C. S. Peirce,
Milford, Pa.
Dear Sir:—

Your favor of Sept. 23d was duly received, and in reply we will enter your name on The Open Court and The Monist lists.

We send you a copy of Dr Carus's "Buddhism and Its Christian Critics", and as we have just published a translation of Lagrange's "Lecons élémentaires sur les mathématiques" we also send you a copy thinking that you might perhaps have it reviewed.

We will hand your letter to Dr Carus.

Respectfully yours,
The Open Court Pub. Co.
D.

---

18 Lagrange's] DeMorgan's *[E] (see Bibliography)*

1900

Peirce and Sacksteder mid-late April 1900 (missing): Sacksteder contacts Peirce with some question of "special interest," to which Peirce replies and in some way criticizes Carus, who opens correspondence 3 March 1900 (see correspondence).

Peirce to Carus 3 Nov 1900 (see correspondence): Asks for type of logic characters used in his "Logic of Relatives," to be employed for his work in Baldwin's *Dictionary of Philosophy*. OCP replies as follows.

OCP (unsigned) to Peirce. LPB: SIU 34.41.640.

November 9th, 1900.

Mr. Charles S. Peirce,
Milford, Penna.

Dear Sir:—

The type for your logical characters were cast from matrices made in the latter part of the year 1896, especially for us, by Messrs. Barnhart Bros. & Spindler, 183 Monroe St., Chicago, Ills.. These matrices are, we believe, our property, and Messrs. Barnhart Bros. & Spindler would doubtless be glad to furnish you what you want at their usual prices. It will be best, however, for you to ascertain beforehand precisely how many characters you want. At the time your articles[10] were printed, we probably had about three pounds of each of these characters.

We enclose three specimens of the type, so that you may use them when you write to the type founders.

Very truly yours,
The Open Court Publishing Co.

Jan 1901–Aug 1904: For the next three years, with the exception of his unusual, incomplete correspondence with OCP compositor Fred Sigrist, no letters between Peirce and OCP are found. It is likely that Peirce simply maintained little contact with OCP during what was then a very busy time for him: he translated for the *Annual Report of the Board of Regents* of the Smithsonian Institution (1901), wrote definitions for Baldwin's *Dictionary of Philosophy* (1901–1902), submitted his elaborate grant proposal to the Carnegie Institute for "Memoirs on Minute Logic" (1902),[11] de-

---

10 "Regenerated Logic" and "Logic of Relatives."

11 Ransdell, "MS L75: Logic, Regard as Semiotic", and partly in *NEM* 4:13–73.

---

17 at their usual prices] *typed above*

livered his Harvard lectures on pragmatism[12] and his Lowell Institute lectures on logic (1903),[13] and all along continued reviewing books for *The Nation*.

He was, however, also gearing up for a new series of *Monist* articles on pragmatism to disentangle his own form of it, pragmaticism, from other contemporary variants, such as those of William James, F. C. S. Schiller, and John Dewey.

1904

OCP to Peirce 10 Sept 1904 (missing): Promises to send check for his first pragmatism article, "What Pragmatism Is," to appear April 1905.

Nattinger to Peirce. LPB: SIU 37.50.106.

September 23, 1904.

Charles S. Peirce, Esq.
Milford, Pa.

Dear Sir:

Please find enclosed check for $50.00 as per our letter the 10th inst.

Yours truly,
Open Court Publishing Co.

JGN.

Robinson to Peirce. TS: RL 77. Stationery headed, "The Monist | The Open Court | Editorial Department | La Salle, Ill. | Dr. Paul Carus" LPB: SIU 36.75.510.

La Salle, Ill., Nov. 22, 1905.

Charles S. Peirce,

12 *EP* 2:133–241, also in full in Turrisi, 1997.
13 *EP* 2:258–299.

21 Peirce] P*i*e*e*irce

Milford, Pa.

My Dear Sir:

In Dr. Carus' absence I am getting your paper on Royce's book[14] ready for the compositor, who you may know from former experience is prone to using large words when it becomes necessary for him to set by hand irregular copy. He admits that there is nothing about this that he can not do, but there is one particular about which I am not quite clear.

As I understand, you have plainly indicated a distinction between the capital Roman E and the Greek, thus: E and Ɛ; but the type for the large Epsilon is exactly the same as the Roman Cap. so I am at a loss how to express the distinction which you need.

In the formulae you use also the characters $\underline{e}$, $\underline{e}'$ and $\underline{e}''$ By the line underneath you mean, do you not, simply to emphasize the fact that the letter is the Greek minuscule ε, in contrast to the italic *e*, and not that the line underneath is necessary to differentiate that ε from any other.

Sincerely regretting the necessity of troubling you in this matter, I remain,

Yours very truly,

(Miss) Lydia G. Robinson.

1906

Robinson to Peirce. LPB: SIU 37.16.190.

La Salle, Ill., July 2, 1906.

Dear Sir:

We are having the diagrams prepared which are to accompany your article on the , and we find that figure 3 is omitted. You began a rough diagram in a sort of honey-comb design, but crossed that out. The place of its insertion is thus introduced on page 37 of the MS.: "In order to avoid the intersection of lines of identity either a selective may be employed or a *bridge* which is imagined to be made of a

14 Likely Peirce's "The Relation of Betweenness and Royce's O-collections" presented that month to the Academy of Sciences on Royce's 1905 *The Relation of the Principles of Logic to the Foundations of Geometry*. Apparently slated for *Monist* but ultimately rejected. See Peirce to Carus ca. 25 Oct 1905.

4 who] who~~m~~ 16 regretting] regre~~e~~\t/ting

ribbon of paper, but will in practice be pictured as in Figure 3."[15] Hoping you will be able to furnish a diagram to insert in this place, we remain,

Yours very truly,
OPEN COURT PUBLISHING CO
R.

Robinson to Peirce. LPB: SIU 37.17.268.

La Salle, Ill., August 9, 1906.

Mr. Chas. S. Peirce,
Milford, Pa.
Dear Sir:

By today's mail we are sending you proofs of your "Prolegomena to an Apology for Pragmaticism", and hope you will find them generally satisfactory. The omission of figure I is due to a mistake which was made by the maker of the cut. We have it now on hand and are sure that there is no question about it. In regard to the other figures we supposed that all your requirements were carefully followed, but find when reading proof that you speak of "a heavy line as a juncture". If it is necessary in order to present your meaning fully that some of these be redrawn with a heavier connecting line, please indicate exactly the change you desire and it will be made, but we hope they may serve as they now stand without further change.

Very respectfully yours,
OPEN COURT PUBLISHING COMPANY.

L. G. R./N.

1907

March–July 1907: Peirce prepares a lengthy "letter to the editor" for *The Nation* and *Atlantic Monthly* on pragmatism (rejected, *EP* 2:398–433) and delivers Harvard Philosophy Club lectures on "Logical Methodeutic" 8–13 April (see *W* and *EP* Chronology). Likely begins working on his imminent "Some Amazing Mazes" series as a respite from his pragmatism series, which he was reportedly too ill to complete at this time (see Peirce to Carus 12 Oct 1908).

15 *Monist* (Oct 1906), p. 531: 
Fig. 3.

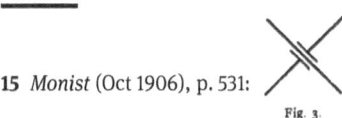

14 are] *above* ~~am~~

Peirce to Carus ca. 25 May 1907 (missing): Asks for a copy of Oct 1906 *Monist* with his "Prolegomena" to be sent to Giovanni Vailati, whose "Pragmatism and Mathematical Logic" appeared in the same *Monist* number. Answered by Lydia Robinson as follows.

Robinson to Peirce. LPB: SIU 37.30.990.

La Salle, Ill., May 31, 1907.

Chas. S. Peirce,
Milford, Pa.
Dear Sir:

Your letter of recent date addressed to Dr. Carus was duly received, and in his absence we have undertaken to comply with your request that certain copies of THE MONIST be sent to Sr. Vailati. However, an article of Vailati's on Pragmatism[16] appeared in the same number of The Monist in which your article on "Existential Graphs" was published. This means that he received at least one copy of that number, being himself a contributor. We shall see that he receives a copy of the errata slip belonging to your article, but will not send him another copy containing the article itself unless we hear from you to the effect that you desire him to receive an extra copy.

Hoping this will be satisfactory in following the spirit if not the letter of your request, we remain,

Yours very truly,
Open Court Publishing Co.
R.

L. G. R./N.

Peirce to Carus ca. 28 Sept 1907 (missing): Sends "The First Curiosity," "Explanation of Curiosity the First," and "A Second Curiosity" (MSS A–C)[17] of his "Amazing Mazes" series.

Robinson to Peirce. LPB: SIU 37.37.238.

La Salle, Ill., October 3, 1907.

Mr. Chas. S. Peirce,
Milford, Pa.
Dear Sir:

16 "Pragmatism and Mathematical Logic."
17 MS denotations explained in note of same date in Peirce and Carus correspondence.

We beg to acknowledge the receipt of your MS. and to inform you that Dr. Carus has been quite ill with a fever for some time and will not be able to attend to business matters for a while yet. At least part of the MS. you sent will appear in the January number of the Monist, and your letter will be called to Dr. Carus' attention when he is once again able to take up the thread of his regular duties.

We remain,

Very truly yours,
OPEN COURT PUBLISHING CO.

L. G. R./N.

Robinson to Peirce. LPB: SIU 37.37.247–248. Bottom of page 1 signed prematurely by Carus, recovering from illness and apparently overlooked page 2.

La Salle, Ill., Oct. 4, 1907.

Mr. Charles S. Peirce,
Milford, Pa.

Dear Sir:

We handed over some of the diagrams in your "Amazing Mazes" to a Japanese artist in our office who frequently does work of this kind for us in preparing illustrations for reproduction.[18] He wishes us to ask you whether the enclosed drawing of Fig. 12 exactly complies with your idea. It will be reduced to about *the size of your own diagram*, and of course the blot will be removed.

He wishes us also to ask you to give some idea as to how exactly the irregularities in outline of figures 11 and 13 are to be followed; for instance whether the outline of figure 13 should be conventionalized approximately as he did in figure 12,

---

18 *Monist* (July 1908), pp. 445–446:

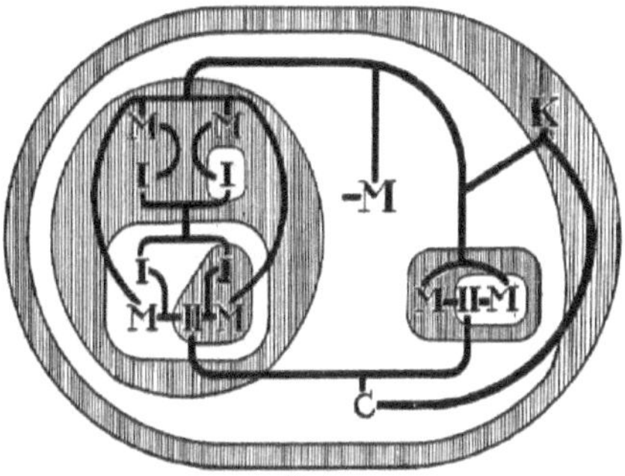

Fig. 11.

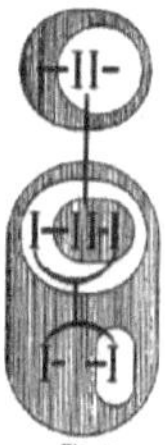

Fig. 12

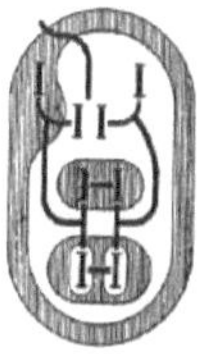

Fig. 13.

---

2 a] *above* ~~the~~

22 figures] figure$^{\vee}$s$^{\vee}$

23 he] *w*<u>h</u>e

and how differently the exact slope of the inner portion must be followed to convey your idea. Also whether the border line of Fig. 11 to the right of K should follow exactly that slope, and whether the line bordering the unshaded interior should be parallel to the outside or should vary from it at just the places and in just the degree in which it does in your sketch. Should the larger shaded body within this center follow exactly your outlines or is that supposed to be conventionally the same width at the bottom as at the top?

In short, if you can explain it in a few words, please let us know what are the vital points about your diagrams which must be followed exactly, and which may be altered to form a symmetrical part of the diagram. If you have any suggestions whatever to make in regard to the tables and figures which have to be reproduced, we shall be glad to know of them.

Enclosed you will find stamps with which to register the sheets when you return them to us.

Paul Carus

Thanking you in advance for your reply, we remain,

Yours very truly,
OPEN COURT PUBLISHING COMPANY.

L. G. R./N.

OCP to Peirce 21 Nov 1907 (missing): Asks for clarification about Peirce's discussion of cyclic and denumeral systems in his "Explanation of Curiosity the First" (at p. 452 of July 1908 *Monist*).

Peirce to OCP. MS: RL 77.

P.O. Milford Pa | 1907 Nov 23

Open Court Co
Chicago Ill
Gentlemen:

I have just received yours of the 21st inst. You can easily imagine the spiral whose equation in polar coördinates is

$$r = e^{\frac{1}{\theta}}$$

$$\text{or} \quad \theta = \frac{1}{\log r}$$

---

3 that slope,] th*e*at slope$^\vee$,$^\vee$
6 conventionally] conveni~~ent~~\tional/ly
10 symmetrical] sym\m/etrical
10 suggestions] suggestion$^\vee$s$^\vee$
15 Paul Carus] *Signature accidentally at bottom of first page.*

It will twist round the origin indefinitely many times when *r* is nearly equal to 1. Draw a radius vector, and travel along it from the origin to a great distance, and your crossings of the spiral will first be an *endless* series, but afterward will be a *beginningless* series.

I enclose a foot-note[19] to meet this difficulty by means of a slightly different example. Will you kindly have it inserted? I thank you very much for pointing out the need of it.

As to the other point, you are of course right 6! = 720.

I don't think beginningless needs any hyphen. To my mind, its meaning is plainer without one.

I give the coördinates for the above spiral. You can take any angle you like as the value of the unit of θ.

Yours cordially
CSPeirce

P.S. Kindly say to Dr. Carus that parts D and E of my "Amazing Mazes"[20] will go on to him during the coming week. The MS of D contains about 70 pages; that of E already exceeds 100 and there will be *at least* twenty more.

Thereupon I shall at once take up my two concluding papers on Pragmatism,[21] which are planned in detail, and the second of which is substantially written out in full. Its principal argument for pragmatism is scientific and strong. But I shall supplement it with what Aristotle would call some "rhetorical" arguments, which may be stronger than scientific ones. The first of the two is in my opinion a Gibraltar, although very counter to the prevailing spirit of the last half of the XIX$^{\underline{\text{th}}}$ century & perhaps still more so of the first half & of the XVIII th.

Also, if you could insinuate that an early remittance would make him a friend indeed, because a friend of more than ordinary need;—if you feel you properly can, it might be of service to me.

---

19 pp. 461–462. Peirce to Carus 26 May 1908 extended the note to p. 464.
20 "Third Curiosity" and "Fourth Curiosity," sent to Carus ca. 1 Dec 1907 (never published).
21 By this point likely "The First Part of an Apology for Pragmaticism" and "Apology for Pragmatism" (R 296–297). See Peirce to Russell 7 Dec 1907; also Peirce to Carus 30 Jan 1907.

---

2 and travel] *over [*?on a*]*
3 your] *over [*?cro*]*
5 slightly] *inserted above*
16 The MS of] *inserted above*
24 first] *over [*?oth*]*

1908

Robinson to Peirce. LPC: SIU 91.24.

La Salle, Ill., February 20, 1908.

Mr. Charles S. Peirce,
Milford, Pa.
Dear Sir:

We enclose the proofs of one of the tables in your Monist article and would call your attention to the fact that in the type of the card indicators which we have procured, the ten spot is represented by the figure 10, and not by the Roman X as you have represented it in your copy. As we have no corresponding type with which to substitute an X for the 10 in the table, we take it for granted that you would prefer us to use the Arabic numeral.[22] This being the case, would you not

22 "First Curiosity," *Monist* (April 1908), p. 235:

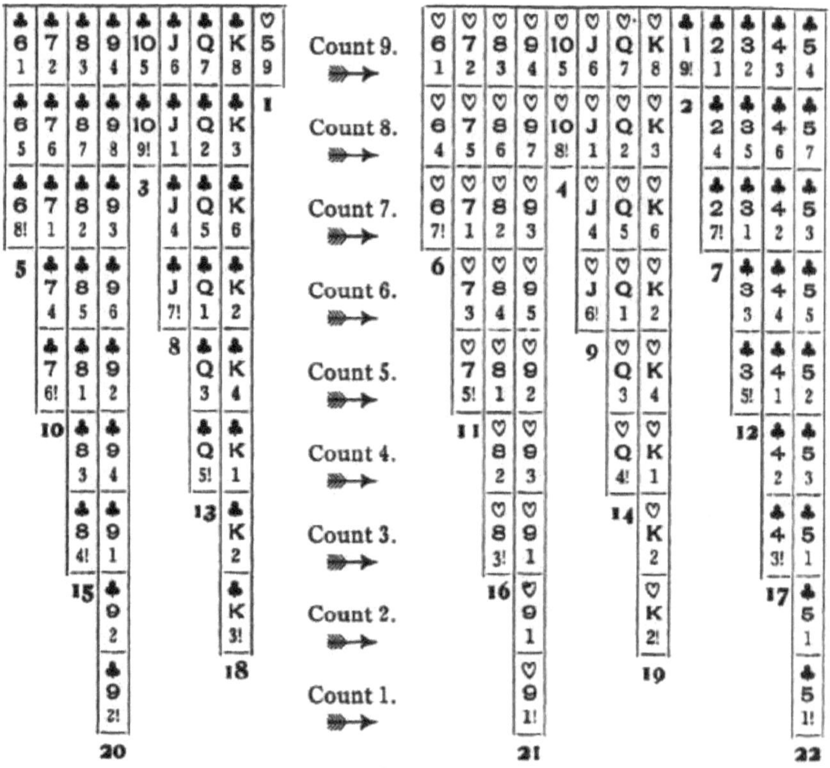

Fig. 4

also prefer the same notation in the figures which we are about to have reproduced, notably Figure 5,[23] where figures and initials are used to represent the cards from one to Queen.

We enclose also your copy for Figures 7 and 8,[24] and would ask you to indicate a little more plainly the distinction between the letters *b* and *f*.

Very respectfully yours,
OPEN COURT PUBLISHING COMPANY.
L. G. R./N.

OCP to Peirce. LPC: SIU 91.24.

La Salle, Ill., March 16, 1908.

Mr. Charles S. Peirce,

---

23 Ibid., p. 236 (Roman X ultimately remained):

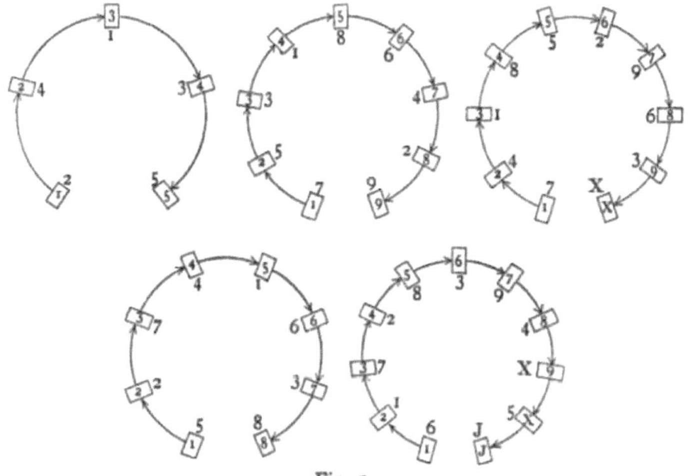

Fig. 5.

24 "Explanation of Curiosity the First," *Monist* (July 1908), p. 441 (no "f" appeared in print):

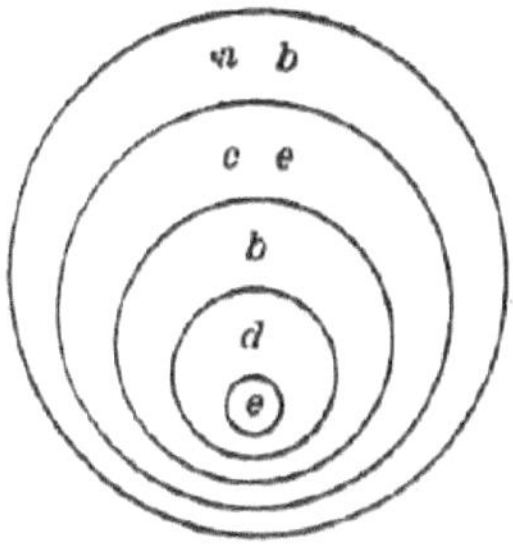

Fig. 7.

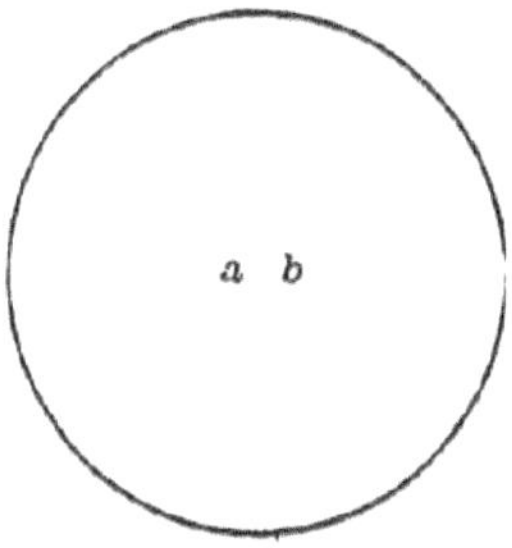

Fig. 8.

Milford, Pa.

Dear Sir:

We have asked the Chicago office to send you at once a copy of Dedekind's "Essay on the Theory of Numbers" translated by W. W. Beman. Will you please let us know immediately upon the receipt of the book whether this edition contains the Dirichlet of Dedekind, or at least that portion of it to which your article makes reference.[25] If so do you mean your personal note with regard to Mr. Beman's edition to be included in the foot note as it appears in the preface.

Very truly yours,

OPEN COURT PUBLISHING COMPANY.

OCP to Peirce. LPC: SIU 91.24.

La Salle, Ill., May 19, 1908.

Mr. Charles S. Peirce,

Milford, Pa.

Dear Sir:

Under separate cover we are sending you the proofs of Part B of your Amazing Mazes.[26] We have made every effort to follow your copy with accuracy, and trust that the proportion of corrections will be small as it is a very difficult piece of composition.

We would like to urge that any changes or additions that are made within the body of a paragraph should be limited to one exact line in length or even multiples of a line. Corrections which would throw the following words out of their position in the line, thus necessitating the resetting of the rest of the paragraph, can not be made, because there will not be time. Of course insertions of any length may easily be made as separate paragraphs or at the end of existing paragraphs.

When we had the diagrams made we thought that figures 29 and 30 could be made in the composition room. We find now that this is not the case, but there will be no difficulty about having them made accurately and suitably. We hope

---

25 "First Curiosity," p. 241, refers to "Dedekind's lucid and elegant redaction" of Dirichlet's *Vorlesungen über Zahlentheorie* (1871).

26 "Explanation of Curiosity the First."

---

6 your] you *[E]*

18 proportion] \pro/portion

23–24 paragraph, … time.] paragraph$^{\vee}$,$^{\vee}$ can not be made*.*, \because there will not be time./

25 easily] *inserted above*

figures 15–17, and 21–28 will serve as they now stand.[27] Though the lines are broken the intention seems to be quite evident. If they will not serve the purpose however, either as they stand or in lines of half the present length, we can yet have them made although at some additional expense.

Very truly yours,
OPEN COURT PUBLISHING COMPANY

27 Ibid., pp. 453–455:

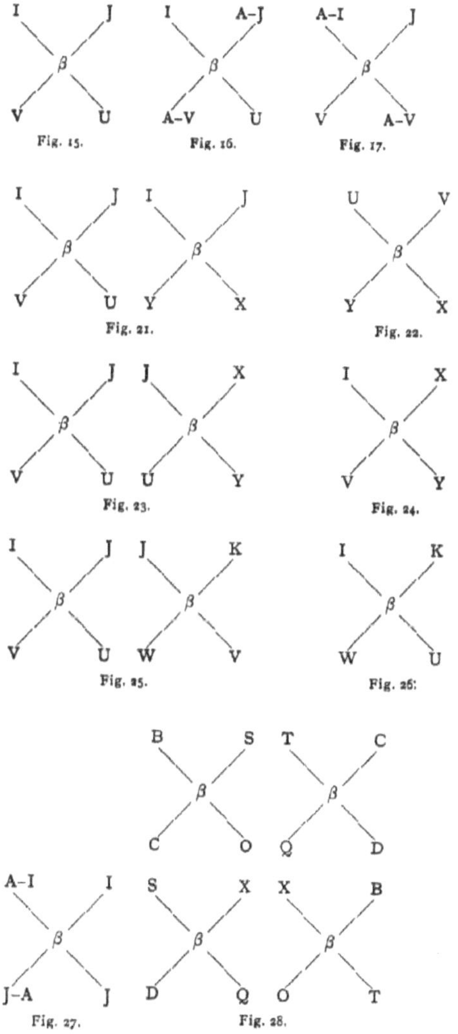

1909

Jan–Dec 1909: Peirce works in earnest on his revisions for the "Illustrations" book and on the concluding articles for his 1905–1906 pragmatism series (never finished).

Carus to Peirce 26 Nov 1909 (see correspondence): Asks for Peirce's permission to include an excerpt of a letter from Peirce to Russell ca. early Oct 1895 in his forthcoming "Nature of Logical and Mathematical Thought" *Monist* article.

Robinson to Peirce. LPC: SIU 91.25. Not received, as reported in Peirce to Carus 14 Dec 1909.

Dec. 7, 1909.

Mr. Charles S. Peirce,
Milford, Pa.

Dear Sir:

Enclosed is the copy of the portion of your letter to Mr. Russell which Dr. Carus has incorporated in his article for the current Monist.[28] We are sending it to you at his request.

Yours very truly,
Open Court Publishing Company.

L. G. R./N.

1911

Peirce to OCP. MS: RL 77.

Milford, Pa. | 1911 June 13

Open Court Publishing Co.
623-633 Wabash Ave
Chicago, Ill.

28 For Peirce's edits to the proof of the excerpt later received, see Appendix, "Peirce and Carus."

21 Publishing] *over* C

Gentlemen:
I want a copy of Prof. W. B. Smith's *Ecce Deus*.[29] Kindly let me know the price.

Yours very truly
C. S. Peirce

Milford Pa.

---

29 *Ecce Deus: The Pre-Christian Jesus*, 1894, or possibly *Ecce Deus: Studies of Primitive Christianity*, 1913 (OCP).

---

2 of] *after* ~~for~~

2.4.2 Part II: Fred Sigrist

Part II of this section contains letters to Peirce from OCP compositor Fred Sigrist, who despised Carus and corresponded with Peirce briefly in 1903 and 1907 to recruit him for a scheme to expose Carus for alleged plagiarism, particularly in his 1902 translation of Kant's *Prolegomena to Any Future Metaphysics* . "[I]f I were a scholar whose contributions had been found too 'high-priced,' " Sigrist wrote Peirce 1 January 1903 (referring to Peirce), "I should have quite a little fun with Dr. Carus before I (or somebody else) would expose him." Although no letters from Peirce are found, he seems not to have gotten involved for fear of anything "squinting at blackmail."[30] Peirce did, however, publish a review in *The Nation* 18 June 1903 of Carus's *Prolegomena* wherein he addressed the extent to which, as Carus's disclaimer states, he "utilized the labors of his predecessors" in the translation.[31] The veracity of the allegations is undetermined.

Late 1902: Sigrist sends multiple letters attempting to open correspondence and Peirce responds ca. 30 Dec 1902 (all missing).

1903

Sigrist to Peirce. MS: RL 77.

La Salle, Ill., Jan. 1, 1903.

Mr. Chas. S. Peirce
Milford, Pa.

Dear Sir:—Your letter received this morning and wish to reply that I am not an autograph hunter, although I was anxious to get a letter from you before I wrote any *more*; but this was simply a precautionary measure,—I wanted to make sure that I am corresponding with the *right* Mr. Peirce, of which I am sure now, as your letter and your manuscript which I set up about five years ago[32] are identical as far as handwriting is concerned. In your mathematical writings I am not interested, for I am not educated enough to understand them, and the reference to them was simply a "blind," intended to prevent me from getting into touch with a "stranger."

30 Sigrist to Peirce 9 Jan 1903.
31 pp. 497–498.
32 "Regenerated Logic" or "Logic of Relatives," *Monist* (Oct 1896 and Jan 1897).

Your little letter has shown me that in writing to you I must not attempt to beat around the bush if I expect an answer at all, but I shall now be as plain as possible.

I am now working as a compositor for the Open Court Pub. Co. for the last six years, and during this time had ample opportunity to study Dr. Carus and his method of "writing" thoroughly; and I must say that he is one of the greatest literary pirates on this or any other Continent. About a year ago, however, he has outdone himself. For a good many years it had been his ambition to translate Kant's "*Prolegomena*," but having neither the ability nor the patience for such a difficult task, he simply took the best *existing* translation (Prof. *Mahaffy's*, published by Macmillans, London)[33] and copied it almost literally, taking great pains, however, to change the *beginning* of every paragraph *to avoid detection*, in which he has so far succeeded to my great astonishment; in fact, he received very flattering reviews from several University professors, *doctors of philosophy*! As a matter of course, Dr. Carus is awful proud of his grand "translation," which he accomplished in less than four days; but poor Mahaffy may have worked on it more than a year! Now, if you are interested in this matter, I shall send you a copy of Carus's "translation"; and if you compare it with Mahaffy's book, you can convince yourself in less than an hour that the two are identical, and that Dr. Carus has committed a piece of literary piracy of which no man of honor, and especially no scholar with the least vestige of self-respect, would be guilty. You can imagine what a stir it would cause among scholars, etc., if this piece of thievery were publicly exposed in one of the leading scientific journals, and especially by a man of your standing and ability; for let me tell you the people here have the utmost dread of your pen when you settle down to review one of their new books, and this is the main reason why I wrote to *you*. Besides, I hear that your manuscript[34] has been rejected some time ago as being too "*high-priced*," from which I concluded that you would hardly be inclined to *betray* me if I should communicate to you that fraud stated above, which latter however is not the only one I know of and can prove; for instance, he also "translated" a Chinese book (Lao-Tze's *Tao Teh King*) without

33 Mahaffy and Bernard, 1889. Carus provides in his translation a disclaimer, however, that he "utilized the labors of his predecessors, among whom Prof. John P. Mahaffy and John H. Bernard deserve special credit" (iv).

34 Likely Peirce's proposed review of Boltzmann's "Recent Development of Method in Theoretical Physics" in Jan 1901 *Monist*, rejected by Carus 16 Jan 1901 due to insufficient funds for "an expensive contributor" like Peirce.

7 has] *inserted above*
17 you^{1}] you~~r~~
21 stir] stirr *[E]*

knowing a word of Chinese:[35] he simply imported with the Old Man's[36] money an educated, very intelligent young Japanese *who did the work*, and Dr. Carus put his name under it,—he is now one of the foremost Oriental scholars! Further, Prof. Gunkel (Germany) offered Dr. Carus some time ago an article on the Origin of Genesis,[37] which was gladly accepted, for Carus saw at a glance that he could steal its main contents, rehash it and so pose also before the public as an original "researcher." Gunkel's article appeared later on, but of course fell flat. I do not know whether Prof. Gunkel ever found out how he had been robbed. I tell you, Mr. Peirce, we publish sometimes articles which are not "*high-priced*" at all! I could fill a little book in telling you similar rascalities of Dr. Carus. At present he writes on mathematics (see "Monist," which I am sending with the present mail),[38] but I am sure he does not know the multiplication table.

I think Carus ought to be exposed, and I have made an attempt at it before, but failed. I wrote (in Sept.) to the manager of the Macmillan Co. in New York, telling him all about the fraud in connection with the Kant book, for which I expected a small compensation; but that gentleman told me that the Mahaffy translation is not copyrighted in the U.S., and nothing could be done. That may be so as far as the *Law* and the *United States* are concerned. But Carus would fear a public exposure ten times more than any Court verdict, for it would knock his (fraudulent) scientific fame into "smithereens," and all of the Old Man's money could not repair it; besides, he could still be prosecuted in England, where they have a branch office, and a law-suit there would give them no end of trouble. I am sure they would be unable to find a respectable publishing house to handle their truck after such a scandal. I am partly at loss to understand why the manager of the Macmillan Co. in New York refused to do anything in the matter. One reason may be this, that he *neglected* to have the book copyrighted in the U. S., which at that time (in 1889) would still have been possible according to the copyright laws existing *then*; and that for this negligence he would be severely censured by the *main firm* in London. Further, the Macmillan Co. gets an enormous amount of free advertising in every number of the "Open Court" (see "Book-reviews"); this of course, is of no value, as the O. C. has no subscribers.

As I stated repeatedly, a public exposure of the fraudulent "translation" of Kant's *Prolegomena* would be most disastrous to the whole O. C. crowd; and if I

35 Oddly, Carus to Russell 12 Sept 1900 admits to his not speaking Chinese: "I am both surprised and pleased to learn that you are contemplating taking up the study of Chinese, which is a language full of beauties. I myself know nothing about it, and am very glad that I do not." (LPB: SIU 34.33.384)

36 Hegeler's.

37 *The Legends of Genesis*, published by OCP 1901.

38 "Philosophical Foundations of Mathematics" (Jan 1903).

were a scholar whose contributions had been found too "high-priced," I should have quite a little fun with Dr. Carus before I (or somebody else) would expose him. I should send him some of my manuscript (if I had any to publish) and charge my usual "high price" for it. I should also send a little letter along, stating incidentally that I had lately read Dr. Carus's translation of Kant's *Prolegomena*, which, however, seems to be an almost literal copy of Dr. Mahaffy's book which had been one of my favorite classics for years. "How does that come, Doctor?" I should conclude.—I am sure, this would work like a charm: my articles would not be too "high-priced" any more, in fact, Dr. Carus would love me like his "long-lost brother,"—as long as he sees any danger ahead.—It is true, such a mode of procedure looks somewhat "French," and I do not know whether you would approve of it. But here we are dealing with a rascal of the highest order, and should be willing to do almost anything to punish him. At present he is in New York and will next attend the convention of the American Association for the Advancement of Science, in Washington, D.C., where he will deliver a lecture on mathematics,—Great Scott!—how did you manage to put such a lot of "gall" into any *single* human being,—it would almost be enough to supply the whole Hebrew race!

Well, I think I have been outspoken enough this time, and you will hardly take me any more for an autograph hunter. You will also have gained the conviction from these lines that I am far from being satisfied with my present "job," and I assure you that I am sick and tired of the O. C. gang and also of this disgusting coal mining camp called La Salle. For years it has been my ambition to get back again to my old home Chicago, but the wages are so low here that I am unable to save anything and support my family decently, especially if doctor bills come in as regularly as during the last two years. I have tried several plans to make Carus pay my moving expenses *indirectly*, but without success so far. I wish I were one of those "high-priced" authors,—and it would be "dead easy," as the slang goes.

Are you going to Washington, D.C., to attend that Convention? I wish you would and have some fun with Dr. Carus. I am sure if you would make a few poignant hints at his "translation," it would strike him like a thunderbolt. He expects to be the "whole thing" among those scientists. Now tell me the truth, Mr. Peirce, is he not considered an all-around fool and fraud by those who <u>*know*</u> him? I can hardly conceive of a more ridiculous, disgusting, repulsive fellow as that little, fat, over-fed son-in-law.

---

4 incidentally] *inserted above*
7 for years] *inserted above*
13 New York] *n*<u>N</u>ew York
15 on] o*f*<u>n</u>
16 how] h*av*<u>ow</u>
25 regularly] regu~~r~~larly
25 plans] planes *[E]*

Now I shall close and hope to hear from you soon. Please use the address exactly as given below and never write "Care of Open Court." Dr. Carus opens every letter addressed thus,—by "mistake" of course, but we poor devils have to stand it; Dr. Carus is always *suspicious*, like every other criminal.

Very resp.
Fred. Sigrist
636 Tonti St | La Salle, Ill

As a matter of course, I expect you to consider this letter as strictly *confidential.*

Peirce to Sigrist ca. 5 Jan 1903 (missing): Replies to Sigrist's unsavory scheme with reservation in doing anything to Carus "squinting at blackmailing."

Sigrist to Peirce. MS: RL 77.

La Salle, Ill., Jan. 9, 1903

Dear Sir:—Your letter received, and I shall make a few additions to my last letter to give you a clearer view of the situation, and I shall also try to correct some misunderstandings which obtain between us.

In the first place, Carus could never *deny* that it was his aim and intention to steal the best translation of Kant's Prolegomena (i.e., Prof. Mahaffy's); for any scholar who would take the trouble to compare the two books for *only one hour*, would undoubtedly reach the conviction that the two are the same thing, and that the few changes which are found in C's. version were evidently made for the mere purpose of concealing the theft. I set up almost the whole book alone, and I know what I am talking about.—Further, when the book was finished, I thought it wise to take care of the "copy" instead of letting it go on the rubbish heap. The "copy" consists of loose leaves torn out of Mahaffy's book with Dr. C's. changes, made by Dr. C. himself in his own handwriting. Now, if such evidence is not sufficient to prove the fraud, nothing else will be. *This "copy" is in my possession*, and if you wish to see it I shall send it to you. Of course I would expect of you to take good care of it,—it might come "handy" some other time. Further, you could not mention it as evidence in your controversy (if you have any such intention), for that would give me away; they all know that I hate Carus like poison and would be willing to do anything against him to show him in his true light; therefore I would

8 As a matter … strictly *confidential*] *This line up left margin.*
19 would] *after* ~~he~~
28 you] *after* ~~it~~
31 to^2] *above* ~~and~~

be the first one to be suspected of having furnished you the "copy," something which I could not risk while I am *here* in this hole; of course, if I were in Chicago, I should not care a snap what they think,—and *prove* they could nothing. Nobody knows (or cares at present) what became of that "copy." The "copy" is complete, that is, the whole translation of Prof. Mahaffy, with the exception of pp. 14–15, which I sent in Sept. last to the manager of the Macmillan Co., N.Y., and which that gentleman failed to return in spite of my urgent request. He acted very queer toward me throughout our correspondence.

With the present mail I send you Dr. C's. "translation," and I recommend the *Preface* to your special attention,—it's a peach.

You say it is quite easy to translate Kant. Why, of course it is if you know how. But I am sure it is impossible for a man who has only a smattering of the English language and is too lazy for any serious work, like Carus. All his so-called articles must be read, revised, re-read, etc. from 6–24 times before his drivel is fit to print, and in his Introduction to the Chinese book we corrected for *three years*, and I am sure every single type costs the Old Man at least a *dime*.—I shall send you one of his proofs as soon as I get hold of a good specimen. Let me tell you: if Carus would translate one of Kant's books *alone*, he would be famous in less than no time,—his English and grammar would do it all.—If Mr. McCormack could be inclined to tell you his tale of woe, you would see how Carus escapes becoming the laughing stock of the country on account of his pitchin-English,—even the poor "Jap" has to improve on it!

As far as his Chinese book is concerned, nothing can be done, I agree with you; and I only mentioned that to show you what frauds C. resorts to to become "famous." However, that fraud also will be exposed some day, and I am sure every scholar in the land will enjoy it.—You do not want to do anything "squinting at blackmailing,"—well, I expected that; but on that point we differ greatly. I believe in beating a rogue at his own game every time!

If you wish further information on any point, please say so; and remember, I am not writing a word which I am not able to prove black on white.

Resp. yours
Fred. Sigrist
636 Tonti St | La Salle, Ill

Do you understand German? I could send you some very interesting sarcastic articles on C. from a local paper,—good writer!

---

14 , etc.] *inserted in-line and above*
19 McCormack] McCormich *[E]*
20 you[1]] *inserted above*
24 show] *after* ~~what~~
24 to] *inserted above*

18 June 1903: Peirce publishes a review in *The Nation* of Carus's *Prolegomena* and addresses the extent to which, as Carus's disclaimer states, he "utilized the labors of his predecessors" Mahaffy and Bernard in the translation (pp. 497–498).

Sigrist to Peirce. MS: RL 77.

La Salle, Ill., June 26, 1903

Dear Sir:—I hear that you exposed that "translation" fraud in the "Nation," and I am very anxious to read your review. Please send me a copy of the "Nation," or what is the price of that paper and where is it is published? I should like to purchase a few copies for future "reference." If you write any more in the matter, please be cautious and do not give me away. You will be surprised to see that I am still in La Salle. Well, "I got left" with that expected job in Chicago. When the machines were ready for operation, my friend (the foreman) was asked to put a relative of the manager in the place he had promised me, and of course the poor fellow could not afford to offend the *boss*. Besides, all my old friends tell me that I shall not get a job in Chicago in a hundred years if I stay in *La Salle* and rely on *promises*,—and I think they are right. I am now trying hard to save a few dollars, and by fall I hope to have enough to move back to Chicago and watch my chances.

Here things are going from bad to worse, and the whole outfit will soon be crippled up completely and be beyond repair: our best man and the intellectual backbone of the enterprise, Mr. McCormack, will leave us in September. He has been elected principal of the La Salle–Peru High School, and I am glad he accepted the position—I am glad for his own sake, for he is a competent man, and in the educational field he has a chance for advancement, while the conditions here in the "Asylum" are simply hopeless,—hopeless for everybody except that *fat pickpocket*.

Do you read "The Monist"? Our pirate has now also entered the ranks of the mathematicians, and he has succeeded in stealing three articles together on the "Foundations of Geometry" (even the title is stolen!).[39] These articles appeared in the last three numbers of "The Monist," and some mean cuss assured me with a twinkle in his eye that they are "*grand*." But since I am no mathematician, and since I do not wish to *exaggerate* this time, I refrain from any personal judgment. However, if you wish to enjoy a hearty laugh I shall send you the three copies of "The Monist."

39 "Philosophical Foundations of Mathematics," "Foundations of Geometry," and "Foundations of Geometry (Concluded)" (Jan, April, and July 1903).

30 I[2]] *above* ~~and~~

It is getting dark and I shall close. May I hear from you?

Very resp.
Fred. Sigrist
636 Tonti St | La Salle, Ill

P.S.—Did you ever read the pamphlet which Mr. Underwood published *in exposé* of Dr. Carus, that is, how he managed to maneuver his *benefactor* (Underwood) out of the editorship and become himself editor and *son-in-law*?[40] I shall send you the pamphlet if you wish to read it (and I assure you, it is very interesting), but you must send it back,—I have only *one copy*.

Peirce to Sigrist ca. 10 July 1903 (missing): Asks for the copy of Carus's translation of *Prolegomena* in Sigrist's hands, which he offered to send Peirce 1 Jan 1903. Also likely sends a copy of *The Nation* per Sigrist's request.

Sigrist to Peirce. MS: RL 77

La Salle, Ill., July 13 *[1903]*

Dear Sir:—Your letter, etc., received. Did you mean to say that you would like to get hold of the *Kant* "translation"? Well, it was my intention months ago to send it to you, but the tenor of your letters led me to believe that you do not like to do anything in the matter at all, and I did not want to bother you too much. However, I resolved to make use of the "translation" myself some day in the way indicated in some of my first letters; the result may be small, but it is better than nothing. I intend to try my game in October when I am settled down in Chicago, but in the meantime I am quite willing to listen to you if you have anything new to say.

With the present mail I send you also a copy of the Dedekind book,[41] and one of our catalogues. I have almost a complete set of the mathematical (and other) publications of the O. C. Pub. Co., for Mr. McC. was in the habit of making us printers a present of a copy of every new book we set up. Some books we were glad to get, others we did not think worth while taking home,—Carus's stuff, for instance.

---

40 *Open Court* (24 Nov 1887), pp. 592–593. For more on Underwood and his dramatic resignation, see Russell to Peirce 2 Oct 1908.
41 Likely *Essays on the Theory of Numbers* (1901).

---

1 I] *inserted above*
7–9 I shall … *copy.*] *down left margin*
19 myself] *inserted above*

Look over the catalogue, and if you find anything in the mathematical line (the only branch you are interested in, I suppose), let me know, and I shall send you the book or books. They are of no practical use to me, and all I could do with them would be to sell them to a hock-shop on a "rainy day"; but that would hardly pay the trouble of dragging them along and pay freight, etc.

Do you read German? If so, I shall send you two numbers of a local German paper in which Carus is being tickled with a pitchfork. Do you wish to read the Underwood pamphlet?

Yours truly
Fred. Sigrist
La Salle, Ill.

1907

Sigrist to Peirce postcard 19 Dec 1907 (missing): Invites Peirce to consider another scheme or some other aspect of the same to debase Carus.

Sigrist to Peirce. MS: RL 77.

La Salle, Ill., Dec. 23, 1907

Dear Sir:—On the 19th inst. I wrote a postal card to you in regard to a matter relating to Dr. P. C. of the "O. C." and "Monist," but I am sorry to say that you did not answer. Kindly tell me if you are not interested in the concern at all, which would be a great mistake, as there is a bushel of fun and perhaps a few hundreds in it for the right party,—a man with the "proper philosophical comprehension."

Please favor me with a reply!

Resp.
Fred. Sigrist
La Salle, Ill

2.5 Peirce and Russell

Sometime in 1887, Peirce began receiving letters from a Chicago attorney and amateur philosopher and logician Francis C. Russell, who admired Peirce's work in logic and would become a longtime correspondent and dedicated disciple of Peirce. "Those who have been scanning the intellectual horizon of the cisatlantic world looking for American leadership," Russell later wrote Peirce 11 October 1895, "little know that not only for the western continent but for the entire earth the great prophet and his gospel is already extant." Although Russell held no regular position at OCP, he was a friend and editorial consultant of Carus and Hegeler and played a key role in Peirce's involvement with OCP. He persuaded Carus to invite Peirce to contribute to the first issue of *The Monist* in 1890 and over the course of Peirce's rocky association with OCP served as a liaison between the two and an advocate for Peirce's work within the OCP organization. At Carus's request, Russell reviewed and edited several of Peirce's *Monist* publications, including "The Regenerated Logic" (1896), "Substitution in Logic" (1905),[1] and Peirce's 1877–1878 "Illustrations of the Logic of Science" series to be republished in book form (never published)—these among many other of Peirce's manuscripts exchanged and discussed between Russell and Peirce. The correspondence reveals candid details about the inner workings and struggles of OCP in its early years, such as the Underwood–Carus affair of 1887,[2] and contains some of Peirce's most valuable informal writings on logic, mathematics, and Existential Graphs, particularly between 1894 and 1896 in the context of Ernst Schröder's 1895 *Algebra und Logik der Relative*,[3] as well as in the summer of 1908 in response to Russell's proposed book on logic. Peirce developed a deep appreciation for Russell and through his friendship found hope for opportunity amidst constant opposition. "It is impossible to tell you the satisfaction your correspondence gives me, and how doubly anxious it makes me to rise above my misfortunes," Peirce noted to Russell 10 March 1896. Russell recognized the caliber of Peirce's genius during a time when few others did, and shortly after Peirce's death April 1914 declared prophetically in the pages of *The Monist*: "*[*Peirce*]* left this world and left also a volume of product the eminent value of which will sooner or later be discovered, perhaps only after it has been rediscovered.... So lives Charles S. Peirce. The Universal Spirit has him and the world that neglected him will care for him—after many days perhaps, but most assuredly."[4]

---

1 Published under Russell's name. See Peirce to Russell 3 Oct 1904 for details.

2 Described at various points, but especially Russell to Peirce 2 Oct 1908; see also "Underwood" in Biographical Register.

3 See "Peirce and the Open Court" for more details.

4 "In Memoriam Charles S. Peirce," July 1914. Also in full in Appendix, "Peirce and Russell".

1887

Russell to Peirce early Sept 1887 (missing): Discusses James J. Sylvester's work on matrices with quaternions, to which Peirce responds as follows.

Peirce to Russell. MS: RL 387.

Milford, Pike Co, Pa. | 1887 Sep 12

F. C. Russell Esq
150 W Madison St.
Chicago

Dear Sir:—

If Sylvester says that *he* has identified any matrix with quaternions, it is another example of his well-known idiosyncrasy. He has not. But it was done by my father in his Linear Associative Algebra, using imaginaries & by me without imaginaries.

I do not know exactly what you mean by setts; but matrices & quaternions are to a certain extent cognate subjects with *groups*, upon which considerable has been done. My father's work & Sylvester's; also Cayley in the Phil Trans for about 1860[5] are the most instructive.

My father's Linear Associative Algebra is published by D. Van Nostrand, Murray St, New York.

Some of my work is in the appendix. The rest is scattered in various places & I have no copies left.

Yours very Truly
C. S. Peirce

5 Likely his "Memoir on the Theory of Matrices" (1858), which Peirce believed substantially anticipated his own algebra of dual relatives. See "Cayley" in Biographical Register for more information.

10 has] has has *[E]*

Sept 1887–Jan 1889: During this period Peirce's time was divided between many familial, professional, and intellectual occupations, including his Coast Survey reports, definitions for the *Century Dictionary*, continued work on his "A Guess at the Riddle" (*W* 6:166–210), Juliette's declining health, and, with an inheritance from his late Aunt Lizzie (4 Feb 1888), the ambitious purchase of the Quick family farm in Milford, PA, soon to be substantially renovated at Peirce's great expense and rechristened "Arisbe."[6]

1889

Russell to Peirce. MS: RL 387. Stationery headed, "Office of | Francis C. Russell | Counselor at Law, | No. 150 West Madison St."

Chicago, Jan 2nd 1889

Prof. C. S. Peirce

Esteemed Sir—

I feel almost like an intruder but really unless men like you are a little gracious I dont know how men like me would get along at all—Will you kindly inform me if your paper on the Logic of Relatives in Vol IX of American Academy is substantially reproduced in the "Studies of Logic"—The latter I have as well as Van Nostrand's edition of the "Linear Associative Algebra" but I find it practically impossible to get at the memoirs of the "American Academy" here in Chicago—No Library has them except that of the "Academy of Science" and that institution has been a tramp for nearly ten years and its library is packed away in a remote corner and not get-at-able—Since I am hereby to bother you let me free myself on a matter or two besides—

You see when your "Illustrations of the Logic of Science" came out the papers initiated in me a new era in my mental history and I am one of a necessary many who recognize in you a master to be followed—so ever since I have gathered in as I could your published works & profited by their study—

You was to publish in the International Scientific Series the "Illustrations" &c but as that has ceased to be advertised in the prospectives I suppose it is given up—Besides the "Studies in Logic", the the Algebra of Logic, and the "Linear A. A." I

6 For a detailed view of this period, see *W* 6:xlv–lxiii and Brent, 186–202.

29 the Algebra of Logic] *inserted above*

have of yours the *three* articles in "The Journal of Speculative Philosophy" Vol. 2, your review & reply in the matter of "Phantasms of the Living" & your contribution to the symposium of the Christian Register—on Immortality—[7]

Now I notice what you say in this latter with respect to the inadequacy of the mechanical method (or theory) with vivid interest and in view of what you here said in the "Illustrations" "The Order of Nature" and also at the close of the "Theory of Probable Inference" was a little surprised—Have you changed your views and if so can you give me just a hint towards where I can get at the considerations proper to be viewed in that regard—I have Stallo's book[8]—Again I ardently desire to know if you still abide by your positions in the three articles in the "Journal of Speculative Philosophy" and if you have modified your views—then the points of change

I assure you it is no idle curiosity that prompts me but the appetite of a plodding student—Dont think I ask for anything more than a word or two—I am moderately well up in symbolic Logic Matrices—Mathematics and speculative Philosophy and a hint or two will go far and put me under more obligation than I can tell—

Very Respectfully & Cordially
Francis C. Russell

Peirce to Russell. MS: RL 387.

Milford Pa | 1889 Jan 8

My dear Sir:

I received yours of the 2nd last night. It had followed me here to my farm. It gives me great pleasure to form the acquaintance of another student of logic. I send you by today's mail, 3 quarto papers on the logic of relatives,—all I can lay my hands on. If they do not reach you, let me know, please. I have pushed the subject considerably further than these papers show, but I have had no time to work at it for the last 5 years.

I am trying to write a book setting forth my theory of the universe & I hope it may appear in the spring, under the title of Reflections on the Logic of Science.

---

7 "Science and Immortality" (April 1887).

8 Likely Johann Bernhard Stallo (1823–1900), *The Concepts and Theories of Modern Physics* (1882).

---

8 at] *over* to

15 Logic] *after* ~~Algebraic~~ *(another* Logic *accidentally repeated above deletion, ommitted here)*

15–16 Philosophy] Phylosophy *[E]*

There you will find the significance of my remarks in the Symposium which surprize you. I could give you no such hint in a few words as you ask for,—there is a long train of thought involved. Suffice it to say that I have not given up any of the more fundamental of my younger opinions so far as I recollect them, but am perhaps more sceptical & materialistic. I am sorry I cannot just now write you more. I am kept busy all day writing various kinds of things & making mathematical computations & have to run my farm, carry on building etc. at the same time. I am consequently not very good company and a detestable correspondent.

Yours very Truly
CSPeirce

Francis C. Russell Esq | 150 W Madison Street | Chicago, Ill.

Russell to Peirce. MS: RL 387. Stationery headed, "Office of | Francis C. Russell | Counselor at Law, | No. 150 West Madison St."

Chicago, Jan 16th 1889

Prof. C. S. Peirce

Esteemed Sir—

On returning from Iowa after the absence of nearly a week I find awaiting me yours of the 8th inst and also three monographs to wit

On the Logic of Relatives, 1870—
" " Algebra " Logic—Vol III. Am J. of Math
" " " " " " 7 " " " "

Your gracious liberality both of your time and papers is recognized and appreciated by me and has proven far beyond my expectations—I hope I am not given to overstepping but you have given me what will conduce to my protracted delight—

You see a solitary student like me has to flounder unaided amongst his books and papers at the very best but when he is debarred from access to important and perhaps indispensable monographs it is very slow progress he is to make—I have only looked over the papers hurriedly but a cursory glance is enough to inform me that I have now got just what I have long wanted

I am delighted to hear that your Scientific Organon is to soon appear. I shall reach for it as a hungry man for his dinner.—It seems to me that the epoch calls loudly, almost vociferously for the organic principles capable of stating the Universe mental and physical with thorough convenience—I look for this to be done on the lines laid down by you—outside of which I can discern no mental salvation

11 150] 120 *[E]*
14 1889] 1888 *[E]*
26 debarred] debard *[E]*
28 hurriedly] hurridly *[E]*

Cordially Yours
Francis C. Russell

Russell to Peirce. MS: RL 387. Stationery headed, "Office of | Francis C. Russell | Counselor at Law, | No. 150 West Madison St."

Chicago, Jan 22nd 1889

Prof. C. S. Peirce

"Beware" says Emerson "How you unmuzzle the valetudinarian" and as I start to write I fancy you in dismay at the possible reward of your graciousness to me in unmuzzling a logical tyro who is presuming on your courtesy.—

So let me say at once that this letter expects no answer—I am too sensible of the favor accorded me, too sensible of the value and engrossment of your time to ask any such thing.—

But as I shall send to you a few copies of "The Open Court" a word of explanation seems necessary.—

"The Open Court" was started by E. C. Hegeler the wealthy zinc manufacturer of La Salle in this State.—

Mr. H. is no mean philosopher in that he besides being *up* in the classical German philosophers he is also *up* in the various German philosophic essays and monographs.—And all these not merely acquired but digested in and by a mind of great vigor and grasp—so persuaded has he become that Science, Ethics, and Religion when recognized in their essential reality are not only not discordant but mutually formed for illustration and reinforcement that he is willing to bear a yearly expense of considerable amount in the conduct and publication of the journal "The Open Court"

When it was first started H. got B. F. Underwood out here from the Index. But soon both found out that they were unsuited in their views to each other. Underwood was a plain agnostic—denying but not constructive—Hegeler had what he deemed a *gospel* to preach—to wit: Monism, Entheism (an Indwelling God) and not Transmigration, but Propagation of the soul—i.e. That we live again in our children, not in a *tropical* but in a *real* sense—discarding all other ideas of Immortality.—

So after a little while Underwood left and Dr. Paul Carus, son of Dr Gustav Carus, Supt. Gen'l of the State Church in Eastern Prussia, was installed as an editor—Underwood has sneered much at Carus but I think that C. will in the long run count for a sufficient substitute.

He has studied under H Grassman, is a very capable mathematician, one of the "Back to Kant"-ers and only needs *as I think* to study the foundations under your lead to acquaint himself well in the work he has undertaken of editing "The Open Court" according to its founder's design and opinions with which he is in thorough accord.—

I have been in the habit ever since your "Illustrations of the Logic of Science" were published in the Pop. Sci. My. of picking up the numbers containing the six parts, taking out the leaves on which they were printed, binding them together, and then of donating such to my *very very* dear and select mental consorts as a gift of a very distinguished value—So after I became intimate with Dr. Carus I gave him one[9]—and I think I can discern the influence its perusal and study has had upon him not only primarily but secondarily by his converse with me.—

In some of the recent numbers of "The Open Court" he has written two essays—one—"Is Nature Alive" and the other a long one continued in four numbers of "The Open Court" entitled "Form & Formal Thought"[10] and I have thought that maybe you might be interested to see a beginning of a style of comment on philosophical topics that under the influence of the mathematical speculative sallies of modern times, the new logical calculi, &c is as we believe to enact a great part in the history of the next and coming generations—so with his consent and in part too as a token of recognition for the papers you sent to me I mail them to you.—

Do not however impart to me the *gaucherie* of asking you to read the essays—I know too well myself the irksomeness of being expected to read things that do not specialy interest me—If you feel like it look at them. If not not.

At any rate even in your waste basket they will have served a good purpose if indicating my grateful feelings for the papers sent me.—

As I have said before this letter does not expect any answer

Cordially Yours
Francis C. Russell

I return you the stamps were on the papers sent to me

---

9 To this booklet (in SIU 91.40) Carus later appended a proposed Preface and table of contents for the (unsuccessful) republication of the papers in book form. See Carus to Peirce 23 Aug 1913 and the accompanying enclosure in Appendix.
10 Nov–Dec 1888.

---

1 H] *inserted above*

Russell to Peirce. MS: RL 387. Stationery headed, "F. C. Russell | Attorney and Counsellor at Law, | 501 Tacoma Building, | La Salle and Madison Streets."

Chicago, Nov 8th 1889

Prof. C. S. Peirce

Esteemed Sir.

I venture to solicit your help again. I should write quite frequently if I took counsel of my own needs alone but the fear of abusing your graciousness restrains me so that I do not apply until I am absolutely "stalled"

My present exigency is my want of a "Boole".[11] I havent been able yet to get a sight into it. I dont believe there is one in Chicago, or its vicinity. I have tried to get it through our booksellers for the last five years and have at last come to believe almost that there is no such book extant.

Now what can I do to get me a Boole? I need it as urgently as can be expressed— If you can offer me any suggestion tending to enable me to satisfy my need it will be of the greatest service to me.

And now since I have for the purpose stated claimed your attention I will take advantage of the occasion and speak of some other things.

You have sent to me.

Your Logic of Relatives 1870

" Algebra of Logic

" On the " " "

I have of yours also

" " an Improvement in Boole's &c

" " New Categories

" " Classifications of Arguments

" " Logical Extension & Comprehension

Your three papers in Vol II of

Jour. of Spec. Phil

" six papers " Pop. Sci. Monthly.

Also Studies in Logic

& the Van Nostrand Linear Associative Algebra, *and I suppose I might say*—

Also, "Nominalism & Realism"

& "What is "determined?" "

in the Journal of Spec. Phil.

11 "The Calculus of Logic" (1848).

Now in your appendix "On the Relative forms of Linear Associative Algebras" you make the reference "See my brochure entitled A Brief description of the Logic of Relatives"—Is what is thus referred to the same as the second appendix to your paper on Induction in Studies in Logic or not. If it is not then I want to find it and have it if possible.[12]

When is your *Suggestions of the Logic of Science* going to appear?.

Do you still abide by your criterion of the compass of a conception which you laid down in your paper "How to make our ideas clear"?.—

How long it does take to get a new and important idea afloat to be sure—That all inference lies between rules, cases, and consequences or results. That Deduction, Induction and Hypothesis are the full tale of distinct simple types of inference I think no competent person can avoid after the matter is once presented. Yet authors labelled logicians go right on in the beaten track.

Now as to Hypothesis and its strength—

Why just now here in Chicago five men are on trial for their lives and presumably will be condemned by proof which confessedly is composed of hypothesis and certain facts called *circumstances*. What our Law Books call "Circumstantial Evidence" is just precisely Hypothesis.—

Whenever you have any errand or commission to be attended to in Chicago consider my gratuitous services tendered in the spirit of heartiest welcome

Very Respy
F. C. Russell

1890

Russell to Peirce. MS: RL 387. Stationery headed, "F. C. Russell | Attorney and Counsellor at Law, | 501 Tacoma Building, | La Salle and Madison Streets."

Chicago, July $1^{\underline{st}}$ 1890

Prof. C. S. Peirce
Esteemed Sir

12 Refers to "A Brief Description of the Algebra of Relatives," a privately printed brochure of Jan 1882, which Peirce would send to Russell ca. 19 Sept 1892 per his repeated requests.

6 *Suggestions of the Logic of Science*] *For "Reflections on" the Logic of Science*

19 to^2] to to *[E]*

26 1890] 18~~89~~$^{\vee}$0$^{\vee}$

I write at the instance of my friend Dr. Paul Carus of "The Open Court" who wishes to inquire if you can be induced to write him an article on the nature and import of logic, to be published in his journal. He has been publishing original articles by such men as Prof. Max Muller, Prof. Rudolph Herring, Prof. Geo Romanes, Prof. Alfred Binet and he has assurances of articles from Prof. Ernst Mach, Prof. Ernset Haeckel and others. It is the intention of the management of the journal to make it the vehicle of such utterances only as shall be competent to the topics treated and they expect to pay for their articles after a measure in some degree fitted to the dignity of the writers and the customary recognition of the value of their productions.—

If you will permit me the suggestion I would say that an article on the lines of your introductory lecture at Johns Hopkins University would be very useful—You have hitherto been addressing only the very specialists themselves and hence are not known to this generation as the importance of your method solicits.

Everybody is talking about scientific method and yet outside of yourself no one so far as I can see has any definite conception as to what that scientific method consists in. And yet that very conception lucidly stated is the crying need of the present times.—You may say of course that you have already stated the case and have outlined at least the main features of that method and hence that there is no need of further expression on your part until the publication of your book—But I have been bold enough to guess that it may yet be some time before we shall have the benefit of that great work: that you are still pushing your researches and according to the signal importance of the same that you intend to perfect them according to your own standard of perfection.

Indeed I am jealous of the next generation and impatient to have the world put in possession of the fruits of your work.

But so profound a system as yours cannot speedily be assimilated and hence *reiterated* anticipatory statements of its nature will but prepare the public for the better reception and comprehension of the book when it is at last given over to us.

Very Respectfully
F. C. Russell

---

4 Romanes] Romaines *[E]*
6 Haeckel] Heckel *[E]*
14 method] *after* ~~results~~
26 the] *after* ~~your~~
28 *reiterated*] *after* ~~sho~~

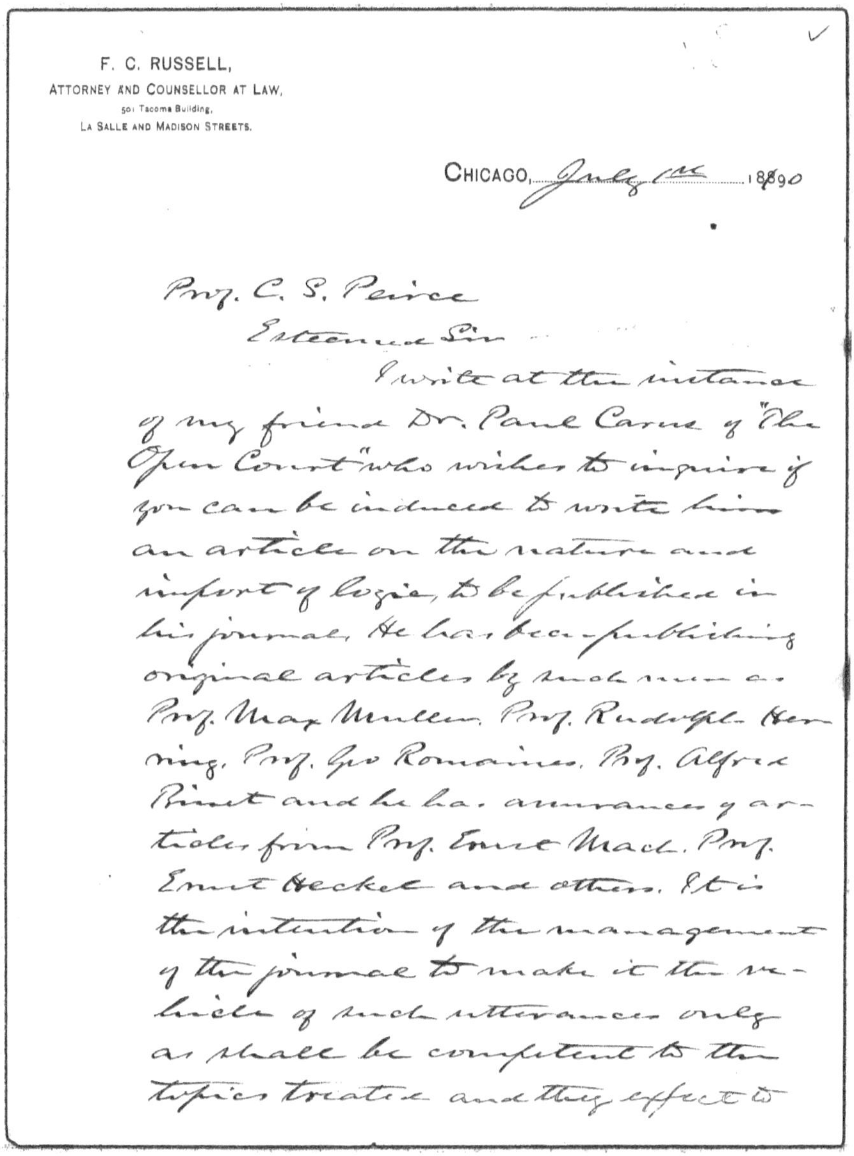

F. C. RUSSELL,
ATTORNEY AND COUNSELLOR AT LAW,
501 Tacoma Building,
LA SALLE AND MADISON STREETS.

CHICAGO, July 1st 1890

Prof. C. S. Peirce
Esteemed Sir
I write at the instance of my friend Dr. Paul Carus of "The Open Court" who wishes to inquire if you can be induced to write him an article on the nature and import of logic, to be published in his journal. He has been publishing original articles by such men as Prof. Max Muller, Prof. Rudolph Herring, Prof. Geo Romaines, Prof. Alfred Binet and he has announced articles from Prof. Ernst Mach, Prof. Ernst Heckel and others. It is the intention of the management of the journal to make it the vehicle of such utterances only as shall be competent to the topics treated and they expect to

Figure 9: July 1890, first page. Invitation to contribute to *The Monist*.

Peirce to Russell. MS: SIU 91.9.

Milford Pa | 1890 July 3

My dear Sir:

Your letter has given me much pleasure. I have not had the fortune of seeing much of the *Open Court*, but what little I have seen, and Dr. Carus's Fundamental Problems (of which I have just written a little notice)[13] has led me to regard it with much respect,—*it*, and also the community in which such a journal can by any means be supported.

As I seldom put my name to what I write for the general public, you naturally suppose I write little. I should really be ashamed to tell you how many articles I do write; for you would begin to think me very superficial. It is true, I am apt to be rather reticent about ideas which I do not think the public in any sense prepared to receive favorably. To give you some idea of the kinds of things I write, I enclose two pieces of which I happen to have copies.

I should be very glad to write for the Open Court. One can profitably put but very little into a single article. I should therefore prefer to write a number.[14] I would write in a general way about the ways in which great ideas become developed, *not* about verification and assurance, to which my Johns Hopkins lectures used chiefly to be directed. If you or Dr. Carus will write and say how many articles the Open Court will take on that subject and how many thousand words per article, I will endeavor to fill the bill. I generally get $ 25 a thousand words.

A philosophy is not a thing to be compiled item by item, promiscuously. It should be constructed architectonically. Elaborate and thorough survey of the ground, consideration of the sources, study of the methods, should be systematically made before one step is taken, even in so much as drawing the general sketch of the structure to be erected.

In this great preliminary task, an acquaintance with the laws which have been found to govern the evolution of the leading ideas of mathematics & physics cannot fail to be useful. This is a subject which I have studied out in my minute way, and I should like to give the readers of the Open Court some general notion of my results.

Yours very truly
CSPeirce

---

13 Review in *The Nation* (7 Aug 1890).

14 What would become his five *Monist* metaphysical series articles Jan 1891–Jan 1893, beginning with "The Architecture of Theories" to which he alludes in the following paragraph.

---

4 Your] *over* I h   10 I[2]] *after* ~~As~~

F. C. Russell Esq.
501 Tacoma Building, Chicago.

1891

Russell to Peirce. MS: RL 387. Stationery headed, "F. C. Russell, | Attorney and Counsellor at Law, | 162 Washington Street, | Rooms 58 and 59."

Chicago Nov 1st 1891

Prof. C. S. Peirce.

Esteemed Sir

In the Linear Associative Algebra (Van Nostrand's ed.) in the Addenda II, by you, "On the Relative forms of the Algebras" and on p. 127, you say in parenthesis, (See my *brochure* entitled *A brief Description of the Algebra of Relatives*)

Now in the "Studies in Logic," by Note B. to your paper "A Theory of Probable Inference" you give a sketch of the Logic of Relatives.

I have often inquired of myself if this latter paper was the one referred to.—But I have had no little doubt as to the real state of the case. The *brochure* would seem to contain information that I cannot find in the Note but I have refrained from troubling you about it.—But now I notice that Prof Schroeder in his new "Algebra Der Logik" in giving a list of your papers[15] finds himself in the same case with myself, so I am prompted again as I have often been before to ask you about the matter.—I have I believe all of your papers listed by him except this one and the one on the logic of number in Vol. 4. Am. Jour. Math.—

I am very much gratified to know that the worth of your work is beginning to become known.—It is singular and not a little unfortunate for Prof. Schroeder that he became acquainted with your papers in the Journal of Speculative Philosophy so late. Does he yet know of the various fragments in the Johns Hopkins Circulars &c. The fact is you "heave off" here and there things of the greatest moment as though they were trifles. For instance how many would suppose in reading your review of "Hardy's Calculus" that they were being treated to a solution of the tough old paradox of the differential method. So I (and I suppose many others) who value your work, get to feeling that for aught they know there may be lying around in unsuspected places such valuable information thus enunciated.—To

15 Vol. 1 bibliography, pp. 710–711.

get at what DeMorgan did is bad enough work but it seems to me you can discount him. It is one of my most earnest wishes that you will be spared until you feel prepared to set forth your system at large.—

Now I am going to give myself away in asking you to enlighten me with respect to § 40 of the Linear Associative Algebra. In order that you may perceive my difficulty (or deficiency) better I will exemplify the language as I understand it.—It is said—

"Take any combination of letters at will and denote it by A."

Well I do so, take, x & y & write $ax + by = A$

"Its square is generally independent of A and its cube may also be independent of A & of A^2"

Here I am stumped. I ask myself How can the square or any power of any thing be *independent* of the same? Will not the former vary in accordance with the variation (if any) of the thing.

But let me go on.—

we have $ax + by = A$

then $a^2x^2 + abxy + abyx + b^2y^2 = A^2$

or $a^2x^2 + ab(xy + yx) + b^2y^2 = A^2$

or $a^2x^2 + ab(xy + yx) + b^2y^2 - A^2 = 0$

& also

$a^3x^3 + a^2bx^2y + a^2bxyx + ab^2xy^2 + a^2byx^2 + ab^2yxy + ab^2y^2x + b^3y^3 = A^3$

or, $a^3x^3 + a^2b(x^2y + xyx + yx^2) + ab^2(xy^2 + yxy + y^2x)b^3y^3 - A^3 = 0$

You see just how I grope & how it is that I am utterly lost when the text goes on to say

"But the *number* of powers of A. that are *independent* of A. and of each other," &c

Either I am very dull or there is something that you can explain to me & if you can do so without too much trouble I wish you would

Cordially

F. C. Russell

Peirce to Russell. MS: RL 387. Stationery headed, "Century Club | 7 West Forty-Third St."

New York 1891 Dec 6

My dear Sir:

---

1 work] *inserted above*

23 that I] that ~~when~~ I

Your letter of Nov 1 only reaches me today. Your difficulty with § 40 of the Linear Associative Algebra is easily resolved. First, by *independent* is meant *linearly* independent. Then, by *letters* are meant not ordinary scalars. Let

$$A = ai + bj + c$$

where *i* and *j* are orthogonal unit vectors

$$A^2 = -(a^2 + b^2) + c^2 + 2aci + 2bcj$$

and is independent of A, that is, is not a *linear* function of it.

But A^3 I think you will find equals

$$[5c^4 - 10c^2(a^2 + b^2) + (a^2 + b^2)^2]A + [-4c^3 + 4c(a^2 + b^2)]A^2$$

I think the notice of Hardy's Calculus was written by Newcomb; and I dont agree with it. Excuse my haste.

very truly
C. S. Peirce

1892

Russell to Peirce. MS: RL 387. On OCP Stationery B. LPC: SIU 91.11.

Chicago, May 11th 1892

Esteemed Sir—

Dr Carus has asked me to write to you concerning the Tale "An Excursion into Thessaly"[16]—We are both much pleased with the prospect of seeing you, and are prepared to do whatever seems advisable in order to promote the writing. As I am more familiar with conditions in Chicago and more constantly on the grind the Dr. relies to some extent on my efforts.—Now in order that I may be a little better furnished with data as to what is expected I will ask you 1st Is the tale to be finished in one *reading*? 2—What is the "*argument*" or view of the story?.—Had you entitled it "An Excursion into Beotia" I could have made a guess, but as it is I am totaly at sea.—3d Are there not quite a many (and who) in Chicago to whom you are known?.—Whatever you produce is of that quality that it ought by nights to be read before an audience capable of appreciating it, and the Chicago fellas that affect philosophy—if your story connects with that—are largely given to "Haygle"

16 Later titled "Embroidered Thessaly," *W* 8:296–340. See also Peirce to Carus 8 May 1892.

5 orthogonal] *above* ~~rectangular~~ 8 But] *above* ~~So~~

and "Hubbut" Spencer?—Still if the persuasion could be raised that a really fine article of philosophical discourse was available there would be no difficulty about the auspices under which it could be delivered nor about the expenses. Even the Chicago average "plug" citizen of the moneyed type is free with his cash whenever his goods are assured to him as realy A.1.—Why dont you come here and be a Professor in our New Chicago University where they are paying $ 7 000. per year

Very Respy
Francis C. Russell

Peirce to Russell. MS: SIU 91.11. Note by OCP in top-left corner, "O. Court | File"

12 W 39$^{\underline{h}}$ St. New York | 1892 May 14

My dear Sir:

My story has no philosophical discussions. It is a story of the adventures of a young traveller in Thessaly about 1862 when the country was pretty wild. It has rather a poetical atmosphere and conveys the impression of being true, but the adventures are quite surprizing. I read it at the Century Club here to some of the very best judges of such things here, and they were much struck and delighted with it. The reading occupies an hour and a half, but it is not at all tedious. Literary people will be sure to like it; as for very stupid persons, they can be assured that it has been pronounced A1, by those who are qualified to judge.

The idea of a professorship in Chicago is new to me, but I confess rather pleasing. I have always felt that Chicago was the real American city. I have, however, fewer friends than enemies there. Would my being an Episcopalian be an objection to my having a professorship?

I have two connected articles[17] for the Monist well under way, and the first will be sent on in a very few days. I hope it will be inserted in the July number.

Yours very truly
C. S. Peirce

Francis C. Russell Esq
Office of The Monist

17 "Law of Mind" and "Man's Glassy Essence" (July and Oct 1892).

13 young] *inserted above*

Russell to Peirce ca. 14 May 1892 (missing): Sends literature about the University of Chicago.

Peirce to Russell. MS: RL 387. Note in top-left corner reads, "Please address | me till further | notice at | 739 Broadway | Cambridge | Mass."

12 W 39$^{\underline{\text{h}}}$ St. New York | 1892 May 17

Francis C. Russell Esq

My dear Sir:

I have been reflecting upon your suggestion that I should go to Chicago and become professor there. It seems to be the thing for me to do, provided they call me. During many years, I felt that for my peculiar powers the world had no use. Hence, I only threw off pieces here and there, and my deeper studies in logic remain today unpublished, and nobody dreams of the things I have found out. But during the last year or two, I have been getting more and more impressed with a prevision of the miserable consequences which must ensue from the prevalent necessitarian conception of the universe. It makes God a limited monarch or *roi fainéant*, acting under law so blind and inexorable as to leave no room for any acts of paternal love, or any listening and answering of prayer. Now whoever will follow out with me the higher logic of relations will see as clearly and as evidently as can be the baselessness of the materialistic-necessitarian fabric. Nor can his eyes fail to be opened to the fearful abyss into which that machine-made doctrine is precipitating society. A return to christian principles, to which a knowledge of my discoveries would lead, is the sole way of salvation. Accordingly, I now feel that if a way is shown to me to teach logic, it is my sacred duty to pursue it.

Such thoughts incline me to reconsider my purpose of reading my tale. So long as I am pursuing a studious life upon my own country place and studying just to learn, with no ulterior design, it is pardonable if not commendable that I should wish to show what I can do in various kinds of literature. But I fear the telling of emotional stories is hardly compatible with the self-abnegation and exclusive devotion to the cause of sound learning and education to which a man who proposes to become a professor must surrender himself. Therefore, if you think there is any prospect that I might receive such a call to Chicago, and if you have not gone too far in the matter of my reading to draw back, I think it will be best to give that up. If I am to be put into a position to do the work I was brought into the world to do, I desire to lay aside all other ambitions & vanities and give myself up to that work exclusively.

9 felt] fe*e*lt

Yours very truly,
C. S. Peirce

Russell to Peirce. MS: RL 387.

Chicago May 19th 1892

Esteemed Sir.

I have got your letter dated on Tuesday and am glad to know that you really wish to come to Chicago as a Professor. I sent to you several days ago some of the literature concerning the new institution. Now I take it that such situations have to be *worked for*. You see I am more or less of a politician and I naturally think that whenever anybody wants a place it is for him to marshal his *influence*. As you are to be in Cambridge when you will get this you will be in position if on other accounts it is desirable to do much.

Prof. Palmer who is there has just declined a chair of Philosophy in Chicago and so he ought to know about the avenues towards such a place so too quite a number of the Clark University professors men that used to be at Johns Hopkins have got chairs here in the Biological line.—

In Mathematics only one appointment has as yet been made, that of Prof. Eaton, formerly of the Northwestern University here.

Of course the Great Mogul in all the appointments is Harper, the President.[18] He came here from Yale where he was a professor of some kind of sacred linguistics. I do not very well know who are his chummies East but that can no doubt be very easily found out. If I were you I should just bone all my friends especially those of influence in the educational world and have them write to Harper on your behalf. That is the way they all do I guess.—Concerning what can be done by us here I cannot assure you of much but still something.—You see Harper is almost a stranger here and has very little heed for mere local influence—How do you stand with Stanley Hall, Pres. Gilman, Willard Gibbs. There must be a host of people who can speak of your distinguished fitness, and who would be glad to do so if only asked

I had not gone far with the reading & shall go no further according to your request. Whenever I can do anything for you freely call upon me

18 William Rainey Harper (1856–1906), American Hebraist and academic selected 1891 by John D. Rockefeller to help organize and preside over the University of Chicago. Also helped organize Bradley University and served as leader of the Chautauqua Institution, New York.

9 politician] politition *[E]* 10 marshal] marshall *[E]*

Very Respectfully
Francis C. Russell

Peirce to Russell ca. 23 May 1892 (missing): Discusses a forthcoming article, likely "Law of Mind." Incorrectly addresses the letter to OCP office.

Russell to Peirce. MS: RL 387. First two pages of letter missing. Estimated around this time based on context. Also, paper type is identical to that of preceding letter.

*[*late May 1892*]*

[…] Your letter was directed to me via. The Open Court office. I therefore send you one of my cards. I have no regular connection with the Monist and Open Court although I am quite intimate with Dr Carus and Mr Hegeler and know pretty well what is going on as regards these periodicals. The Dr. is a prime good fellow that you will be sure to like.

He is going to Europe in June if he can get prepared so to do. He is therefore obliged to settle the make up of the July and October numbers of The Monist. The July no. is already settled, and so your article cannot appear in that, which is to be regretted.

Will write to you farther *in. re.* the "Trip into Thessaly" as soon as I have somewhat to write. I really dont know what can be done, I am rather green in the line of getting up affairs.

Very Respy
Francis C. Russell

Peirce to Russell June–Sept 1892 (missing): Sends "several letters," at least one of which suggesting that Russell use his influence with Hegeler to get Peirce in better favor with him.

Russell to Peirce. MS: RL 387. Stationery headed, "Francis C. Russell, | Attorney and Counsellor at Law, | 123 and 125 La Salle Street, | Bet. Madison and Washington Sts., | Suite 54, 55 and E."

Chicago Sept 10th 1892

Dear Mr. Peirce.

I write more as a notice that I am receiving your several letters and that I will address myself to the suggestions made therein with all the dispatch possible than

---

29 dispatch] dispach *[E]*

as in any wise any answer. The fact is the Courts here are now just starting up after vacation and a multitude of little preparatory things so press for attention that I have been more constantly occupied than I am wont to be. Still I shall get time at once to move in your affairs.

I however distrust my influence with Hegeler. He & I are on good terms enough but he is a partly case hardened subject to influence, or rather he is a man of *contradictory* susceptibilities. You aim to stir him with considerations addressed to his Scylla side & bang you run against the Charybdis rocks. I never saw a man harder to "get at" than he. I see that in my ignorance of the true facts I have misjudged the case between you and him in divers particulars. Now I am one of those men who as it were makes it kind of a religious principle to "stick by my friends."—I dont ask for friends that are perfect.—I will take them as they are. Hence I would shun saying anything to the prejudice of either Mr. Hegeler or Dr. Carus. But you may be sure that a lawyer whose very trade it is to study human nature generally "tumbles" to the foibles &c of even his friends.—I may however I think go so far as this without any disloyalty—Mr. Hegeler had and has had the idea that he personally was a superior philosopher.—Hence the foundation of the Open Court to advocate what Hegeler understands (or misunderstands) by "*Monism*". He got Underwood (B.F.) at first as an editor.—Then Carus turned up.—Then followed a quarrel with Underwood (in which I think Hegeler had the right) & Carus was made editor. Now Carus also thinks himself a superior philosopher, but he is full of adroitness in dealing with Hegeler.—I firmly believe Carus is friendly to you (in a qualified way) i.e.—He will first take good care of Carus (why shouldnt he) & then exercise his good will *as opportunity shall offer* towards you. But as for either of them voting themselves subordinate to you in philosophical competence, that is an idea that does not even occur to them as proper to be done.—

So you see that my way is not "an easy one." But I will do my best ("en suite")

Very Respy

Francis C. Russell

Russell to Peirce. MS: RL 387. Stationery headed, "F. C. Russell, | Attorney and Counsellor at Law, | 162 Washington Street, | Room 60."

Chicago Sept 13th 1892

Esteemed Sir

---

7 susceptibilities] susceptabilities *[E]*
7 considerations] *after* ~~con~~
12 for friends] for men for friends *[E]*
13 or] of *[E]*
16 has] *after* ~~had~~
20 (in which] *after* ~~I~~

The perusal of your article in the Open Court—Pythagorics[19]—prompts me to write to you principally to express my great satisfaction that you and Dr. Carus get on so well together. Do you know I promise great things in the reasonably immediate future from this acquaintance. Not that the way is *all* clear and everything booming but because of your merits and the Dr's character. When you come to know Dr. Carus as I do you will know him for one of the most excellent men alive, a man of a truly philosophical bent of mind and of great philosophical ability.—Still he and you are not upon the same philosophical basis and of course that will be an occasion for preventing that thorough coalescence between you that would complete the satisfaction that is to be desired. I would not dare to be so bold as to say to you that you can accommodate your system to fit his nor indeed to say to him that he can fit his philosophy to fit yours. But I think that it is permitted to me to inform you as to the situation here in order that you may counsel yourself as you are advised out of your own judgment and after the best information available. You see I account myself as a sort of match-maker. It was myself who first "*steered*" the Dr. to the invitation to you, and if the connection proves as fertile as I give myself leave to dream, I shall give myself no end of credit for the part I, (*that is I myself*) have played in the history.—Since I am on the offensive view I may as well make that confession to you that I have many times wanted to make. You see I regard you as almost my intellectual creation. I had dabbled in philosophy after the usual way year after year without any more result than becoming more and more befogged.—Then I fell upon your papers in the Popular Science Monthly. Then for the first time I found a clue that "made things clear and reasonable." So I was put to looking for more and more of your instructions. I believe I have now all that you have published except your "Logic of Number" and that paper (supposed to exist but which I have sometimes doubted) referred to by you in your second appendix of Van Nostrand's edition of the "Linear Associative Algebra"—"On the Relative forms of the Algebras"—as your "brochure" entitled "*A brief description of the Algebra of Relatives*"—I wrote to you once inquiring for this paper[20] but you in your answer forgot to tell me about it. (I see also that Prof. Schröeder has been inquiring for it)(By the way Prof Schroeder does not seem to have your six papers in the Pop. Sci. My.). So I became in a way a disciple of yours as much as I can be a disciple of any one for I am a natural born "*kicker.*" I follow just so far and no

19 8 Sept 1892.
20 Russell to Peirce 1 Nov 1891; also 8 Nov 1889.

26 by] *inserted above*

farther than I can see my way clearly. My great hobby is that there is just one way and one way only to become truly educated and that is to exercise one's own "*observation*" that the so much be-lauded and so little understood "*scientific method*" is purely and simply the method of *observation*, that experiment is simply for the sake of observation, that we experiment with our mental "*constructs*" just as much and perhaps more than we do with the objective entities and all in order that we may just open our senses external and *internal* to see—intuit—or *at-look* the results that are to be perceived by any one who will take the pains to study to see them.—*and* that there is no intellectual life that is of real worth for anyone who is contented to take *anything* for granted on *anyone's* mere *say so*.—

But I find some of your papers perfectly awful to study out and comprehend. You are so brief and I am so ill furnished with subsidiary knowledge that I am repeatedly reminded of that saying that is credited to Bowdich regarding La Place's Mechanique Celeste,[21] viz: that he found that when La Place made the remark "thus it plainly appears" he had learned to know that he had a long study before him to find out how it was that it appeared at all. I am just mad to compass the logic of Relatives but I make very slow progress. I am sure that it is the key to much that I want to know. I have been studying to see how I could spare the money to take lessons of you by correspondence but I have not as yet seen the time when I could spare even the small sum needed. A student is apt to be not a money maker and I have just at present a daughter and a son who are completing their schooling and are quite expensive. Within a few years more they will become self supporting and I will be able to do more for myself provided I do not get so old as to have become intellectually dulled.—

But pardon this long digression over my own self and my affairs. I am really frightened at and ashamed of it. When I began I had no intention to bore you in that way. I wanted principally to inform you about the state of affairs here anent the Open Court & The Monist.—

The substantial (financial) basis of these publications is Mr. Edward C. Hegeler of La Salle, Ill.—He is *the one* on whom they depend in *every* sense of the expression. What he says, nay even what he wishes in regard to them is *all important*.—Unless you fully understand the character of Mr. Hegeler you will be sure to misapprehend the situation and leave the way open for some mistakes.—

21 *Mécanique Céleste* (1829); trans. Nathaniel Bowditch (1773–1838), American mathematician and pioneer in maritime navigation.

7 intuit] intuite *[E]*
21 schooling] *after* ~~scooling~~
26 ashamed] *after* ~~as~~

Mr Hegeler is a peculiar man in many ways, in fact *a character* in every good sense of the word.—He was born, reared, and educated in Germany. His father was a merchant in Bremen and a staunch Lutheran of the old school, a thoroughly "practical" man who intended to, and did give to his sons a scientific rather than a literary education. Edward received his college education at Freiberg and while there became intimate in the family of Prof. Weisbach of Weisbach's Mechanics.[22] Mrs. Hegeler is a daughter of Prof. Weisbach.—Prof. Weisbach made a great impression on Mr. Hegeler and many of the dicta of his teacher with regard to the teaching of mathematics and other things are irremovably planted in Mr. Hegeler's body of doctrines.

Whether it was on the death of his father or not I do not now just remember but about thirty five years ago Mr. Hegeler and a college chum of his a Mr. Matthiessen came to this country with about \$ 75,000.00 between them, about equally divided I believe. While they were in college together their minds turned to this country as the field of their lives, because this was the land of freedom &c and of opportunity &c—They studied as a specialty metallurgy and in particular the manufacture of zinc, and they came here intending to establish a manufactory of that metal.—

So purposing and after looking the country over for about a year they established themselves at La Salle.

For several years they had a tough struggle to save themselves from ruin but at length they became successful and Mr. Hegeler is now a very rich man, worth several millions at least and with an income (he singly) of at least \$ 100,000.00 per year. The works are very profitable, are run by a superintendent, and require from him very slight attention. He has however for (I guess) the main part of his estate a large number of farms, homes, &c.—

Mr Hegeler is emphatically a *ruler*. He was never yet in anything that he was not ruler over. It is his innate constitution to be this. At the same time he is an eminently reasonable man in his intentions, a man who is on all things governed by his regard for right, a man to whom his conscience is a *supreme law*.—He would do anything to avoid what would seem to him to be wrong, or even to savor of

22 Julius Ludwig Weisbach (1806–1871), Mathematics and engineering professor at Freiberg University of Mining and Technology during Hegeler's studies there 1853–1856, and later Hegeler's father-in-law. Reference is to his 1845–1847 *Lehrbuch der Ingenieur- und Maschinenmechanik* (*Principles of the Mechanics of Machinary and Engineering*).

4 his] *inserted above*
16 metallurgy] mettallurgy *[E]*
18 over] *after* ~~of~~
23 superintendent] superintendant *[E]*
24 has] *after* ~~have~~
28 on] *i*on
30 what would seem] ~~doing anything that~~ \what/ would ~~be~~ seem

wrong doing. In other words he is punctilious to a singular degree. So again if he finds that he has made any mistake of an ethical nature he is never quiet until he has repaired the *[?ruin]* to the last iota. This extreme conscientiousness is a marked feature of his character. But it is by no means a conscientiousness that is born of weakness. Mr Hegeler is preeminently a *strong* man in every phase of his character.—Many find him overbearing and combative, and such would indeed be the case if his conduct were not at the same time to be estimated in the light of his conscientiousness and his *real solicitude* to be *liberal*.—He is a man of *great ability* and in judging of his conduct it must never be left out of sight that he has time after time been on opposite sides of an issue from others with whom he was associated and carrying through his opinions by force of an indomitable determination, has seen the results justify him in his course.—It is a cardinal item of his principles that while a man should always keep his mind open to fairly consider all opposing opinions and ingenuously to accept whatever instruction opposing contentions may afford still he should while he remains unaltered in mind maintain and *fight for* his own ideas.

The "struggle for existence" and the eminent morality of the same seem to him the plainest lessons of modern knowledge. So as the saying is he is "ready to fight at the drop of the hat" for the opinions he deems valid. Indeed he regards it as a duty for him to do so.

Now it is just this trait of his character that has led him to establish "The Open Court" and "The Monist." They are instituted and maintained by him for the purpose of promulgating what he regards as a new gospel. They are not business ventures in the ordinary acceptation. While Mr. Hegeler is a thorough *business* man and has shaped the business features of his publication schemes, so as to permit and so far as might be to invite, whatever business success might fairly be attracted, still this question of the profits, indeed even as to the question of remuneration is wholly subordinate to the missionary character of the design.—He would do precisely as he is doing even if the returns were wholly insignificant.—

He harbored this design (as is his wont) for a long long while and pondered and ruminated over it and when he puts his hand to any plough he never looks back. You can rest sure that so far as depends upon him the O. C. & M. will be published right along, pay or no pay.

---

1 is punctilious] \is/ punctillious *[E]*
2–3 quiet until he has repaired] ~~quiet~~ quiet until he has repared *[E]*
5 preeminently] *after* ~~a~~
8 *real solicitude*] *real~~y~~ solicitude*
12 the] *after* ~~his~~
14 ingenuously] ingenously *[E]*
15 remains] *after* ~~thus~~
15 in mind] *inserted above*
26 and] *after* ~~the f~~

His chief difficulty in the beginning was to find a suitable editor. The Open Court would have been established several years before it was, except for this difficulty. At last in despair of being able to do any better he made overtures to Underwood (who was never very acceptable to him)

Now the Open Court had scarcely been established under Underwood's direction when Dr. Carus became known to Mr. Hegeler and on comparison of views it proved that he came very near to the views of Mr. Hegeler, so much so that he seemed almost a godsend for him. As Mr. Hegeler had expressly stipulated with Underwood that he might have either in person or by another his monistic views advocated in opposition to the agnostic views of Underwood he put in Dr. Carus to the editorial staff to exercise that office. So Underwood rebelled and made the condition of his remaining that Dr. Carus should keep out. Had he known Mr. Hegeler better he might have known that this was only to commit *hari kari*. So Underwood went out & Dr Carus became editor. Meanwhile Dr. Carus married Mr. Hegeler's favorite daughter.—The editorial department is you know at La Salle in Mr Hegeler's house, and as soon as other things will come convenient there is to be built there a building especially devoted to the convenience of that department.

Dr. Carus and Mr. Hegeler keep in very good accord as to their views, and their mutual esteem is unbounded. Yet I am persuaded that not even the relations that so exist would avail to overrule Mr Hegeler's missionary design in the O. C. & M. should the case arise of any serious difference between them. Mr Hegeler *is wholly inflexible* as to the substance of his designs. Now having said that I am right away going to qualify it. Mr. Hegeler has in spite of his determined character and combative disposition an important margin of accommodativeness. His disposition to be fair and liberal is quite as marked as the seemingly opposed disposition.—I can see that Dr. Carus has modified the sharpness of his views considerably and has in fact instructed him far more than he Hegeler is aware of. As to Mr Hegeler & myself we are in a great many things of opinion wholly unconformable.

Now I regard your philosophical views as altogether the most important ones that have ever been broached—This is stating the case very very mildly.—I want you and Dr. Carus to agree so well together that you will become a necessity for him and Mr. Hegeler. They can give you opportunity and you can give to their efforts that solid foundation that will command the respect of the world. I am satisfied that with tact you and they can find a permanent *modus vivendi*. Please tolerate from me an expression that I fear (without really knowing much about it) that

18 as to their views,] *after* ~~an~~
21 any] *after* ~~diff~~
26 the] *after* ~~his~~
31 agree] *after* ~~joint join~~
33 command] *after* ~~comm~~
35 fear] *before* ~~that~~

you are somewhat angular or irritable or impolite or something of that sort. I can imagine how it must torture you to find others dull witted, ignorant, or incapable to apprehend things that you see readily, but a man may be slow and yet on the whole very capable. I fear very much that you would at first think that Mr Hegeler could cut only a poor figure at having a philosopher. But opportunity is not a thing to be lightly held and I regard this one here as one that is yours in a very special manner. I am so sanguine that I even think that right here in Philistine Chicago there is already started a movement in the intellectual and moral world which with your *politic* co-operation will cut a *leading figure in the history of philosophy.*—

I have written this right along in a garrulous way not stopping to weigh my words and I submit it in the trust that my real feeling will be apparent and will be set off against all the arrogance, impudence, &c that will plentifully appear.

Cordially
F. C. Russell

Peirce to Russell. MS: RL 387.

Milford Pa 1892 Sep 17

Mr dear Mr. Russell:

You have put me under a deep obligation by your letter of the 13th. You amuse me with your antithesis of my "merits and the doctor's character." I am fully ready to believe that Dr. Carus has a noble character and philosophical spirit. Nor can I think that differences of opinion would outlast daily contact long. Obstinate disputes in philosophy are maintained by *life* presenting itself under diverse aspects. At the same time, as you imply, it is not permitted to a scientific man to "fit" his philosophy to anybody's teachings. With a strong, positive, business man, I should be less certain to get on. My great word is that the thing to go your bottom dollar on should not be a *doctrine* but a *method*. For a vital method will correct *itself* and doctrines too. Doctrines are crystals, methods are ferments. Practical men, who value acquisitions so much and sneer and mere powers, will seldom take this view. Make your pile, and then die, if you haven't vitality enough to begin again.

"Disciples" are *ipso facto* on the wrong track.

That all thinking is perception I have often said. See Century Dictionary under *inference*. Also that observation *of* nature is an evolution of thought.

If you will kindly point out to me passages in my screeds which require elucidation you will do me a great favor.

You have a vivid and strong pen for character-sketches.

7 Philistine] Phillistine *[E]*

Chicago is coarse and calculated to make a man with nerves miserable I don't doubt. But that it is "Philistine" I can hardly think. Even New York hardly rises to the light and sweetness of Philistinism. I know what I am saying. I have seen the $2^{\underline{\text{nd}}}$ Edition of the Philistines,—the Philistines of Berlin from within, and intimately,—and I have seen New York.

This valley of the Delaware where I live will compare with any country I have seen,—and I have seen many,—for picturesqueness. Moreover, the roads are set down by the bicyclist's guide-books as the best in the land. Moreover, there are many French settlers here, & especially *chefs*, and they have disseminated good cooking wonderfully. My wife and I own over two square miles in one piece nearly square; and I think that will be of value eventually. Outside of that we have two places. One is the small farm where we live, stretching a third of a mile along the river and reaching a quarter of a mile back. It is a most lovely spot, and never do I look out the window without refreshment. The house is a 2 story frame house of my design. The width of the larger part of it is 36 feet, (in front 48 feet, and in the ell only 27 feet, in part 38 feet). The total depth is 76 feet or with the verandah 84 feet, namely 16 of greatest width, 38 or 36 and 38 feet and 22 of 27 feet. It is very pretty and very original, but the farthest possible from pretentiousness. The floors are mainly whiteash waxed. The rest Georgia pine. There are eleven open fireplaces, some large; one you can stand up in. There is also steam heat throughout the house. A good cellar under the whole. Running water and all conveniences. The rooms are as follows:—

You enter at a *conservatory* or glass-gallery, $17\frac{1}{2} \times 7$ft. There you have three doors at Right (A), In Front (B) to left (C). Door A leads in to my study 16×26 feet, 3 windows, large open fireplace and an outer door. From the study you pass into the music-room 16×20 feet, open fireplace, long windows opening upon verandah 8ft $\times$ 34ft. (I do not mention steam heat except where it is *not*.)

Door B goes into parlor $17\frac{1}{2} \times 15$ feet, where is the big fireplace. This opens by wide door into the Garden Room $17\frac{1}{2} \times 12\frac{1}{2}$ feet with large fireplace. Two doors into music room and 9 feet glass door into flower garden. It opens by folding doors into Dining Room $19 \times 16\frac{1}{2}$ feet with large fireplace & broad windows.

Door C goes into corridor (with large closet for coats etc) leading into dining room and also having the staircase, and leading into servants dining room. The latter is $10\frac{1}{2} \times 9\frac{1}{2}$ feet and has an outside door. It leads into the cook's room which is 9×11 feet has 2 closets. Has no steam heat.

---

6 Delaware] *d*Delaware
8 land] *after* ~~count~~
19 eleven] *after [?~~Do~~]*
25 large] *inserted above*
29 door] door~~s~~
30 glass] *inserted above*
32 (with … etc)] *inserted above*

From servants dining room you pass into kitchen $11 \times 15\frac{1}{2}$ feet with outside door range, sink, closets, and cellar door.

From kitchen you pass into larder $7\frac{1}{2} \times 14$ feet with inner closet etc and 3 windows. No steam heat.

From kitchen you pass into Laundry $10\frac{1}{2} \times 15$ feet with set tubs, large closet for preserves etc. but no steam heat.

There are two large covered porches about kitchen.

The pantry between dining room and kitchen is $5\frac{1}{2} \times 14$ feet and has two *[?rows]* cupboards. There *is* steam heat in pantry, along one side, porcelain lined sink with hot & cold water etc.

Going upstairs you find yourself in a corridor but the staircase turns away from the length of it toward the sitting room door at the end of it. The corridor goes straight back to the end of the house. This corridor has no steam heat upstairs, that below being quite sufficient.

Sitting room is 15×17 feet, fireplace, balcony with beautiful view of river, 2 closets.

From it you pass forward to a lobby with two doors and no steam heat. One leads into a chamber over music-room, with fire place and opening upon upper verandah, & closet. The other into chamber over study with fireplace.

From sitting room you pass into Mrs. Peirce's room 17×18 feet with window–seats extra space. Fireplace. From there into dressing room with 2 closets.

Thence into my room with fireplace $10\frac{1}{2} \times 22$ feet and closet. I go out of my room into corridor and opposite is a room we call the studio $12\frac{1}{2} \times 19$ feet with fireplace (large).

The corridor has 3 linen closets, and two large store closets and rooms. There is also a bathroom 6×12 feet with four closets. And a maid's room 8×10 feet with 2 closets, and no steam heat.

Besides this, there is a cottage nearby, very pretty with two rooms. We use it for servants. There are two barns, and various little buildings.

The 60 acres of land are worth \$ 6 000, the house, insured for \$ 4 500, is worth \$ 6 000, the other buildings \$ 1 000. My library, insured for \$ 2 500 is worth $\frac{\infty}{\infty}$. Say in all \$ 15 000. Now I propose to put up three pretty cottages of about 4 rooms each, and make the house a sort of Casino for fashionable people of "cultured" tendencies, to spend the summer, have a good time, and take a mild

2 and] *e*and
6 but no steam heat.] *inserted below*
9 pantry] *after* ~~laundry~~
15 fireplace] *after* ~~large~~
17 and no steam heat.] *inserted above*
19 & closet] *inserted in-line*
23 we] we~~l~~
34 "cultured"] "cultur*al*ed"

dose of philosophy. There is now no railway through the valley. When it comes, as it will in a few years, values will be greatly enhanced, and my place, with the business I shall have built up will be worth considerable. My ultimate aim is to set agoing an institution for the pursuit of pure science & philosophy which shall be self-supporting.

Now I need to start $ 5 000. There is no bank in this country, and it is so cut off from the rest of the state that there is no use trying to raise a loan in Philadelphia on a mortgage. In short, I cannot raise a loan in my prudent way.

I therefore propose to turn the thing into a company with $ 5 000 of preferred stock receiving 8 per cent. I can pay that $ 400 and earn a living here and have 7 months leisure to write books. In a few years, the stock holders will find their stock at 133 or more.

I am a chemist by profession, and when I get this working right, I intend on another piece of land I have here to start a little works for a process of my invention. I want $ 5 000 to be taken by *very very* eligible parties who will be more or less interested in my remoter purposes. I write to you hoping you may know such persons. Do you think I could induce Mr. Hegeler to come on early next month with a view of looking at the place & perhaps of making one of my stockholders?

Peirce to Russell ca. 19 Sept 1892 (missing): Sends his "Brief Description of the Algebra of Relatives" per Russell's repeated requests. Received by Russell in the course of writing his response that follows (see final paragraph).

Russell to Peirce. MS: RL 387. Stationery headed, "F. C. Russell, | Attorney and Counsellor at Law, | 162 Washington Street, | Room 60."

Chicago Sept 21st 1892

Dear Mr. Peirce.—

Yours of the 17th inst. is just come. I think you must have intended to send me another sheet for after using up two sheets, your letter ends at the bottom of the page abruptly and without signature.—

In regard to your plan for a Casino I think it altogether a most inspiring prospect and will do anything in my power to promote it. I should judge that

---

1 through the] through the~~r~~
2 as] *after* ~~val~~
2 values] *over* pri
4 pursuit] *s*pursuit
15 want] want to *[E]*

Mr. Hegeler would be quite likely to be favorably inclined towards it. Some address in proposing it to him would however be advisable. He is very much inclined to want to have things center around himself at La Salle and vicinity. The valley of the Illinois in which is La Salle is not by any means void of charms. Still he and his family go a good deal and a lodging place where they would feel as it were at home, where they could be not too far away from New York and Philadelphia, where they could meet persons of consequence, and where Mr Hegeler, Dr. Carus, and his wife could talk philosophy &c would I think be alluring for them and him. That you are specially a *chemist* is an additional attraction. That is the very branch of science in which all the family are most interested, the manufacture of zinc spelter having substantially a chemical process.—

I suggest that you invite Mr. Hegeler to visit you. Perhaps not as a special invitation (and perhaps yes). You can at least invite him to visit you on his next journey East. You would find him better company perhaps than I do for I believe you talk German.

If he was inclined to the Casino plan and especially if he thought it likely to prove a good investment he would take all the stock you would be inclined to let him have. He has *always* money on hand seeking investment.

You say in your letter "If you will kindly point out to me passages in my screeds which require elucidation you will do me a great favor." This encourages me to *put at you* a matter or two that gives me much trouble. But I would not have you understand that I claim that your account is unclear. I only say that *I as yet* find them indigestible.

In your Algebra of Logic in Vol 3 Am. Jour. Math. under "Classification of Relatives"[23] after distinguishing between the two types $A:A$ & $A:B$ you divide relatives into those that are composed 1st entirely of those of the form $A:A$ (concurrents) and those composed (a) either of those of pure $A:B$ form, or (b) of those of mixed forms.—Those non concurrents are "opponents."

Then you divide again into Alio relatives and Self Relatives. Alio's being those composed entirely of $A:B$ forms and Self Relatives being the concurrents (sibis) and those of mixed forms.—This can be tabulated thus

23 Beginning at p. 47.

18 investment.] *above the following deleted paragraph (papers received during writing—see final paragraph of letter):*
~~You havent told me as yet about the "Brochure entitled 'A brief description of the Algebra of Relati- ves" which you speak of in the Van Nostrand edition of the "Linear Associative Algebra"~~

23 indigestible] indigestable *[E]*

$$
\begin{array}{ll}
\text{self relatives} \left\{ \begin{array}{l} \underline{\text{All } A{:}A \text{ forms}} = \text{sibis or concurrents} \\ \left(\begin{array}{l} \text{Some } A{:}A \text{ forms} \\ \underline{\text{And } '' \; A{:}B \; ''} \end{array} \right) \end{array} \right. & \left. \begin{array}{l} \\ \\ \end{array} \right\} \text{Opponents} \\
\text{Alio relatives} = \text{All } \; A{:}B \; '' &
\end{array}
$$

So far all is clear enough =

But then you define *negatives*

Negative of Concurrent = All $\overline{A{:}A}$ form

″ ″ Opponents = Have $\overline{A{:}B}$ ″

What are to be taken as negatives of Self Relatives & of Alio Relatives you do not state, but you do state the relations of the different classes these made.

Now I cannot but appear to see that taking S. to represent elements of the $A{:}A$ form, and A to represent elements of the $A : B$ form there are fifteen sorts of relatives, viz:

sibis.?. (9)	$\ldots\ldots\; S$	(1)	= sibis or concurrents
	$\bar{A} \;\ldots\ldots$	(2)	
	$\ldots\; \bar{S} + S$	(3)	
	$\bar{A} \;\ldots + S$		
	$\bar{A} + \bar{S} + S$		
	$A \;\ldots\ldots + S$	(4)	(8) Op?
	$A + \bar{A} \;\ldots + S$		
	$A + \bar{A} + \bar{S} + S$		
	$A \;\ldots + \bar{S} + S$		
	$\ldots\; \bar{A} + \bar{S}$	(5)	
	$A + \bar{A} + \bar{S}$		
	$A \;\ldots + \bar{S}$		
	$A + \bar{A} \;\ldots$		
	$\ldots\ldots\; \bar{S}$	(6)	
Alio Relatives =	$A \;\ldots\ldots$	(7)	

$\bar{A}$ is put in accordance with your remark that Every $\bar{A}$ is at once a Self Relative and the negative of an opponent and $\bar{S}$ in pursuance of the corresponding remark—Now since class (9) are all self relatives (they *containing* elements $A : A$ form) they cannot (as it seems to me) be *negatives* of self relatives. This would

4 All $A{:}B$ ″] *before* ~~= Alio Relatives~~

leave classes (5) (6) and (7) only available as such self-relative-*negatives*. Correspondingly classes (1) (2) & (3) only are available as negatives of opponents. And what shall be said of class (4) and the relative $\overline{A} + \overline{S}$.—Being in this *fix* of mind I cannot at all divide up relatives into mutually exclusive classes as you would indicate. Of course it is a confusion of mind on my part but so much the *more need of getting straightened out.*

Suppose one should take notice not only of the *converse* of a relation but its *perverse*, that is suppose, say, lover and its converse say *sweetheart* for the *perverse* say *hater* where converse would be *hated*. These *perverses* are somewhat numerous in language and as it seems to me especially prominent in mathematics.

In such a case the *negative* of a relation would be its perverse plus all relations neutral.

In a *formal* way we could impute a perverse to any relation and we could imagine all relations to intersect by means of their neuter regions just as any point may be taken to be common to a three-fold infinity of right lines. If anyone could devise a notation and discover and express the relations of all kinds to one another and also discover a (so to speak) *geometry* of relations so as to find a *calculus* wouldnt that be worth while!!!. Of course this is only a dream, but I like to be audacious in thinking of what may come to pass in the future when the Pythagorics have got in their work. By the way Mr. Sacksteder who carries the clerical end of The Open Court & Monist here said to me to day that many letters were coming in asking for your address *in order to learn further regarding the Pythagorics*. They take the matter in real earnest and *want to join*. I say *real earnest* not meaning at all to imply that there is not a *very solid foundation* in reality for your hint.—In fact I have had an essay *begun* long ago in which I took the ground that the law of natural selection should imply development in the future in an analogous manner to evolution in the past, that we had good reason to forecast a superposition over man of a superior race of beings generated out of us, that casting about for the probabilities concerning the nature of the generation and its results it would not be unreasonable to suppose that such generation would be occasioned by the general inability in relation to mathematics, that mathematical science has sprung so much ahead and promises so much, that in the development it will soon reach only a

---

3 $\overline{A} + \overline{S}$] *after* $\overline{A} + \overline{A}$
9 say *hater … hated*] \say/ *hater* \where converse would be *hated*/.
11 relations] *after* ~~neutral~~
16 the] *after* ~~and~~
17 *geometry*] a *geometry [E]*
18 audacious] *after* ~~adu~~
20 Sacksteder] Saxtetter *[E]*
20–21 who carries … here] *inserted above*
25–26 natural selection] *after* ~~evolution~~
26–27 evolution] *after* ~~the~~
31 science] *after* ~~ability~~

few of earth's elite *[who] will* be able at all to compass its resources, that at any rate the great bulk of human cattle (in which I include the great *bourgeois* world and all the merely "literary" folk) will be barred out of its advantages by sheer and insuperable *inability*. That therefore the "*mathematically competent*" will become possessed of powers of a special kind that will enormously ensure to their militant power in the struggle for life, that they will come to see that the *masses* cannot be lifted up to a higher plane by the *mere* instruction and persuasions of "the competent," the masses being wholly *unable* to comprehend what "the competent" see clearly, that "the *competent*" after due efforts to have the masses accompany them in an upward stride will despair of that result, comprehend the implications of the law of natural selection will *refuse longer to* forego for themselves and their children their proper development, will gain conciousness of the solidarity of "*the competent*" and their ability to suppress all rebellion on the part of the masses, will assume the direction of earth's affairs, cut loose and become a *superior race*. By *mathematical competence* I mean especially the mathematical logic and its developments.—Ha! Ha! Ha! How is that for wild dreaming.!

I have just received your package of papers. I cannot thank you enough for them. I will go and strike out the passage above on page 3.[24] Pardon my length. I will never repeat the offence

Very Very Cordially
F. C. Russell

Peirce to Russell. MS: RL 387.

Milford Pa 1892 Sept 23

My dear Mr. Russell:

In the Century Dictionary, under *relation*, I have somewhat changed my nomenclature of classes of relations. To this nomenclature I shall adhere. Dual relations may be divided in 9 classes, according as they contain under them *all*, *part*, or *none* of the individual relatives of the form $A : A$ and of the form $A : B$. Of these 9 classes, four are special logical relations. Namely,

24 See critical note for line 18 on page 400

2 *bourgeois*] *burgeois [E]*
6 militant] millitant *[E]*
7 the] *after* ~~their their~~
11 for] *after* ~~the~~
18 length] leangth *[E]*
23 1892] 189*3*2
26 Dual] *over* T

coexistence	embraces	all	of form $A : A$ and	all	of form $A : B$
identity	″	″	″ ″ ″ ″	none	″ ″ ″
otherness	″	none	″ ″ ″ ″	all	″ ″ ″
incompossibility	″	″	″ ″ ″ ″	none	″ ″ ″

The other 5 classes are

1. *Differences* or *alio-relations*, which contain *none* of the form $A : A$
Thus, it is absurd to say A is greater than A, for
A is *not greater* than A.
2. *Sibi-relations* or *concurrencies*, which contain *none* of the form $A : B$
Thus, it is absurd to say A is self-conscious of not A; for
A is *not self-conscious* of not A.
3. *Agreements* which contain *all* of the form $A : A$. These are *negatives* of differences. "not greater" is an example. Another is "similar to" A is similar to A. But *dissimilar*, the negative of similar, is a *difference* or *alio-relative*.
4. *Distances* which include *all* of the form $A : B$. These are negatives of concurrencies. Thus, *not-self-conscious of* is a distance.
5. *Variform relations* include a part and a part only of the $A : A$ form and of the $A : B$ form. Thus A *kills* B.

The negative of a *dual relative* is a dual relative which includes every individual dual relative the former does not include and includes none the former does include. Thus, *lover of* and *not a lover of* are negatives.

		Containing of individuals of form $A : B$		
		all	some	none
Containing of individuals of form $A : A$	all	∞	agreements	1
	some	distances	variform	concurrencies
	none	$\overline{n}$	Differences	0

What you called *perverse* is by logicians called the *contrary*. In order to give every character a contrary, it is necessary to adopt a limited universe of marks.

A has *all* the marks of loving B
A has *some* of the marks of loving B
A has *none* of the marks of loving B.

19 *dual*] *inserted above*
19 includes] *after* ~~emb~~
23 individuals] *above* ~~elemen~~

See my note on limited universes of marks in *Studies in Logic*.[25] I worked this out many years ago. It has nothing particular to do with relatives. I never printed anything because it was too remote from any use. There are special cases, as in mathematics, where contrariety comes in largely; but then it is just like any other relation.

Probably I must have forgotten to finish my last letter. Thanks for your suggestions about Hegeler. That talent is already somewhat a matter of breeding Galton[26] certainly proves.

Yours very truly
CSPeirce

1893

Between 28 Jan and 6 Feb 1893: Peirce meets with Hegeler and likely Carus at Hegeler's home in La Salle to discuss larger publication projects, including Peirce's "A Quest for a Method" (or "Search for a Method")[27] and several math books. Hegeler seems also to have asked Peirce to begin investigating into printers to contract with OCP on the books, especially his arithmetic book.

7 Feb 1893: Returning from La Salle, Peirce meets with Russell at Richelieu in Chicago[28]—details undetermined, though an unsatisfactory meeting for Peirce due to Russell's "not showing me the side of him that I wanted to see" (Peirce to Hegeler 11 Feb 1893).

Peirce to Russell. Telegram: RL 387. Form headed, "The Western Union Telegraph Company."

8 Feb 1893 | Niagara Falls Ny

Judge F. C. Russell
125 LaSalle st Chgo
Kindly express umbrella to me
buckingham hotel New York

C. S. Peirce

---

25 "Note A," pp. 182–186.

26 Francis Galton (1822–1911), English sociologist, psychologist, and eugenicist; coined the phrase "nature versus nurture" in his 1874 "On men of science, their nature and their nurture" (*Proceedings of the Royal Institution of Great Britain*, 7:227–236).

27 (R 592–594). For details, see Peirce to Hegeler 24 Feb and 7 Mar 1893. Also de Waal, 15–17, and Brent, 221–224.

28 Carus to Russell telegram 6 Feb 1893: "Peirce will arrive tonight about seven invite you to lunch tomorrow Richlieu" (RL 387).

1894

Aug 1894–May 1896: For the next two years Peirce's correspondence with OCP goes silent and only Russell remains in contact. By this time Peirce had fallen out of favor and contact with Hegeler due to consistent complications and misunderstandings with agreements, which consequently affected his standing with the OCP, including Carus. The situation elicited the following response from Russell and would be a recurring subject in many of the remaining letters of their correspondence.

Russell to Peirce. MS: RL 387. Stationery headed, "Francis C. Russell, | Attorney and Counsellor at Law, | 123 and 125 La Salle Street, | Bet. Madison and Washington Sts., | Suite 54, 55 and E."

Chicago Aug 29th 1894

My dear Peirce.—

Since our meeting in Chicago now more than eighteen months ago,[29] I have heard from you not a word directly, although considerable in an indirect way.—Had I believed that all went well with you I should probably await overtures on your part. But since I am fearing that your lives are not easy at present, and because in spite of whatever bones I have to pick with you, I care more for the prosperity of your mission in the world. I venture to make this first offer. And a venture, it is, I cannot help but feel. To be of any considerable service to you I must say or hint several things that I fear will be apt to arouse one or more of several very untoward excitabilities that I must impute to you.—My hope is that you will see in my lines and in spite of what I shall say my real and hearty good will towards you personally and my great solicitude that what you no doubt care most for, to wit: your burden of precious discoveries shall not be balked of revelation.—For I really fear that such is the case that your message to the world is in great danger of being in great measure lost or at all counts that the same shall remain in such *obscurity* that they will have to be virtually re-discovered before the world can profit by them. You no doubt comfort yourself with the reflection that the merit of your work is already recognized by several of the élite among the savants of the world, and that in their verdict your future fame is assured.—But I tell you that the adapters and interpreters of your work will manage to appropriate to them-

---

29 At Richelieu in Chicago 7 Feb 1893, on his return from meeting with Hegeler in La Salle. See Peirce to Hegeler 11 Feb 1893.

---

20 untoward] *above* ~~irksome~~
21 and hearty] *inserted above*
28 savants] savans *[E]*

selves the greater part of the credit. At least they will do so for the present and the proximate future. The fact is that you have almost buried your oracles in a style of utterance so technical and condensed that only the very able and bright can interpret them into their own ideas and then only after the closest and most continuous study.—Now this is not because of any inability on your part for the (so called) popular style of exposition. Indeed you have when you choose a style of discourse that is singularly easy and luminous.—But I judge that you have little or no patience with inferiority or mediocrity of comprehension, and fling out your expressions, as who should say, There that is *in fact* clear and comprehensive to an adequate instruction and faculty, and let the dullards grope and be hanged.

Now has a man of *genius* no obligations to fulfil?. Is *he* privileged by his gift to separate himself from his duller fellows, or to amuse himself with contempt and perhaps scorn for such as find his oracles too difficult?.—I say No.—As the parable of the talents prescribes. As the precèpt "noblesse oblige" inculcates, so likewise we ought to have generally accepted some such precèpt as *génie oblige*.

Besides when genius does too constantly contemn the dullards with conduct to correspond to its contempt, inferiority and mediocrity in some way usually manage to "*get even*," at least during the time when genius is able to enjoy the good things of life.—It is so ordained that that cast of soul that is inclined to *brooding* is not one, and by no charm can become one, that is fit to win those material goods that men have need of. Indeed the *brooder* has more urgent need of an abundance than any other sort of a man. The very fruition of his laboring mind may depend on his circumstances as to plenty or its contrary. The consequence is that the brooder has most usually a pretty "rocky" time. His contributions, so precious, are born in anguish of mind and body. Needing the *utmost* leisure and tranquility for the prosperity of his mission. Needing also unrestricted resources, so far as the same are fit to assist his laboring soul, he is nevertheless forbidden that line of conduct by which these can be secured. If he were abundantly fit to be a money gainer (so far as his general character and inclinations are concerned) he would be prevented by the necessities of his "call." Brooding & moneymaking are absolutely incompatible. The monasteries of old furnished after a fashion a refuge for the brooder. But modern life seems to have no corresponding appliances. Oh! that we had some recognized place and fit nursery for those whom the world ought to *search out* and put to the single hearted and single minded fulfilling of their "call."

---

2 proximate] ~~ap~~proximate
13 the] the the *[E]*
17–18 manage] manage~~s~~
31 monasteries] monestaries *[E]*

Huxley[30] (I believe it is) has called attention to the mere money value to the world (so great as to be altogether beyond computation) of the discoveries of Faraday[31] and says (in substance) that society would reap an enormous profit could it only by any investment and régime select out such men and put them about their own proper function. There *ought* to be some such institution and the expense of the same on the *most liberal scale* would be wholly *unworthy* of *consideration if* we could only make the right selections.—Indeed it seems to me that society could *abundantly* afford to maintain such an institution even though it turned out a *real success* only now and then. Now if instead of being very poor I were *rich* like, say Mr. Hegeler and it was my good fortune to find out such a man as you I would be only too glad to say, "Just make yourself easy, I will furnish all your needs, and on a liberal scale only devote yourself to the working out and production of those results which God seems to have specially fitted you for"

But so the conditions are *not* cast at at present, and the question is what is the best thing to do now. You have "got out" with Mr. Hegeler *or* he has "got out" with you. *I* judge this is partly your own fault (which I am very very willing to attribute to the "eccentricities of genius" although that does not alter the fact) and partly his fault that 1st he does not adequately appreciate your genius, and 2d that he acts on certain principles good enough for the most part, but not adapted for cases like that one in hand.—I fear too that you suspect Dr. Carus of being jealous of you and to that extent unfriendly.—Perhaps he is jealous but if so I am quite sure that he is not so *consciously*.—Both Mr. Hegeler and Dr Carus are too liberal minded and too liberal hearted to indulge *consciously* in any pusillanimities I am quite sure that Mr. Hegeler had a sincere desire not merely to obtain your work as a *business* venture, but also to be useful to you in buying and marketing such work as you might produce at what (to him) seemed full prices. He was also willing to make occasional and temporary advances. I ought to say right here that it is one of the salient points of Mr. Hegeler's code of conduct that whatever aid one extends to another ought to be so ordered that any accommodation will never put the one benefited in an ungracious position towards the other. He didnt want your

30 Thomas Henry Huxley (1825–1895), English biologist and opponent of Darwin, credited with coining the term "agnostic."
31 Michael Faraday (1791–1867), English scientist and magnetist, pioneered the concept of the electromagnetic field in physics.

1 Huxley] Huxly *[E]*
14 at present] *inserted above*
15 now] *inserted above*
17 eccentricities] excentricities *[E]*
18 genius] genious *[E]*
23 pusillanimities] pus*c*ilanimities *[E]*
29 so] *inserted above*
29 will never] will be never *[E]*

library and dont want it and cares little what it is worth.[32] He wanted to accommodate you with some extra funds. The library was taken as a saving to your pride. Still, since the transaction was for the reason mentioned put in a business form he wanted to have it governed according to the customs of business.—He has got into his head as the upshot of his transactions with you, 1st That as to money you are extravagant and heedless, a "bag without a bottom" into which more money than he would be willing to spare could be poured and then would remain as empty as ever. He would probably say "I would have liked to have helped Mr. Peirce but I found myself unable to do him any *permanent* service."—2d, That you lack appreciation of the expectations that business men feel injured in not finding satisfied. He expected you would edit "Watts Logic," would collect and edit your scattered papers, would write a certain textbook, &c, &c.[33]—He feels injured at being disappointed.—Now I can see (I think) how you are not very much to blame. I take it that you have been harassed to get along from one week to another and in a very real sense *obliged* to so work one expedient and another that your time has almost been spent to no prosperous purpose. But he feels this sense of injury and some resentment and has told Dr Carus to deal with you as with all the rest that have work to dispose of. He will take anything from you that is desirable paying you for it (irrespective of any standing account) at the regular rates he has before paid for your work. Farther than this he will not go.—But Mr Hegeler is a man that does not cherish displeasure and I hope fortune will so order matters that you and he will "let up" on one another. As for Dr Carus I *know* that he desires more of your work. If I were you I would just continue right along with those papers of yours in The Monist also those papers, The "Critic of Arguments" in The Open Court.—I wouldn't spend much care on controversy over "tychism." Every system must stand on its merits *in its integrity*. Yours is not yet before us so that we can judge of it in its proportions. I do not think that Dr Carus has as yet at all "tumbled" to the rôle that *chance* is made to play in your philosophy. Scarcely any one except one used to mathematical habits of thinking is likely at present to do so. Maybe I do not myself for I take "pure chance" to be the analogue in philosophy to zero in continuous number, a purely ideal "limit" antithetical to *rigid law* another "purely ideal limit."—

32 Meaning, Peirce's personal library sold to Hegeler spring 1893. For details, see footnote in Peirce to Hegeler 1 April 1893.
33 Referring to Isaac Watts's *Logic, or The Right Use of Reason* (1724), Peirce's "Quest for a Method," outlined in Peirce to Hegeler 7 Mar 1893, and his proposed arithmetic book.

31 continuous] *after* ~~number~~

I have several times thought to write for The Monist my own ideas of your philosophy say an article on "The Philosophy of C. S. Peirce" as a means in the first place of perhaps starting you up again in The Monist and also of clearing up various misconceptions of your ideas (according to me) that have been published.

Doubtless such an article would be crude enough (from your standpoint) and stir your scorn to the bottom. But what of that? Just as the boy immigrated to America wrote back to his father, "Pap come over. Mighty mean men get office here" so in philosophy (that is in current philosophical literature) mighty mean stuff passes as respectable.—You see we lawyers are not so thin skinned as secluded scholars like yourself are. We "sass" one another right readily and are none the worse friends. And this leads me to remark how fortunate we lawyers are in that when we make fools of ourselves we suffer the discipline (so salutary) of our brethren. I cant help but imagining that if *you* had someone who to your knowledge or full confidence loved you so well that you would *stand* his chastening, such a one would many a time keep you from commissions and omissions that afterwards bring you to grief.—Nothing is worse for a man than to be too well shielded from criticism. It takes a pretty good man to be long a minister without being full of mean streaks. I must now close this long letter. If you have become offended at my officiousness or other faults I beg to protest that I have nothing but good will for you and an urgent desire to be in some way of service to you. What you think of me or my lucubrations I will readily tolerate if I can contribute in any way to your prosperity. If #genius# dont care to "oblige" why then #dullness# accompanied by magnanimity *will* "oblige" if it can get a chance. Perhaps I ought to say in closing that neither Mr. Hegeler nor Dr. Carus have or have had the slightest idea of my writing to you or that I thought of doing so. It is all entirely out of my own impulses and suggestions that I have written.

Yours Cordially
F. C. Russell

Peirce to Russell. MS: RL 387. Letter sent in place of that ca. 1 Sept 1894 which follows.

Milford Pa 1894 Sep 5

My dear Judge:

My letter was so long that I think I had better try to write a shorter one. I was very much struck with Hegeler when I saw him & I am such a fool about men that I extended my admiration of him much too far, and that was why we could not get along, perhaps. To me, there is nothing so utterly incomprehensible as a man's not

---

6 immigrated] emigrated *[E]* 12 suffer] *above* ~~enjoy~~

wanting to be set right when he is wrong. I have formerly kept a worthless fellow in my employ for years for no other reason than that he was very impertinent & sometimes spoke home truths I found useful. It is difficult for me to conceive the state of mind of a person who wants to have their errors go uncorrected. Still, I see the phenomenon on every hand; and one of my greatest difficulties is to steer my course so as not to offend people's absurd susceptibilities in that respect. But when I see persons of that absurd temper pretending to be *philosophers*, my contempt is very deep. Now when I had a very exalted idea of Hegeler, I could not conceive that setting up to be a bit of a philosopher, he should look upon philosophy as a *contest*, where your triumph consisted in not having your errors exposed. Consequently, I soon mortally offended him, by pointing out that he was wrong about something, a thing which anybody doing for me has always endeared himself to me. Of course, in that respect I have a small opinion of Hegeler. I came here and found all my pipes frozen, the house uninhabitable, the railway blocked, & in this neighborhood one never can get work done in time. There was a great delay during which it was impossible to advance my work much, & Hegeler became also offended *not*, as you suppose, because I did *not* prepare my papers for publication, but because I *did* so. True, he at first wanted to publish them; but he changed his mind, and though he had promised to do so, flatly refused to keep his promise. My arithmetic was, however, by that cause, and also because of certain domestic events upon which I could not calculate unavoidably delayed. If Hegeler would have advanced me a small additional sum, that would have been all right; but as he would not, & *I had to live*, the only thing I could do was to give it up, go to New York, & live on my earnings. Even then, I should have been all right, if I had not been awfully swindled; and then I should have been all right, if it had not been for certain treacherous friends who continued to play me a bad trick. But still I should have come out all right, if it had not been for the terrible times. And even as it is, though it has been a terrible squeeze, I shall soon be in condition, I believe, to repay all I owe to Hegeler, pack up the books & send them to him, and go about my proper business of making the exposition of my philosophy. You may readily imagine that I did all I possibly could to avoid a rupture with Hegeler; for I saw it coming. A short time ago, I wrote to Carus to see if I could arrange to get any writing to do for the Open Court (though on reflection I *first* wrote on another subject, & then *Carus* opened *that* subject) and Carus said he would like to have me write *and that Hegeler assented to it*, while wishing to have business details

---

1 formerly] *inserted above*
21 upon … calculate] *inserted above*
35 me] *above* ~~my~~

referred to him for final decision.[34] My business proposition was that they should pay the prices they had formerly paid me, unless they thought them too high in in which case they could pay what they saw fit, and 2/3 of the amount was to come to me and 1/3 be put to my credit unless Mr. Hegeler saw fit to keep back more. They were to have a weekly article from me in the Open Court and one in the Monist unless they thought that too much when they were to say how much they would take of mine. Mr. Hegeler said nothing, but just let me lose ten days preparing (rewriting etc) a number of articles,—they to reject what they did not like,—when he suddenly said he did not want anything accepted from me & gave Carus a $5 bill to send me![35] As he has helped me, I will say nothing against him; but when you are talking of what you would do if you were rich, you forget the greatest privilege of wealth, that of making yourself more despicable than another man would dare to do.

Of course, the *Open Court* is so badly conducted at present that I really believe Carus couldn't do worse if he were trying to scuttle it. It will be a damned shame; for the paper is a splendid monument to Hegeler. If I had hold of it, I would make it rival The Nation on the literary side and on the scientific side be the best thing in the country. *That* would really further Hegeler's ideas, which are not furthered at all by it at present. I would have ten to fifteen thousand subscribers at $3 for an increased size of the paper & valuable line of advertisements. But Hegeler dont care for money. What he wants is obstinately to continue mistakes he has fallen into. Of course, I would be glad of employment and pay. But as before he will only have me on to lose a lot of time and money and behave just as before. Of that I am confident.

C.S.P.

Peirce to Russell. MS: RL 387. Unsent letter replaced by that of 5 Sept 1894.

[Unsent, ca. 1 September 1894]

My dear Judge Russell:

34 See Peirce and Carus 9 July to 15 July 1894.
35 Carus to Peirce 16 July 1894.

2 had] *above* ~~saw~~
2–3 in which] *above* ~~and~~
4 saw] *p*saw
4–7 They were … mine.] *inserted above*
14 is so] is \~~at present~~/ so
22–24 course, I would … C.S.P.] *up left margin and along header of first page of letter*

I am very much obliged to you for your kind letter received today. The reason I did not write at first was partly because I was very much occupied & partly that I am an extremely bad correspondent especialy toward those I most care for, because I am not satisfied to write such letters as I *can* write. But afterwards came two special reasons for not writing. One was that I seemed to be drifting into an unpleasant state of things with Hegeler & I did not think it fair to draw you into it & the other I did not wish to say much about my relations with Hegeler because I did not comprehend them. I dont know now what the matter with Hegeler is. It is quite clear to me that his state of mind is entirely different from what you suppose it to be, that for some reason or other he is violently inimical to me. I dare say I may be greatly at fault in some way, & probably more than you seem to think, or more than you say. But how I am so, I really dont at all comprehend. (As I continue writing & reading your letter I begin perhaps to see.) That you dont rightly understand the matter is clear from this, that Hegeler has told Carus that he dont want anything from me accepted for the Monist or Open Court whether to be paid for or not. I have a system of philosophy which I was very anxious to sketch out in its several parts & show the connection of them. I should have been glad also to show the reasons that support it, but that I was less anxious to do. I wanted at least to say what it was. I regret that I could not do so. Hegeler never wanted me to publish my papers. He reluctantly consented for a while then objected & finally talked to reporters about "heading" me "off." As for Carus, I think if it were not for Hegeler, he would give me some aid to say my say,—at least, he wishes me to think so, and I haven't very closely considered whether he is sincere about it or not. I think he has very little regard for the proprieties of debate. For instance, his citations of Scotus are simply faked. Anybody who knows anything about Scotus can see that. And I could mention a lot of other things. But I dont care a straw for all that.

I am just as hard at work as I possibly can be. I have one volume of my logic complete, the deductive part. I was induced to finish that first because I had a positive order for it from a reputable firm.[36] But when it was ready instead of taking it, they put it into the hands of a reader whose judgment of it was simply ridiculous. But his judgment was entirely *favorable*, and he recommended the publication. The publishers thereupon promptly *rejected* it & wrote me saying they had not

36 Ginn & Co., regarding Peirce's "How to Reason: A Critick of Arguments," the logic book that evolved from Peirce's "Quest for a Method." See Peirce to Hegeler 11 July 1894 and Peirce to Russell 8 Sept 1894.

6 draw] *after* ~~drift~~
12–13 (As I ... to see.)] *inserted above*
25 citations of] *above* ~~quotations from~~
26 And] *after* ~~But~~

the slightest doubt it would be a success! All these inconsistencies can only be accounted for by supposing them to be very much affected by the state of business. I intend to offer it to other publishers but am holding it back in order to make some alterations which I have not time to make at present.

That same publishing house had already urged me to write a geometry and I was thus decided to do it, or rather very radically to revise and rewrite my father's book before I finished the Logic.[37] I thought it would only take me a few weeks. But it has proved a much harder task than I supposed. And I doubt it being successful as a text book, although I have had the matter under advisement for many years, have tried my method with a class & had written considerable parts of it.

My arithmetic has been finished also all but the examples and part of the advanced book.[38] But it has been rewritten & rewritten in the effort to adapt it to children & is now in the hands of a lady for criticism. I am rushing everything to try to get that to press. But my idea is to finish the geometry first & I hope it wont take me much longer.

No person can understand what the nature of my difficulties are. Besides the "eccentricity of genius" which seems to consist in being a perfect idiot in all dealings with human beings, in my case, I have to contend with certain peculiarities in the health & mind of some of those I have to think of, which I certainly ought not to explain, I ought not even to hint at them as I am doing, but which would account for certain things for which I am violently blamed.

The only objection I should have to you or anybody writing about my philosophy is that my philosophy has never been stated even in the most general outline. It is absolutely impossible to get the slightest comprehension of it until certain sides of it not yet touched upon are sketched & the way the different parts fit together to form one unitary conception are explained. I defy the greatest intellect,—I would defy Aristotle himself,—to see in a general way what I am driving at from what I have said so far. As I cannot get any help to say it, I mean to publish it my-

---

37 Peirce's "New Elements of Geometry" (*NEM* 2:233–474), to be a revision of Benjamin Peirce's 1837 *Elementary Treatise on Plane and Solid Geometry*, in concert with his brother James Mills.
38 His Arithmetic outlined in Peirce to Hegeler 22 May 1893.

---

6 and] *above* ~~my~~
17 being] *after* ~~I~~
25 sketched] *after* ~~do~~

self. My Principles of Philosophy has only been advertized *once* in the Nation.[39] That brought me in about 100 subscribers & I certainly think it encouraging. I have sent out a few circulars,—very few. It is pretty clear to me the publication will pay for itself; but I must first be at leisure. If I can finish my geometry & arithmetic, perhaps I shall have enough income to manage it.

As to personal credit for my system I am not thinking of that. It is a mere flatus vocis, except so far as prestige means power to bring my philosophy before the public & gain their ear. I am one of the men who do not inspire confidence of my kind & people calmly assume I have various radical incapacities which I have not & which they have no reason to suppose I have.

You speak of my obscurity. I think my "obscurity" arises from this, that having an enormous amount of thinking to bring before the public, it is a hard task at best for a person to see what I mean. But then I cant even get myself expressed in the merest outline,—and an outline of such views *must* be obscure. I shall be clear enough when I am spread over 12 volumes. My theory of the universe is a good deal more difficult intrinsically than the Theory of Perturbations. How can it be otherwise considering the immensely greater complexity of the problem? Now how foolish it is to try to popularize the Theory of Perturbations Herschel's Outlines shows,[40] which is at once far too hard to be popular, and far too popular to be correct even in the outline of the subject. To my mind it is futile to undertake to give people any philosophy of whose value rises as high as *zero* without requiring them to think as hard as a man has to think to read a book on differential equations.

The duty of a man who has valuable thoughts is to get those thoughts stated. That is as much as he can generally do. If he has such wonderful power that he can also state them so that he not only imparts them to those who alone can receive

39 Vol. 58 (11 Jan 1894). Peirce's proposed 12-volume work to serve as "a Working Hypothesis for use in all branches of experiential inquiry" (printed circular ca. 1894, under "Peirce and Russell" in Appendix). There were also two other printed announcements, one privately printed brochure from 1893 consisting of a detailed outline (*CBPW* P00552), and another estimated in *W* and *EP* chronologies as an announcement by Henry Holt & Co. in 1894 (SIU 91.27, that in Appendix cited above). An unpublished announcement with endorsements from William James, Josiah Royce, and others is also found in R 1581.

40 *Outlines of Astronomy* (1849).

7 my] *after* ~~out~~
9 have[2]] *over* do
13 for] *over* to
16 Theory of Perturbations] *t*Theory of *pu*Perturbations
21 whose] *above* ~~any~~
24 those] *after* ~~thoughts~~

them,—namely those who think,—but also excites the applause of those who are incapable of thought and therefore cannot get those thoughts, perhaps that may be his duty. He will I suppose state them as clearly as he knows how. However, that is not the case with me. I have far less chance of being heard or listened to if I pretend to make the idle understand than if I do not. Were I to write twaddle no person to whom what I have to say could be intelligible would pay the slightest attention to me. Would you have Klein write his Icosahedron[41] so that the Sunday*World would find it acceptable? I can write in a light and agreeable way, and can get a little work of that kind from time to time. But it is strictly on condition I never reveal or even hint at anything which I have to contribute to the world's thought or knowledge. Nobody would publish it if I did,—certainly not Hegeler,—and if it were published no thinker would go to such a writing with the idea of thinking about it. If I wish to contribute to human knowledge I must do so, as everybody else does, in the forms which real thinkers like.

But as I am on very short commons & my wife too, I should be extremely glad of any opportunity to earn a little money either by writing, teaching, lecturing, or in any way whatever. I am a tremendous worker and very adaptable. I dont at all insist on airing my own views. On the contrary, prefer not to do so. Never would do so in anything intended for general readers except with a view to earning more money or more easily that way than another way.

I dont know that it ever occurred to me Carus was jealous of me. I dont rate Carus very high certainly. I don't think him or Hegeler either at all "liberal," if you mean broad and tolerant. I think them very narrow, as all the Büchnerite sect are for the most part. But I dont think Carus has sufficient understanding of what thought is to be jealous of my power of thought & what he could be jealous of I can't see.

I have a very high respect a sort of veneration for Hegeler. I think his ideals are admirable & the whole morale of the man high, though he is too arrogant and too little aware that truth is not something to be pushed like a business. I think he says to himself, "Now here is Peirce. Judge Russell and others say that I could do the world some good by helping him. Besides, he is rather an interesting man, I can see. I will see if I can't help him." But for some reason or other, perhaps because he has made up his mind my philosophy is pernicious, he is only anxious to put me down.

---

41 *Lectures on the Icosahedron* (1884).

---

9 strictly] *after* ~~stretly~~

16 lecturing,] *inserted above*

19 intended] in*d*tended

29 says] *after* ~~h~~

That I am extravagant and heedless about money to put the thing in its mildest terms I must acknowledge. Neither do I for a single instant feel anything but a heavy load of obligation toward Mr. Hegeler. I certainly have not the slightest desire to have him do anything more for me in a financial way. On the contrary, the thing which is driving me crazy is how I can ever get clear of the situation.

But Hegeler is not desirous of having me edit Watt's Logic or collect my papers. The former was never broached by him. The latter he was persuaded to consent to publish; then expressed his reluctance; finally when I had broken with another publisher flatly refused.[42]

You are mistaken in saying "he will take anything from you that is desirable paying you for it at the regular rates he has before paid for your work." On the contrary I sent him on a number of things and said if they weren't liked I had a variety of other things and if none of them were liked would write on any subject not committing me against my conscience, and would make an effort to please him. Would consent also to taking only a portion in cash the rest to go to my credit on his books. The answer came back that Mr. Hegeler did not wish any of my work be accepted.

When you say you hope I will "let up" on Mr. Hegeler you do me a very serious injustice. I have never attacked or accused him or complained or been wanting in any way of the highest esteem for him. I think he has some peculiarities; but nothing in comparison with his grand and noble character. He seems to me to be changeable; but what I have said all along was that I could not comprehend what was the difficulty. My horrible business habits I suppose, from what you now say. Certainly I am very culpable in that respect.

Among the papers I sent to Hegeler & which he rejected were some articles on the Critic of Arguments. I much prefer to keep out of philosophy altogether in writing for Hegeler, because they absolutely refuse to allow me any reply to Carus, though I think he is absolutely slanderous in talking of my holding that there are two truths etc. doctrines that no philosopher holds or ever did hold except that "*Damned Averroes*" as Scotus calls him.[43]

I entirely agree with you about avoiding controversy. I detest it & with you think it helps very little. I only replied to Carus because he having devoted 2 long articles to me, it seemed courtesy required me not to pass them by with the appear-

42 See Hegeler to Peirce 16 March and 15 May 1893.

43 *Ordinatio* II, d. 3, Q. 6, n. 164.

21 He] *after* ~~What~~
22 but] *after* ~~and~~
26 altogether] *after* ~~all alo~~
33 not] *after* ~~to~~

ance of indifference I really felt.[44] As for the tychism, Carus is entirely mistaken in supposing that that is the chief feature of my philosophy. It is only the one that had to be brought forward in the first instance in order to explain what my philosophy was.

If you can talk Hegeler round, why you may be sure, that I shall be as grateful as I can be. In point of fact there are some features of my philosophy which chime in remarkably with his views of immortality, though I dont think they are quite satisfactory. But what I would like would be to have an Article in every Monist (or every other one) until I can sketch my whole system. After that I should be glad if I could have further space to *argue* it; for I have really *argued* nothing except the tychism. That argument will rest on observations, some new & some old.

I should also be glad to be allowed say one page or two in each Open Court (or every other one) on Logic or preferably the History of Science, or any other subject or subjects that may be agreeable. I dont object to articles being returned they dont like provided the reasons can be made so definite that I can avoid the objectionable features.

This is what I tried to bring about through Carus some time ago; but for some reason it did not work.[45] Hegeler was just as insulting and disagreeable as Prussian resources in that line enabled him to be. If you can manage it, as I shouldn't wonder, it would be a great help; for till my books are out I am quite strapped. In fact, there is hardly enough food to keep body & soul together & I am considerably weakened by it. I dont suppose this state of things will last very long; but it dont take very long to finish a man.

very truly
C. S. Peirce

Peirce to Russell. MS: RL 387. Another letter in response to Russell's of 29 Aug 1894, and one of two letters sent on the same day (see next letter).

"Arisbe" Milford Pa. | 1894 Sep 6.

My dear Judge

Your thinking I have no capacity for making money because I am a "brooder," has this mistake about it, that I am not the least a *brooder*. There is nobody less so.

44 Peirce's "Reply to the Necessitarians" (July 1893) in response to Carus's "Mr. Charles S. Peirce's Onslaught" and "The Idea of Necessity" (July and Oct 1892).
45 Peirce and Carus 9 July to 28 July 1894.

3 had] *after* ~~was~~
7 they are] *that*they ~~is~~ are
8 have] *over* give
18–19 Prussian] *after* ~~a~~

I think the idea of trying to advance thought in any such way utterly wrong. When I have had anything to do with business men so that they could understand my course, they have never failed to be struck with my business ability. Besides, I have always made from three to five thousand a year outside of any salary, until the last few years when I have had to contend against a fearful combination designed to ruin me.

It is true I have always and continually had losses owing to my excessive confidence in men, & I have never failed to excite the enmity of every stupid man I had to deal with, because jokes slip out of my mouth & such men dont like jokes.

If you know any way of inducing Hegeler to give me work for a while, do so. Of course, I could make his concern a very effective means of furthering what he has at heart & especially the *Open Court*. But of course I know very well that he is the kind of man who wistfully sighs after the *end* & refuses to take the *means*. Carus is not jealous of *me* that I know of, certainly not at present; But his intense jealousy of *Hegeler* and bitter hatred *of* his brother in law sticks out obtrusively. Carus will never allow the *Open Court* paper to be a success; and there would be no doing anything with it with him working the other way. I could also help the publishing concern. I offered them my logic, the first volume of which (a complete work in itself) is ready for the press. I am awaiting the return of a publisher to send it to him.[46] But Hegeler would not listen to it. I repeat that I have not the least hope that you will be able to arrange anything. He will act as before, pretend to agree to it in the hope of getting me into further difficulty & turn round at last & show his animus. "Kiss it," he'll say. Kiss what? I shall ask. My animus, he will reply.

I have seen several rich men whose great amusement & pleasure was to get poor men into difficulties & who would rub their hands & chuckle when they heard somebody had committed suicide by their machinations.

Hegeler has elevated & noble elements; but it is clear to me his mind is somewhat diseased & he has something of the trait which makes a man the most contemptable of all traits. Those of us who are poor may thank our stars that whatever the balefulness of poverty may be, it would be better to suffer it a thousand fold

46 His "How to Reason," sent to Ginn & Co.

1 utterly] *over* ab
14 intense] *over* je
19–20 I am awaiting … to him.] *inserted above*
23 "Kiss it," … reply.] *inserted in-line and below*
24 several] *above* ~~many~~

to all eternity, than be contaminated with the soul-poison which can effect even as good a man as Hegeler.

Come, virtue, in an earldom's cot!
Go, vice, in ducal mansion![47]

C.S.P.

I should anyway be glad to sell a few articles for immediate relief. I have a number which were written for them & which Carus says were *accepted*. But Hegeler dont seem to think himself bound by anything. (I easily understand how he gets the courts down on him.) He simply refuses to take them & lets his engagements go to hell—I am badly in need of cash—

Peirce to Russell. MS: RL 387. Another of two letters sent on the same day (see previous letter).

Milford Pa. 1894 Sep 6.

My dear Judge Russell:

I happen just now to be off the *Nation* for the time being, owing to my striking, although they really pay me at a very high rate as newspapers go. The thing will be adjusted soon and a new agreement made. Meantime I am free to make a suggestion which I shall not remain long free to make.

The *Nation* only has about 10 000 subscribers. The reason is partly that they have *no* subscribers in NY. city, because the *Nation* is only a weekly edition of the *Evening Post*; but it is chiefly because the *Nation* is rather decidedly Democratic, while most of the people who would naturally subscribe for such a journal are Republicans. The *Nation* is also regarded as rather fogey and rather dilettante.

Now if the *Open Court* were made as large as the *Nation*, a first rate critical journal, advocating Mr. Hegeler's views of Religious matters, & moderately but distinctly republican in its tone, giving attention also to Science, to Intellectual Amusements, theatres, chess, whist, etc. and paying also more attention to *the tastes of women*, there is no doubt in my mind its subscription list would equal or surpass the *Nation*'s very soon. Meantime, it would gain advertisements which would also swell the treasury. The *Nation* gets as much from advertisements as from Subscribers. The *Nation* is not conducted in a very economical manner, in

47 William S. Gilbert, "The Periwinkle Girl" (1868).

6–10 I Should anyway … need of cash] *Paragraph on a card pinned to the top of the last page of the letter*

30 in^2] *above* ~~if~~

various respects, which I dont care to speak about. In my opinion it could be better done for less money; in fact, I guess the editors are quite aware of that. But various things are done for the sake of prestige. I know a number of rich men who would put their hands in their pockets to establish a journal like that only of a Republican complexion. Condense the printing and make 18 pages a week of reading matter, i.e. As much as the Nation. Have the bulk of it done by a staff of writers rigidly drilled & trained to write in one set way. Let the opinionated part be confined to one page of Politics and two pages of Religion. Let the rest be distinguished by an anxiety to give all sides a fair hearing (*reversing* in this respect the course of the *Open Court* which is much laughed at for the contrast between its liberal professions & its intolerant practice). Not that the editors should allow anybody to advocate any *cause* or opinion whatever, but only to direct attention to both sides & treat both sides of all questions where there really are competent men on both sides with perfect respect and impartiality. Another thing: the *Nation* notices what it finds *convenient* to notice. But a new journal should notice *every* work of real merit within its range. It should record the whole weekly history of the intellectual life of the world, passing by all quite unimportant things and explaining the bearings of things noticed. It should attempt to pass no judgments on the substance of disputes until the moment came to express the mind of the learned world, using great caution in this.

If I were offered a responsible position on such a journal, I would endeavor to bring capital into it, & no doubt should succeed in doing so. Of course I would accept no other than a responsible position; and I should think a place where any opinions or sentiments were to be inculcated entirely derogatory to my standing. I would, if about 4 pages weekly as long as the *Nation*'s were devoted to critical matters & to science, accept the editorship of that part, and write from half to three quarters of it myself, if desired. Or, I would add a page on *Games*; or, in fact, run up to taking care of as much of it as they saw fit to shove on me. I know we could make a success at it & nothing could further Mr. Hegeler's aims half so much, as well as setting up a proud monument to his philanthropy. My writing of advertisements, by the way, has been repeatedly praised by men at the top of *that* profession.

very truly
C. S. Peirce

---

2 in] *after* ~~of~~
4 only] *inserted above*
11 the editors] *above* ~~we allow~~
14 respect and] *inserted above*
17 quite] *inserted above*
23 than a responsible] *inserted in-line and above*
26 part,] *inserted above*
27 *Games*] *g*Games
29 could] *w*could

I should then be in the best possible situation to get out my philosophy. My two volumes already complete would speedily find a publisher. But I know the stupid German ire too well to expect anything from the suggestion.

Peirce to Russell. MS: RL 387. Several gaps and conjectures are present due to torn-off portions of the MS, the most severe being the bottom left corner of the first leaf.

Milford Pa 1894 Sep 8

My dear Judge:

I ought to be ashamed of myself for inflicting such long epistles upon you all about myself. My extreme need is my only excuse.

You suggest that Hegeler has the idea I indulge in idle pleasures. This is a mistake. If for a time I was given to seeking society & trying to forget myself, it was owing to perturbations of spirit from causes sufficient to drive a man mad. I had better not dwell on that. It must be remembered too that I had a right to suppose I was on the way to making a great deal of money; and in ordinary times so I should have done. One has to feed one's family & in the last year I suppose I have written about 60 papers for the Nation, of which the most laborious have never been printed, though a great many have *[?ap]*peared.[48] I have not confined myself to Scientific & Philosophical *[?subjec]*ts but have written on Hale's New England Boyhood, Charles *[?Leland's]* Memoirs, Scott's Letters, Vathek, etc. Such things as *[?Durege's Elemen]*ts of the Theory of Functions etc occupied *[?my time, although Ernest]* Naville's Définition de le philosophie, Ward's Psychic Factors, etc were still more elaborate. Also, a great literary & scientific effort on Aeronautics which the Nation would not take because it went too far from criticism & which will open people's eyes. I have a great method in hydrodynamics never published by which I have solved many things & have views on Airsailing which I set forth as if pure fiction,—as if a sort of Jules Verne affair,—but which is really a deeply studied matter. Besides that, I have done a great deal of hard work on Petrus Peregrinus, have read between 50 & 60 works of medieval science, have written an elaborate analysis of the character of Napoleon which in abridged form was published in 4 columns of the Independent & excited much interest among Napoleonographers, have done a great deal of hard mathematical work on the development of

48 See Bibliography for publication details of the works mentioned in this paragraph.

421.29–422.3 much, as well as … suggestion.] *up left margin, along header, and up right margin of first page of letter*

25 by which] *above* ~~from~~

my philosophy, have written two volumes, and have another more than half done. My volume on Logic is completely ready for the press; though I am anxious to make some alterations in it. Yet it could go to press & in the opinion of publishers would meet with moderate pecuniary success. The reason the publishers gave for not taking it was that my own judgment was against it as a textbook (which I tried to *[?]*. School-Book-publishers have to give away their books to all the teachers & *[?that would]* kill my market.

Now I have an idea. I must have some immediate cash or *[?]* to smash. Suppose I assign you the copyright of this work as collateral secu*[?rity]* for a loan of $ 250. You see if you can get it for me. If you can smooth down Hegeler's ruffled fur, why should he not advance that sum? The security is good. There are several chapters that would make excellent Monist articles & the book would pay his publishing house. I will send you the MS if required though I had rather keep it by me for reference & to work on occasionally. But I will proceed to describe precisely its present condition & how I would like to improve it.[49] If you will as lawyer send me the right kind of assignment to execute I will do so & return it to you. I enclose an agreement to do so.[50]

Present State of a Volume entitled
The Art of Reasoning.

Chapter I. *What is a sign?* My old theory, but entirely newly written & one of the clearest pieces of analysis I have ever made. Likenesses, Indices, Symbols, more clearly discriminated than ever before, especially the last, which had been a little hazy.

Chapter 2 The Materialistic Aspect of Reasoning. My old theory newly written out, & somewhat elaborated.

Chapter 3. What is the Use of Consciousness? New. Questions the assumption that it is one thing to *see* red or green & another to *look* red or green. Empedocles a little pirated here.

Chapter 4. The fixation of belief. My old piece. Improved in parts, adhering to the same inflated style. The *a priori* method much better treated & enforced by criticisms of Descartes, Kant, & Hegel.

Chapter 5. Of Inference in General. This is a new piece of work, largely occupied with discussion of terminology.

49 Most of the outlined material that follows is contained in R 397–424, "Grand Logic (How to Reason: A Critick of Arguments)." See Bibliography for details.
50 Enclosure not found.

2 completely] *inserted above*

"Simple Apprehension, Judgment, & Reasoning"

"Term, Proposition, Argumentation"

Understanding, Verstand, *intellectus*.

That most languages have no nouns, at least not so radically separated in the minds of those use them from verbs as are ours. The old Egyptian mode of conceiving the analysis of a sentence.

The Egyptian conception is the most convenient in Logic.

Onslaught upon grammatical terminology in general & upon the Procrustes bed into which European grammarians force languages. Prof. Nicholl's absurd criticism on Palmer's Arabic Grammar—Pronouns are not "words used in place of nouns." The parts of speech are to considerable extent only true of Indo European languages. "Root" & "stem" good designations, but "inflexion" poor & "declension" downright silly. "Oblique" cases, a phrase that goes with "declension." "Case" belongs to the same foolish metaphor.

"Gender" for sex is stupid. "Accusative" a bad transference of αἰτιατική, itself utterly inappropriate. "Genitive" bad, worse than "possessive," but "correlative" is the true word. "Instrumental" poor. The English temper shown in expressing this by *with*, which really means *against*. "Ablative" not good; it ought to be "linguitive."

"Comparison" of adjective utterly inconsistent with terming the dual degree "the comparative."

The origin of the term "voice." The "middle" voice a weak designation. Tenses in different languages. "Aorist" & "imperfect" not good. Our Indo Germanic *moods* are modified tenses. But this is not a logical way of looking at the matter & many barbarous languages are far better about this. The *eskimo* languages very interesting in this respect. Protest against calling the *assertoric* mood the "indicative." Meaning of *indĭcare*. Comparison of the meanings of the roots DIK and MAN. Two meanings of shew.

The word *infinitive* is a tranfer to Latin of ἀπαρεμφάτως, without παρέμφασις, that is, without variation by number or person. The reason not so varied is that Indo European languages do not attach persons to nouns, (for the infinitive is a noun) as many other languages do. *Gerund* would have been better called *gerundive*, so imitating the excellent method of terminology of the Arabian gram-

13 "Oblique"] "Obique" *[E]*
16 "Genitive"] ~~Gete~~ "Genetive" *[corrected spelling mine]*
20 utterly] *above* ~~perfectly~~
23 Indo Germanic] *inserted above*
24 way] *above* ~~con~~
31–32 (for … as] \(for the infinitive is a noun)/ ~~which~~ \as/
33 excellent] *inserted above*

marians. *Supine* goes along with "Case," "Declension," "Oblique." It is a form so oblique as to quite lie down.

"Demonstrative" pronouns vilely designated. Utter want of nice sense of the use of words shown by grammarians. *Demonstrative* is just precisely what demonstrative pronouns are *not*. They are only *indicative*.

The Eskimo wealth in demonstratives owing to their mode of life.

That there are many more *relative* pronouns than grammarians recognize, and especially the very useful ones A, B, C, used by lawyers in stating hypothetical cases, by geometers, etc. *Reflexives* and *reciprocal* pronouns stupid words. *Intensives* valuable. The "indefinite" pronouns ought to be called *selective* and distinguished as *universal selectives* and *particular selectives*. Careful study of the value of these requisite for the logician.

Prepositions. How they probably originated. The original meaning of several Egyptian prepositions are traceable and appear in their hieroglyphs. *En*, meaning "to" is a relative pronoun, "which." *Em*, meaning same as German *von*, is probably "who." *Er*, "for" or "fore" is probably "mouth," and that is its hieroglyph. Shown by sentences. Perhaps originally "breath" (an imitative in sound) Compare Greek εἰς or ἐνς from root AN, *an*imus, breath. So Sanskrit *ana*, etc. Egyptian *Ap*,—with hieroglyph of bearded head, which is the ideograph of primacy,—means "upon." It is quite equivalent to Greek ἐπί whose root is seen in *op*timus, *op*us, *op*ulent, where the idea is probably that of mastery. Egyptian *Hr*, meaning *above* is the same word & hieroglyph as the noun meaning "face." Egyptian Xg meaning *under* and *with*, also means *fight*.

Gr, meaning *with*, also means *taking*.

Ha, meaning *behind*, also means *the back of the head*.

Nesw, meaning *belonging to*, *concerning*, has the pronunciation and hieroglyph of *tongue*.

Zr, meaning *since*, also means *prison*, *end*, *stop*.

Armen, meaning *until*, also means a man's *arm*.

Xeft, meaning opposite, also means *face*.

In fact there are only three prepositions whose meanings are not apparent (Probably if I knew more Egyptian I might explain these.)

Conjunctive conjunction a pretty awkward phrase but not so bad as *disjunctive conjunction*.

The kind of language that reasoning demands. How each kind of sign is required shown more clearly than ever before. The psychology of language.

---

2 quite] *l*quite 25 *the*] *inserted above*

The terms *conclusion*, συμπέρασμα, *syllogism*. *Premiss* a word introduced to translate an Arabic word. *Consequence, antecedent, consequent.*

Our *hope* that the question which happens to occupy our minds is capable of final decision. The conception of *perfect knowledge. Sure knowledge. Practically perfect belief.*

The *necessary*, *possible*, etc. Their primitive conceptions, & various modifications. The *possible* and the *future*, the *necessary* and the *past*. Absurdity of jumping to the conclusion that nothing in the real world does correspond to possibility and necessity, simply because we can see how there *might* be nothing. The razor of Ockham a good rule, but not to supersede inquiry into the facts. Whether there is or is not any real necessity and possibility is a real question to be decided by positive investigation.

Onslaught on "conceivability" and still worse "inconceivability" as tests of truth. Application of the doctrine to the question of our own personal existence. Here is something the conceivabilitarians consider the most certain of all truth. Yet certainly it is to a great extent a delusion; and in the light of modern psychology those theologians who say personal existence is a sham have the best of the argument. A little rhetoric about the April (*aprilis* = opening) morning of the other world when those who believed in themselves will be found April fools, and they discover that neither selves nor neighbor selves were anything more than *vicinities*, while the Love they rejected was the *sence* of every *sent*. This lesson the most balsamic of all the sweets of philosophy.

What good reasoning must be. Hegelians rate only that inference A1 which setting out with falsity lands us in the truth. But it is the Compulsion of *experience* that ought to be worshipped. Rules of inference. Tradition: what a wise respect for it is, what a childish clinging to it is. Descartes and his *je pense; donc je suis*. False reasoning in Elementary Geometry.

The *reductio ad absurdum* and *Induction* compared. Stoical views of inductive reasoning.

Necessary and probable inference.

Diagrams and experiments with them.

Examples of *Insolubilia*—mostly novel.

Mathematics purely ideal.

Chapter 6. The Algebra of the Copula

This difficult subject is here perfectly cleared up and systematized for the first time, and made to lead directly to the Boolian algebra.

History of the principles of identity, contradiction, and excluded middle.

---

7 Absurdity] *after* ~~How far~~

36 , and made … algebra.] *inserted in-line and below*

Chapter 7. The Aristotelian Syllogistic. Aristotle's formal logic shown to rest on his doctrine that we cannot know anything unless it be a uniformity. In place of the *Quantity* of propositions, I speak of their *Lexis*; in place of *Quality* of their *Phasis*. This chapter presents the ideas of my paper on the classification of arguments in greatly improved form. I show that it is just as logical (only from a different point of view) to make two negatives mean a negative as to make them mean an affirmative. And I now triumphantly make it clear that *Some* may be so conceived that *Some-some* is universal again. This I tried to show in my early paper; but could not make it out clearly. But it is now plain. I show how you can take some in this sense, or else take *not* so that a double negative shall mean a negative; but you must in logical consistency do one or the other.

Chapter 8. The Quantification of the Predicate.

Refutation of Hamilton.[51]

I then consider DeMorgan's propositional scheme from two points of view, which I call the *Diodoran* and the *Philonian*, with reference to a celebrated controversy between Diodorus and Philo. I myself advocate the Philonian view, DeMorgan the Diodoran.[52] I first treat his scheme from his own point of view and show that putting his propositions at the vertices of a cube, every face and edge is significant. I then take the Philonian standpoint, and put his eight forms upon an anchor-ring; and I show the relation to Aristotelian syllogistic more clearly than before. I next treat *Spurious Propositions* and show that they are of extreme importance. In fact, all mathematical or at least arithmetical propositions are spurious. I then take up a *Limited Universe of Marks*, and end with a *General Canon of Syllogism.*

Chapter 9 Logical Breadth and Depth. This is my old paper worked over. Breadth & Depth *in relatives.*

Chapter 10. The Boolian Calculus. I show that the rules of the non-relative algebra may be immensely simplified by dropping all connection with arithmetic & I show that arithmetic itself dictates this if we regard the Boolian equation as an arithmetical congruence with modulus 2. I then pass to the *Logic of Relatives*

51 William Rowan Hamilton (1805–1865), Irish physicist and mathematician, founder of Hamiltonian mechanics (reformulation of Newtonian mechanics), and inventor of quaternions (1843).

52 For details about this controversy and Peirce's views on it, see his 1898 "Types of Reasoning," *NEM* 4:169–170; see also Zeman, "Peirce and Philo," *Studies in the Logic of Charles Sanders Peirce* (1997), 402–417.

3 *Lexis*] ~~*Meta*~~*l*Lexis

3 *Phasis*] ~~*Meta*~~*p*Phasis

4 chapter] *after* ~~pa~~

22 or at least arithmetical] *inserted above*

26 Breadth & Depth *in relatives.*] *inserted in-line*

which I now bring to a much more perfect and interesting form than before & greatly simplify the proceedure. The student will work it practically with facility and advantage. I show that it is necessary to learn to think with it, and that is the chief use of it.

I also give enough of the Logic of Second Intentions to show that it is a vast subject, of the utmost precision, and that it is really an "Objective" logic,—or rather a Material Logic,—that is, that it is something like Hegel's logic, one idea developing into another. I then put it aside as requiring a volume in itself. I may say that I have done an enormous amount of work on this: I have a formula which embraces Hegel's as a particular ease. I do not pretend, however, to be able to work it out from my *Ichheit*. I dont believe any mortal can. Yet theoretically it might be done.

Chapter 11. Graphs and Graphical diagrams. Shows the value and limitations of the geometrical way of thinking.

Chapter 12. The Logic of Mathematics. This is I think the strongest piece of logic I have ever done. It analyzes the reasoning of mathematics by means of the Calculus of Relatives. I dont see what loop-hole there is to escape my conclusion as to the nature of Mathematics, that it is merely tracing out the necessary consequences of hypotheses.

Chapter 13. The Doctrine of Chances. The worst chapter in the book. A mere revision of my Popular Science Monthly article. This chapter demands amplification.

Chapter 14. The Theory of Probable Inference. My old essay. Remarks on this I shall make presently.

Chapter 15. How to Make Our Ideas Clear. Old article.

I intended to add two more chapters. One on the Law of Continuity and the other on Logical Recreations & Exercises & the last I will throw together. Then there was intended to be a Glossarial Index which would be made in reading proof sheets.

The book could perfectly well go to press just as it is, though there are some little things I should desire to do to it. I should also want to insert a chapter on putting questions into Logical Form.

But I have commenced a more Radical revision of it. The idea of this is to restrict this volume to Necessary Reasoning and make a separate book about Probable Reasoning, and a third about Objective Logic.

---

15 I] *after* ~~on~~
20 *The Doctrine of Chances.*] above ~~*The Theory of Probable Inference*~~.
28 a Glossarial Index] an ~~Index~~ Glossarial Index *[E]*
31–32 I should … Form.] *inserted in-line and below*

For this purpose I should make the plan of the present volume as follows:

The Principles of Strict Reasoning. By C. S. Peirce

Book I. *First Principles of Logic.*

Introduction. The Association of Ideas—[Ready.]

Division I. The Formal Aspect.

Chapter 1. *The Categories*—[Ready.]

Chapter 2. *Signs*—[The chapter I have.]

Division II. The Transcendental Aspect.

(By transcendental I mean in Kant's sense that is the epistemological aspect)

Chapter I. Materialistic Aspect of Thought [The chapter I have]

Chapter 2. What is the Use of Consciousness [The chapter I have, a little revised.]

Chapter 3. The Fixation of Belief [The chapter I have.]

Division III The Substantial Aspect.

Chapter Sole. The Nature of Reasoning with notes on Terminology [The very entertaining chapter I have, except part to be shoved into next part.]

Book II Demonstrative Reasoning.

Introduction. Analysis of Propositions [Part of chapter I have, with additions.]

Division I. Stecheology.

Part I. Non-relative Logic.

Chapter 1. The Algebra of the Copula [The chapter I have.]

Chapter 2. The The Aristotelian Syllogistic

[The chapter I have, with considerable additions giving a full view of Aristotle's whole theory of logic and a sketch of his philosophy & a history of Logic

Also: Treatment of Fallacies & *Examples*]

Chapter 3. Hamilton, DeMorgan, etc. [The chap. I have]

Chapter 4. The Boolian Calculus

[A new chapter more elementary and with all my examples, which are a fine collection.]

Part 2 Logic of Relatives.

Chapter 1 The Algebra of Relatives } The chap. I have, split in two

Chapter 2 The Algebra of Dual Relatives }

Chapter 3 Graphs etc. [To be improved.]

---

22 The Aristotelian Syllogistic] *before* ~~[The chapter I have with additions about fallacies]~~

23–26 [The chapter … *Examples*]] *below* ~~Chapter 3~~

31 Logic of Relatives.] *R*Logic of Relatives.

Division II. *Methodology*

Chapter 1. Breadth & Depth [The chapter I have with additions.]
Chapter 2. Clearness of Ideas. [The chapter I have]
Chapter 3. Formal Grammar and Formal Rhetoric.
Chapter 4 Definition and Division. [Nearly ready.]
Chapter 5. Applications [Nearly ready]

Glossarial Index.

This would make a valuable treatise of this & would leave room enough to treat the great subjects of the "Logic of Science" and "Objective Logic" in separate volumes.

My dear friend. The crops have utterly failed. No hay. No oats. Not even the seeds back. Three magnificent horses I have advertized all around, but though willing to sell for a song cannot because of the ruinous price of hay & the failure of oats etc. In Europe last year's drought, & this year's wet, and certain special circumstances reduce my wife's income to *below her taxes.*

My Geometry promised to be so profitable (& I dont think it can fail) is *nearly ready*. My arithmetic is *out of my hands*. If I could only borrow enough on the copyright of my logic to carry me along a while I should be all right. I dont want to force myself on Hegeler who *don't want* my writings. He has shown that clearly enough. But perhaps you can show him that he is harsh in his judgment of me. I *can't*, because I can get no inkling of what he thinks of me. I only know he loses no opportunity to increase my difficulties. I know I have one of the most powerful men in Europe as enemy who will leave & has left no stone unturned to make people act in a way really ruinous to me—Contrived to cause people to give my wife such advice as nearly landed me in a madhouse—& how can I guess what has been said to Hegeler? I know Rood and others (how influenced I don't know) have joined in a regular hounding of me. I believe I shall come out on top; but I dont know *[?when]*. I guess you can influence Hegeler if anybody can; but his animosity is nasty. Try to get me scholars; to have me invited to give lectures or readings.

um|26|||||C. S. Peirce

Above all *[?dont]* lose time. Starvation dont wait. If you cant raise all at once, raise a part and send in haste. I need it & my wife needs it even more. Need cannot be

---

2 with additions.] ~~with~~ \with additions./
19 harsh in his judgment of] *above* ~~unjust to~~
20–21 I only … difficulties.] *inserted above*
22 leave] *d*leave
30 Hegeler if … C. S. Peirce] *up right margin*

more extreme. My two brothers are both abroad. There is nobody to whom I can make my condition known.

Russell to Peirce ca. 8 Sept 1894 (missing): Reports on his efforts to persuade Hegeler to accept Peirce's work.

Peirce to Russell. MS: RL 387. Undated, in reply to Russell of ca. 8 Sept 1894. Date estimated per Peirce to Russell 22 Sept 1894.

*[*ca. 10 September 1894*]*

Dear Judge

I have received yours in reply to some of mine. Dont try too hard! I wouldn't have you get *yourself* into bad odor. That I could never forgive myself. Besides, if a cool presentation don't work, nothing will. Nor do I care to be a suppliant. I shouldn't wonder if one had the silly notion one could starve me into acceptance of a given philosophy. I will write for a paper without writing on philosophy. That is of course the farthest I will go. I will endeavor not to express my opinion of Hegeler and Carus as philosophers, but if the occasion comes directly across my path I shall do so.

C. S. Peirce

Russell to Peirce. MS: RL 387. Stationery headed, "Francis C. Russell, | Attorney and Counsellor at Law, | 123 and 125 La Salle Street, | Bet. Madison and Washington Sts., | Suite 54, 55 and E."

Chicago Sept 19$^{\underline{th}}$ 1894

Dear Mr. Peirce

I have not as yet received a single word from Mr. Hegeler in answer to my representations in the matter of your affairs. I hardly know what to think. I cannot believe that he will flatly ignore and contemn my inter-locution. Perhaps I ought to argue favorably from his deliberation. He is an awful slow man to make up his mind sometimes. But at any rate I did what I could and rest (as rest I must) hoping for the best.

What do you do with in-transitive verbs as relatives. Take say "Not all that glitters is gold" *or* Not all glitterers are gold *or* $\overline{\Pi}$ glitterers $-\!\!<$ gold. What is the converse of *glitterer*.

---

1 My] *B*<u>My</u>

11 cool] *after* ~~slight &~~

28 in-transitive] \in-/transitive

Also take the relation “between”. What are the *two* converse relations. Also take the plural relation “among” and then the relation “in”. I have little doubt but that my questions “give me dead away” but please understand. I make no pretensions whatever only to those who are just as big hunks as I am myself.

Cordially—
Francis C. Russell

Hegeler to Russell 19 Sept 1894 (see copy in Russell’s letter that follows): Replies to Russell with indifference to Peirce’s affairs and refuses to work with Peirce further.

Russell to Peirce. MS: RL 387. Stationery headed, “Francis C. Russell, | Attorney and Counsellor at Law, | 123 and 125 La Salle Street, | Bet. Madison and Washington Sts., | Suite 54, 55 and E.” Russell includes with this letter a copy of Hegeler’s response refusing to work with Peirce.

Chicago Sept 20th 1894

Dear Mr. Peirce.

Mr. Hegeler’s answer to my letter came this morning and I send you a copy. His challenge of your attack contained in the next to the last clause ought to have led me perhaps to have kept that back, for I feel sure that should you really care to “*go for*” the favorite doctrines (and the ideas that are supposed to support the same) of Mr. Hegeler and Dr. Carus, you would at least give them full occupation. You have however without doubt enough to attend to otherwise. It would seem that there is no further use to try and make Mr. Hegeler listen. May I presume to counsel you that should you give time to Mr. Hegeler & Dr. Carus’ doctrines that you will *punctiliously* suppress every *personal* element.—Dont be offended at my saying this. It is *so hard* to give up a good chance to strike home under such circumstances that it takes a grade of human nature to resist that is not often met with.

I am still not merely willing but desirous to be of service to you. Would it accord with your inclinations if I should try and find a publisher for you here?.—How do you stand with *Halstead*, of the University of Texas. He and I are pretty good friends and I take it that he is in pretty fair shape financially, not rich but he gets a pretty good salary and has small expenses. He knows about the grade of work that you furnish. When I go to my friends here who are well to do & talk about security upon a copyright of a work on Logic, they say Oh! There is no money in one book out of a thousand & as for a book on logic!!!. In vain I try to show them that

---

4 am] *after ~~any~~*

22 *punctiliously*] *punctilliously [E]*

27 *Halstead*] *For George B. “Halsted” (1853–1922). See Biographical Register.*

this is not an *ordinary* work. They dont understand the situation at all & I dont seem able to make them. Oh! that I myself were (not rich I wouldnt need to be that) but only just not driven and pushed and harassed with demands that keep me drained dry all the time, I would actually give $ 250.[00] myself (were I able to raise it without robbing those who depend on me) for the privilege of studying your manuscript.—But there is no use I can tender you only good will & service. But command me to that extent.—There are a good many publishers here and I am told that manuscripts find a better welcome among them than at the East. It is true that they have been publishing mostly books of a low grade of merit, fiction &c. It is also undoubtedly a great advantage in all ways to have one's book put out and put on the market under the auspices of an old and established concern of large trade and connections. But it seems to me that when you once get before the public in a *book* such as you can write then you are forever put in a condition to *demand* rather than ask. You are as yet comparatively unknown outside of the ranks of a few experts, who as myself can only give you praise and good will but no *shekels.*—

Cordially
Francis C. Russell

(Copy)

La Salle Sept 19th 1894

Dear Mr. Russell

Your esteemed letter in regard to Mr. Peirce has been in my hands several days. When it came I felt that I could only decline the proposition but delayed writing in the thought that some idea might occur how the objects you have in view might be reached. But such is not the case and I have to positively decline. I will not enter again into any kind of business relation or correspondence with Mr. Peirce.

Should he unfavorably antagonize the work of my publications I shall be glad rather than otherwise.—

Hoping yourself and family are well I remain

Yours Truly
Edward C. Hegeler

Peirce to Russell. MS: RL 387.

Milford Pa 1894 Sep 22.

3 harassed] harrassed *[E]* 16 *shekels*] *sheckels* [E]

My dear Judge:

Yours of the 19$^{\underline{h}}$ reached me yesterday afternoon at 5PM. The previous letter from you[53] which I answered briefly only reached me because a casual person happened to pick it up on the road & that is a specimen of the way my mail is served here. I wrote you a very long letter[54] to which you make no allusion & I conclude you either did not receive it (it was written a day or two before I received your previous one) or else you have written me some letter I have not got.

I will say in a general way that in that long letter I told how I have the MS. of a book on Logic ready for the press. It has been favorably reported to a firm of schoolbook publishers but they do not want it, because they say it is a teacher's book & they have to give away a whole edition of every book they publish to teachers. They express the decided opinion the work would sell well. But although the book is ready for the press, I am not altogether satisfied with it & want to make some changes in it, which I indicated to you. I gave a full description of the work, & of the proposed changes. I also enclosed an assignment of copyright & suggested that you might be able to raise me a loan of $ 250 on that security. It would be valuable security to Hegeler, because there are several entertaining chapters fit for the *Monist*. (By the way, your notion that my articles can only be ready by specialists does not agree at all with the letters I am continually receiving.)

Of course, I want either to raise the loan through you or get my assignment of copy right back again.

As for Hegeler, you know I never at all expected any success in that quarter, though I thought it might be worth trying. It is plain his feeling is very bitter, intensely so, but I cannot imagine why, I cannot even imagine any creditable possible explanation. If you find out why it is, I beg you will let me know. I do not feel at all sure it is not something in which I am greatly at fault, though my conscience informs me of nothing. But when my difficulties have been at their height,—& my present plight of simple starvation is a mere bagatelle in comparison,—I have been nearly perhaps quite out of my head & may have done something of which I am not aware. I want to *know what is* the matter with Hegeler. I cannot find out but I think I am entitled to be informed. I am extremely grateful to you for all you have done; but I shall be more so if you can find out & tell me what it is Hegeler is so wroth about.

53 ca. 8 Sept 1894 (missing).
54 8 Sept 1894.

13 not] no *[E]*

Neuter verbs are equivalent to classnames with copula; other intransitives only differ in making the classification depend on temporary conditions. (This by the way makes me think you did not get my description of the contents of my logic.) Considered as relatives, I take all the objects of the universe as correlatives as explained in Studies in Logic. Thus the converse to "man" or "man that exists" is "coexistent with a man that is". Every negro is a man. That is everything is coexistent with some man or other. That is any given negro.

The relatives of "among" and "in" differ but little. Among is simply the relation of the copula of inclusion. All men are mortal and all men are among mortals are pretty much the same. "In" differs as collective from distributive. This relation is the subject of careful analysis in my book on Logic, being of great importance in mathematics. Between is a complication of a transitive relation not equiparant. Take any such relation. If then A < B B < C, B is in reference to that relation between A & C. These things all receive full treatment in my book.

Yours faithfully
CSPeirce

Peirce to Russell. MS: RL 387.

Milford Pa 1894 Sep. 23.

My dear Judge

I have received yours enclosing copy of Hegeler's. I have no thoughts of pitching into them.

1. I am not so basely ungrateful as to wish anything but well to Hegeler & as for Carus he substantially said to me that he would have helped me if Hegeler would have let him, & I have no reason to disbelieve him.

2. If I wished to hurt them, I could not "antagonize" their work, because I eminently approve of it. True, in some details I think they are quite wrong; but those, considering the degree of culture of those they influence, are of little moment compared with the truths they inculcate.

3. Considered as philosophers, I think them both beneath criticism. Carus, who alone has any pretensions, is so confused so in leading-strings, that he never can make any impression on real thinkers.

---

7 some man or other] ~~a~~ \somet/ man \or other/ somet *[E]*
8 The relatives of] *This paragraph below deleted new paragraph:* ~~"Between" is an equiparant relative like "cousin" etc. It is its own converse.~~
9 mortal] *comma deleted*
10 collective from] *inserted above*
12–14 Between is … my book.] *inserted in-line and below*

4. If they were much stronger than they are, my business is to present the aspects of truth I think I see, & leave refutations to those who are less full of positive suggestions of new truth.

I really should like, as a matter of curiosity, to know what it is Hegeler has against me. If ever you learn, I depend on your informing me. I am inclined to think that it is because I did not finish my arithmetic within the time specified. But that is most unjust. First if that is the reason, he was bound to say so and hear what my defence was. In the second place he like many such men forget that it is necessary to eat in order to do anything. However, if I ever can learn what his grievance is I shall be glad.

I dont think it would do at all to get money from Halstead or any such person. True if he had some money he wanted to loan & thought the security good the thing might come in such a shape it would be unobjectionable. But it would be apt to be compromising. I have known Halstead to do a very indelicate thing. But I have had some friendly dealings with him about his Lobatchewsky etc.[55]

My special business is to bring mathematical exactitude,—I mean *modern* mathematical exactitude into philosophy,—and to apply the ideas of mathematics in philosophy. I am anxious to get up a sort of Academy at my place here for pursuing those studies. My house is large & I mean to make it larger & to build other houses. I want to get an autumn gathering here & when I can make some money to make a colony here to study philosophy in an exact & scientific way & to be self-supporting. Of course, I am in great danger for the moment of going to wreck; but if I escape I am likely in a few years to make a great deal of money. My bleaching process is of great value. You are one of the people whom I hope to attract here.

You speak of wanting to see the MS of my logic. I will send it to you shortly. The same mail that brought your last letter brought me a letter from a publisher about my preparing a small book on logic. What I would like to do would be to get your aid in revising my present book & in writing others. I want to complete this one in two more volumes & write two small ones besides.

One of my reasons for not sending you my MS at once is that I want to write to you about it & would like to have it by me for that. But I dont know exactly what I want to say & haven't time just now to consider it much. Perhaps I might as well dispatch it. I will try to think it over.

---

55 Peirce reviewed Halsted's 1891 translation of Lobachevsky's *Geometrical Researches on the Theory of Parallels* in *The Nation* (11 Feb 1892), and leading up to this review held correspondence with him on related matters (see *NEM* 2:x; also "Halsted" in Biographical Register).

---

9 ever] *1*ever

I really havent the least idea where food is coming from owing to the failure of crops etc. But I am rushing my geometry so I have not time to think of anything else. Perhaps I might like a Chicago publisher. I would like to deliver some lectures. I would like to do the literary & scientific columns of a newspaper. I would like pupils or boarders. I have a certain preference for not dying of starvation just now,—not a very strong one.

Your questions about relatives, so far from giving you away, show you are on a line of thought which may be pursued very profitably. If I can only get two or three profitable books launched & my process at work I shall see if I cant make it profitable for you to come here & help on the geometrizing of philosophy. I dont mean to shackle anybody with any condition other than that they shall work at the rendering of philosophy mathematically exact & scientifically founded on positive experience of some kind.

I write just before going to bed, pretty sleepy & pretty dull no doubt.

very faithfully
CSPeirce

P.S. In the general line of your questions, I may say that in my mathematical chapter, I discuss first "relations of covering," which are relations expressible in the form

$$\ell \mathfrak{f} \breve{\bar{\ell}}$$

To say that *A covers B* in reference to a given relation is to say that *A* is in that relation to everything to which *B* is in that relation.

Everything covers itself.

If *A* covers *B* & *B* covers *C* then *A* covers C.

If *A* covers *B*, then in a negative sense *B* covers *A*. That is, if *A* loves everything that *B* loves, then *B* non-loves everything that *A* non-loves.

I next consider "relations of superiority" which are relations of the form

$$(\ell \mathfrak{f} \breve{\bar{\ell}}) \cdot \ell \breve{\bar{\ell}}$$

that is relations complicate of a relation of covering and the negative converse of that relation. "To cover and not be covered by," is a relation of "superiority."

A can lift everything *B* can lift and some thing besides.

To say that *A* is superior to *B* in reference to a given relation is to say that taking any third thing, *C* either *A* is in the given relation to *C* or *B* is not in that relation to *C*, and *C can* be so chosen that both alternatives are true.

If *A* is superior to *B*, *A* is not *B*.

Superiority is a transitive relation.

If *A* is superior to *B*, then *B* is superior to *A* in reference to the negative relation.

Everything is $\ell ɟ \breve{\bar{\ell}}$ to itself that is

$$1 \prec \ell ɟ \breve{\bar{\ell}}$$

In a universe of more than one individual nothing is $1ɟ1$ to anything

$$1ɟ1 = 0$$

But it is *possible to find such a relation* ℓ that *everything* is

$$\ell ɟ \breve{\bar{\ell}} ɟ \ell ɟ \breve{\bar{\ell}}$$

of *itself*. Such a relation ℓ I call a "foundation of subsequence." For such a relation

$$\ell\breve{\bar{\ell}}$$
$$\therefore \ell\breve{\bar{\ell}}(\ell ɟ \breve{\bar{\ell}} ɟ \ell ɟ \breve{\bar{\ell}})$$
$$\therefore \ell(\breve{\bar{\ell}}\ell ɟ \breve{\bar{\ell}} ɟ \ell ɟ \breve{\bar{\ell}})$$
$$\therefore \ell(\breve{\bar{\ell}} ɟ \ell ɟ \breve{\bar{\ell}})$$
$$\therefore \ell\breve{\bar{\ell}} ɟ \ell ɟ \breve{\bar{\ell}}$$
$$\therefore \ell ɟ \breve{\bar{\ell}}$$

The relation $\ell\breve{\bar{\ell}}$ where everything is $\ell ɟ \breve{\bar{\ell}} ɟ \ell ɟ \breve{\bar{\ell}}$ of itself I call a *subsequence*. Every subsequence is a superiority.

The converse of a subsequence is a subsequence in negative reference.

The negative of a subsequence is a relation of covering in reference to the negative of the foundation.

For a foundation in reference to which things which cover one another are considered as identical, a relation of subsequence is said to be *linear*.

More and *less* are examples of linear subsequence.

In all cases, $(\ell ɟ \breve{\bar{\ell}})(\ell ɟ \breve{\bar{\ell}}) \prec \ell ɟ \breve{\bar{\ell}}$

For relations of subsequence $\ell\breve{\bar{\ell}}$ $\ell\breve{\bar{\ell}} \prec \ell\breve{\bar{\ell}}$

But if $\ell\breve{\bar{\ell}} \prec \ell\breve{\bar{\ell}}\ell\breve{\bar{\ell}}$ the relation of superiority is said to be *concatenated.*

A relation of superiority for which everything superior to another is superior to something without being superior to anything superior to that thing is termed a *deuterosis*. (Aristotle has a passage in which he attempts to analyze these ideas. It occupies several chapters of the *De Caelo*. Of course he was at an immense disadvantage but it is truly wonderful to see how far he succeeded in seeing.)

The above are the main conceptions from which my discussion sets out. We see that Numerical propositions are *spurious*

Some consult is not some consul

---

28 *deuterosis*] *deutereusis [E]* 32 Numerical] *after* ~~Sp~~

i.e. the consuls ≧ 2.
These are inferences from

Some consul is overbearing.
Some consult is not overbearing.

An antispurious proposition is like

Every Sun is every Sun

or Suns ≦ 1.
Of course we easily get higher spurious propositions, by inference from spurious & particular premises etc.

Multitude is the complicate of the possibilities and impossibilities of the forms of relationship between things of a class.

All differences of multitude are subsequences.

If one class has possibilities of forms of relations which another has not (and therefore necessarily has all those the latter has) it is said to be *greater* and its members *more*, the other *less*, and its members *fewer*

I guess this is all you care for today. Gooby!

Russell to Peirce. MS: RL 387. Stationery headed, "Francis C. Russell, | Attorney and Counsellor at Law, | 123 and 125 La Salle Street, | Bet. Madison and Washington Sts., | Suite 54, 55 and E."

Chicago Sept 27th 1894

Dear Mr. Peirce

Your last two letters of 22d & 23d inst respectively reached me together yesterday morning. I got your long letter of Sept 8th all right and my lack of acknowledging same must have been an oversight for it is a letter that I highly prize on account of its contents. In it you give a syllabus of your book on Logic &c. You also enclosed a power to me to use in pledging your copyright &c. This power I now return.

When I said that your writings were difficult, I meant *only* such as, say, your paper of 1870 in Memoirs A. Acad of A. & Sci. & the more technical of those in the Proceedings of same Society, and your papers in the Am. Jour. of Mathematics, your "brochure" on the logic of Relatives & the appendix B, in the Studies in Logic. As for your other writings, when you write discursively, you are charmingly easy consideration being had for the often very recondite topics you discuss. It may be: indeed it is quite likely, that were I as expert in mathematical reading as one should be who pretends to approach such papers I would not find any just cause of complaint. But so it is that I could never assimilate say the ideas of relative

6 Sun] *s*Sun   19 1894] 1890 *[E]*

composition (relative multiplication, progressive & regressive involution, transaddition infinitesimal and elementary relatives) until (having the good fortune to get a copy of DeMorgan's five papers in the Camb. Phil. Trans[56]) I studied DeMorgan's paper No IV, and until I made a diagram thus-like

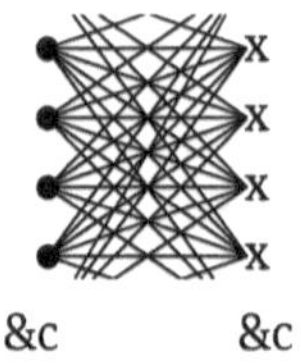

so that I could see what say lover (in whatever way he loves at all) of a servant &c meant.—

I will make every endeavor to find out what ails Hegeler in the matter of him and you and so far as I find out let you know. My own idea is that it is more than half simple misunderstanding and the rest some great big bugaboos spun out of his imagination. For instance in respect to the review of Mach he construed your letter (or one of your letters) to mean that if you got $ 200. for your review you would make no attack on it but if you had $ 100. only you would feel at liberty to pitch into it.[57] I dont believe at all that you even thought or dreamed of such an idea but he certainly did. Perhaps I had better say in the bluntest terms (which of course exaggerate but which make plain) that he has put you down as unscrupulous, unreliable, designing, and tried to "play him for a sucker". That is my idea of his state of mind and of course it wrongs you in the most shameful manner. Still I may not be on the right track. Except in casual allusions and quite rarely he has never said a word to me on the subject and it is hazardous to try and "pump" him. All I know I know through Dr Carus and his assistants at the office. If you should write to the Dr. on the subject so that he would feel at liberty to do so, I have no doubt that Mr. Hegeler would become acquainted with what you would say and if he could be *made to see* that he had done you wrong I am sure he would make amends, for to do right and not to do wrong is the very strongest impulse in Mr. Hegeler's nature.

56 "On the Syllogism" series, vols. 8–10 (1846–1864).
57 Letter or letters addressing this not found.

16 exaggerate] exaggerate *[E]*

Write to the Dr and tell him to tell you fully what the matter is. Tell him to put it without the *slightest extenuation*, that you want to have the matter put just as men put the same when they want to *abuse* one another; in all its particulars and in faithful compliance with the customs of vituperation. Tell him you ask this as a favor that one friend might render to another. Tell him you want an account in full of each and every item, instance, and particular of your offences and offencivenesses. It is certainly due you or anyone else that no one shall flatly refuse to reply to a civil communication without assigning some reason good or bad for such conduct.

I very much approve of the sentiments you express in your last letter and fully believe them to be the natural outpouring of your character. I imagine that your worst fault is to say things not intended to be so but which do in fact bite and rankle. You are so clear headed and able in every way that you hit the mark at every little chip leaving a blister.

When I expressed my burning desire to pursue your Logic I did not mean it as an out-feeler towards that end. I think I realize how much you would be inconvenienced by having it away from you for any great length of time, but if in the course of good fortune I only could get a chance at it if only for a short time, I should almost regard it as chance of a lifetime. I should sit up nights to devour it (in so far as I could devour it.)

I am very very grateful for the insight which what you give in your long letter (containing the Syllabus of your Logic) and in the letter with the long Postscript (about the relations of covering, superiority, subsequence &c) has given me into the nature of the subjects treated of. I feel more than ever that the advent of your methods and systems will mark an epoch in the history of philosophy, in the history of mathematics, in the history of physics, in the history of philology, in the history of morals, religion and politics. Pardon me for this long letter

Truly your friend
Francis C. Russell

Peirce to Russell. MS: RL 387.

Milford Pa | 1894 Oct 17.

My dear Judge

It occurs to me that I might write a number of articles about logical points & send them to you. Let you add & subtract till you were willing to father them (I am not the least bit sensitive about how much you alter) sell them to Carus as yours & send me what you might deem my share. I enclose one.[58] I would have

58 "Achilles and the Tortoise" (R 814–815).

had it copied in another hand; but I thought if you didn't alter it, it would be recognized as mine. Feeling so hampered, it's a rather poor thing. As to the questions, of course neither is logical. The distance of Achilles from the tortoise varies as the cube of the time. Argument 2 is only a different form of Zeno's that since things cannot be separated while together they cannot be separated if together. An object may be moving past a station while its velocity is zero. There is therefore an ambiguity. Velocity zero & not moving are different, for the former refers to the limit, the latter to the times coming to the limit.

very truly
C. S. Peirce

Alter just as much as you think best. I absolutely dont care or can only be pleased. Upset the whole doctrine, if you like, you are the sponsor.

Russell to Peirce. MS: RL 387. Stationery headed, "Francis C. Russell, | Attorney and Counsellor at Law, | 123 and 125 La Salle Street, | Bet. Madison and Washington Sts., | Suite 54, 55 and E."

Chicago Oct 27th 1894

Dear Mr Peirce.

I received yours enclosing the "Achilles & Tortoise" paper several days ago I ought to have noticed sooner. But a pressure of little affairs, visitors from the East &c &c have reduced me into delay.

I have altered the paper not a little mostly in the way of expansion. Very possibly you will be vexed to use it in the shape in which it will appear and think it a pity to spoil a good thing in such a way but I thought I had to alter it in order to avoid a "dead give away."—I have little doubt but that it will appear soon.

I will be very proud and happy to follow up this kind of work. I am sure I shall have the best end of the affair. Whatever is paid I will religiously remit to you but you must, of course you will, remember that the rates they pay me are very much less than those you can command.—For what little they have of late had occasion to pay me they have done so semi-occasionally within from a month to two months after the matters appeared. So you see how it will be.

I have of late as I had leisure been at work over what I would call the DeMorgan-Peirce Inference the "Syllogism of Transposed Quantity". I shall pretty soon be ready to submit some observations. It is quite possible that you will find them "all-wrong" and then I shall find out something that I do not at present

8 coming] *m*coming
3–10 Achilles from ... C. S. Peirce] *up right margin and upside down across header*
11–12 Alter ... sponsor.] *down left margin*
30 what] the what *[E]*
32 will] *inserted above*

see at all. It is a firm habit of my mind not to take what I do not *see* to be thus or thus on *anyone's authority*. That others see the matters otherwise is of course according to the quality of the authority more or less good reason to suspect that there are points one does not "tumble to". But until that illumination ensues I hold it the only *profitable* precept of mental economy to *stay along* with what one does *really see*. In the start of Emerson's Nature he asks "Why should not we also hold an original relation to the Universe?."[59] and Kant in his Doctrine of Method[60] points out that mathematics have the unique virtue that while fit to be taught in a dogmatic way they also are specially adapted to bring the mind into contact *with the very matters themselves*—so as to induce a *direct and original* communion. This *original relation* to the objects of knowledge is what I prize, and for the sake of it I am willing to appear crude and ill-informed and perhaps conceited. I take it that it is this very point that the scientific (in the sense of natural science alone) world has been so much insisting on.—

Well the present state of my mind in respect to the DeMorgan-Peirce Inference is to challenge the perfection of the accounts given of it.—

Cordially
Francis C. Russell

Peirce to Russell. MS: RL 387. Undated, in response to Russell's of 27 Oct 1894.

[ca. 1 Nov 1894]

My dear Judge:

I dont care a straw what you do about Achilles. But of course if the pay is going to dally like that, that must prevent any further work of that kind.

I dont know where anything to eat is to come from tomorrow. In the last 24 hours one cracker & a little oatmeal—My wife will go very soon.

I have now an elementary geometry all done & ready for the press, but no way of getting it to a publisher. It is no doubt worth enough cash down to make me all right. This place the agents say is worth $ 18 000. If I could get some photographs of it I could sell it—However, I did not mean to inflict my woes on you. What I was about to say was that if DeMorgan's Syllogism of Transposed Quantity is only objected to from a formal point of view, I have myself in my MS. improved on that, &

59 p. 5.
60 *Critique of Pure Reason*, Section II.

12 ill-informed] *after* ~~of~~
16 perfection] *after* ~~ways~~
26 an elementary geometry] $\text{a}^{\vee}\text{n}^{\vee}$ \elementary/ geometry

dont doubt more still is feasible. But if you mean to attack the distinction of finite & infinite, I am confident you will make a mistake.

My wife cannot live unless we get away at once. Now if the University of Texas would ask me down there to give some lectures *at once* & pay expenses, i.e. journey & hotel bill for self & wife, that would suit nicely! Do you think Halstead could manage that? My geometry being done now, one objection to getting his good offices if he has any disappears.

very truly
CSPeirce

Peirce to Russell. Postcard: RL 387. Recto headed, "Postal Card with Paid Reply | United States of America" with portrait at right.

[recto]

Nov 6 | 1894 *[stamped]*

Judge Francis C. Russell
123 La Salle St
Chicago Ill.
Suite 54, 55, E

[verso]

I will send you four or five chapters of my logic if you like, including what I say about the Syllogism of Transposed Quantity.[61] You to pay expressage both ways in full; and return to me within a week from date of sending.

C. S. Peirce

1895

Peirce to Russell. MS: RL 387. Undated, estimated based on context.

[ca. Summer 1895]

My dear Judge:

61 From "The Art of Reasoning" or "The Principles of Strict Reasoning," outlined in Peirce to Russell 8 Sept 1894.

5 Do] *over* You

I dont hear a word from you or from Carus. I suppose it is hot in Chicago.

I wonder if my article[62] about which I wrote to you ever came to hand to Carus. I am awfully in want of cash notwithstanding I have two great companies trying to get my patent & various other things that promise well. If I could get,—say x dollars—why even that would be something provided $x > 0$.

very faithfully
CSPeirce
Care Albert Stickney | 35 Nassau St. | N.Y.

Peirce to Russell. MS: RL 387.

New York 1895 Sep 20

Francis C. Russell Esq
125 La Salle St
Chicago

Dear Sir:

I beg you to step slightly out of your professional line (if it be such a step) in order to do me such an office as one lover of philosophy may hope another will be willing to do.

I expect in a few days to sign a contract for the publication of a schoolbook which is expected to bring in some income for a good many years. I also have understandings with publishers concerning three other books, all of which will be profitable. In addition to all these, I have my Logic which has been completed & largely rewritten, & which before I get through with it, will, I hope, be a work to do a great deal of good & to receive a due reward.

Now, what I want to do is, to make all the moneys payable to you, as a sort of trustee, or receiver (you know the legal title, but I dont) to pay off all my creditors until they are completely paid, with interest, before I receive anything. I cannot see that there is any moral or legal objection to my doing this.

I should want you, upon receipt of each dividend from the publishers, first, to pay your own commissions & fees, & then to proceed upon the following principle (subject to special exceptions in the interest of my creditors); namely, taking the names in alphabetical order, to pay the first man, next after the last who has received anything, the whole amount of the dividend, if so much is due him. If

62 Possibly "Achilles and the Tortoise," sent to Russell 17 Oct 1894.

28 dividend] *before deleted comma*

something remains over due to him, he is to wait till his turn comes round. If there remains a balance of the dividend, you pay it to the next man.

The exceptions are, 1st, cases in which the immediate payment of a given creditor will release something which will go to increasing the payments to other creditors. 2nd, I am much tempted to put one or two persons first who have brought themselves into embarrassments by helping me. I should also like very much, if I could, to put Hegeler ahead, because of his peculiarities of one kind & another, & also because that would perhaps induce him to name the amount due him.

The reason I ask you to undertake this is, first, that it would insure my creditors being paid, whether I live to pay them or not; and this would conduce infinitely to my peace of mind, and consequently to the objects of my life; second, because I am quite decided to have some lawyer do it, & I would rather it should be a man who has some comprehension of my work; third, because I think the announcement of a perfectly fair & impartial arrangement would afford a pleasure to my creditors. I wish you to be adequately remunerated for your services & my rule given above is intended to make your task as little troublesome as possible.

Of course I should send you a list of claims & if any were accidentally omitted they could be subsequently inserted. I shall be very grateful if you consent.

very truly
C. S. Peirce

Russell to Peirce. MS: RL 387.

Chicago Sept 23d 1895

Dear Mr Peirce.

I am *very* greatly pleased to hear from you again and to hear that material prosperity again turns its face towards you.—I saw the newspaper reports of the seizure of your library and that day or the next Dr. Carus was up to Chicago and called on me. I was so full of sorrow and a kind of resentment that when the Dr treated the matter in an "all-to-be expected"-fashion, I let my tongue fly on the burning shame it was that American Science and American Philosophy and American Scholarship should let such a thing occur. This led to some German Chauvinism on his part and to a reply proper thereto on mine. I also told him that he himself had altogether too soon forgotten in his *good fortune* his own former estate and had apparently hardened his appreciation and sympathies to the "*business*" (How I

3 immediate] *inserted above*
8 name the amount due him.] *above* ~~come to some terms.~~
14 pleasure] *above* ~~satisfaction~~
15 adequately] *after* ~~full~~
17–20 Of course … C. S. Peirce] *up left margin of first page*

hate that epithet) standard.—He must have taken some offence at my freedom for he has not cultivated my society as formerly and I have not felt justified in offering any thing in the nature of an apology so for many months we have grown more and more apart.—Indeed I have not even met him for several months.

I have always felt that if Dr Carus was really and fully friendly to you or even if he were able to appreciate your measure he would and could have "steered" Mr Hegeler into the mood useful to both of you and especially useful to Mr. Hegeler in his (as I must believe) magnanimous purposes.—

In reference to the matter of business broached by you I have only to say that I feel greatly honored by your proposal and that I will gladly serve you in the way you mention or in any other way you desire if it is within my power.

You may rely on me to stand your friend in every exigency that I can forecast. That I have not been of much if any good to you heretofore has not been for lack of any good will or effort but solely because this odious age is blind drunk (or daft) over the rags in the gutter and will not or cannot give just one little upward look to see the glorious crown that all the good angels are tendering.

I am *very* cordially
yours
Francis C. Russell

Russell to Peirce. MS: RL 387.

Chicago Sept 25th 1895

Dear Mr. Peirce

Today Dr Carus called again on me and we had a general overhauling of our words and conduct. I feel sure that he has little relished our *strained* relations. He is I guess sometimes puzzled how to cut in between Mr Hegeler and thou against whom the latter has become offended. At any rate the Dr & myself wiped the slate and took salt together. He wants me to give him some more reviews. In particular he wants a review of Schröeder's last volume of the "Algebra der Logik."[63] I told him I was totally incompetent to any such a task and that there was only one man in the whole world who could write any kind of sense on the Logic of Relatives. Of course he knew I meant you. To this and with that infernal Chauvinistic spirit (Ger-

63 *Algebra und Logik der Relative* (1895).

13 heretofore] heretofor *[E]*
13 for lack] *after* ~~of an~~
23 again on me] *inserted above*

mans of course) which you must have noticed he as much as said that Schröeder seemed to be the only one as yet who *had any doctrine of relatives and its Logic.*—

Now *I urgently wish* you could find some way to *enable* me to write such a review or something that will pass for it without my making a *fool* of myself.

Cordially
Francis C. Russell

Peirce to Russell. MS: RL 387. Stationery headed, "Express Address. | Port Jervis, N.Y. | Wells, Fargo Exp. Co. | C. S. Peirce."Arisbe" | P.O. Milford, PA."

1895 Oct 4

Dear Judge Russell:

Returning home last night I found your first letter, & today received your second. I will answer this last first.

I think it was the summer of last year that I wrote to you asking if you would like to see certain chapters of my logic, but never got any reply.[64]

It seems to me the best way to enable you to review Schröder will be to send you those chapters. When you have read them you had better send them to him. If I can get a publisher, I should, before going to press, prune them of all sorts of personalities & of most of the side remarks. The chapter on the Logic of Quantity has been since repeatedly rewritten. As sent to you, I don't know whether you can make head or tail of it.

Schröder himself does not agree with Dr. Carus.

See, for instance, what he says of my work in the Mathematische Annalen.[65]

He is very generous & scientific. Dr. Carus is the reverse, narrow and theologic,—though his theology is negative. However, he is rich.

Schröder's chief criticism of my work is that I have left undeveloped the methods of the algebra. I acknowledge that his contributions in this direction are of much value. But it is easily exaggerated. The principal use of an algebra of logic is not to solve problems with it, but to make use of it to *analyze reasoning* & discover its essence. A regular method is not necessary for that; yet you will see I have one.

In short, I am delighted with Schröder. Yet I dont think his work on relatives compares in importance with my own.

---

64 Peirce to Russell 6 Nov 1894.
65 In "Note über die Algebra der binären Relative," vol. 46, pp. 144–158.

---

13 the summer of last year] ~~last~~ \the/ summer \of last year/
15 will be] *above* ~~is~~
18 Quantity] *after* ~~Numb~~

Had Hegeler kept his positive promise to allow me six articles in the Monist to develope my philosophy, in which I do not count my defence against Carus[66] (to whose subsequent attack no reply was permitted) the 6$^{\underline{h}}$ Article, which would have been the keystone of the whole, would have related to that branch which I variously call Second Intentional Logic, Objective Logic, & Pure rhetoric.[67] By this I mean the doctrine of the Evolution of thought. For instance, take the Theory of Functions. Here is a branch of mathematics severed from the real world, & working out hypotheses. Yet the development has been dominated by an order; and if the mathematicians of Mars have turned their attention that way, they must have made pretty near the same hypotheses, and worked the subject up in somewhat the same way. Now is there any law by which a given idea calls up another. Hegel says there is. It is his movement of dialectic. *I* say there is, too; and I grant that Hegel's formula is a case of the true formula. But it is nothing more. How does an idea in the mind call up another not in the mind? Always, according to me by attraction,—by evolutionary love. But in numerous cases this attraction takes certain degenerate forms. Of these Hegel's formula is one. Another is what is ordinarily called Generalization. Look at Generalization from the point of view of the Logic of Relatives. What is it which in the logic of relatives fulfils the same function as Generalization in non-relative logic. It is the passage from an idea to the recognition of that idea as belonging to another idea as its complement which with it goes to form a *System*. In that sense, all suggestion is generalization.

I give this as a specimen of the sort of use which it seems to me the Logic of Relatives chiefly is. It widens our ideas.

It seems to me Schröder is too mechanical and cares too little for such applications. But mind, it is merely a question of a little more or a little less between us—Schröder leans chiefly toward investigations whose value I concede. I lean more toward other applications of the theory of relatives, whose value *he* concedes. Real *philo-sophers*, or sophophiles, do not desire that all should look through their spectacles, as the theologue insists on men's doing, or else he will try to take away their bread & butter.

---

66 "Reply to the Necessitarians," followed by "The Founder of Tychism" (July 1894).

67 The sixth was to be "A General Sketch of a Theory of the Universe," as outlined in Peirce to Hegeler 7 March 1893.

---

3 to whose] to ~~which~~ \whose/ ~~of and no re~~ *[(first revision)* to ~~which~~ *(second revision)* ~~of~~ *(third revision)* ~~and no re~~ *(fourth revision)* to ~~which~~ \whose/*]*

22 specimen] speciment~~t~~

23 It] I *[E]*

I shall soon sign a contract for my Elements of Mathematics,[68] putting in your name; and then shall send the list of creditors. This does not mean I am flourishing but means I am arranging for my starvation. I shall get my creditors paid. My ideas are immortal. They are all right. My wife will find stronger friends.

very faithfully,
C. S. Peirce

Peirce to Russell. MS: RL 387. Stationery headed, "Alexander Stewart, Prop. | Stewart House, | Bachelor Hotel, | Broadway & 41st St."

New York, Oct 7 1895

My dear Judge:

I have written to a farmer neighbor of mine to go to my house find & send to you 5 chapters on the Algebra of the Copula

Aristotelian Syllogistic
Modifications thereof
Boolian Algebra
Logic of Quantity.

I want you to remember that this was written last summer & all about relatives has been written for many years; and this I can *prove*.

I want to call your attention to the definition of negation by an infinite series.

There are some other things of interest about the algebra of the copula.

In the Aristotelian syllogistic, observe my *two* pairs of definitions of *not* & *some*.

According to one

not not makes *not*
some some makes *some*

According to the other pair of definitions, *perfectly clear*,

not not makes affirmative
some some makes *all*.

In the Boolian algebra, observe my code of rules for non-relative addition & multiplication. Observe the importance of multiplying a premise into itself.

In the Logic of Quantity, a badly written chapter, (since twice rewritten to great advantage) observe the necessary evolution of the ideas.

68 Never published by Peirce; in *NEM* 2:1–232.

24 *not not*] *above* ~~t two nots~~ 25 *some some*] *above* ~~tw~~

I will now call your attention to some points (all small) in Schröder's 3rd Vol.[69] The first two chapters (about the others I have not time to write tonight.)

p. 1. "Hauptförderer" is pleasant. Schröder is a gentleman, for all he is a German.

p. 2. He might have known his accent was going to make trouble

p. 3. "Binary" is a better word than "dual".

p. 4. I dont much like the word "elements" as synonym for individual

A, B, C, etc are *indices*. As such, can have no generality. Hence, two cannot be identical & every one exists.

p. 5. Spurious propositions, as I show in my Logic of Quantity are arithmetical. Or arithmetical propositions are nothing but spurious logical forms.

But in general in the absence of any special qualification, the individuals of the universe are,—that is, are possibly,—that is all the "are" can mean,—*innumerable* (not merely inenumerable). Now I think a good sign to denote the aggregate of all the individuals of a possibly innumerable logical universe is ∞. I dont see the beauty of calling it 1.

p. 8 Observe that there is a world of difference between aggregating two given classes or individuals, and describing a symbol as signifying that something is to be aggregated for every one of the innumerable individuals of the universe. Addition gives no result till completed. The summation of an infinite series cannot be completed; and the final result, or limit, is evaluated by reasoning about what it would have to be, *not* by performing the addition. Therefore $\sum_i$ is not, strictly speaking, a sign of logical aggregation nor is $\prod_i$ a sign of logical combination. They are signs of the limit of endless series of summations & combinations. Here and again p. 17 Schröder seems to overlook the fact that they are not signs of logical addition & multiplication.

p. 9. No doubt it is better a sign should be symmetrical or otherwise according as the relation signified is so or not.

I believe it was I who first urged this; though I never proposed to have it override all other considerations. Hence, it would be better, instead of the colon to use Stokes's solidus

$i/i \quad i/j \quad i/k$
$j/i \quad j/j \quad j/k$
$k/i \quad k/j \quad k/k.$

p. 10 The word "array" is taken from Scott's treatise on Determinants.[70]

69 *Algebra und Logik der Relative* (1895).
70 Robert F. Scott, *Treatise on the Theory of Determinants* (1880).

2 two] *above* ~~four~~   7 individual] idividual *[E]*

p. 12. In Latin, *idem* is not used as a prefix and *ipse* hardly ever used. *Se* is the more proper one; but that is rare too. We make up Greek words freely to suit. And that is quite in the genius of the Greek language. Even such words as "thermic" perhaps would not greatly shock a Greek, though it might make him laugh. But Latin is different, and when one reads latin every day, *idem-relative* is too jarring. *Aliorelative* is just supportable.

Now as to relative multiplication. I write it without a sign. The sequence is in itself asymmetrical. But there are two mighty strong reasons for my practice, quite overweighing those of Schröder.

1st. It is a psychological law that men in all languages write one thing directly after another to indicate the correlate of a relation.

2nd. It is the universal practice of mathematicians to do so. We have not only functional multiplication which is precisely the relative multiplication, but we have also the usage of quaternionists, writers on the theory of matrices, etc.

As to my comma, I gave it up and prefer the dot. But observe that my comma is explained in my paper of 1870[71] as meaning precisely what his apostrophe after 1 means

m,b is "man-that-is black."

p. 15. *Uninary* (not *urinary*) is a sweet pretty word. Let's adopt it.

p. 25 Observe that his apostrophe means two separate & distinct,—nay opposed— things

1' is "what is *that is*"

0' is "what is nothing *unless different from*"

p. 26. This 0 = is a case of Schröder nodding. He ought to write, of course, $0 = \prod_i \prod_j i/j$, that is what combines all the relations that anything has to anything, i.e. what is *inconsistent with.*

p. 33. I always curve † in writing it, as Schröder ought to have seen in my MS. I dont make the little curl he prefers, but a longer one

$$x \dagger y.$$

p. 39 His first equation (y) is wrong. It is not true that

$$\sum_u \sum (A_u \prec B) = (\prod_u A_u \prec B)$$

All that is true is

$$\sum_u \sum (A_u \prec B) \prec (\prod_u A_u \prec B)$$

71 "Description of a Notation for the Logic of Relatives."

1 ever] every

20 two] to *[E]*

His (𝔣) is *correct*, though in his table of Errata he says it is not.
p 45 A good example.
p. 50. I maintain that ∞ means precisely "coexistent with" and *not* thought together with. Of course when I say "coexistent" I mean "compossible" if we have to do with a universe of possibles.
p 69. I think the second convention ill-advised.

very faithfully
CSPeirce

Peirce to Russell. MS (copy by Russell): SIU 13.9. LPC (copy of Russell's MS by OCP, TS missing): SIU 91.25. Excerpt of a missing letter copied by Russell and sent to Miss Cook of OCP 10 Nov 1909 to be used by Carus in his "The Nature of Logical and Mathematical Thought" (see Carus to Peirce 26 Nov 1909). Carus sent proofs of the excerpt to Peirce 17 Dec 1909, which he edited and returned ca. 25 Dec (see Appendix, "Peirce and Carus"). The excerpt was included in Carus's paper in *The Monist* (Jan 1910), p. 45.

The text that follows comes from Russell's MS copy made from the original. Peirce's edits as shown in the Appendix are not reflected here because they are believed to be emendations preparatory for publication and not restorations of the original text. Although undated, the excerpt is estimated to be from around this time, when Peirce began discussing Schröder's 1895 *Logik der Relative* with Russell.

[early October 1895]

[…] Schroeder then passes on to relations characterized by such formulae as
$\bar{x}\,\bar{x} \mathrel{-\!\!<} x$, $\breve{x}\,\breve{x} \mathrel{-\!\!<} x$, $\breve{x}\,\breve{\bar{x}} \mathrel{-\!\!<} \breve{\bar{x}}$, $\bar{x}\,x \mathrel{-\!\!<} x$

I studied these and others before I first took up the general study of relations, in the course of investigation into the consequences of supposing the laws of logic to be different from what they are.

It was a sort of non-Aristotelian logic in the sense in which we speak of a non-Euclidean geometry. Some of the developments were somewhat interesting, but not sufficiently so to induce me to publish them. The general idea was of course obvious to anybody of sufficient grasp of logical analysis to see that logic reposes upon certain positive facts, and is not mere formalism. I only mention it because a writer of another character afterward suggested such a false logic, as if it were the wildest lunacy instead of being a plain and natural hypothesis worth looking into and which as the investigation turned out lacked little of being worth printing. *[ ]*

---

22 $\bar{x}\,\bar{x} \mathrel{-\!\!<} x$, $\breve{x}\,\breve{x} \mathrel{-\!\!<} x$, $\breve{x}\,\breve{\bar{x}} \mathrel{-\!\!<} \breve{\bar{x}}$, $\bar{x}\,x \mathrel{-\!\!<} x$] In the left margin is Russell's instruction regarding this formula:

To the typewriter.

You had better write these *xx* (*x* &c using the first horn of the parenthesis ~~and~~ in place of the $\mathrel{-\!\!<}$ and then with the pen write over the proper *x*'s the macron's or the breve's or the combined macron's and breve's.

Russell to Peirce. MS: RL 387.

Chicago Oct 11$^{\underline{\text{th}}}$ 1895

Dear Mr Peirce

I stop in the midst of some hurry to send you an acknowledgment of the receipt of chapters VIII to XIV & chap either XV or XVII (I cant just make out what is intended but it is the chapter on "Quantity") of your Book on Logic—I cannot tell you how *very very* much gratified I am. I sat up last night till nearly three o'clock this morning making a cursory dip into the whole. I think it is a work that runs over with new truths & views of the most valuable sort. I never cease to wonder at the masterful hold you have got upon the very foundations of our mental economy. How you can bear to defer the day when through you the world shall begin to know something is also wonderful. Those who have been scanning the intellectual horizon of the the cisatlantic world looking for American leadership, little know that not only for the western continent but for the entire earth the great prophet and his gospel is already extant.

very cordially
Francis C. Russell

Peirce to Russell. MS: RL 387.

158 W 50^{h} St New York | 1895 Nov 5

My dear Judge:

In writing about Schröder's first four lectures, I forgot to mention that the propositional forms at the top of p 149 are spurious and anti-spurious, concerning which I have given sufficient hints in the Johns Hopkins Bulletin.[72]

His fifth lecture reminds me of a question I sometimes put to children: "If you had your choice of two things, which should you like best?" His problem, to find a general solution of $Fx = 0$, means: Given a problem (no matter what), required to find the general rule of proceedure for solving it. The solution he gives is

$$x = \sum_u \left[a(\infty F_u \infty) + u(0 \text{ɟ} \overline{F}_u \text{ɟ} 0)\right]$$

72 Likely "On the Relative Forms of Quaternions," abstract in *JHU Circulars* (1882); in full *W* 4:334–335.

5 XVII] XVII~~I~~
9 new truths & views] *above* ~~novelties~~
13 the cisatlantic] \the/ ~~trans~~\cis/atlantic
13 that] *before* ~~already~~
26 Given] *g*Given

This means: Let *a* be any proceedure which will answer the purpose. Then you either proceed in that way or in any way whatever which will answer the purpose. This may teach something about the system of algebra (and *does* teach something important) but it teaches nothing about logic. So a child might reply to my above question that he would prefer the one best adapted to please him.

His sixth chapter is good; but is susceptible of simplification. Thus on p. 212, he has a time proving

$a\dagger 0 = (a\dagger 1)T\dagger 0.$

Now I fully approve of his pentagrammatic method. But in this case, ordinary methods will answer. Let c be any relative such that $\infty c\dagger 0$. Then starting with

$(a\dagger\breve{\bar{c}})c\dagger 0$ we have

$(a\dagger\breve{\bar{c}})c\dagger 0$

$\prec a\dagger\breve{\bar{c}}c\dagger 0$

$\prec a\dagger T\dagger 0$

$\prec a\dagger 0$

On the other hand since $\infty c\dagger 0$

$a\dagger 0$

$\prec(a\dagger 0)(\infty c\dagger 0)$

$\prec(a\dagger 0\infty)c\dagger 0$

$\prec(a\dagger 0)c\dagger 0$

$\prec(a\dagger\breve{\bar{c}})c\dagger 0$

Thus $a\dagger 0 = (a\dagger\breve{\bar{c}})\,c\dagger 0$

We might generalize further, and say that if $qc = \breve{\bar{p}}$,, then $a\dagger p = (a\dagger pq)c\dagger p$

Now *T* is such a relative that $\infty T\dagger 0$ That is, everything is other than some thing. In like manner because every person has a mother, to be the man-servant of every person is the same thing as to be to every person the man-servant of every person mothered by some mother of that person.

It was partly because of the complexity of the formulae in the Sixth Lecture that I long ago preferred to return to the other mode of writing. Namely

For $a\dagger 0\prod_j a_{ij}$ and for $\infty c\dagger 0\prod_g\sum_h c_{hg}$

Multiplying $\prod_g\sum_h\prod_j a_{ij}c_{hg}$

And increasing $\prod_g\sum_h\prod_j(a_{ij}+\bar{c}_{hi})c_{hg}$. What could be simpler.

3–4 (and … important)] *inserted below*

4–5 So a child … please him.] *inserted in-line and below*

16 $\infty c\dagger 0$] *above* ~~multiplying $a\dagger 0$by~~

21 $\prec(a\dagger\breve{\bar{c}})c\dagger 0$] *above* ~~$\prec(ag)$~~ $\breve{c}e$

23 that if $qc = \breve{\bar{p}}$,] that if ∞ ~~$\prec qc\dagger p$ and $pq\prec\breve{\bar{c}}$ then~~ \ $\breve{\bar{p}}$ ~~$\prec qc$~~/ qc = $\breve{\bar{p}}$,

25 person] *above* ~~man~~

26 person] *above* ~~body~~

27 person] *above* ~~thing~~

28 was partly] *above* ~~is~~

28 Lecture] *after* ~~Chapt~~

On the other hand, starting with the last, and diversifying j (which we do by squaring the whole)

$$\prod_g \sum_h \prod_j \prod_k (a_{ij} + \bar{c}_{hk}) c_{hg}$$

Now identifying k with g

$$\prod_g \sum_h \prod_j (a_{ij} + \bar{c}_{hg}) c_{hg}$$

Whence

$$\prod_j a_{ij}$$

Nothing could be much more obvious than the general formula of which his (21) is a special case. Still, if handled awkwardly, it might no doubt be puzzling; so it is just as well to write it. But I think it illustrates the superiority of the algebra which I prefer. Still, I admit the other has its charms.

As I write, I hear the newsboys calling out that Tammany has won by 40 000. Naturally. They were trying to make New York too good. The majority of the voters think it for their interest in their business to have it a hell of a place. Besides, it is cruel to shut off the Sunday beer. The gambling houses were never running more openly than under Teddy Roosevelt. In that respect, it is a perfect Chicago. I rather like his pluck, and of course naturally sympathize with efforts to stop corruption. Still, they really are such hypocrites, I cant say I feel *very* bad.

What I want to say is, that I am to fill 2 columns a week in the Pike County Press, of which one will be on the Logic of Relatives in its *general* and *philosophical* bearings. Subscription $ 1.50. Cant you get me some subscribers? Will any dealer take a few?

very truly
C. S. Peirce
158 W 50$^{\underline{h}}$ St | New York

P.S. Schröder's Sixth Lecture relates to a subject of the very first importance, *if we are to use the algebra he prefers*. It is a complicated doctrine necessitated merely by that preference. I shall copy from my old notes some formulae to show he has not carried his analysis back to first principles. In giving them in your review, you can say they are mine; but need not say where you find them. If asked, you might say you have had access to some of my MSS. If further pressed, refer to me. I have a book which I have kept for 30 years with every page dated. It affords conclusive evidence of my long priority.

---

8 much] *inserted above*
12 newsboys] news*p*boys
27 necessitated] *after* ~~wholly~~

On p. 210 formulae (8) to (12) come simply from

$$a \dagger 0 \prec a \dagger T \prec a$$
$$a \prec a1 \prec a\infty$$

One advantage of writing ∞ for his 1, is that it suggests

$$\begin{array}{ccccc} & & 1 & & \\ & \curlywedge & & \curlyvee & \\ 1 & & & & \infty \\ & \curlyvee & & \curlywedge & \\ & & T & & \end{array}$$

In the other notation

$$(a \dagger 0)_{ik} = \textstyle\prod_j a_{ij} \qquad [a\infty]_{ik} = \textstyle\sum_j a_{ij}$$

Now obviously $\prod_j a_{ij} \prec a_{ik} \prec \sum_j a_{ij}$.

The other formulae on the same page strongly illustrate the imperfection of the algebra he prefers. Thus (13) is a special case of

$$(a \dagger 1) \cdot b \prec (a +\!, b) \dagger 0$$

With the other notation this is instantly seen. For

$$\textstyle\prod_j (a_{ij} +\!, 1_{jk}) \cdot b_{ik} \prec \prod_j a_{ij} +\!, b_{ij}$$

But in the algebra he prefers new fundamental formulae are requisite; and these cannot be put in their most general shapes, because that algebra does not permit us to express them so. For example $(ab \cdot c)d$ equals something which we might write $\underbrace{a \cdot \overbrace{c\,\breve{b}}}\,\infty \cdot d$. That is,

$$\begin{aligned}[(ab \cdot c)d]_{i\ell} &= \textstyle\sum_k \sum_j a_{ij} \cdot b_{jk} \cdot c_{ik} \cdot d_{k\ell} \\ &= \textstyle\sum_j \sum_k a_{ij} \cdot c_{ik} \cdot \breve{b}_{kj} \cdot d_{k\ell}\end{aligned}$$

This is not expressed by $\{(a \cdot c\breve{b})\infty\} \cdot (cd)$ nor by $(a\infty) \cdot c(\breve{b}\infty \cdot d)$

In short, the algebra does not allow us to express it with the letters $ac\breve{b}d$ in that order. To deduce the formulae in the lower part of p. 210 we ought, first, to use the symbol ℓ' for lover of itself. Then we have $\ell' = (\ell \cdot 1)\infty$

For $[l']_i = \ell_{ii}$ and $[(l \cdot 1)\infty]_{ik} = \sum_j \ell_{ij} \cdot 1_{ij} = (\ell_{ij} \sum_j 1_{ij}) = \ell_{ii}$

We next have $(ab \cdot c)\infty = (a \cdot c\breve{b})\infty$, of which a corollary is,

$$a\infty = (a \cdot a)\infty = (1a \cdot a)\infty = (1 \cdot a\breve{a})\infty \prec a\breve{a}\infty \cdot 1\infty = a\breve{a}\infty \cdot \infty = a\breve{a}\infty$$

and $a\breve{a}\infty \prec a\infty\infty = a\infty$ whence $a\infty = a\breve{a}\infty$. In the other notation this is evident at a glance.

$$\textstyle\sum_j a_{ij} = \sum_j \sum_k a_{ij}a_{kj}\ ;\text{ for } a_{ij} \prec \sum_k a_{kj}$$

That $a \cdot 1 = \breve{a} \cdot 1$ is evident in the other notation; for $a_{ij} \cdot 1_{ij} = a_{ji} \cdot 1_{ij}$ But in the algebra used by Schröder, it is quite difficult to prove it.

Another corollary of $(a \cdot c\breve{b})\infty = (ab \cdot c)\infty$ is $(a \cdot \infty\breve{b})\infty = ab\infty$

Another is $\{(a \cdot \breve{b})\infty\} \cdot 1 = \{(ab \cdot 1)\infty\} \cdot 1 = (ab)' \cdot 1 = (ab) \cdot 1$

We have $(\bar{a} \cdot \bar{b})\infty = (\bar{a} \cdot \bar{b})(T \mathbin{+\!\!,} 1) = (\bar{a} \cdot \bar{b})T \mathbin{+\!\!,} (\bar{a} \cdot \bar{b})1 \prec \bar{a}T \mathbin{+\!\!,} \bar{b}1 \prec \bar{a}T \mathbin{+\!\!,} \bar{b}$

The contraposition of this is $(a \dagger 1) \cdot b \prec (a \mathbin{+\!\!,} b) \dagger 0$ Putting $b = a (a \dagger 1) \cdot a \prec a \dagger 0$

To show the converse

$$(a \dagger 0) = (a \dagger 0) \cdot (a \dagger 0) \prec (a \dagger 1) \cdot a$$

These two give his (13); and (14) and (15) are similarly deducible. To get (16), on the one hand, by $\dagger 0 \prec x$, we have

$$a\infty \dagger 0 \prec a\infty$$

and on the other hand

$$a\infty \prec a(b \dagger \breve{\bar{b}}) \text{ Put } b = \infty \text{ and this becomes}$$
$$a(\infty \dagger 0) \prec a\infty \dagger 0$$

Equation (17) rests upon the principle that

$$ab \dagger c \prec a\infty \mathbin{+\!\!,} (0 \dagger c)$$

And this again upon

$$(p \mathbin{+\!\!,} q) \dagger c \prec p\bar{c} \mathbin{+\!\!,} (q \dagger c)$$

And this again upon

$$p \dagger c \prec p\bar{c} \text{ or else } \infty \prec c$$
$$(p \dagger c)_{ik} = \textstyle\prod_j p_{ij} \mathbin{+\!\!,} c_{jk} = \prod_j p_{ij}\bar{c}_{jk} \mathbin{+\!\!,} c_{jk}$$
$$\prec \textstyle\sum_\ell p_{i\ell}\bar{c}_{\ell k} \mathbin{+\!\!,} \prod_j c_{jk}$$
$$\prec \textstyle\sum_\ell p_{i\ell} \mathbin{+\!\!,} \prod_j c_{jk}$$

Having thus got

$$ab \dagger c \prec a\infty \mathbin{+\!\!,} (0 \dagger c)$$

we put $c = 1$ Now $0 \dagger 1 = 0$
Hence $ab \dagger 1 \prec a\infty$
and $a\infty \dagger 1 \prec a\infty$
On the other hand $\infty \prec (\infty \dagger 1)$

---

21 $p \dagger c \prec p\bar{c}$ or else $\infty \prec c$] *below* ~~$p \dagger c \prec p\bar{c} \prec (0 \dagger c)$ i.e.~~

Hence $a\infty \prec a(\infty \mathfrak{f} 1) \prec a\infty \mathfrak{f} 1$

All this is perfectly evident with the other notation. For

$$\textstyle\prod_k \sum_j a_{ij} b_{jk} \dagger c_{k\ell} \prec \sum_i a_{ij} \dagger \prod_k c_{k\ell}$$

Equations (20) are tough even for the powerful notation. There his pentagrammatic method shows well. I will first show that

$$(a\mathfrak{f}1)\infty \cdot (\bar{a}T \dagger a) \prec (a\mathfrak{f}1)T$$

This is strictly speaking not true. It is necessary to be quite accurate to assume that there are at least two objects in the universe

$$\infty T \cdot (a\mathfrak{f}1)\infty \cdot (\bar{a}T \dagger a) \prec (a\mathfrak{f}1)T$$

Without that assumption, a conclusion results; but it cannot be expressed in the form of algebra preferred by Schröder, except by writing

$$(a\mathfrak{f}1)T \dagger (0\mathfrak{f}1)$$

We start with

$$\begin{aligned} &[(a\mathfrak{f}1)\infty]_{i\ell} = \textstyle\sum_q \prod_u (a_{iu} \dagger 1_{uq}) \text{ . We square this} \\ &\textstyle\sum_q \prod_u \prod_v (a_{iu} \dagger 1_{uq})(a_{iv} \dagger 1_{vq}) = \\ &\textstyle\sum_q \prod_u \prod_v (a_{iu} \dagger a_{iv} \dagger 1_{uq} 1_{vq}) \\ \prec &\textstyle\prod_u \prod_v (a_{iu} \dagger a_{iv} \dagger 1_{uv}) \end{aligned}$$

This may be expressed in the other algebra

$(a\mathfrak{f}1)\infty \prec \{(a\mathfrak{f}1) \dagger a\}\mathfrak{f}0$. This is the first step. In ordinary language whatever loves everybody except one individual stands to every individual in the relation either of loving it or of loving everybody else.

We now go on

$$\begin{aligned} [\bar{a}T \dagger a]_{i\ell} &= \textstyle\sum_r (\bar{a}_{ir} \cdot T_{r\ell} \dagger a_{i\ell}) \\ &= \textstyle\sum_r (\bar{a}_{ir} \cdot \bar{a}_{i\ell} \cdot T_{r\ell} \dagger a_{i\ell}) \end{aligned}$$

Multiply this by the result of the first step.

$$\textstyle\sum_r \prod_u \prod_v (\bar{a}_{ir} \cdot \bar{a}_{i\ell} \cdot T_{r\ell} \dagger a_{i\ell}) \cdot (a_{iu} \dagger a_{iv} \dagger 1_{uv})$$

Identify u with r and v with ℓ, and we get

$$\textstyle\sum_r (\bar{a}_{ir} \cdot a_{i\ell} \cdot T_{r\ell} \dagger a_{i\ell}) \cdot (a_{ir} \dagger a_{i\ell} \dagger 1_{r\ell}) \prec a_{i\ell}$$

---

7 This is] *above* ~~The truth~~

7–8 assume … universe] above and below ~~note~~

10 results] *above* ~~is poss~~

20 everybody] body *above* ~~thing~~

In the other algebra,

$$(\bar{a}T \dagger a) \cdot \left[\{(a \mathfrak{f} 1) \dagger a\} \mathfrak{f} 0\right] \prec a. \textit{ Second Step.}$$

Multiply $a_{i\ell}$ by the identical proposition

$$\textstyle\prod_w (T_{w\ell} \dagger 1_{w\ell}). \text{ We have}$$
$$\begin{aligned}\textstyle\prod_w (T_{w\ell} \dagger 1_{w\ell}) a_{i\ell} &\prec \textstyle\prod_w (T_{w\ell} \dagger 1_{w\ell} a_{i\ell}) \\ &\prec \textstyle\prod_w (T_{w\ell} \dagger 1_{w\ell} a_{iw}) \\ &\prec \textstyle\prod_w (T_{w\ell} \dagger a_{iw})\end{aligned}$$

In the other algebra

$a \prec a \mathfrak{f} T$ which is a regular formula.

We have thus found

$$(a \mathfrak{f} 1)\infty \cdot (\bar{a}T \dagger a) \prec (a \mathfrak{f} 1)\infty \cdot (a \mathfrak{f} T)$$

That is

$$\begin{aligned}&\textstyle\sum_q \prod_p (a_{ip} \dagger 1_{pq}) \cdot (a_{iq} \dagger T_{q\ell}) \\ &= \textstyle\sum_q \prod_p (a_{ip} \dagger 1_{pq} \cdot \bar{a}_{iq}) \cdot (a_{iq} \dagger T_{q\ell}) \\ &\prec \textstyle\sum_q \prod_p (a_{ip} \dagger 1_{pq} \cdot T_{q\ell}).\end{aligned}$$

This is the conclusion; and as I said it cannot very well be expressed in the other algebra.

We will multiply this by ∞T

$$[\infty T]_{i\ell} = \textstyle\sum_j T_{j\ell}$$

It will be seen that $\sum_q \prod_p (a_{ip} \dagger 1_{pq} \cdot T_{q\ell})$ expressed in ordinary language is: I can select q so that, whatever p you take, you will either take my q and it will not be ℓ, or (even if owing to their being no individual but ℓ in the universe I have been forced to take ℓ for my q) you will find your p is loved by i. But when we multiply in $\sum_j T_{j\ell}$ which means that there is something beside ℓ in the universe, I can *always* take my q so that it shall not be ℓ and we have $\sum_q \prod_p (a_{ip} \dagger 1_{pq}) \cdot T_{q\ell}$, which is $(a \mathfrak{f} 1)T$.

The converse is easy enough. Since $T \prec \infty$ it follows that $(a \mathfrak{f} 1)T \prec (a \mathfrak{f} 1)\infty$

That $(a \mathfrak{f} 1)T \prec (\bar{a}\, T \dagger a)$ is merely a special case of the formula

$$\ell \mathfrak{f} s \prec (\ell \mathfrak{f} 0) \dagger \bar{\ell} s$$

whoever loves everybody but servants either loves everybody or fails to love some servant.

---

21 if] *inserted above*

24 so that it shall not be ℓ] *above* ~~sha~~

26 converse] *above* ~~last~~

$$\textstyle\prod_j(\ell_{ij} +\!\!, s_{jk}) \prec \prod_j \ell_{ij} +\!\!, \sum_j \bar{\ell}_{ij} s_{jk}$$

This is so because between these we have $\prod_j(\ell_{ij} +\!\!, \bar{\ell}_{ij} s_{jk})$.
Hence putting $\ell = a \quad s = 1$

$$\begin{aligned} a\dagger 1 &\prec (a\dagger 0) +\!\!, \bar{a}1 \\ (a\dagger 1)T &\prec \{(a\dagger 0) +\!\!, \bar{a}\}T \\ &\prec \{(a\dagger 0)T +\!\!, \bar{a}T\} \\ &\prec (a\dagger 0) +\!\!, \bar{a}T \\ &\prec a +\!\!, \bar{a}T \end{aligned}$$

I now proceed to show how far it is true that $a\dagger 0 \prec (a\dagger 1)T$. We have

$\prod_r a_{ir} = [a\dagger 0]_{i\ell}$ We must assume $\prod_u \sum_v T_{uv}$.

This gives $\sum_v T_{v\ell}$. Multiplying

$$\textstyle\sum_v \prod_r a_{ir} T_{v\ell} \prec \sum_v \prod_r (a_{ir} +\!\!, 1_{rv}) T_{v\ell}$$

The last is $[(a\dagger 1)T]_{i\ell}$.
As we have already seen that

$(a\dagger 1)\infty \cdot (\bar{a}T +\!\!, a) \prec (a\dagger T)$ à fortiori
$(a\dagger 1)\infty \cdot \bar{a}T \prec (a\dagger 1)T$. Hence,
$(a\dagger 0) +\!\!, (a\dagger 1)\infty \cdot \bar{a}T \prec (a\dagger 1)T$

I now have to show the converse. We start with $(a\dagger 1)T$. We square it.

$$\begin{aligned} &\{(a\dagger 1)T\} \cdot \{(a\dagger 1)T\} \prec \{a\dagger T\} \cdot \{(a\dagger 1)T\} \\ &\prec \{(a\dagger 0) +\!\!, \bar{a}T\} \cdot (a\dagger 1)T \\ &\prec (a\dagger 0) +\!\!, (a\dagger 1)T \cdot \bar{a}T \\ &\prec (a\dagger 0) +\!\!, (a\dagger 1)\infty \cdot \bar{a}T \end{aligned}$$

In my lectures in the Johns Hopkins Univ. I introduced $\sum^2$ and $\prod^{-1}$
$\sum_i^2 m_i$ means there are *two different m*'s
$\prod_i^{-1} m_i$ means everything but one is an m.

In my memoir of 1870,[73] I wrote ℓ_w^2 to denote lover of two women. I also wrote ℓ^{w-1} to denote lover of all women but one.

So $\ell_\infty = \ell^\infty + \ell^1 \cdot \bar{\ell}^{\infty-1} + \ell^{11} \cdot \ell^{11} + \ell^{\infty-1} \cdot \bar{\ell}^1$ where ℓ^{11} means lover of two or more $\bar{\ell}^{11}$ non-lover of two or more. This might also be written

73 "Description of a Notation for the Logic of Relatives."

2 between] *over* the ~~jaf~~ $\bar{a}$
5 $(a\dagger 1)T \prec \{(a\dagger 0) +\!\!, \bar{a}\}T$] ~~$\ell\dagger 1 \prec (\ell\dagger 0)f\ell$~~ $a\dagger 1 \prec (a\dagger 0) +\!\!, \bar{a}1$ $(a\dagger 1)T \prec$ ~~$(af\bar{a})T$~~ $\{(a\dagger 0) +\!\!, \bar{a}\}T$
18 the] th*at*e

$$\ell_\infty = (\ell \mathfrak{f} 0) + \ell \cdot (\bar{\ell} \mathfrak{f} 1) + \ell^{11} \bar{\ell}^{11} + (\ell \mathfrak{f} 1) \cdot \bar{\ell}$$
$$\ell_T = (\ell \mathfrak{f} 1) + (\ell \mathfrak{f} (1+1)) \cdot \bar{\ell}_T^1 + \ell_T^{11} \cdot \bar{\ell}_T^{11} + \ell_T^1 \cdot (\bar{\ell} \mathfrak{f} 1 \cdot T)$$

and with this we might work somewhat as Schröder's pentagrammatic notation. Thus, the principle of that is in my first paper.

Again we might divide loving in various different kinds; that kind that if you love that way you love everything, that kind that if you love that you may love everything but some one individual whom you do *not* love that way, loving of that kind in which you love you love more things than one and fail to love more things than one, loving of that kind in which you love one only. Write $\ell = \ell^\infty + \ell^{\dot{\infty}} + \ell^s + \ell^1$ There is also that kind of loving in which all beings love whatever is so loved ${}^\infty\ell$. That kind in which all beings but some one or another love whatever is so loved ${}^{\dot{\infty}}\ell$, Thus there are 16 kinds of loving. But if we include non-loving, as we should, what? The advantage of this over Schröder's notation is that it extends to any number of letters.

Thus

$$\begin{aligned} \ell + b = {} & \ell^\infty({}^\infty \bar{b} + \bar{b} +^s \bar{b} +^1 \bar{b}) \\ & + \ell^{\dot{\infty}}({}^\infty \bar{b} +^{\dot{\infty}} \bar{b} +^s \bar{b} + \dot{\infty} \cdot^1 \bar{b}) \\ & + \ell^s({}^\infty \bar{b} +^{\dot{\infty}} \bar{b} + s \cdot^s \bar{b}) \\ & + \ell^s({}^\infty \bar{b} + 1 +^{\dot{\infty}} \bar{b}) \end{aligned}$$

I don't know as I have these right. I dont particularly interest myself in matters of this kind.

Let us take his (21) on p 212.

$$(a \mathfrak{f} 1) T \mathfrak{f} 0 \prec (a \mathfrak{f} 1 T) \mathfrak{f} 0 \prec (a \mathfrak{f} T) \mathfrak{f} 0 \prec a \mathfrak{f} 0$$
$$a \mathfrak{f} 0 \prec (a \mathfrak{f} T) \mathfrak{f} 0 \prec (a \mathfrak{f} 1 T) \mathfrak{f} 0$$

Then if there are more than one object in the universe $(a \mathfrak{f} 1 T) \mathfrak{f} 0 \prec (a \mathfrak{f} 1) T \mathfrak{f} 0$

Please understand that notwithstanding my criticisms I think that Sixth Lecture a great piece of work.

Schröder's Sixth Lecture

$$10000 = a \mathfrak{f} 0$$

$$\begin{cases} 0\alpha000 = (a \mathfrak{f} 1)\infty \cdot \bar{a} T \\ 01000 = (\bar{a}\infty) \cdot (a \mathfrak{f} 1)\infty \\ 0\bar{\alpha}000 = (a \mathfrak{f} 1) \cdot \bar{a} \end{cases}$$

---

12 But if we include non-loving, as] *above* ~~The advantage of this~~

$$\begin{cases} 00\beta00 = aT \cdot a \cdot (\bar{a}T\mathfrak{f}0) \\ 00100 = (aT\mathfrak{f}0) \cdot (\bar{a}T\mathfrak{f}0) \\ 00\bar{\beta}00 = \bar{a}T \cdot \bar{a} \cdot (aT\mathfrak{f}0) \end{cases}$$

$$\begin{cases} 000\gamma0 = (a\infty) \cdot (\bar{a}\mathfrak{f}1) = (\bar{a}\mathfrak{f}1) \cdot a \\ 00010 = (a\infty) \cdot (\bar{a}\mathfrak{f}1)\infty \\ 000\bar{\gamma}0 = (a\infty) \cdot (\bar{a}\mathfrak{f}1)T \end{cases}$$

$$00001 = \bar{a}\mathfrak{f}0$$

Russell to Peirce. MS: RL 387.

Chicago Nov 21st 1895

Dear Mr. Peirce

I have so many things to say that I know scarcely where to begin.

First let me thank you again for the loan of your Ms.—I rise from its study every time with more and ever more wonder at the immensity of the work you have been able to accomplish—You have really shown a way to the conquest of not only the field of logic but of many other things besides. For one thing, take the science of language. Philology is now a mere science of word assorting and word pedigree. The philologists have got down to the *roots* and their broad way has run down to a squirrel's track ending in a chipmunk's hole. They have buried themselves with the mere *matter* of language to the almost entire neglect of the vital *form* for whose behalf the spirit of man as convenience has allowed or necessity has required has caught up that heterogenous stuff with which they bury themselves.—It is plain to me that there is *no thoroughfare* for any *real science* by the methods they are trying to work.—

I have now on the way an article which I have entitled "Language from the Point of View of Modern Logic" in which I first explain as best I can what modern logic is, by who it has been elaborated. Its characteristics. Its points of difference from the old word logic. The problems it has had to work out and claiming for it that it promises a complete solution of the aims of those who desire to comprehend the philosophy of language.—

I even claim that in the light of Modern Logic the old dream of an *invented* and truly *architectonic* organon of expression is now only a question of work in detail.

16 word] *inserted above*
16 word] *inserted above*
17 their] *after* ~~now~~
18 ending] *after* ~~and~~
20 as] *after* ~~has~~
21 heterogenous] heterogenis *[E]*
26 characteristics] charicteristics *[E]*
28 who] *inserted above*
31 now] *after* ~~on~~

I have given every hour I could save or steal from business and often from sleep to the study of your Ms.—That "*Algebra of the Copula*" is simply stupendous as is also the chapter on the "*logic of quantity*"—In fact one can only pick out a specially fine gem among a lot of brilliants any one of which surpasses any other in the world.—I shall return you the Ms. before long but unless you have need of it let me keep it a little while longer. I notice in several places various *errata* which I have noted on a separate piece of paper. Why do you insert "The Aristotelian Syllogistic" & DeMorgan's System between the Algebra of the Copula & the Boolian Calculus. I suppose you have a good reason for it but it naturally strikes me as a break.— Since you use the sign Ψ for $\left(\begin{matrix}\text{and}\\\text{or}\end{matrix}\right)$ why not use the sign ᛦ for ʃ.—*Vel* is an excellent name for Ψ. In calling " · " ac did you have in mind "ace"—I have been in the habit of calling any sign of composition blet (from "blend")

I should make you laugh I know in reading formulae for I am a great hand to invent my own words when I need them. For instance in reading formulae that involve brackets of various sorts (more than three sorts are rarely used) I call "( )" curbs "[ ]" bracks "{ }" braces—The first of either I call *urcurb* (or urbrack or urbrace) *when I have need to call it at all*—The last one I call curbful (or brackful or braceful). I extend also the names to the contents so that any formula say this one

$$a \,\Psi \left\{ b \,\Psi\, c \cdot [d \cdot e \,\Psi\, (g \cdot h) \cdot i] \,\Psi\, j \right\} \Psi\, k$$

I would read thusly to wit:

"*a* vel *the brace b* vel *c* blet *the brack d* blet *e* vel *the curb g* blet *h* curbful blet *i* brackful vel *j* braceful vel *k*"

Now you as I notice have much reverence for established words but as for me if I can find them to suit all well and good, but as I have said I never allow my thinking to be impeded by any such easily cured need as the lack of a word

Cordially & Admiringly
F. C. Russell

---

26 thinking] thin*g*<u>k</u>ing

26 cured need] *after* ~~found need~~

1896

Russell to Peirce. MS: RL 387.

Chicago Feb 6$^{\underline{\text{th}}}$ 1896

Dear Prof. Peirce

I am really quite shocked when I consider what a long while I have retained your manuscript. But I hope you will take into consideration the great temptation I have been laboring under. It has been such a priceless advantage to me to be allowed to study your altogether uniquely masterful productions, that I could not bear to think that any of the items of any of them should get away from me. Indeed I have been forced to remark that the more I gained mastery of your ideas the more crowded with significance they became. If your sheets shall show marks of much handling it will be at least a true mark of the degree of attention I have given to the text.

However I am now willing to return the chapters and will send them as soon as I get a word from you to any address you shall designate, expressage paid.

I hardly dare to suggest a thing as being something like "riding a free horse &c" but I must protest that I have the most consummate craving to read and study the other chapters especially Chapters 1$^{\underline{\text{st}}}$ (Signs) Chap 3$^{\underline{\text{rd}}}$ (Use of Consciousness) Chap 5$^{\underline{\text{th}}}$ (Grammar &c) and the chapter on Graphs.—

Of course I recognize that it is much more likely to be for your interests to have any book of yours published by a firm that has the facilities for and the business habits that permit them to *crowd* the sales of their books. Except for that I believe I could interest some one of our publishers here to publish your great 12 volume work[74] at as good terms you could get anywhere else. The first volume "The review of the Ideas of the 19$^{\underline{\text{th}}}$ Century" I am perfectly sure would at once seize the attention and call to your aid (in all ways pro & con) the whole thinking world & of course per consequence the reading world with large sales &c. This would give a splendid send off for the logical volumes which while more difficult it would be the *fashion* to be supposed to be read by all the same ones who had talked about the first volumes. Then the remaining volumes would go off like hot cakes &c.—

74 "The Principles of Philosophy." See footnoted description in Peirce to Russell ca. 1 Sept 1894; see also announcement included in Appendix, "Peirce and Russell."

20 interests] *after* ~~t~~
22 habits] *after* ~~in~~
28 it] *inserted above*

My review of Schroeder was not far enough advanced for the Jan. no of the Monist in fact it is not yet done but it will be ready for Apr.

With a heart full of thanks I am

Yours most Cordially
Francis C. Russell

Peirce to Russell. MS: RL 387. Undated, estimated per Russell's response 12 March 1896.

[Undated, ca. 10 March 1896]

My dear Russell:

I consider myself a well-educated man. Some branches have been neglected. I do not play on any instrument, except a nocturne passionata of my own composition on the pianoforte in which while ten fingers play the treble, five toes do the bass. My efforts in the way of counterpoint are not quite up to Bach,—not Sebastian. If you ask me if I sing, I must reply with Florence in the play that my friends assure me I do not. Though I draw incessantly, I have never drawn a prize. But I have covered a pretty wide field with my studies. There is, however, no part of my education that I value more than that of the last few years which have made me by rights a Doctor of Misery. I could write a book on the subject which would teach people much. If there be time, after I have written about 18 other books on subjects fewer persons are acquainted with, I will do so. Among the compensations for my experience,—if I ought to speak of compensations for anything so valuable,—not the least is that of discovering some most genuine and disinterested good-feeling in the world. It is impossible to tell you the satisfaction your correspondence gives me, and how doubly anxious it makes me to rise above my misfortunes. I have not written in answer to your last, because I was considering it. I cannot get into my house in Milford, for reasons I will explain later; and most of my MSS. are there. I have with me only some things that I have written since; and none of them are in their final state. Still, I shall send you some of them.[75] Among them is a piece upon the classification of the sciences.[76] I began to rewrite the part relating to the moral sciences; but it would be a long job, and I had better not undertake it. Further, the classification of the arts is very rude; but I do not think that much matters. The list of arts, though very far from complete, is I think sufficient to af-

75 These articles are not precisely determined, but estimated as follows based on Peirce's descriptions and on Russell to Peirce 2 June 1896.

76 Likely related to or part of R 1335–1347, precursory to his more developed classification delivered 1902 before the National Academy of Sciences or his "Outline" included in the syllabus of his 1903 Lowell Institute Lectures (*EP* 2:258–262).

ford data for a sketch of the true classification, when anybody wishes to attempt such a thing. Next, there is a pretty carefully written paper about the three elements of consciousness, nearly corresponding to Feeling, Will, and Cognition.[77] Here again, what I have said about *vividness* of ideas is not quite self consistent throughout, and some repetition is involved. Still, I think my doctrine can be understood. The part about Thought is slurred over, and evidently requires further study. Another paper is a sort of introduction to logic, and particularly shows the importance of my division of signs into Icons, Indices, and Symbols.[78] It is my best statement of this point. Other things in this paper are slurred over in order to bring this point into prominence. Another paper is about the Association of Ideas,[79] and is intended to show that a general conception is a sort of composite photograph of a number of images associated according to Resemblance, although William James and others deny that there is any association by resemblance. The paper is so long that its main idea is not well brought out. There are also sundry papers about the philosophy of relatives. The thing which has most occupied me since I have been in New York, my Geometry, is not among the papers I send you.

It is now necessary for me to inform you about some personal matters. I have proof that several very rich men are my bitter and active enemies. Some of them are well acquainted with one another, in fact, particular friends. They may very likely be acting in concert. At any rate there is positive and almost conclusive evidence of at least one conspiracy to ruin me. Be that as it may, incredible obstacles are put in the way of every attempt to better my situation, or even getting any employment. Besides that, it turns out that (though I have always considered myself the most affable and unpretentious of men) a large number of my former "friends" are perfectly delighted to see me humiliated, and are ready to do anything easy to them to increase my humiliation. Some of these are offended at my having, however quietly, conceited myself the peer of the greatest philosophers, and still more because some eminent men have admitted such claims,—perhaps all the more

77 Likely one of or related to his many "One, Two, Three" writings on his cosmological categories, especially what is included in *W* 5:242–247 and 292–308.

78 "Of Reasoning in General" (*EP* 2:11–26, R 595), identified in Russell to Peirce 2 June 1896. The first part of an incomplete "Short Logic" begun 1895 for Ginn & Co. following their rejection of his lengthier "How to Reason" (a.k.a. "Grand Logic").

79 Likely part of or related to R 400–403 carrying the same title, introductory portion of Book I of "Principles of Strict Reasoning" (outlined in Peirce to Russell 8 Sept 1894).

9–10 Other things ... prominence.] *inserted above*

13–14 The paper ... brought out.] *inserted above*

because I have always by my manners shown my ridicule of a man because of any gifts or powers, apart from any official position calling upon him to lead and command, putting on airs of superiority. In fact, I have always laughed at all talk of "authority" not derived from a very definite power of enforcement. Anyway the fact is, that many old "friends" and my brother[80] have shown a disposition to do anything not too conspicuous or troublesome to aid my enemies in their most wicked plots.

I have been indicted by the Grand Jury of Pike County, Pennsylvania, for assault upon a woman with aggravating circumstances,—intent to kill, or something, —I don't know precisely what.[81] It is easy to show that the only time, within months of the alleged date, at which I could have gone near the woman, was one time when my wife, a servant girl, and a servant lad were looking on .and nobody else. The boy has enlisted in the navy, and has sailed away. But his testimony would fully confirm that of my wife and former servant girl, that I only stood at the door of the kitchen, and very calmly and gently uttered these two sentences: "Laura, I understand you refuse to do what Mrs. Peirce has ordered done. Now I want you and your husband to understand that you must either both work or both quit." I made no gestures nor faces to frighten her. I immediately retired, closing the door without anything like a bang. In short, there was not the least approach to an "assault." Nevertheless, the return of the Grand Jury showed there were two witnesses against me; and as I am told the woman did not dare to appear, both must have been persons who were not there, at all. My trial is to come off on the 16th, and as I have no money for lawyers, nor even to get my witnesses there, it seems quite likely I shall be sent to the penitentiary. My brother says he is "sorry," but that he thinks I richly deserve to go to prison for turning the woman away; and he shall do nothing to prevent it. I believe he has sent the other side money to help. At any rate, it is said in Milford that he aids them. I shall go up and stand my trial. But I will not go to the penitentiary, because I am persuaded my poor wife would ruin herself and kill herself in efforts to get me out. Hence, if I am sentenced I shall then and there take sudden poison. I have fully considered every

80 James Mills (Jem), identified in the next paragraph.

81 Commonwealth v. Peirce, beginning April 1895 and involving Laura (house servant) and Leo Walters, as described by Peirce below. Charles and Juliette had been staying in New York since August 1895 to be closer to Juliette's doctors, and as this lawsuit escalated, there they remained to evade Charles's warranted arrest until the case's resolution and Charles's temporary return ca. Dec 1897. He and Juliette returned permanently May 1898. (Brent, 244–245)

4 from] *above ~~in~~*
10 within] *above ~~for~~*
12–13 and nobody else] *inserted above*
15 and gently] *inserted above*
24 penitentiary] penetentiary *[E]*

aspect of my duty, and am quite sure I am right in this resolution. The reason I write it to you is, that I want you and Schröder to have my MSS. I wish you, so far as you can, to use them so as to pay my creditors. I am under no sort of obligation to mankind at large which has emphatically told me to get out & begone. As for "fame," without caring much, I rather prefer that nothing I have done should be praised, not set down to the honor of the country and age which have wronged me. I should like to have posterity benefit by my work, but would slightly prefer my name should not be attached to it. What I care much more for is to have my creditors paid. If some such person as Carus is willing to pay me roundly to have my work go by his name, that would suit me immensely. If such a person can conceive of what a man of honor is, he may be assured I have prepared no trap for him. He can say, "Oh yes, there was one Peirce who is said to have had some confused notions of this kind. But he was nothing but a beggarly ne'er-do-well." Anyway you and Schröder do what you think best; and be careful my brother James has no say in the matter. I owe him about $ 250 *at most*. He employed a lawyer named Sykes, employee of Taylor, firm of Hornbeck or Horn something (proposed for U.S. Supreme Judge) who made me sign a note for $ 2300 as a "mere form," and for my interest. I declare solemnly it was a perfect fraud. He never would give it back.

My only staunch friends throughout whom I need name to you have been my wife, Mrs. Juliette Peirce, Dr. W. K. Otis, 5 W 50$^{\underline{h}}$ St, New York, to whom I owe about $ 450; Comte d'Aulby, who is by rights Prince Borghese, rather an effete strain, and much worn out by wrongs of his own, but what little there is of him, true, honorable, and ideal-worshipping, and who is under obligations to my wife; Wendell Phillips Garrison, Editor of the Nation, P.O. Box 794 New York, to whom I believe I owe $ 100 (but it may have been paid, for he had my authority to withhold it by degrees from payments to me); Dr. Francis Ellingwood Abbot, Cambridge, Mass. to whom I owe $ 5; Dr. John Milton Dexter, a young physician of no influence, 160 W 87$^{\underline{h}}$ St (and I ought not to have put in his name, because he could do nothing.) There is no sort of appeal that I would ever consent to being made to these people; but I mention them as persons who, should they appear to take an interest in my publications, may be thoroughly trusted. Perhaps I ought to add to them Prof. W$^{\underline{m}}$ James (always thinking of his own ambitions, and ready to shore another man down if tempted in that way), Mr. Theodore L. DeVinne the printer, to whom I owe $ 50, Prof. B. E. Smith, Century Co., Dr. Abraham Jacobi, 34$^{\underline{h}}$ St, New York, to whom I owe $ 10, of whose wife I have been a friend, but I really es-

---

7 would] *above* ~~not~~
17 who] *after* ~~to s~~
19 whom I need name to you] *inserted above*
19–20 my wife, Mrs. Juliette Peirce,] *inserted above*
22 and much … his own,] *inserted above*
30 take] *above* ~~have~~

teem him much more than her. It is to be supposed that my scholars would have some regard for me. There are doubtless many others. Some I know of, but do not mention, because it would be to no purpose.

It is my opinion that my Geometry, or "Elements of Mathematics," as I call it, is a good book, that will at least pay all my debts. My arithmetic, if my wife can get the MS. and get it completed, ought to belong to her after she has paid Hegeler out of it, if he is not paid out of the Geometry. But, remember, he would rather go without his money than have the vexation of knowing my creditors were honorably paid. As for Carus, I have the lowest opinion of him all round. You think him a good fellow; and I do not seek to shake your faith; only I need not say that you have not the right to put in jeopardy the payment of my creditors, because *you* trust him, when *I* do not.

I do not wish to load you and Schröder down with an obligation which, if you accept it, may be burdensome. I do not give you any instructions whatever. I just give you and Schröder my MSS, and hope, if you can, you and he will make them contribute to pay my creditors. Only the arithmetic, I do not give you. I give it to my wife. And the Geometry or "Elements of Mathematics" I only give to you and Schröder, on condition of your getting it published and applying the profits to paying my creditors. If you do not accept this trust, I give it to my wife on the same conditions. If Schröder accepts the trusteeship of the Geometry and you do not, I substitute my wife for you. If you accept, and Schröder does not, I substitute my wife for him. My wife is Mrs. Juliette Peirce of "Arisbe," Westfall Township, Pike Co, Pa. I shall draw a will to this effect. Of my MSS at the house, you will find a great quantity "of value only to the owner."

I trust I shall beat my enemies at my trial in one way or another, and then you will be saved this trouble.

At the same time, I stand in such a position that life appears to me like a spectacle upon which the green curtain is descending. If, just as I disembark upon the shores of eternity, the cabman who drives me to my mansion in the skies says to me "What did your Honor think of Man, anyway?" I shall say, "Well, Cabby, he promises to develope into something very noble. But it is curious how many of them think of power only as a glorious opportunity to make themselves more con-

---

2–3 Some I know … no purpose.] *inserted below*
5 at least] *after* ~~be~~
6 the MS.] *above* ~~it~~
10 and I … say that] *above* ~~but~~
11 put] *above* ~~jeop~~
15 and Schröder] *inserted above*
15 you and he will make] you \and he/ will ~~may~~ make
18 and Schröder,] *inserted above*
23–24 I shall draw … owner."] *inserted above, at top of last page of letter*
28 just] *after* ~~arrived d~~
30 Honor] *h*Honor

temptable than they could ever otherwise hope to be." I shall reserve for a more lofty spirit the remark that such a man as Dr. Ernst Schröder is a sufficient excuse for the existence of all the rest.

Of course, I shall immediately drop you a post-card, if I am acquitted, or the trial is postponed.

In any case, without waiting for that, I wish you would send my chapters of logic now in your hands at once to Prof. Dr. Ernst Schröder, Gottesauerstrasse 9., Karlsruhe in Baden, Germany.

very faithfully
C. S. Peirce
160 W $87^{\underline{h}}$ St | *New York City*

I have not explained why I cannot get into my house in Milford. I cannot. Nor can my wife immediately, should I die.

Russell to Peirce. MS: RL 387. Stationery headed, "Francis C. Russell, | Counselor at Law, | 145 La Salle St., Suite 217 & 218 | Chicago."

Mar $12^{\underline{th}}$ 1896

My *dear dear* Mr Peirce

I have just received your chapters last sent and have read your letter accompanying. I am deeply grieved over your troubles and never felt the curse of material impotence so deeply as now when all my soul bids me to fly to your aid. I *will* not admit that your straits are really so desperate and you *must not give up in any event*. Forgive me if I put the matter in this way that you *ought not* to abandon the mission God sent you to fulfil. No *personal* reasons of *any sort* even the *most desperate possible* must prevail. You do not *belong to yourself* but to *humanity* and *to the future*. You are the repository of a *sacred trust* of the *most exigent kind*, and *take hold on that sublime horizon* that is fit to match with that trust. All will come out in the end happily, *believe me*, no matter what tribulations and anguish you are compelled to endure. I swear to you that I intend and for a long time have intended to become able to help you, and to do it. My prospects are good. But wait for me a little longer. Please dont even if I must say so dont for my sake give up

Yours very Affectionately and sympathizingly
Francis C. Russell

---

4 acquitted] *after* ~~aq~~

Peirce to Russell. Postcard: Recto headed, "Postal Card—One Cent | United States of America" with portrait at right.

[recto]

New York | Mar 19 | 7 PM | 1896 *[stamped]*

Judge Francis C. Russell
125 LaSalle St
Chicago
Ill.

[verso]

I got my case continued.
C. S. P.

Russell to Peirce. MS: RL 387. Stationery headed, "Francis C. Russell, | Counselor at Law, | 145 La Salle St., Suite 217 & 218 | Chicago."

Chicago Mar 21st 1896

Dear Mr Peirce

I got your postal card this morning and am rejoiced to see that we have at least a breathing spell. I want to tell you in all frankness that I could not but feel that your cry of despair, although perfectly genuine, would be seen to be uncalled for, by the nature of the jeopardy as viewed by the lawyer's eye. Still I could not tell for sure and so I wrote you in the strain I did as the best appeal I could then make to you against anything desperate. Now we have a little spell for calm thought I want to say that endorsements such as yours ought to and would could you only comprehend them aright in their relations to your own personality arm you with a dignity and sense of wealth and general overlordship that would make you proof against the utmost of the slings and outrages of fortunes.

But although I tell you this, do not for a moment infer that I am inclined to extenuate or underestimate what I have always considered as one of the most *shameful* systems of instances of the way which one generation deals with those whom succeeding generations recognize as the most illustrious of the benefactors of mankind. Nor do I fail to appreciate and sympathize with you in all the various and plentiful torments that fortune seems perversely bound to pile upon your head and heart. The *simple plain truth* is the world is all unprepared to recognize your real measure. In this I am not referring to the ordinary plug citizen, of course

---

18 would be seen to be] *above* ~~was~~

19 by the nature … lawyer's eye.] *inserted below*

20 then] *inserted above*

22 to²] *inserted above*

such specimens are not, but rather to the professorial hierarchy *damn them*. The dilettante look up to them and regard them as the repositories of pretty much everything that is worth knowing. They are flattered and fawned upon, feasted and showered with material favor. Some of them are really rather bright fellows and capable in some measure of taking and holding an *original* relation to divers objects of knowledge. The rest are ordinary stuff crammed with stock traditions of knowledge and perhaps with a limber tongue that enables them to make a good show. But *success* (how I hate that word) spoils them all. They all become imbued with all the pursiness, big headedness, little low petty jealousies, intolerances, little-souled-nesses of oriental despots. From hearing everybody else (except these auspices) say so they soon begin themselves to really believe they know it all. Furthermore they soon come to recognize that they belong to the *in's* and that to be so is a rather a soft snap both by way of honor and profit. Most of them are lazy too, they have *finished* in the main their education and have no inclination at all to begin to learn again their alphabet and multiplication table. They also are well aware of the professorial solidarity and its united power and disposition.

Now when a man like you comes into the world, by the very nature of the case and however much you may strive to soften your incidence you come to their comprehension ((i.e. as to the way it effects them) as a revolutionist as even perhaps a horrible *anarchist*!! in the world of science and philosophy. At first they are a little puzzled to know how to trust you. Their first essays may be somewhat equivocal. They are all animated by the spirit of the hunter who said in a certain dubious case that he took his aim "so to hit if it was a deer but to miss it if it was a cow". *They want to be with you if you win*. But if they venture either expressly or tacitly to put you down or to crowd you to one side or to ignore you it is astonishing to observe what they will compass to that end. A conspiracy of silence is of course adopted. Then all sorts of exertions to prevent you from gaining a hearing. You find yourself balked everywhere most often by some hand that stabs in the dark. Having got you discredited among "*the hierarchy*" and the tacit understanding having become generally operative that you are not to be accounted "*one of us*" or a candidate for that, it is the easiest thing in the world to humbug the lower grade's minds. Such a one say, some puffy editor of some great newspaper or magazine says to Prof. X. What is the matter with Peirce? I thought once that he was like to cut a big figure among you learned people, and Prof X replies, Oh. Well, I dont exactly know, he dont come exactly in my line but he dont seem to get any

2 dilettante] dilletante *[E]*
9 petty] pety *[E]*
9–10 little-souled-nesses] little-soul\ed/-nesses
16 the] the~~ir~~
19 (i.e.] *after* ~~as~~

one to take any stock in his notions. And so on with such like all round the whole outfit. Now what does Mr Editor think. Why! *he* reasons thus, These professors *say* they welcome all new light with gladness and honor. It must be there is some fatal laconism in Peirce's make up, in his science and in his philosophy. He is *afraid* and publishers subject to the same influence are *afraid*, and so the ban proceeds.

Now what is the practical lesson. It would be for me if I were you, "You poor little pusillanimous cusses. I of mine own plenteous exaltation and grace *can afford* to despise and overcome your machinations. I can do it and I will do it though all Hell assails me. I will not be driven to despair by *Any* fortune. I'll take whatever good fortune that comes my way and enjoy it, and if bad fortune comes I'll stand it. Though poverty comes that will I stand. Yea even though the Prison yawns even that will I endure."

But dear friend, I feel quite convinced that your terrible case will turn out not very bad to manage, perhaps it will prove a very tame affair. It is your financial status that is undoubtedly the worse of the two. As to that I propose to give it some attention before long if my own affairs turn out as well as I have a right to expect.—I have myself been under a pretty bad financial racket for four or five years last past, and maybe you will see that I have been much apt to go to jail many a time and for long spaces of time subject to that danger than you have ever been. But I wasnt phased and didnt intend to be, and had it come I should have said & *felt*, "Well it is only part of a lifetime"

Cordially
Francis C. Russell

Carus to Russell Feb–April 1896 (TS: SIU 91.7): Apparently influenced by Schröder's insistence that Peirce was the only one suited to review his *Algebra der Logik*,[82] Carus asked Russell 25 Feb 1896 to invite Peirce to contribute, and under his own name. "As matters are," Carus adds, "I cannot ask him to write the review, but if he would send it I would not refuse it.... Mr. Hegeler does not wish me to be in any business connections with him, but that would not exclude the acceptance of a contribution which he might offer at a cash price, the bill to be settled at once ... without further complications."

After much delay, possibly due to a misunderstanding of whether or not the review would in fact appear under Peirce's name, Russell contacts Peirce as follows.

---

82 Brent, 256; also *Peirce Project Newsletter* (1994).

20 phased] phazed *[E]*

Russell to Peirce. MS: RL 387.

Chicago April 22, 1896

Dear Mr. Peirce.

I am authorized to obtain from you either a review of Dr. Schroeder's work or *better* an article from you on "Modern Logic" commenting on the subject with especial reference to Schroeder's three volumes. For this purpose I have withdrawn my review which I have always been ashamed of. Just how Dr. Carus and I will have your article published whether under my name or without any name is not as yet settled but it is plainly understood that you are to get paid for it at *your old rates* and as soon as the article is written. This will give you perhaps a little ready money and so be welcome to you just about now-a-days. I have been trying to get some of our Chicago men of affluence interested in you and in your work. *Melville E. Stone*, a former editor and co-proprietor of The Chicago News, a vice president of The Globe National Bank and the Western General Manager of The Associated Press, here, has been listening to me. I have given to him in writing a sort of a synoptical view of your field of labor and success and the prospects and bearings of the same and I am encouraged to hope that he will aid me in interesting someone of our millionaires.—

As for myself I am preparing an article for The Monist in July on "The Logic of Geometry". I am going to tell you (even though I can see you laughing at me and saying "another paradoxer") that I have succeeded in framing a definition for the straight line that consists of a universal proposition concerning the distribution of its *points* along its curve.

I say "A straight line is one such that every point-triad that may be assigned of its points is a straight point-triad." "A straight point-triad is one whose several points are so placed that the *ad-spheres retentive* of any one of said points have no other points in common"

You see every pair of points determine two conjugate spheres taking commutably each point as a centre and the other point as a sample point of the surface. Every point-triad distributes into three pairs and determine six spheres. Spheres thus determined by the points of an "ad" (i.e. pair, triad, tetrad &c &c) I call ad-spheres. The six spheres determined by a point-triad, always pair as follows with reference to any certain point. Two of them centre at said point. The other four may for the moment be called with reference to the point in question the *opposite ad-*

15 in writing] *inserted above*

28–29 two conjugate ... each point] two \conjugate/ spheres taking \commutably/ each ~~sphere~~ point

475.34–476.1 the *opposite ad-spheres*] \the/ *opposite* *ad-*/*spheres*

spheres. Two of these opposite spheres are always conjugate to one another (that is, either one is a centre while the other is a surface point). These spheres do not *in general* contain any other points of said triad. The remaining pair always include the point in question, so I call this pair the *ad-spheres, retentive* of said point.

If the *ad-spheres retentive* of any point of a point-triad have no other points in common then the both cases of *ad-spheres retentive* of the two other points of said point-triad have no points in common (This statement is unclear but I guess you will tumble)

With this definition it is easy to prove the equivalent of Euclid's 10th Axiom. I make scarce any use of the plane. After proving that "two right lines that have one point in common can have no other point in common" I define a Dialelle as "the figure consisting of two right lines that have a common point" then the half rays of a dialelle ending at the common point as the *arms* of the dialelle, then "An *angle* is the figure determined by any two arms of a dialelle". I make a distinction between a triangle and a trialelle which I need not point out to you. Then I define the diameter of the sphere and the poles of the diameter and also the radius. Then after a number of propositions relative the dialelle the triangle and the complete quadrilateral (tetralelle) I prove that, "The adjacent angles of every dialelle whose common point lies in the surface of a sphere and two of whose arms pass each through one of the poles of any diameter of that sphere are equal" from which follows as a mere corollary, that all four angles are equal and then defining such angles as *right* angles the Euclidean axiom "All right angles are equal" is proved. But further, we can then go on easily and (after defining a right angled triangle) prove first that "the angles of a right angled triangle are together equal to two right angles" and then that "the three angles of *any* triangle are together equal to two right angles," and so on to all *that is involved.*

Now the article that I am preparing will treat of the data and logic of geometry in view of these results, showing, 1st We need nothing but definitions, explanations (if what we mean by our words), indications (which we cannot do without), diagrams (or *signs* other than words), *immediate* inferences, *mediate* inferences, and experiences that *are common* to all reasoning beings (Euclid's *common no-*

---

2 a¹] *inserted above*
2–3 These spheres … said triad.] *inserted below and up right margin*
6 both cases of] *inserted above*
6 the two other] *above* ~~every~~
7 This statement is unclear] This \statement/ is ~~equival~~ \unclear/
10 right] *inserted above*
11 Dialelle] *For "diallel"*
15 trialelle] *For "triallel"*
18 tetralelle] *For "tetrallel"*
19 each] *inserted above*
21 a] *inserted above*
21 corollary] corrollary *[E]*
25 that] *inserted above*

tions—common sense in a new sense i.e. a sense that is common to all reasoning beings—and propositions in the nature of his postulates, i.e. propositions as to what may be *done*)

Let me hear from you at your earliest convenience about the article we want from you

Very Cordially
Francis C. Russell

Peirce to Russell. MS: RL 387.

New York 1896 Apr 26 (Address will be forwarded.)
Stewart's Hotel. Broadway and 41st St.

My dear Judge:

Yours of the 22nd reached me today. I shall be happy to write the article in question, leaving it to you to decide whether it shall appear under your name, or anonymously, so long as it does not so appear as to be taken for the work of any other person. I shall keep quiet, simply saying that the article represents my views, but telling Schröder about it, confidentially, perhaps. I am under a promise to write two articles for the *New York Herald*. I have also promised a friend, in case he pays my expenses, to go on to Washington and do certain work for him there.[83] The weather is very cool here, at present; and I have no money to pay for a lodging, so that I am forced to walk the streets all night. This fatigue makes my work in the daytime, which has to be confined to the hours during which the libraries are open, slow. Then, I get very little food. I am strongly urged for an excellent situation—salary 25 000 fr.—in Paris, which, were it to go to the most efficient man, I should be sure to get. I have an invitation to pass the summer at a château in France. If I could do so, I should be able to finish my Arithmetics during the summer, and should find assistants there for another profitable work, which could be commenced.[84] If I could get an appointment as newspaper correspondent in Paris, with an advance of money to take me there, I should be able to go, and make a proper appearance. And making a proper appearance, I should be able to get my

83 See unsent draft that follows for details.

84 A *dictionnaire raisonné* (a.k.a. *Peirce's Logotheca*) intended to replace *Roget's Thesaurus*. For details, see unsent draft below and Peirce to Russell ca. 10 Oct 1896.

1–2 *common sense* … reasoning beings] *inserted up left margin; en dashes mine*
15 any] *after* ~~Carus~~
23 salary 25 000 fr.] *inserted above; en dashes mine*

Arithmetics published to advantage. I have succeeded in getting in an application for a patent for an indispensable adjunct to a new machine for domestic lighting by acetylene.[85] I have great confidence that that machine is going to revolutionize domestic lighting, and it has to be put in every house where the new mode of lighting is used. They will be forced to use my invention; and were I to be able to make a decent appearance, no doubt, I could make them pay me a dollar or more royalty; which would come to hundreds of thousands. But unless I can get some start, what is to prevent their stealing it? I have other inventions whose real value is immense; but this is the only one I have taken steps to patent. So far, I am sole proprietor of it. You see, therefore, that if you were to get me a position as Paris correspondent of a newspaper, or any other good start, such that it would enable me at once to make a good appearance, it would come to very much more than itself, and would put me in a situation in which I could turn money your way, in return. And you may be sure my gratitude would be genuine and loyal. Of course, I do not describe the thing; because to *prove* to you what its value is would require many pages. But I would part with 1/4 interest in the patent to a good business man who would exert himself actively to do the necessary business with either of the two parties who might buy it, namely, the owners of the machine for which mine is indispensable, and the great company with ten millions capital who would like to gobble that machine, but can't without me. For that 1/4 I would require $ 5 000 cash and would give you another 1/4 to find me such a party. I should require you not to part with *all* your fourth; so that, we could protect ourselves against dishonest deals. Now if you want a description of the matter with an argument as to its value, let me know. I say that, supposing the patent passes, it is worth in the neighborhood of a million.

I dare say that after you have assumed certain properties of a circle not true in non-Euclidean geometry you can prove the 3 angles of a triangle are equal, by your definition of a straight line which makes the straight line the limiting case of a circle or of part of a circle. But that would not suffice to show that you can prove geometry without postulates. This would be as much as to say that two different

85 Peirce had invested and would continue to invest substantial time and expense on this invention (ultimately to no avail), in collaboration with wealthy partners including painter Albert Bierstadt and stockbroker Edmund Stedman (see Biographical Register). For Peirce's writings on acetylene and its commercial use for lighting, see *The Nation*, "Acetylene and Alcohol" (26 Dec 1895) and his later review of V. B. Lewes *Acetylene* (27 Sept 1900). Other related papers are found in *HP* 2:1048–1094 and R 1030–1035.

18 for which mine] ~~of~~ \for/ which *it*<u>m</u>ine

21 cash] *inserted above*

constitutions of space are not possible. I think that erroneous; and would ask how you prove that space has three dimensions in any such way; how you prove, that space has no points, lines, or surfaces from which there are more or fewer ways of motion than from ordinary points; how you prove that space is all one piece, that every ring in space can shrink to nothing in space without breaking (if you hold this to be true, or if not, how you prove whatever you do hold true.) I call your attention too, to the fact that all projective geometry rests on the assumption that every plane returns into itself, but so that if an object having three dimensions rests on the plane and sliding upon it passes through infinity and so back to its original position, it is now on the *other side* of the plane; so that the plane of projective geometry is what I call a *perissid* surface; and some writers call a *unifacial* surface. But the theory of functions deals with a plane which is a simple *artiad* surface. All the pairs of straight lines upon it meet twice unless they are parallel, and in that case they have two coincident points at infinity, and when an object passes along such a line through infinity, it returns to its original place just as it was. There then are two conflicting hypotheses about a plane with which mathematicians are working, without inconvenience. If therefore you can prove the propositions of geometry without postulates,—that is, without any initial hypothesis about the constitution of space,—you ought to be able to prove that one or other of those two conceptions is absurd. And since both branches of mathematics make continual use of these very properties, you will utterly overthrow one or other of the two greatest branches of mathematical doctrine.

I have just completed a memoir I intended reading to the National Academy of Sciences in Washington this last week.[86] But I was unable to get there. In the introductory part of this memoir, I undertake to state in general terms what is logically possible, and what not. This assumes that some things are impossible although they do not involve any contradiction, such for example as that there should exist only two or only three things, that there should be a relation which could not exist between a certain set of things although there were no contradiction involved, etc. etc. Having thus described the logically possible, I go on to consider the multitudes of collections. I succeed in that way in proving that of two collections not

86 Undetermined.

3 points,] *above* ~~sur~~
8 every] *after* ~~in~~
9 back] *f*back
13 pairs of] *inserted above*
14 object] objects *[E]*
27 should exist] \should/ exist~~s~~
28 should] *is*should

equal one must be greater than the other; although there is no contradiction in supposing that in every possible way of setting them off into pairs, one object of each pair belonging to the one collection and the other to the other, there should always remain unpaired objects among both collections. I also show that greater than the collection equal to all finite whole numbers there are a series of possible collections each next greater than the last, but infinitely greater than that last; and these collections are equal, in the multitude of them, to the finite whole numbers; and greater than them all is a possible collection, than which no collection can be greater. Having thus disposed of collectional quantity, I propose in subsequent papers to consider other kinds of quantity.

very faithfully
C. S. Peirce

P.S. None of the collections mentioned at the end of my letter has the properties of an infinity of the first order. For in all cases, beyond the finite collections, the square of the multitude equals the multitude. The maximum collection equals its own exponential. The lowest infinite multitude has no logarithm. The rule of possibility would be worth telling. It is very simple. But I have no space for it, this time.

[Unsent, incomplete draft]
160 W 87$^{\underline{h}}$ St. New York 1896 Apr 26

My dear Judge:

Yours of the 22$^{\underline{nd}}$ reached me today. I shall be happy to write the article in question. I have nothing to say as to whether it appears under your name or anonymously, further than that I should suppose the latter would be more agreeable to you. I should kick if it so appeared as to be taken for the work of Carus. Otherwise, I will keep quiet, merely saying that it represents my views. I shall tell Schröder just how it is, confidentially. Two things, one will & the other may, cause some delay. Namely, I am under a promise to write certain articles for the *New York Herald*. I have also promised a friend, in case he pays my expenses, to go on to Washington and get through a bill which will cause the repayment to him of some $ 18 000 duties collected from him on pictures & which cannot be rebated on the pictures being taken out of the country because the law imposing the duties has been re-

2 object] *inserted above*
3 the other] *above* ~~one~~
5 possible] *inserted above*
7 the[1]] *inserted above*
13–18 P.S. None … this time] *Postscript added at top-left corner of first page of letter.*
13 mentioned] *above* ~~end~~
26 Schröder] Sch\r/öder

pealed. I understand how to do that business, & can do it. I have succeeded in getting in an application for a patent, for an indispensable attachment to a new machine not yet brought out for domestic lighting by acetylene. I have great confidence that that machine is going to revolutionize domestic lighting, and hundreds of thousands will be put in. I hope to be able to force them to pay me a royalty of $ 1 on every one. Anyway, I am all alone in it so far; and whatever my patent may be worth belongs, so far, to me. I have an invitation to pass the summer with some friends at their château in France. I have also been strongly recommended for a place as Paris correspondent about scientific and literary matters of a syndicate of English newspapers. The salary is 25 000 francs. If I go to France I shall spend the summer in putting my Arithmetics into their final shape and also shall start up work, with assistants, upon my *dictionnaire raisonné [ ]*

Peirce to Russell. MS: RL 387. Undated, estimated per Carus to Peirce 9 May 1896 and Peirce to Carus 11 May 1896.

[ca. 30 April 1896]

Dear Judge

I enclose my article on Schröder.[87] I have counted the words on 6 pages as follows

page	words
2	301
5	286
10	333
15	317
20	227
25	278

The mean is 290 words to the page. (There are really 300, page 20 being exceptional) Since there are 29 pages there are 8410 words at that rate. The pay should be $ 210 and the need is pressing. Please endeavor to have a New York draft sent to me at once.

very faithfully
C. S. Peirce

It is fearfully long; but I would have gladly written much more.

87 "The Regenerated Logic," *Monist* (Oct 1897). Some of the material here sent may have been unused or abbreviated in the final article due to space restrictions. (Compare what appeared in *The Monist* with Peirce's commentary about the article in the letters that follow.)

2 indispensable] indispensible *[E]*

---

ca. 1–6 May 1896: "The Regenerated Logic" was delayed in Russell's hands until it reached Carus 8 May; hence Peirce's urgent telegram below and to Carus 11 May. For details, see footnote to this article in Carus to Peirce 9 May 1896.

Peirce to Russell. Telegram: SIU 91.14. Form headed, "The Western Union Telegraph Company."

May 6th 1896 | New York

Francis C Russell

145 Lasalle st | Suite 217 Chicago

Please remit immediately as promised where upon will rewrite to suit wife imminent danger appendicitis hence urgency.

C S Peirce

Russell to Peirce telegram ca. 6 May 1896 (missing): Reminds Peirce about his Logic chapters still in his possession, likely re-asking for Peirce's stable address for their return.

Peirce to Russell. MS: RL 387.

1896 May 7.

Dear Russell

I had to have my poor wife carried to a hospital & engage the most skillful physician for her. The hospital is damning me; and my wife being so ill responsibility for certain payments falls on me. I have a few pictures of value, I would like to sell or hypothecate. Among others a splendid landscape with a snowy mountain by Bierstadt. It is worth $ 2500. If any Chicagoan wants to loan me $ 1000 to $ 1200 on it till Oct 31st, I may sell it before that time. It ought to be easy to get $ 1500 to $ 2000 for it. If I have some old pictures of greater value.

As to the notice of Schröder, having another of importance to write for the Bulletin of the American Mathematical Society, I thought all interests would best be subserved by choosing the topics I did for the Monist article. For the Logic of Relatives, as Schroeder has treated it, is nothing but my work with mathematical developments by Schröder. As for the philosophical aspect of that subject, he has, as yet, said little or nothing about it. It seemed therefore that the best way to divide what I had to say was to put the comments on his development of the logic of

---

10 Peirce] Pierce *[E]*
14 May] *above* ~~Apr~~
17 so] *inserted above*
24 thought] though *[E]*

relatives mainly into the Bulletin. However, that is held back; and I stand ready to substitute other topics for those chosen, if desired. As to the frequent recurrence of my name, the book cannot well be criticized without that. It would not have been fair dealing on my part to have written otherwise. Still, I have no objection to suppressing partly or wholly such references.

If my invention turns out as valuable as it looks, which will soon be determined now, my first effort will be to make a sacrifice of a portion in order to square accounts with Hegeler, which is the thing that weighs most heavily upon me, & then to recover the MS. of my Arithmetics which was seized long ago,[88]—the very first attack upon me did that,—& I am afraid Hegeler has it and is determined not to let me have it. However, it looks now as if I was going to right myself in spite of all the powers against me & be able to settle down to the completion of my life-work after paying all obligations. The latter I am confident of doing sooner or later; but it may not leave me life and energy for the former.

Things can hardly go right without going so exceedingly right that I can do justice to your golden endeavors in my behalf. It would be ridiculous for a man who thinks a day is particularly fortunate if it brings him a sausage to be making promises; but you shall see whether I am at bottom worthy of all you have done, or not. That is I *hope* such opportunity may come.

Your telegram received.

Have you ever sent Schröder those chapters? If not, please send them here & I will forward them.

very faithfully
C. S. Peirce
Stewart House | W 41st St & Broadway | *New York City*

Peirce to Russell. Telegram: RL 387. Form headed, "Night Telegram | Postal Telegraph-Cable Company" with company seal at left.

New York, N.Y. May 12, 1896

Francis C. Russell,
145 LaSalle St.,
Suite 217.
Chicago, Ills.

88 ca. 8 Sept 1894, when Peirce reported to Russell that the arithmetic MSS were "out of my hands." Details about this seizure and "first attack" are unclear.

18 all] *over* ~~it~~

This broken promise loses me twenty thousand to save me something telegraph fully acknowledging you were authorized to promise immediate remittance.

Peirce | 7:45 p.m.

Peirce to Russell. MS: RL 387. Undated, estimated based on context. Whether sent or not is undetermined. This and the next two letters were likely written around the same time, and one or more of them were likely alternate drafts.

[ca. 12 May 1896]

My dear Russell:

You have had those chapters of mine a long time. Please send them *at once* to me at 5 W $50^{\underline{h}}$ St New York City.

Can you not contrive to communicate in some way with Carus and have *something* sent to me at once, telegraphically if possible.

Here was a plain promise made to me that I should be paid "as soon as MS. was sent in" It was sent in just 13 days ago. If it had not been for that I should have done some other work which would have brought in what was absolutely essential. Also trusting in that promise,—which it never occurred to me was a mere trap,—so innocent am I of German ways,—I sent my wife to a hospital.

The result will be that they will kick her into the street to die. I have had nothing to eat this week except a sandwich which was given me yesterday morning. I shall have no place to sleep after last night, except the park.

I could have made your fortune, if I had been properly backed up. Or even independent of any backing.

I feel now that the whole business of writing for the Monist was a trap, that Carus never intended to keep his promise & merely used my credulity to injure me.

The last copy sent I will incorporate with the rest as soon as I am informed what the form of the article whether anonymous or what is to be.

The problem which I use eight letters to solve can be solved with 6 (but not fewer) thus

$$\begin{array}{ll} H_n \prec xH_{n+1} & H_n \prec \overline{x}H_n \\ H_n \prec xH_{n+3} & H_n \prec \overline{x}H_{n+2} \\ H_n \prec xH_{n+4} & H_n \prec \overline{x}H_{n-1} \end{array}$$

18 had] *after* ~~n~~

$$
\begin{array}{ll}
H_n \mathrel{-\!\!\!<} xH_{n+1} & H_{n+1} \mathrel{-\!\!\!<} \bar{x}H_n \\
 & H_{n+1} \mathrel{-\!\!\!<} \bar{x}\,H_{n+1} \\
 & \underline{H_{n+1} \mathrel{-\!\!\!<} \bar{x}\,H_{n+3}} \\
H_n \mathrel{-\!\!\!<} xH_{n+3} & H_{n+3} \mathrel{-\!\!\!<} \bar{x}H_{n+2} \\
 & H_{n+3} \mathrel{-\!\!\!<} \bar{x}\,H_{n+3} \\
 & H_{n+3} \mathrel{-\!\!\!<} \bar{x}\,H_{n+5} \\
\hline
H_n \mathrel{-\!\!\!<} xH_{n+4} & H_{n+4} \mathrel{-\!\!\!<} \bar{x}H_{n+3} \\
 & \qquad \mathrel{-\!\!\!<} \bar{x}H_{n+4} \\
 & \qquad \mathrel{-\!\!\!<} \bar{x}H_n
\end{array}
$$

Make a great effort and get some money sent me. Owing to the delay, I am unable to get you something which would have made you some money

very truly
CSPeirce

Peirce to Russell. MS: RL 387. Undated, estimated based on context. Whether sent or not is undetermined.

[ca. 12 May 1896]

My dear Russell

I have no time to explain the terrible effects which have resulted from the breach of Carus's promise. For one thing my wife is turned out of the hospital in a state of imminent danger to her life as the Doctors themselves say having given up the room where she was and with no place to go & money in advance is required everywhere. She will simply have to die in the streets. If this cursed treacherous promise had never been made,—which Carus no doubt chuckles at the idea of my believing,—I should have been decently off by this time. But I cannot believe you will not take steps about it. I will have my lawyer write to Hegeler if something is not done soon.

The problem $a \mathrel{-\!\!\!<} x\bar{x}$ can be solved and a eliminated if there are more than 3 objects in the universe. Namely 1, 2, 3, 4 denoting 4 individuals and 5 any 5th whatever

$$x = 1{:}2 \mathbin{\psi} 1{:}4 \mathbin{\psi} 4{:}2 \mathbin{\psi} 4{:}4 \mathbin{\psi} 2{:}1 \mathbin{\psi} 2{:}3 \mathbin{\psi} 3{:}1 \mathbin{\psi} 3{:}3 \mathbin{\psi} 5{:}1 \mathbin{\psi} 5{:}2$$
$$\bar{x} = 1{:}1 \mathbin{\psi} 1{:}3 \mathbin{\psi} 4{:}1 \mathbin{\psi} 4{:}3 \mathbin{\psi} 2{:}2 \mathbin{\psi} 2{:}4 \mathbin{\psi} 3{:}2 \mathbin{\psi} 3{:}4 \mathbin{\psi} 1{:}5 \mathbin{\psi} 2{:}5$$

It does not matter whether 3:5, 4:5, 5:5, 5:3, 5:4 go in x or in $\bar{x}$.

Obviously $1{:}1 \mathrel{-\!\!\!<} (1{:}4)(4{:}1) \mathrel{-\!\!\!<} x\bar{x}$ $1{:}2 \mathrel{-\!\!\!<} (1{:}2)(2{:}2) \mathrel{-\!\!\!<} x\bar{x}$ $1{:}3 \mathrel{-\!\!\!<} (1{:}4)(4{:}3) \mathrel{-\!\!\!<} x\bar{x}$

1:4 ⤙(1:2)(2:4) ⤙$x\bar{x}$ 1:5 ⤙(1:2)(2:5) ⤙$x\bar{x}$
2:1 ⤙(2:1)(1:1) 2:2 ⤙(2:3)(3:2) 2:3 ⤙(2:1)(1:3) 2:4 ⤙(2:3)(3:4) 2:5 ⤙(2:1)(1:5)
3:1 ⤙(3:1)(1:2) etc like (2:1) etc (4:1) etc like (1:1) etc.
5:1 ⤙(5:1)(1:1) 5:2 ⤙(5:2)(2:2) 5:3 ⤙(5:1)(1:3) 5:4 ⤙(5:2)(2:4) 5:5 ⤙(5:2)(2:5) ᛘ (5:1)(1:5)

Whoever works up the copy sent into an article, I or you, which can be done only when it is decided just what form the article is to take will insert the above. The solution for 3 objects though Schröder seems to think it complicated is really very easy.

Dont fail somehow to remit at once, so that I may possibly still save my life from the consequences of that lie or if that is too late, may at any rate do something for my creditors. I intend the thing shall be known all over the world.

very faithfully
C. S. Peirce

Peirce to Russell. MS: RL 387. Undated, estimated based on context. Whether sent or not is undetermined.

[ca. 12 May 1896]

Dear Judge Russell

For God's sake send me *some* money if you can communicate with Carus. It is a matter of life and death!

I send a lot more stuff. Part is a rewriting of part of what I sent before. Part is a pretty thorough overhauling of Schröder's Vol III. It will raise the amount to *about* $ 300. I need it. I need *some* say $ 100 the worst way. If I had it, I could at once put you in the way of earning anywhere up to $ 500 in a few days by a little operation. But I have to have something to start on.

On p. 28 near the bottom of this new copy, you will see an allusion to yourself.

very faithfully
C. S. Peirce | 5 W 50^{h} St. | New York.

Peirce to Russell. Telegram: RL 387. Form headed, "Telegram | Postal Telegraph-Cable Company" with company seal at left.

New York, 13 May, '96.

Francis C. Russell,

---

1 1:4 ⤙(1:2)(2:4) ⤙$x\bar{x}$ 1:5 ⤙(1:2)(2:5) ⤙$x\bar{x}$] 1:4 ⤙(1:2)(2:4) 1:5 ⤙(1:2)(25) ⤙$x\bar{x}$ *[E]*

12 known] know *[E]*

No. 147 La Salle St.,
Suite 217, & 218,
Chicago, Ill's.
Remittance received.[89] Thanks. Letter to-day.

The Peirce. | 12:37 p.

Peirce to Russell. MS: RL 387. Undated, estimated per Peirce to Carus of same date.

[13 May 1896]

My dear Judge Russell

I this morning received a Chicago check so that I shall next week, or week after, get some money. I beg you to express to Carus my warm thanks. Its being too late we will leave in the background. You may think I make a row over little. But that is very far from the truth.

Please say that owing to the vagueness of the account of the form in which the article was to appear, I have been unable to give final form to it as yet. But that I stand ready to do as soon as I learn what is wanted and what modifications are desirable.

In my second copy a better form is given to some statements & new ones are added. Of course, I have *loads* to say of the book beyond what is in either copy. I think you and Carus had *[*better*]* go over the thing, decide what modifications (in a general way) you would like made & return to me when I will give the whole its final shape. My $ 25 a thousand of course includes the last touches.

very faithfully
C. S. Peirce

Address me at 5 W 50$^{\underline{h}}$ till I find a room.

Peirce to Russell. Recto headed, "Postal Card—One Cent | United States of America" with portrait at right.

[recto]

New York | May 16 | 7 PM *[1896] [stamped]*

Judge Francis C. Russell
145 LaSalle St

89 Carus to Peirce 9 May 1896.

11–12 You may ... the truth.] *inserted in-line and below*

13 of the form] *inserted above*

15 what[1]] *after* ~~in~~

Suite 217
Chicago, Ill.

[verso]

Communicate the substance of this to Carus without giving the card. I know already his love of unfair dealing.

I have restatements of several things which are vast improvements. I do not send them though they are all written out as I expect all the copy to be sent to me to make up into its final form. I am to publish something in another prominent journal of philosophy about the subject;[90] but my most comprehensive statement will go to Chicago if wanted there. Otherwise, its inferiority will be plainly stated elsewhere.

C. S. Peirce

Peirce to Russell. MS: RL 387. Undated, estimated based on context. Whether sent or not is undetermined.

[ca. 20 May 1896]

My dear Judge:

I hear nothing at all either from you or from Carus, which disturbs me; especially, as I am dreadfully afraid I have offended you, which I wouldn't do for the world. But where are those MSS? And why are not my pieces returned to me to put into shape according to the form decided upon by Carus since they were written? Observe that Carus has twice before dealt with me in a fashion which would ruin his reputation as an editor, if it were known; first, in printing a private letter giving a somewhat jocular account of my views of prayer & although consenting on my emphatic protest to suppress the letter, yet refusing to substitute a careful statement, which was unfair.[91] Second, in insisting on printing some hasty remarks of mine about Venn's book on Logic of Chance though I sent him, in ample time, a revised version resulting from reëxamining Venn's book.[92] This was a disgrace-

90 *Bulletin of the American Mathematical Society*; see Peirce to Carus 13 May 1896.
91 Peirce and Carus 5 and 9 May 1893, regarding the printing of Peirce's letter of 9 April 1893.
92 Peirce and Carus 24 and 27 Feb 1893, regarding the remark about Venn in "Reply to the Necessitarians" (July 1893), p. 528.

4–5 Communicate … dealing.] *This paragraph inserted at top of verso.*
6 vast] *above* ~~great~~
7 though they are all written out] *inserted above*
20 decided upon by Carus] *above* ~~given~~
20 written?] written$^{\vee}$?$^{\vee}$ ~~by Carus?~~
23 although] *above* ~~with~~
25 unfair] *after* ~~most~~
26 on Logic of Chance] *inserted above*

fully treacherous performance. Now if he does the like this third time, I shall act in a way which will injure his reputation for fairness,—if he has any,—and show that the code respected by all American editors of standing condemns him. I won't yet say he *will* act so; but considering his past performances, I have a right to consider the *possibility* of it. Philosophical discussion does not consist in stretching strings across an opponent's path to trip him up. He is a theologian in his ideas of controversy.

At the same time, having been handsomely paid by the Monist, I wish to offer him the best I can do, if that is what he wants. I have been writing something very careful, clear, and cogent about the questions raised by Schröder; and it will be published. That is settled. The Monist can have it, in place of the article written in a state of starvation, if Carus & Hegeler want it.[93] If they don't, it won't hurt me materially; but it will hurt Carus. I shall deal honorably with him, & if he thinks that he will gain an advantage by tricky behaviour to me, it will not be my fault. I don't say he does. I only say it would be in line with his previous performances.

My wife was turned out of the hospital, as I supposed. I paid them the next day. But she is now in imminent danger. I do not leave her bedside for over half an hour at a time; & as I am the only nurse, am pretty thoroughly done up. Still, I write my logic, just the same. I lost some 300 acres of bluestone quarries, owing to want of cash; and I fear I have lost my house. But I have had no positive information. I had arranged another way of getting the money in time; but being obliged to say whether I wanted it or not, said no, counting on the Monist money, which I was not able to get till yesterday.

The prospect is however that I shall soon be in receipt of large sums from an invention. It is not yet certain; but it looks that way.

I have a picture by Albert Bierstadt about $2\frac{1}{2} \times 3\frac{1}{2}$ feet with a handsome frame outside of that measure. He gets $ 2 500 for such a picture. There is a mountain with snow in the background, towering up high. It is big, solid, cool, calm. In the foreground large trees, rocks, etc. I came within an ace of selling it to a bank for their ladies' room. I shall be content if I get $ 1 250 for it. I think I shall probably sell it before long; but being unable to leave my wife's chamber, have not done so yet. Do banks in Chicago hang up such things? If you can find a purchaser for $ 1 500 or more, I will give you half of the excess above $ 1 000. It can be seen here. Of

93 This proposition also discussed in Peirce and Carus 13 May–2 June 1896.

6 an] *above* ~~one's path~~
17 imminent] immanent *[E]*
26 Albert] Alfred *[E]*
33 the] *after* ~~exe~~

course, this only applies if you find me a purchaser before I sell it independently. It is a fine picture, with great technique, and very natural. But there is no passion or sentimentality about it. It is a thing calculated to make people who look at it, especially women, calm and reasonable, and to suggest ideas of greatness, reliability, etc. That is why I thought of a lady's room in a Bank; & I should have sold it. Only the President said, "You are so persuasive, I don't dare to trust myself."

very faithfully
C. S. Peirce
5 W 50$^{\underline{h}}$ St New York City

Peirce to Russell. MS: RL 387. Undated, estimated per Russell's response 2 June 1896.

[29 May 1896]

My dear Judge:

I hear nothing from you, and conclude you are either out of town or are offended. The latter hypothesis is hardly possible if you reflect upon the consequences which befell me in consequence of not receiving the money as I had calculated upon. My wife was turned out of the hospital; and I have been working day & night to pull her through her imminent danger; and I have lost a valuable piece of property. Besides, I cannot think that even if you were offended, you would retain my MSS. against my will. I am forced to give an account of this incident in an article which is left imperfect in consequence. I have attributed it *[to]* your supposed absence.

Do you think you could sell for me in Chicago, a picture by Albert Bierstadt worth $ 2 500 for half price? If so, I would give you a good commission. It is suitable for a Ladies' Room in a Bank, or for a private house. About 2 1/3 × 3 1/2 without the deep frame.

very truly
C. S. Peirce
160 W 87$^{\underline{h}}$ St | New York City

Russell to Peirce. TS: RL 387.

Chicago, June 2nd, 1896.

Dear Mr. Peirce:—

---

1 find] *after* ~~sell~~

Yours of the 29th ult. is at hand. I am not in the least "riled," only I have been more than ordinarily busy, and so have neglected writing to you as I ought to have done.

I have not been able as yet, to get Dr. Carus to take your additional manuscript, although in my opinion it is much the more important matter in any review of Schroeder's book. He said it would make the article too long, but leaves some ground of expectation that perhaps he might buy it for the October number. I hold the same however, subject to your orders, although I would like to have you leave it with me a little while longer so I can study it. What put it into your head that I wanted to hold your manuscript? Such an idea has never entered my head. I drew the inference in some way or other, of which process I can give little account, that you were somewhat uncertain as to what you wanted done with the manuscript, and in the mixed condition of your affairs in and around your eastern home, you quite as leave have the manuscript out of harm's way.

The five chapters you first sent me[94] I will forward to you in two or three days, unless I hear something to the contrary from you very soon. The other manuscript I would like very much to study somewhat further, but of course your rights and wishes in the matter ought to prevail.

The chapter on the association of ideas, is such a marvel of comprehensive penetration, that I feel almost as though I wanted to get it by heart. The chapter on Reasoning in General, indices, icons, speculative grammar, etc., is also very precious in my eyes, and so also is the beginning of the chapter on the classification of sciences.[95]

I do not know at all what I could do with your "Bierstadt" picture. I have quite an acquaintance among young and struggling artists, but they have pictures to sell, rather than are buyers of pictures. Of course nothing could be done unless the picture was sent here, and I wouldn't like to have you do that until I had found some opportunity where it would be put on view, and then as to how much it would bring would depend upon circumstances.

I hope you are well and your wife improving, and I close with assurances of my deep esteem and not a little gratitude.

Cordially,
Francis C. Russell

Address hereafter | #1231–1235, Stock Exchange Building

94 Peirce to Russell 4 Oct 1895.
95 Peirce to Russell ca. 10 March 1896.

32 Cordially,] Crodially, *[E]*

Russell to Peirce ca. 5 June 1896 (missing): Returns Peirce's chapters from "The Art of Reasoning" sent 4 Oct 1895.

Peirce to Russell. MS: SIU 91.14.

1896 July 16 | Address: Care Albert Stickney Esq
35 Nassau St | New York City

My dear Judge:

We have been so knocked about from pillar to post, that I believe I have omitted to thank you for the return of the chapters of the Logic Book. You still have some of those I sent later. Having the time and inclination to write more, which I cannot later be in condition to do, I wrote a continuation for Carus[96]; since he was plainly disposed to take it if Hegeler would consent. The paper I have sent him is enough for two long articles.[97] It ought to fetch $ 400; but of course I would take less to get it printed. I told Carus in my letter I had other articles for the Mathematical Society etc. to write on the same subject, & so that I could know what I could put in those articles I should like to know at his earliest convenience whether he would accept what I sent. I fear he may not appreciate that what I have sent to him is a most brilliant contribution to the subject. I rely on you to expound it to him. In the first half I set forth a graphical method with the logic of relatives. This has not been altogether inspired by Kempe's work; for I had ideas on the subject before. I show that Kempe's method could never satisfactorily represent a proposition in the logic of relatives. The reason is that Kempe attempts to make the *spots* of his diagrams represent the objects and the *lines* connecting them the relations. I show that the only way is to make the *spots* represent relations and the lines the correlates. This is the illuminating, and altogether original, and far from obvious key to the problem. I also in the first half of my new paper give a brief account of the algebra of dyadic relatives which so enchants Schröder. In my second half I first explain my general algebra of relatives, which Schröder uses, and give brief rules for working it, which though much abridged are an improvement on those I printed ten years ago.[98] I also explain my theory of logical *involution*, or multiplication of the premises into themselves, which I consider one of the most fundamental dis-

96 Peirce to Carus ca. 15 July 1896.
97 "The Logic of Relatives."
98 Likely "On the Algebra of Logic: A Contribution to the Philosophy of Notation" (1885).

7 knocked about] *above* ~~abused~~ 24 key] *above* ~~solut~~

coveries I have ever made in logic. This though made many years ago, has never before been printed. It is a great and well-matured logical discovery going down to the heart of the subject. I then explain Schröder's general methods of procee-dure, praising them as they deserve, but sharply criticizing the limitations of his conception of problems and of their solutions. This was vaguely said before in my first article[99]; but now appears in fuller shape. I then press on to give a specimen of another kind of work to which the logic of relatives may be applied by investigat-ing the question of possible infinite collections. A more difficult logical inquiry it would not be easy to select. In order to treat it, I am obliged to create a new branch of logic the doctrine of substantive logical possibility. I give a theoretical view of this doctrine & proceed to enunciate three propositions from it which I apply to the theory of collections. In this way I demonstrate, by most ingenious reasoning very far from obvious that there are an infinite series of grades of infinite collec-tions. I fear Carus may think that these are nothing but the well-known "orders" of infinity. But they are nothing of the kind. In *collectional quantity*, there is nothing like an infinite of finite order. All the series of infinites which I develope are rather infinites of zero order. For an infinite of finite order multiplied into itself gives a higher order. That is not the case with the collectional infinites. On the contrary, each is the *exponential* of the preceding. Calling the first, which is the multitude of whole numbers, E_0, the second is $E_1 = \mathfrak{S}^{E_0}$, the third is $E_2 = \mathfrak{S}^{E_1}$ etc. This is absolutely new, except that Georg Cantor makes out the two first; and it is *demon-strative*. It is the breaking of the ground for a wonderful new logic of mathematics. I shall show in following writings that in other kinds of quantity there is nothing of the sort. You will see, therefore, that the paper is quite the most brilliant that has ever appeared upon the logic of relatives & will be a feather in the Monist's cap. Make Carus see it will help him if you can

C. S. Peirce

Of course, you understand I would like at least half the amount as soon as I can get it.

Peirce to Russell. MS: RL 387. Undated, estimated based on context.

[late July or early August 1896]

---

99 "Regenerated Logic."

---

23 following] *after* ~~sub~~

28–29 Of course ... get it.] *This line up right margin.*

My dear Russell

I sent Carus a second article (long enough for two) on the Logic of Relatives.[100] I think he seemed to wish to take one if Hegeler would consent. But I have not heard whether the MS came to hand.

With a moderate amount of help at this crisis, it looks strongly as if I should be all right in future & be able to pay all debts. But just now I fear all will be lost for want of small amount. If you can persuade Carus to advance $ 100, I beg you to do so.

very truly
C. S. Peirce
Care A Stickney, Esq | 35 Nassau St | New York

P.S. I have a more formal and exact statement of all the properties of the graphical method, if Carus prefers such a statement.

Russell to Peirce. LPC: RL 387.

Chicago, Ill., August 11, 1896.

Prof. C. S. Peirce,
35 Nassau Street,
City of New York.

Dear Mr. Peirce:

I am quite shocked to remember my negligence in answering your letter. The fact is, soon after I got your letter, Dr. Carus wrote to me in respect to your manuscript. I don't know why I haven't written on the occasion this made, to you, but the fact is, I haven't, and plead peccavi.

Now the Dr. is a good enough friend of yours, but has his limitations, like all of us, and he has to cut in between yourself and Hegeler. He wrote me that he would not feel justified in paying over $ 250 for your last most improved form of your article. I ought to have immediately told you that he would pay that amount. Now, if you are willing to settle for that, I am sure the Dr. will send you a check at once. In fact it would be the paying by the Dr out of his own funds, taking the risk of making himself whole on Hegeler's account.

100 Peirce to Carus ca. 15 July 1896.

12–13 P.S. I have ... statement.] *Postscript written at top-right corner.*
21 Carus] Caius *[E]*
25 Hegeler] Hegler *[E]*
30 Hegeler's] Hegler's *[E]*

There is a matter about which I scarcely know what to say to you, because I am sure that I entirely agree with the Dr. and don't know how you will be apt to take it. Both the Dr., and especially Mr. Hegeler, are excessively punctilious about their reputations as paying for their dues. A telegram, whatever may be the pretended secrecy, is more or less a publication and when, at LaSalle, they get a telegram demanding money, it puts both the gentlemen in a rage.[101] If you will make me, as much as you can, a mediator, I will promise no further neglect, and my judgment is you can market a good deal of your manuscript at LaSalle.

Hoping you are nearing your deserts in some measure, I remain as ever

Cordially your friend,
Francis C. Russell

Peirce to Russell. MS: RL 387. Stationery headed, "Stuart House | Broadway & 41$^{\underline{st}}$ St. | New York | R. A. Stranahan" with company seal at center.

Sep. 21 *[1896]*

My dear Judge:

I sent to Carus a chapter of my Logic on Graphical Methods of treating relatives.[102] I prefer to have this substituted for the corresponding part of the MS sent him, provided it be deemed suitable for the Monist. He thought it was & I asked him to let you see what you thought, if you have time.[103] But really, it strikes me that if he takes this which is a long article in itself, in addition to the rest of the MS sent which is another very long article, he must feel that he ought to send me some more money. Of that he sent, one half had to be paid away at once to a friend who had advanced me that sum & who was grievously in need of it. The result is that I want some money the worst way now.

I don't want to influence your judgment about the availability of the last MS. But if it is taken, don't you think you could say a word to Carus about the money. I can't.

I enclose you a little piece.[104] There is no chance of the Open Court wanting it; but perhaps there is some paper you know of that would take it. If not, no matter. Ce n'est pas grand'chose.

---

101 Exact telegram not known. Possibly Peirce to Carus 11 May 1896.
102 "On Logical Graphs," Peirce to Carus 31 Aug 1896.
103 Peirce and Carus 8 and 11 Sept 1896.
104 Undetermined.

---

3 Hegeler] Hegler *[E]*
5 or] a *[E]*
23 sum] some *[E]*

A hundred dollars just now would be—

very faithfully
CSPeirce

Peirce to Russell. MS: RL 387.

Stuart House. 41$^{\underline{st}}$ St & Broadway, New York | 1896 Sep. 28.

My dear Judge:

I dont get any of your delightful letters of late. How is it? Are you so taken up with politics, and if so on which side of the fence are you ranging your forces? It is needless to say that living in New York I am in favor of Palmer & Buckner.

Come; have you not time to drop me a few thousand words?

Dr. Carus seems to be better disposed toward my logic than he formerly was. Partly, he is influenced by what you and Schröder and others tell him; partly, he does not now think that there is any danger of my injuring him. Don't you think you could now bring it about that the Open Court Co. should publish my book of logic? You have read a number of the chapters. I should like the whole of the formal part to go to press substantially as you have read it. The philosophical introductory portion, which you have not seen, is also perfectly satisfactory to me. Those two parts would make a volume which might be called Qualitative Logic. The chapter on *Quantitative Logic*, as you saw it, is very bad. There are several years of hard continuous work on the subject not represented in that chapter. A few months more of terribly hard work are needed to complete it; and then in my opinion the result, which will take a small volume, will be epochmaking in the philosophy of mathematics. Upon this would follow a treatise on probabilities, containing much that is new; and then my theory of inductive reasoning, which I could make essentially stronger and also clearer and with better illustrations than in the Studies in Logic. The subject of Induction would be treated from the point of view of the logic of relatives, which is absolutely new. Here would come in, too, practically important discussions of the conceptions of Cause, Burden of Proof, Sufficient Reason, Fact, Explanation, etc. This would together make a pretty large volume to be called Quantitative Logic. This would still leave my "Objective Logic," untouched. Thus, of the three volumes, one is quite ready. Another would be ready by the time the printers wanted it, if I got an advance of money to give me time to do the work. The third would have to wait.

I would issue the three as separate works with distinct titles.

---

28 practically important] *after* ~~very important~~

32–33 if I got … the work.] *inserted above*

If you will negotiate the sale of the first volume to the Open Court they to pay me a sum down and the rest in the future, I will give you a commission of 10%. I am awfully in want of money. I must have $ 100 or $ 150 at once or go under.

You still have some MSS. of mine which I should propose to make use of in revising the introductory chapters, which you have not seen. I believe you are aware that my Arithmetic MS. was seized two years ago;[105] and if I don't have the money to get that back soon, it will be gone forever. However, the sum I now speak of would not enable me to do that. But if it came promptly, it would enable me to sell my place, which is just now in demand. I cannot now go near it or do anything; and it has been broken open & things are disappearing fast.[106]

very faithfully
C. S. Peirce

It would have been far better if I had made way with myself as soon as my troubles began. Foolish procrastination—

Russell to Peirce. TS: RL 387.

Chicago, Sept. 30, 1896.

Dear Mr. Peirce:

Yours of the 28th inst. has just come. As to politics, I am a Bryan[107] man, not because that I have great faith that the free coinage of silver will bring about a millennium, but because I am a democrat, with socialistic proclivities, and it seems to me that whatever may be done to curb the plutocratic tendencies of the times, is well done. I am somewhat puzzled to solve the problem of how it is that a person with the principles I know from your writings you must possess has been led to favor the gold democracy, but "*chacun à son goût.*"

105 See Peirce to Russell 8 Sept 1894.

106 Peirce had been evading arrest in New York since April 1895 due to the Walters lawsuit for alleged assault. For details, see Peirce to Russell ca. 10 March 1896.

107 William Jennings Bryan (1860–1925) American politician and U.S. presidential candidate during the 1896 election, which he lost to William McKinley (1843–1901) on 3 Nov 1896. Bryan was a proponent of the Free Silver Movement that opposed the gold standard, as Russell references. The 1896 presidential campaign resulted in a political realignment in the U.S. and approached the end of an economic depression that had begun in 1893 (the Panic of 1893). Peirce refers briefly to this election and its political and economic impact in his responses to Russell ca. 10 Oct and 5 Nov 1896.

4 MSS.] *above* ~~things~~
6 two] *above* ~~some~~
13 if] *after* ~~for~~
19–20 millennium] millenium *[E]*
24 *chacun à son goût*] *chacun a son gout [E]*

On May 1st I entered the office of Story, Westover & Story as a partner, thus resuming active practice which I had suspended since I went out of office several years ago. My time is therefore very fully occupied, much more fully than I have a taste for, for I am naturally lazy. However, I am in for it and so have to do the best I can. My suspending my letters to you is a part of the things I have been obliged, by circumstances, to do.

I wish above all things that Dr. Carus would take hold of the matter of the publication of your Logic. I need not stop here to express to you what I have so often said, that I regard the work as one altogether singular in its excellence and importance. But, it so happens that no one but an expert somewhat used to the topic in general can form any adequate idea of how great a work it is. Even Dr. Carus, I fear, only regards me as a mere enthusiast when he hears me recite my commendations. Had not Schroeder published his volume, and had not Schroeder been a German professor of high standing in his country, I fear that Dr. Carus would have let what I had to say in at one ear and out at the other. Even now the doctor is very much inclined to believe that Schroeder overlaid you in a marked degree, even though Schroeder on almost every page of his great volume refers to you as the "hauptfoerderer" and great inventor in the line of research on which he writes. I do not believe it will be possible at present to interest the doctor as we desire. The attention of the doctor is at present largely occupied with a new edition he is getting out of the "Tao te King" and really he is making a very interesting volume; one which will be of great use to me in a speculation which, as I have time, I am indulging in, namely: the exemplification of what would really be an improved language. Instructed by you and by your manuscripts, I have a mass of ideas upon this subject which, when I have time to put them into order will, I believe, be of considerable interest.

I shall, however, open up the subject of your letter to the doctor as soon as I get any opportunity. Your long article on the Logic of Relatives cannot appear in the Monist before the January number, and perhaps then only by its first half. The doctor is still undecided as to whether he will embody in the article your manuscript on "Logical Graphs". He says if he does so the whole article will have to be divided into three parts, and he does not like to take so much space. In speaking to the doctor in commendation of your work, I have occasion for the use of all my tact, and it is absolutely impossible to hurry matters. I am quite sure that had your late articles already been published and had time to draw to them the attention they are certain to gain, so that the doctor could observe the interest they excite, the way would be quite open for him to take hold of the Logic. I will speak to the doctor and let you know as soon as possible what the result may be.

5 the] the the *[E]*

Cordially,
Francis C. Russell

Peirce to Russell. MS: RL 387. Stationery headed, "Stuart House | Broadway & 41st St. | New York | R. A. Stranahan" with company seal at center. Undated, in response to Russell's of 30 Sept 1896.

[ca. 3 October 1896]

My dear Judge:

Thanks for your letter. I feel I owe it to you to answer your question as to why I favor gold. You say you are a "democrat with socialistic tendencies." I suppose I am, too, if such a thing is not a contradiction *in adjecto*. A democrat believes in individual freedom & as little government as possible, while a socialist proposes enormously to increase the functions of government. I am a democrat in my feelings while convinced that socialism must and ought to be extended even further than at present.

My natural sympathies are with the poor rather than the rich; and I favor what will elevate the masses. For the same reason I should be very sorry to see them commit an act of great folly which will reduce them to the condition of the poor in silver countries. Besides, while honesty is the best policy in almost all cases, it is especially so on the part of those who need help. I do not think the Free Silver idea is honest.

I think it would be the most disastrous thing possible for all classes; but it would almost enslave the poor, while producing terrible want & many deaths and unspeakable woe of every kind in the coming winter. I think it would be the final knell of the cause of intellect in this country,—if there is any life in that cause now.

All this, my dear Judge, makes no pretence to being argument. It only explains to you, as my friend, what you wanted to know, namely, how my favoring gold coheres with my declared sentiments.

very faithfully
C. S. Peirce

PS I will add one thing more. My wife has been a great sufferer physically & among the numerous physicians who have been consulted there have been several who upon talking to me have been so impressed with my knowledge of medicine that they have asked my opinion. Each time that has happened, as soon as the Doctor left I have sat down & written for his bill. For I want my wife treated scientifically and not in an amateurish way; and if a doctor thinks my opinion of

20 disastrous] dis\as/trous
22 in the coming winter] *inserted above*
23 cause] caus *[E]*
33 the Doctor] *above* ~~they~~

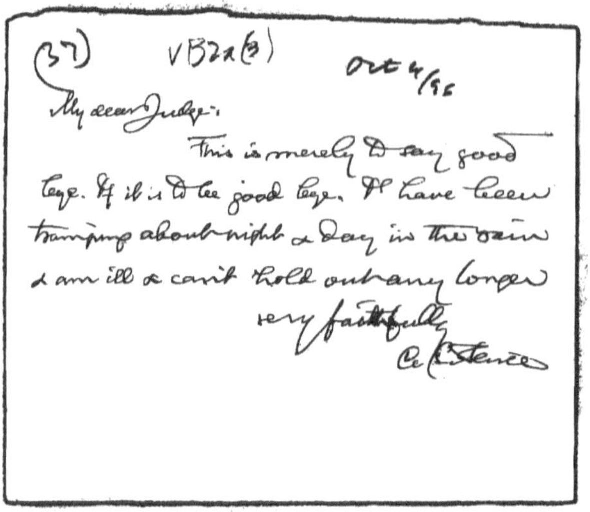
(37) VB2a(3) Oct 4/96
My dear Judge:
This is merely to say good bye. If it is to be good bye. I have been tramping about night & day in the rain & am ill & can't hold out any longer
very faithfully
C. S. Peirce

Figure 10: Peirce to Russell ca. 4 Oct 1896. Death note.

any consequence, I think he is a dam quack. It is the same with problems of government. Out West your voters seem to think their business is to have opinions about questions of the utmost difficulty. Our voters East think,—as it seems to me with good sense,—their part is to decide questions of honor and also in what doctors they will confide.

Peirce to Russell. MS: RL 387. Undated death note, estimated date noted on MS. Whether sent or not is undetermined.

[ca. 4 Oct 1896]

My dear Judge:

This is merely to say good bye. If it is to be good bye. I have been tramping about night & day in the rain & am ill & can't hold out any longer

very faithfully
C. S. Peirce

4 their part is] *above* ~~is~~

Peirce to Russell. MS: RL 387. Undated, estimated date noted on MS. In response to Russell's of 2 June 1896 (belated due to mail discrepancy).

[ca. 10 Oct 1896]

My dear Judge

Owing to a letter you wrote in June having been put into the wrong bag it never was answered & I never knew your right address which was given in that alone. You asked me why I thought you wanted to keep my papers. I never did. I simply could not find any way of inducing you to take the trouble to send them back; and I was finally driven to bawling & yelling for them. I had particular need of some of them.

I think I told you that I had made an invention relative to acetylene.[108] A company of promoters well provided with money,—not excessively rich but with a couple of hundred thousand, finally agreed to give me one fifth of all their European profit & they proved themselves extremely loyal. We have not yet started up the thing either here or abroad, because of the Bryan derangement of business.

But meantime in looking about for means to prepare the carbide of Calcium cheaply, Ca_2C_2, they have fallen upon something the most stupendous,—almost an Arabian Night's cave of wealth. They might have said that it had nothing to do with European profits & kept me out. But far from that they are going to give me, I think, one fortieth of the profits. It will be enough to ensure the completion and publication of my whole philosophy.

It amounts to this. We already have the most extraordinary charter from the state of New York which never would have been given if it had not been supposed the object (as it *was* with the original company) was to supply a town with water!

What it gives us is the means, by the expenditure of less than a million (as the very first engineers have calculated) of getting over 200 000 horse power at all seasons, with several square miles of land, the very most accessible and everyway suitable for a city imaginable. As soon as the election is over, if it goes for McKinley, the money wanted (which we could almost make up ourselves) will be got without difficulty. The 200 000 horse power can be rented *at once* for $ 14 and it is safe to say the profits will be $ 10.

Now if you want to go in, I will take you in on the ground floor, that is, at the same rate I shall pay for anything above what is given to me & which I can't sell. There will be four millions of shares & two millions of 5 percent bonds.

---

108 Peirce to Russell 26 April 1896.

---

12 of promoters] *above* ~~very~~

The very ablest engineers in the world hydraulic, electrical, chemical, have reported on it.

Not a share will ever be given to the public. It will be a close corporation. I want you to be in it. But dont talk about it, either now or later. You shall see all the evidence but must keep quiet.

This brilliant prospect does not prevent my being on the verge of starvation for want of cash. I havent a roof over my head, but am wandering penniless about the streets. So if you can induce Carus to advance something for my logic or to take my Dictionary *raisonné* do so. This last is a new Roget's Thesaurus, arranged on the principles of my philosophy and with indications of the precise meanings of the words & illustrative quotations.[109] These are not of the vague and conjectural description of ordinary books of synonyms; but positive facts. The illustrative quotations are as familiar as possible; because familiar phrases & quotations do much to determine the shades of meanings of words.

I have done a good deal on this book, which I began in the sixties, & have worked on from time to time ever since, & always had its problems simmering in my brain. Roget's Thesaurus in the American reprint still brings in about $ 15 000 a year. So I think there is no doubt mine will pay. I will *now* sell a share of it, not over one half at a valuation of $ 6 000 for the whole, to be paid $ 500 cash down and a second $ 500 on the completion of a hundred "numbers." By a "number" I mean a division of which there are to be about a thousand; and the introduction and scheme of numbers is to count for 200 numbers, that is, is to be valued *now* at $ 10 000. If a person agreed to aid to the extent of $ 2 000 that would pay for 400 numbers; and that is all the help I should need for by that time I could borrow from publishers on it, even if my ship were not in. I can do about a number a day. In my opinion a person putting $ 2 000 to $ 3 000 in it (I would prefer the former sum) would get their money back in 4 years or 5 years and after that would get cent per cent for many years.

Perhaps you can talk this into somebody

very faithfully
C. S. Peirce
Stuart House | 41st St & Broadway | New York

---

109 a.k.a. *Peirce's Logotheca*, which Peirce tried (and failed) to publish in May 1894 with New York publisher George A. Plimpton and in August 1897 with *Century Dictionary* editor Benjamin E. Smith (Brent, 237–238 and 261).

---

9 *raisonné*] *raisonnée*   23 aid] *after* ~~take~~

Peirce to Russell. MS: RL 387. Undated, estimated date noted on MS.

[ca. 5 Nov 1896]

My dear Judge

There will be no great boom in business for two reasons, first, because the Bryan vote was so large, and second, because there is a danger of another extravagant tariff. But there will be sufficient revival to make voters for Bryan feel differently, I think. Meantime, everybody can see this, that when the labor and brains of the country have anything to propose which would be honest and honorable and at the same time would be genuinely and truly for their interests, they will carry the country fast enough. That being the case, there is no occasion for hysterics. It isn't by any means the worker with brawn or brain who is in danger of being trampled on by any public measure, but the rich man. But it is not likely that a majority of such workers will think it for their interest to curtail accumulations of capital in our day. That will not come until they are far worse off and less intelligent. Consequently, and in view of the vast power which rich men wield, I don't think they can regard themselves as in serious danger either.

The real danger of our affairs is that such ignorant men are sent to congress; also, that when we make some headway against protection, we are thrown back and have all our work undone by the extreme folly of the democrats. When I review the history of that party from the time I was a boy, say from Pierce's[110] election to this day, the way they have set back their own cause, time, times, and half a time, by some error strikes me forcibly.

My grand schemes for making money are all going to come to nothing, because I am so infernally strapped hand & foot for want of money. That waterpower scheme will certainly be such a vast success that my share in it will be worth several millions of dollars. It just eludes me. Never was there such an opportunity. But it will escape me because I cannot go to London and exercize the influence which I am sure I could do to make the present arrangement succeed. Unfortunately I naturally cannot prove to anybody that I have such influence. I know it myself; but I cant raise money on it. I can easily convince any engineer of the immense value of the scheme; but I cant sell any of my share, because I am out of the business unless the man we have sent to London floats the bonds. I know he would, if I were at his elbow. For he has the very best of acquaintances in Lombard Street, and I have the very best in the Scientific and Engineering Worlds. And the

110 Franklin Pierce (1804–1869), U.S. president 1853–1857.

8 have] *after* ~~propose~~
9 be] *over* g
31–32 I am out of the business] *above* ~~it is worth nothing~~

project is easily proved to be immensely valuable to any engineer while the Lombard Street men will act according to what the engineers say. I went down to the promoters today & offered to bring in one of the first engineers in the country (a Chicago man, too) if they would guarantee him a small share in the stock, he to advance me $ 1 000 or $ 1 500 to go to London and help secure the placing of the bonds. They replied, our interest does not lie in having those bonds so placed; for we could make a much better deal and leave you out. We felt bound to give you the chance to place them on the same terms as the acetylene machine of which this was the offshoot. But it is not for interest that you should do so. You will have our loyal aid so far as we are bound to give it, but we shall not give any stock to help you beyond what we have agreed to, and certainly we wont guarantee that any engineer or anybody shall have stock if you do not succeed. We don't even mean to sell shares; far less give them away. This was the general idea they conveyed in a few words & pretty forcible ones.

Not being able to raise money in that way, I naturally cast about for some other way. Now this idea occurs to me; and if you can get me the money I want I will give you 10% not only of that but of all the money it leads to my making from the waterpower up to a limit of $ 5 000 a year paid to you. That is should I sell out my rights you would be entitled to 10% up to $ 100 000 and until I sold out to 10% of dividends up to $ 5 000 a year. I fully expect it would amount to that very soon, and go above it. Of course, I may get the money elsewhere if I can before you get it.

The idea which occurs to me is this. Bierstadt has invented a car which opens out into a room 27 feet wide. The Russian & German governments have taken it up; the N.Y. Central people are about to go into it. It is a very practical thing. It goes about like a car, and then can be transformed in a few minutes, by lifting the roof and drawing up sides for part of the roof and by letting down inner sides for a floor and other movements, into a chapel, or a theatre, or a picture gallery, or a shop, etc.

I believe a car costs $ 8 000. It is quite cheap. Now let Hegeler pay me $ 2 000 and I will get him the right to build a number of such cars for Sunday or other lectures on the Religion of Science, which being sent about the country & free sermons & lectures given, & would distribute the Open Court and raise that to a satisfactory paying basis. The lecturers (preferably two) would sleep and eat in the car, and their expenses would be light. If it would add to the inducement I will

2 will] *after* ~~without such~~
7 We] *over* ~~The~~
13 the] the the *[E]*
19 and] *after* ~~or~~
21 Of course, ... get it.] *inserted below*
23 out] *u*out
25 lifting] *after* ~~letting down~~
29 I] *after* ~~It~~
33 and eat] *after* ~~in the e~~

give 100 lectures for my bread & butter during the trip in case I dont succeed in placing those bonds; and if I do, will contribute 10% of all I make, without limit, to the support of such lectures and of the publication of books by the Open Court Company, without asking for any vote in regard to how it is to be spent. Of course, I should expect any recommendations that I might make to be civilly considered.

I would put in more; but that my intention is, after giving my wife the third to which she is justly entitled, to devote the rest to a self-supporting brotherhood to include, if possible, a first class logician, mathematician, physicist, chemist, biologist, psychologist, and then for the sake of the earnings practical scientific men, who are to be sworn to devote their lives to the investigation of the evolution of the universe. You see were my wildest dreams anticipated, it would need all I could make by all my inventions and otherwise to get this thing properly started. It would be a great civilizing agent, far more than you would think until you deeply consider it in all its aspects.

It is strange I should have such vast powers almost within my grasp when I am often for days & days at a time without food or shelter. But there is not the slightest doubt of the value of the enterprise. Upon that the greatest engineers in the country have reported favorably. Nor is there, to my mind, the slightest doubt the men I am in with are exceptionally upright & honorable men, and fully mean to and *will* give me my share of a twentieth in case those bonds are placed.

If they are not placed, I have only indefinite claims to a share in the thing; and it wouldn't amount to much.

I am so confident of the thing that if Hegeler would supply $ 2500 so that I should have a little more to leave my wife for preparations for the winter I would throw in MSS of Logic, Elements of Mathematics, both prepared for the press, though I want to alter both in some respects, and would do so for nothing if I succeed & if not for the least sum possible, together with all the MSS of lectures I could find, of which there are a lot on the History of Science, the History of Logic, theory of Logic, etc.

Anyway if you think it is worth while trying the thing in the way I suggest or in any other, the commission I offer remains the same—

Yours faithfully
CSPeirce

If I go over, I shall go in the steerage; but after I get there I must consult appearances.

---

6 but that] *above* ~~But~~
8 to] *after* ~~for the~~
34–35 If I ... appearances.] *This sentence at bottom left corner.*

If you were only to witness the horrible sufferings of my poor wife from starvation you would make haste to help.

P.S. I enclose a view of the church car open. The chancel at the end shows the size of the section when shut up for travelling, that is 9 × 9. It folds up and the roof lets down. The benches all fold up flat. Of course, there are as many benches as you please. If not wanted, or if nothing is done, please return the picture to me Care A. Bierstadt Esq 1271 Broadway New York, which is my address for the present.

Dec 1896–Sept 1904: No letters between Peirce and Russell are found for this period. For an overview of significant events elsewhere in the correspondence and other related context during this time, see the Peirce and Carus correspondence and the Chronology for this edition.

Peirce and Carus late Sept 1904: Peirce proposes to write a review (later titled "Substitution in Logic") of Hippolyte Taine's 1870 *De L'Intelligence*. Carus in a missing letter replies accepting the proposal but suggests that Russell sign it in his name.[111] Peirce complies and opens correspondence with Russell as follows.

1904

Peirce to Russell. MS: RL 387.

Milford Pa | 1904 Oct 3

My dear Mr. Russell

Our friend Carus suggests that you might sign a communication to the *Monist* that I at first thought of making anonymously.

---

111 Apparently because Peirce quotes his own "Natural Classification of Arguments" (1868) in the review. Furthermore, "seeing that Mr. Peirce cannot sign the communication himself," Carus wrote Russell 7 Nov 1904, "and as the editor should not publish it under his own name it seemed to me best to have some one of Mr. Peirce's friends who would unhesitatingly endorse his views, sign the paper." (RL 387)

1–2 If you … help.] *This sentence at bottom right corner, below signature.*

3 a] *over* in

I have rewritten it to that end; but should be more pleased if you would state the matter in your own language and from your own point of view.[112]

However, I cannot ask you to take that trouble. If you dont object to signing it, then as I have rewritten it, I shall be greatly obliged to you. If you prefer to word your own statement I shall be still better pleased; and in any case you will not hesitate to make any alteration you see fit

very truly
CSPeirce

Russell to Peirce early Nov 1904 (missing): Accepts Peirce's proposition to submit "Substitution in Logic" in his name.

Peirce to Russell. MS: RL 387.

P.O. Milford Pa 1904 Nov 15

My dear Russell:

I receive your witty and valued letter as I am in the midst of things getting ready a long paper[113] for the Natl. Academy of Science which I am told a quantity of people have been invited to come in to hear. For that reason I cannot now finish reading your letter. I only read as much as I could read while drinking a cup of tea. I have been at work on this for eleven consecutive hours today & shall work till I go to sleep.

As to what you say about me, partly seriously, strictly *sub rosâ* I hold that a man of 65 well read in philosophy & a thinker himself must be a precious fool or be able to place himself better than anybody else can do, and I place myself somewhere about the real rank of Leibniz. Of course, Leibniz had the advantage of coming to a field into which no reapers had come. But what I want to say which is more practical, is that I am by nature most inaccurate, that I am quite exceptional for almost complete deficiency of imaginative power, & whatever I amount to is due to two things, first, a perseverance like that of a wasp in a bottle & 2nd to the happy accident that I early hit upon a METHOD of thinking, which any intelligent

112 Russell's rendering of the article appeared in the April 1905 *Monist*. For a transcription of Peirce's MS, see Appendix, "Peirce and Russell."
113 "On Topical Geometry," delivered that day or the next.

1 pleased] *over* s
3 ask] *over* ex
4 rewritten] *over* w

person could master, and which I am so far from having exhausted it that I leave it about where I found it,—a great reservoir from which ideas of a certain kind might be drawn for many generations.[114] It is a pity that necessities have prevented my having a scholar to take up this method. From my point of view, it seems like awful stupidity & waste. But that is the way of things, they get done, but they get done in the least economical way imaginable.

As to what you say about being willing to put your signature to anything,—if it were a question of a mathematical demonstration, that would be very kind, but it would be possible. But when it comes to statements of fact, I cannot ask any friend, a lawyer least of all, to testify to what he does not know of his own knowledge.

I will send you my latest on the Classification of the Sciences. I had forgotten completely all those papers & do not recall anything about them. But I must go back to my job.

very affectionately
CSPeirce

P.S. Add to the elements of whatever success I have had that I have always unceasingly exercizing my power of *learning new tricks*—to keep myself in possession of the childish *trait* as long as possible. That is an immense thing.

1905

Peirce to Russell. MS: RL 387.

Milford, Pa. 1905 July 3

My dear Russell:

Professor Vailati sent me the other day from Florence (where pragmatism flourishes, and I noticed not long ago that one of them, Calderoni, was accused of being "più Peirciano che Peirce," more Persian than C. Shahal Perce, I suppose it is to be translated) a very little paper upon a very large subject, "Sull'arte d'interrogare,"[115] and among the great number of letters I am receiving in these days, I note how much more to the point are the questions of the logicians than those of the psychologists; and I was particularly struck with this in regard to your way of putting the question of the connection of soul & body, "What is the use of consciousness?" Yes, that is just the question. The XIX th Century gave such

114 For various theories about the "METHOD" Peirce might be referring to, see Brent, 322–347.
115 *Rivista di Psicologia* (1905).

prestige to science that there are quantities of people who seem to think that the very highest grade of certainty about everything is to be attained by laboratory experiments; and because physiological experiments, from the very nature of them, cannot bring to light any real efficiency in consciousness, they conclude that it has none. It seems to me the very lowest depth of pedantry. That my cogitations as to whether I shall be more comfortable during a given day with my windows open or with them closed determines the physical fact of their remaining closed all day or being opened in the morning is far more certain than all the laboratory experiments in the world. A man must be deficient in *logica utens*[116] not to see that. It is true that there may be another way of stating the matter which is equally true. An insect which lays its eggs with the utmost care in a place where the progeny will have a supply a food, if asked what made her lay them there, would say it was the intense delight she takes in laying eggs in such a fine place. For she has no means of knowing (we will suppose,) that the eggs will ever produce anything. Now if we were to undertake to judge of this case by itself, the apparent mechanical regularity of the insect's conduct, would make it hard to say whether there was any truth in her view or not; but judging by the analogy of the higher animals and man (how remarkably like insects these Japanese are, by the way!) it seems to me proper to say that the insect is right. But that does not make the Darwinian wrong in saying that she lays her eggs there for a purpose she knows nothing about, the preservation of her race. On the contrary, this evidently goes deeper into the question asked. Nominalism introduced the notion that consciousness, i.e. percepts, is not the real thing but only the *sign* of the thing. But, as I argued in the Pop Sci Monthly for Jan 1901,[117] these signs are the very thing. Reals *are* signs. To try to peel off signs & get down to the real thing is like trying to peel an onion and get down to onion itself, the onion per se, the onion *an sich*. If not consciousness, then *sciousness*, is the very being of things; and consciousness is their co-being. When one suffers from rheumatic fever,—for which I am keenly sorry for you,—or *gout*, isn't it,—too much beer, perhaps?—the kidneys not equal

116 As opposed to *logica docens*: "In all reasoning, therefore, there is a more or less conscious reference to a general method, implying some commencement of such a classification of arguments as the logician attempts. Such a classification of arguments, antecedent to any systematic study of the subject, is called the reasoner's logica utens, in contradistinction to the result of the scientific study, which is called logica docens." (*CP* 2.204–205)

117 "Pearson's *Grammar of Science*."

13 the] *over* her
18 it] *over* I
19 Darwinian] *above* ~~evolution~~
20 lays] *over* do
22–23 consciousness ... thing] ~~conce~~ consciousness$^{\vee}$,$^{\vee}$ ~~and~~ \i.e./ percepts$^{\vee}$,$^{\vee}$ ~~are~~ \is/ not the real thing
28 rheumatic] rh*u*eumatic

to the work put upon them,—I have never had it, myself, but only slight twinges of gout, but I have seen intimate friends suffer fearfully with it, so that my heart and sympathy are with you,—why under these circumstances one would like to shut the doors, & not have so much "cobeing." One puts in the willow-acid to dissolve the stuff & sometimes dissolves parts of the body that one cannot well get along without,—teeth & bones & such like. I have had several relatives on my mother's side whose lives were ruined by rheumatic fevers. One of them, a most charming man, took to laudanum, which he would drink by the tumbler-full. The result, of course, was that he lost all power of doing work; but he was a most delightful companion and a true gentleman to the last. In this respect, he did not forget that the true consolation is more cobeing, not less. In my miseries, which have not been light, I have found philosophy really did afford consolation & especially that recommendation of Paul in his fine epistle to the Philippians. "Whatever things are true, whatever things are fine" etc.[118]

So it seems Teddy is thinking of Judge Taft for Secretary of State. That is Teddy, all over. In my opinion it would be better to send a fleet to capture Iceland and make Taft the viceroy. Because there there would be nothing in the climate to heat the blood. In Washington, as summer comes on some hatchets fly off their handles. I think Wallace had a perfect right to resign. I think it is a symptom of insanity to suppose that Alexander or even Hyde have done anything plainly criminal. I haven't thought much of the Barrens since the Beecher scandal; but I think there was some excuse for this one. I think Loomis has been punished (and will be) enough. Roosevelt is one of the best presidents we have had. It is his theatrical charlatanism that makes him popular.

I doubt if the next one will be any better. We dont deserve a better. Government is certainly the worst managed of all human affairs.

Lately when I was suffering at every mouth through which a man can drink suffering, I tried to beguile it by reading three books that I hadn't read for a long time, three religious books; Bunyan's Pilgrim's Progress, Boethius's Consolations of Philosophy & Hume's Dialogues concerning Natural Religion. The last did one most good owing to the utter blindness of the man. Man can naturally get but a vague idea of the *all* of things; and a vague idea is always open being driven into contradictions. But man will never find a doctrine of the all nearer truth than

118 Philippians 4:8.

7 whose] who *[E]*
11 consolation] *above* ~~remedy~~
25 Government] Goverment *[E]*
31 get] *over* ha

theism. People say there are no miracles nowadays to convince us of the truth of christianity; but it seems to me that a person for whom the very slight extent to which christian pastors are addicted to rogering the females of their flock is not miracle enough is simply insatiable

ever faithfully
C. S. Peirce

Russell to Peirce. MS: RL 387. Pages 5 and 9 only of letter. Estimated to be the letter to which Peirce responds 14 July 1905. Also, the versos of these pages contain logic notes, formulae, and graphs by Peirce that closely resemble those in Peirce to William James Aug 1905 (RL 224; Pietarinen, forthcoming), possibly dating this letter around the same time.

[early July 1905]

[…] Now the sovereign sway of the Life appetency is not perceived as it ought to be because *Pleasure* so generally (but not always) attends upon Healthful Life and just because Pleasure the more special fact is more *intensely* noticed than is Life, its source and matrix. Purblind writers upon Ethics have erroneously resolved that the superlative stress that rules the conduct of man has Pleasure for its final course, and to this resolution they adhere in spite of the common knowledge that all direct pursuit of Pleasure proves abortive while the same arises into being upon all sorts of occasions without any plan or conscious effort for its realization. The substitution of this merely subjective condition all perfectly uncompassable as it is to any direct effort towards its realization *[…]*

[Pages 6–8 missing]

[…] to do and which proposed shirking you yourself condemn. Dont be absurd! Do'nt! Do'nt!. Do'nt!. What you want with supreme craving and will to have is Life, more Life and richer Life. The shirking you propose is as you yourself confess neglectful of the interests of Life. How irrational and base for you to default!. You ca'nt have your cake and eat it too.—

Such like is the case with all the details of conduct as well in regard to acts of commission as to those of omission.

In matters of conduct as well as in matters of cognition we proceed from facts known or believed to be such. These facts entail consequences which in matters of cognition lead the intellect to conclusions of cognition and which in matters of practice lead the will to conclusions of practice. Hence I would define Ethics as the Logic of Conduct. I am greedy to hear from you again.

Your friend and disciple

Francis C. Russell. 1268 Jackson Boul. | Chicago

Peirce to Russell. MS: RL 387.

Milford Pa | 1905 July 14

My dear Russell:

Decidedly I must send you my article of Jan. 1901.[119] Your *summum bonum*, "life", is probably at bottom about the same as mine, though I view it more concretely. I look upon creation as going on & I believe that such vague idea as we can have of the power of creation is best identified with the idea of theism. So then the ideal would be to be fulfilling our appropriate offices in the work of creation. Or to come down to the practical, every man sees some task cut out for him. Let him do it, & feel that he is doing what God made him in order that he should do.

It appears to me that given any man there are certain propositions as to the truth of which he has no doubt whatever, or no discernable doubt. He would act upon them with no misgiving, not the slightest. And roughly speaking these are in the main the same for all men. At any rate there are propositions we all believe, every grown man. But all these propositions are *vague*, and as soon as we attempt any precise definition we fall into conflict with one another and oftentimes into selfcontradiction. Now there is, I think, no department of logic in which new work is more desirable than in working out the whole theory of vagueness. Take a regular tetrahedron *per se*. Any two of its vertices make up an object of which there are six. They are all just alike in every respect and yet each is different from the other five. Of these five there is one to which it is peculiarly related. Some pair of indistinguishable pairs of vertices is formed of three vertices only. Some pair is not so formed. The indefiniteness of the *some* makes the principle of contradiction inapplicable. How far is this, or something corresponding to this true of all that is vague?

A *law* fails to confer existence on its subject. That is, nothing perhaps fulfils the condition of the law. If it does, it does so independently of the law. A general then lacks existence.

---

119 "Pearson's *Grammar of Science*."

511.32–512.1 and which in matters of practice … | Chicago] *inserted below and up left margin*
6–7 concretely] conc*cr*retely
11 in order] *after* ~~to do~~
12 given any man] *inserted above*
20 make up an] *above* ~~as a pai~~
21 each is] *above* ~~all~~

Now take the principle that in every contest some contestant must fail. This principle *does* confer *existence*. What is it, then, which it fails to confer. What element of being is it that the vague lacks. You cannot say it lacks *determinacy*; for this is equally lacking to the general. In fact, determinacy is lacking to whatever is not a complete reality. The vague seems to lack what we may call *whichness*, self-identity, that which makes the principle of contradiction pertinent. Is that so?

C. S. Peirce

1906

Russell to Peirce. MS: RL 387. Emended stationery headed, "\1268 Jackson Boul/ Francis C. Russell, | Attorney and Counsellor at Law, | ~~123 and 125 La Salle Street, | Bet. Madison and Washington Sts., | Suite 54, 55 and E.~~"

Chicago Nov 27th 1906

Dear Prof. Peirce:

When I write to you I always have an itching to make that address "*Dear Master*" but I long ago laid down a law to myself that I would call no man in that way however exalted he were in my reverence.

The *special* occasion of this is to get you to tell me where to find a certain quotation, to wit:

"Viewing all things unremittingly

In *disconnection*" &c &c[120] I want to use it extensively in the course of a certain series of Sabbath School conferences that I have been roped in to undertake—(I send enclosed my opening sheet)

Emerson says somewhere

"Whilst thus the world *will be whole* and refuses to be disparted, we seek to act partially—to sunder—to divide—to detach—" &c &c[121]

The way I regard an important matter is like this. The *All* is in truth and reality *continuous throughout* and *thoroughly* but because we are finite beings we can only proceed step by step by means of diverse sorts of devices the characteristic of all

120 Wordsworth, *The Excursion*, "Book Fourth" (1814), ll. 960–961.
121 *Essays*, "Compensation" (1841), p. 85.

16 reverence] reverance *[E]*

which is their *discrete* nature and that it is only when and in so much as we learn to pass *truly* from our discrete means to the continuous realities that we truly “get there”. The error that infects the world seems to me largely connected with this source and in matters of religion *now* in this opening twentieth century nothing worthy or serviceable will emerge unless this source of error is largely eliminated. I shall try as best I may to instruct my class in this regard although the attempt to interpret to the ordinary *plug* church goer the results of the researches of yourself, George Cantor, Weierstrass, Dedekind and the other pioneers may be so Quixotic as to be comical in the extreme.

¶.—Are you going to let yourself *pass on* without explaining to the world your discoveries in the matter of “*Methods of Study*”?. It is perfectly evident to me that you have turned up methods and practices of study of the most fertile sort, and in my confirmed opinion after numerous recurrences to the subject you will have been guilty of a most cruel neglect if you fail to let the world into your confidence in this regard. Of all other services you were born to render this is the one that is supreme and paramount in its consequences. *Dont fail us.*

Cordially
Francis C. Russell

P.S. Since we are in touch with the subject of Religion I am going to send you enclosed[122]

---

122 The enclosed MSS consist of the following:
(a) Unpublished TS (5 pages) “in defense of religion,” its power and necessity, especially religion that instills the love of God, “a love that should obtain and prevail as a real vital deep-seated motive and passion ever moving men to conduct in true expression of its aims and intent.” Signed “Agapetos” (beloved).
(b) MS note (1 page) reading as follows:

> *Note*—The foregoing article was sent to a publication that bears upon its title page and cover that it is “An Organ of the Independent Thinking of America”. It was prepared and sent in compliance with a solicitation of its editor and author who is the head of an “Independent *Religious* Society”.
>
> The solicitation did not however indicate in any way what sorts of articles might prove unacceptable. This article was found to be such, one of its causes assigned for its non acceptance being that its publication would invoke danger of displeasure in the part of the promoters of the publication.

---

1 *discrete*] *after* ~~several~~
5 worthy] *before* ~~can be ren servi~~ *[(first revision)* can ~~be ren~~ *(second revision)* can ~~be ren servi~~
(third revision shown in letter)]
5 serviceable] servicable *[E]*
8 the other] *after* ~~other~~
13 confirmed] *above* ~~firm~~
13 numerous] *after* ~~at~~
13 to] to to *[E]*
15 that is] *inserted above*

(a) A paper entitled "Religion"
(b) *note*, attached to the foregoing
(c) Draft of letter to Christian Register (over)

Perusal of these papers will explain their relations to one another. I finally concluded *not* to send the letter &c to the "Register". I had to say too much about *myself*, &c &c

But I can send them to you and perhaps you may find some of the amusement that I did in the experience.

I have another copy of (a) but as to (b) and (c) I should like to have them back again *after a while*.

Bye By F.C.R

Peirce to Russell. MS: RL 387.

8 Prescott Hall | 472 Broadway | Cambridge, Mass. | 1906 Dec 6

My dear Russell:

Owing to my not knowing before what my address would be, I have only this minute received your letter here, where I expect to remain for the winter. But as I have overdrawn my bank account owing to the apparent breach of understanding of a newspaper manager & have only a loaf of bread and \$ 1$^{\underline{70}}$ cash to go on until I can make further deposits, I shall be forced to postpone reading your letter, because it will certainly interest me too much to allow earnings to go on. It may be a week before I get round to it, unless I can induce Carus to advance the vast sum that he pays me per article. I have an article about half written for him & the rest is a mere question of writing it down.[123] But I don't think Carus hankers after my work. Still he may for the sake of friendship make the advance & if he does I shall be able to take up your letter at once & surely will do so.

Very faithfully
CSPeirce

Αγαπητος *[*Agapetos (beloved)*]*

(c) Not found.

123 Undetermined, possibly "Phaneroscopy" (R 298); see note for Peirce to Carus 30 Jan 1907.

---

1 (a) … "Religion"] *below* ~~1st <A p> A paper~~

7 some] *after* ~~written~~

Peirce to Russell. MS: RL 387.

8 Prescott Hall | | 1906 Dec 28

My dear Russell:

Believe me I haven't forgotten the article on religion. But I really have not had any time. When I have been forced to take a little rest, I have been so overwhelmed with worry that I am in danger of serious illness,—very serious, & the temptation to give up & commit suicide when I find my mind weakening under worry & starvation is greater.

I had seen Roman Hamilton's prayer before. He was a nice fellow.

I will take up the article the first moment I really have a right to.

most faithfully
CSPeirce

1907

Russell to Peirce 21 Oct 1907 (missing): Criticizes Peirce's unclear and complex style of expression in his "Prolegomena to an Apology for Pragmaticism" and in his two preceding pragmatism articles. Also asks for Peirce's help in some regard (which Peirce discards discards in his reply that follows).

Peirce to Russell. MS: RL 387.

P.O. Milford Pa | 1907 Dec 7

My dear Russell:

On Oct 21 last you wrote me a letter, to which I devoted a day or the larger part of a day, to writing an answer. But thinking that my letter would be of no help to you & was poor stuff, I never sent it. I at first kept it by me, hoping to get an idea of a better answer; but my mind being very hard at work on another matter, it slipped out of my thoughts.

Now I revert to your letter; not, I am sorry to say, because I see my way to helping you, but because you can help me. Namely, one purpose of your letter was to remonstrate with me about my style of exposition. I am just about to write

---

3 Russell] Russel *[E]* 21–22 or … day] *inserted above*

out an explanation of the last sentence of my Monist paper of Oct. 1906.[124] It is a very difficult task & I wish you would kindly revert to that and to the two others of April & Oct. 1905[125] and point out some of the worst passages & especially those in which, as you say, I have made too great leaps in my reasoning, in order that I may profit by the admonitions. You know that for over 40 years I wrote pretty constantly for the *Nation* & many articles of mine were rejected with contumely & whenever my definite fault was found, so that I could discover what was the matter with them, I would always rewrite them; and many times Garrison, the editor, said that my power of profiting by criticism was something astonishing. So you see if you will kindly take the trouble to do as I ask, you may have the satisfaction of having a good chance to do some good. The article to be written is the most difficult to present of any I have ever written.[126] It is the first part,—the lemma,—to my proof of the truth of pragmaticism. I know you will try to help me if you have time; but I fear lawyers are much occupied along now, and I don't want to have you do yourself any wrong in helping me.

very faithfully
CSPeirce

1908

ca. 1 July 1908: Russell publishes his "Hints for the Elucidation of Mr. Peirce's Logical Work" preceding Peirce's "Some Amazing Mazes: Explanation of Curiosity the First" in *The Monist*. Peirce responds to the matter as follows.

Peirce to Russell. MS: RL 387.

P.O. Milford *Pa.* | 1908 July 6.

124 "Prolegomena to an Apology for Pragmaticism." Last line reads: "In my next paper, the utility of this diagrammatization of thought in the discussion of the truth of Pragmaticism shall be made to appear." Peirce might also be referring to the penultimate line of the paper: "Of course, the explanation of the structure of the Conditional gives the explanation of negation; for the negative is simply that from whose Truth it would be true to say that anything you please would follow *de inesse*."

125 "What Pragmatism Is" and "Issues of Pragmaticism."

126 "The First Part of an Apology for Pragmaticism" (R 296; Bisanz, 366–374), never published by OCP. One of several articles intended to be the fourth of the *Monist* pragmatism series (others include "Phaneroscopy" and "The Bed-Rock Beneath Pragmaticism"). See footnote in Peirce to Carus 30 Jan 1907 for details.

My dear Mr. Russell:

I must thank you for your very ingenious defence of me. Surely, loyalty could not go further.

When I sent Carus the MS., I did not expect it to be accepted for publication; because I did not think that readers of the Monist would care for an explanation of the trick, seeing that the only explanation possible would obviously be mathematical, although of the simplest and most elementary grade. But if anybody *did* care for that, I thought they would care still more for the really valuable reflexions concerning the nature of mathematical demonstration that sprung out of it. Accordingly, I recommended that its publication be at any rate delayed until two or three such tricks had been shown. I only wrote it because I was too ill for real work.

But now you can do me a service even greater, as it will influence my following articles on Pragmatism very much, and I have put off the completion of my next until I could get the information I desire. For what I have to beg of you is a bit of information. I want to know whether Carus or anyone of the Hegelers (and which ones) or anybody else you can name, or you yourself ever actually did the trick so as to be able to show it. Kindly inform me specifically about that. Of course three who did not see the phenomenon would not care for its explanation.

ever very faithfully & with renewed thanks which I shall never forget,
C. S. Peirce

Russell to Peirce ca. 8 July 1908 (missing): Asks for Peirce's counsel on a proposed book on logic Russell is to undertake.

Peirce to Russell. MS: RL 387. Followed by 12 additional MS pages (divided into 2 sets) of the letter that form part of the "more than double" of what Peirce intended to but ultimately did not send to Russell (mentioned in final paragraph).

P.O. Milford Pa 1908 July 10

My dear Russell:

I highly approve of your writing a logic & will give you all the counsel you desire; and I think I may promise that in some form my name shall go on the title page if you desire it. But in every undertaking some one person must have the final say & responsibility; and that must be you; because I have by no means given up the hope of getting out a book of my own. The two wont in the least interfere with

---

1 Russell] Russel *[E]*   31 in] *inserted in-line*

each other's sales: quite the contrary. My book will inevitably be harder to read than yours.

You must decide on the plan & everything else. But you may be sure that I will criticize without gloves. There is no time for formalities of manners in business of coöperation. I don't like the idea of making it historical. The reason for doing so must be that the development of the individual must go along the same lines as that of the race. That is true; but there are two things to be said: $1^{\underline{st}}$ that this is only true in a sketchy way and $2^{\underline{nd}}$ that there is a reason for the truth of it & it is better to follow that reason direct. Mach had a most exceptional subject which was developed more rationally than any other owing to the great superiority of the leaders. Yet he filled a thick book with mere outlines easily stated in 50 pages, & better stated for most purposes. So with Muir's unfinished work on Determinants. Logic has pursued no rational course of evolution not only because of the wretched class of minds who have occupied themselves with it, but also because every considerable addition to such a subject implies falsity in views already entertained. The reason why the historic development and the individual development of such a doctrine must in a measure coincide is because both are determined by Reason, by the nature of the human mind, etc. and it is better to follow these guides directly in making your plan.

In my view, logic has three parts, $1^{\underline{st}}$ the *Elements* which makes analysis of what one has to deal with; Arguments &c. $2^{\underline{nd}}$ *Critic*, which examines the conditions of the validity of arguments, and $3^{\underline{rd}}$ *Methodeutic*, which shows how any inquiry ought to be conducted.

I dare say you might choose to begin by giving an idea of what logic is; and that you might do by the historical method. The science was originated by no other people but the Greeks; and by them only after about two centuries of cosmological speculation. Their motive was to find some way of distinguishing between a futile and fruitful line of thought. At that time nothing was known of psychology, and thought and language were for the Greeks much the same thing. Accordingly they paid no attention to what took place in the mind but only to the substance of the linguistic expression of it. Now what proves a thought to be true, as well as what the thought proves to be true, neither of these depends at all upon the shape in which they are thought, but only upon the reality and truth of that which the thought represents. One may think that the heavens turn daily round the earth or that the earth spins upon its axis in every 24 hours, and whatever follows from one thought equally follows from the other, since they are but two ways of thinking the same fact; although it ultimately turned out that if there is any reality in space, it is the latter shape which represents the fact which harmonizes with the

Foucault pendulum experiment[127] as well as with the general principle of action and reaction. We may define logic as that science of mental operations which takes account only of the outward realities of which thoughts are the sign.

It was necessary in taking the first steps toward investigating this matter to have a small number of forms of thought into some one of which and into but one every sufficiently small item of thinking could be put. The Greeks, however, knew nothing or next to nothing of the ways of thinking that antecede the expression of facts in sentences; and accordingly their forms almost necessarily became these four

Any thing that would be an *S* would be a *P*
Nothing that should be an *S* would be a *P*
Something that is an *S* is *P* and
Something that is *S* is not *P*.

On this, Aristotle's system was built, which he claims at the end of the Sophistical Elenchi was entirely original. But he was obliged to take some account of the inductive method of Socrates. He overlooked many things. First the simpler forms

If there be an *S* there is a *P*
There is an *S* and yet there is not a *P*

and these forms are believed to have been studied by his successor Theophrastus. At any rate, Boethius, about A.D. 500, knew something about them, without perceiving that their essential structure was simpler than Aristotle's forms which added to them the assertion of identity; thus:

If there be an *S*, that same *S* is *P*
There is an *S*, and yet that same *S* is not *P*.

They also overlooked more complex forms such as

Any *S* is in the relation *R* to every *P*
Any *S* is in the relation *R* to some *P*
Some *P* is in the relation of being *R*'d by every *S*
Some *P* is in the relation of being *R*'d by some *S*

Also such still more complicated forms as

A abases *B* to *C*

or *A* is in the triadic relation *R* to *B* for *C*.

---

127 Jean Bernard Léon Foucault (1819–1868), French physicist whose pendulum experiments demonstrated the earth's rotation clearly without minute cosmological analysis.

---

4 taking] *m*taking
5 of²] *i*of
5 some] *above* ~~wha~~
8 almost] *over* n
19 believed] *above* ~~supposed~~
28 *P*] *over S*
28 being] be *[E]*

These more complicated forms were recognized during the middle ages but not really studied until the latter half of the XIX th century.

On the death of Theophrastus, Aristotle's writings were removed from his school, and in consequence his logical doctrine was neglected.

About a century after A's death Euclid, a professor of Alexandria, writing upon Geometry gave illustrations of the true logic of demonstration far superior to Aristotle's in that he made it manifest to any student capable of reflection, that a pictorial image of the state of things reasoned about was essential to any real progress in demonstration. But this does not seem to have made any impression on logicians.

It was not until the century before Christ that Aristotle's MSS were recovered in a damaged condition (see the story in Strabo[128] and others) and were brought to Rome and published. Meantime two other important schools of logic had grown up, that of the Epicureans, which was almost the counterpart of J. S. Mill's modern system of Logic (see the very important [Περὶ σημείων καὶ σημειώσεων] of Philodemus, Περὶ σημείων καὶ σημειώσεων brought to light in a sadly mutilated state from a papyrus found in Herculaneum & published by Gomperz in 1865) and the logic of the Stoics. The Epicureans contended that the only reasoning of any value is induction, which they founded on the knowledge of certain uniformities. The stoics and all the other ancient schools of logic held that induction is utterly fallacious and thought the only way was to find some form of *reductio ad absurdum* by which progress could be made.

However, ultimately the Aristotelean system obtained universal favour for the reason that it well represented the thought of the Middle Ages. It was very greatly developed in the Western Monasteries—Or rather we must suppose that there was some considerable development in the Roman schools as exhibited in the Summulae Logicales which Prantl[129] idiotically fancies originated in Constantinople

128 Likely his *Geographica* (ca. 7 BC), an encyclopedic history of significant peoples and regions around the world known at the time.

129 Karl "Ritter" von Prantl (1820–1888), German philosopher and philologist. Reference is to *Summulae logicales magistri Petri Hispani*, a.k.a., *Tractatus*, supposed today to have been written by Petrus Hispanus (13 th century), later Pope John XXI. (See "Peter of Spain," *Stanford Encyclopedia of Philosophy*.)

5 writing] *after* ~~pro~~
8 real] rea~~l~~l
12 Strabo] *before* ~~)~~
16 Περὶ σημείων καὶ σημειώσεων] Περὶ σημεῖων καὶ σημειοσεῖων *[E]; On Signs and Inferences* (ca. 110–30 BC).
27 Prantl] *after* ~~Pratl~~

for no better reason than that a Greek version, as manifestly a translation from the Latin as could be, & greatly influenced by the Latin Grammar of Priscian is attributed in one MS to *Michael Psellus*. Now it happens there had been a Byzantine writer of that name; but he did not write the kind of latinized Greek of this treatise; and it was probably a tolerably common name in Constantinople; for there were certainly two of that name & if this work was translated by another, that makes three. Just as there are three voluminous English writers of some note named Thomas Browne.

In the centuries from Scotus Erigena to Roscellin arose the notion that *Ideas* are the breath of the voice. At least, so we must believe, if we accept the statements; and some remains partially bear them out. This gave rise in the 12$^{\underline{th}}$ century to the great confused free fight over Realism & Nominalism. Abelard is the great figure.

About the beginning of the XIII th century Aristotle's writings as complete as we have them now (except for the work on the Constitution of the Government of Athens) were brought to light & translated both from the Greek & from the Arabic translations. And the philosophy of Avicenna and other Shemites had considerable influence on the further development of logic. Duns Scotus at the end of that century held a very carefully qualified realistic position. Ockham in the first half of the XIV th took the position that all thoughts, are of the nature of words, except that words derive their meaning from convention & concepts theirs from natural dispositions.

This was a very erroneous position, although it undoubtedly is true that all thoughts take the shape of signs of some kind and *some* of them that of linguistic signs though others differ essentially. For some are *images* and others such as the local signs by which we know what part of one's person gets a poke, are of a third kind which represent the the objects they do represent simply by virtue of being in fact connected with those objects,—which is not true either of images (which represent their objects simply by resembling them) or of words, (which represent their objects imply by virtue of the fact that they will be so understood.)

But what was more unfortunate than the error of this "terministic" logic of Ockham than its falsity was that while it was plausible enough to make a tremendous controversy, yet it was not enough so to win the day. Consequently when the new learning came in (owing to the fall of Constantinople and the consequent exodus of learned Greeks to Italy,) it happened that the Scotists—or Dunces—as

---

1 Greek] *after* ~~boo~~
7 voluminous] *inserted above*
13 About] *over* D
13 writings] *after* ~~com~~
24 *images*] ~~I~~ *inserted above*
26 the objects] *above* ~~what~~
28 or] *after* ~~nor~~
29 so] *over* u
32 so] *t*s̲o

they were called (and the word Dunce was by no means understood as meaning a dullard but on the contrary as a man so skillful in logic that the literary men were quite vanquished by him, but who was a stubborn opponent of the new light.) The result was that the new people who knew nothing to speak of philosophy or logic naturally opposed Scotism and fell into a blind worship of the doctrine of Ockham, which was identified with Nominalism in general. For these new men could not distinguish between one kind of nominalism and another. All they thought in their crude way was that whatever is *general*,—such as Ideas of right & wrong, laws of nature, & the like,—are *mere words*. This has been the doctrine of all modern philosophers almost without exception, though some of them have held it with a happy admixture of inconsistency.

Because Italians perhaps were by race superior to the Germans in intellect, and at any rate had been surrounded with the monuments of Roman civilization, and because the capital of the church was Rome, it certainly had always been several degrees less barbarous than Teutonic Europe. Only the Irish & Gaelic monks had long upheld learning north of the Alps, with those who had been educated directly or indirectly by Celts and Italians;—with a slighter influence of Shemites. The Constantinopolitans who now came were no great shakes. Still they were brighter minds and freer minds than there were in the West & the result was seen when I think in 1545—or thereabouts Coppernik, of Polish extraction, though partly German, but educated in Italy, brought out his *De Revolutionibus* which gave the initial impulse to modern science. After Galileo, Keppler (a German), Harvey, Gilbert (who stole largely from Petrus Peregrinus probably partly a Celt,) Francis Bacon gave a literary form to what may be called the Morals of Science in the first book of his Novam Orgamem.

Descartesand all those philosophers were essentially Aristotelean in their views of Logic; and Leibniz the greatest of them was a regular Ockhamist. Kant considered the science of logic to have been completed by Aristotle.

In fact there was no essential & fundamental change until Hegel's total bankruptcy in philosophy caused Fechner and Lotze (and we may add Herbart before Hegel) to start up modern psychology.

From that time to this philosophy has been dominated by psychology; and the effect upon logic has simply been to castrate it completely. All the new German logicians are thinking of how we think, most of them protesting that the great evil of logic is its confusion with psychology, but unable all the time to extricate their

---

4 of] of of *[E]*
12 Italians] *after* ~~Italy~~
14 capital] *above* ~~head~~
15 Teutonic] *after* ~~the~~
15 & Gaelic monks] *inserted above*

own systems from it for the simple reason that they are too indolent to study the successful sciences and to learn from them what logic really is.

I have written more than double what I send.[130] But I guess this is quite enough for one installment.

very faithfully
CSPeirce

[Omitted set 1 of 2: Pages continuing from penultimate paragraph above]
[. . what logic really is.*]*

I have surmised[131] that the above was the sort of suggestion you wanted just now, but of course I can't tell.

---

130 See draft material that follows this letter.

131 I have surmised … distinctions to be considered.] *another draft of these 2 MS pages reads as follows (variants recorded in-line):*

[Page 1] In writing the above, I have gone on the conjecture that that sort of thing would be the sort of suggestion you would be wishing for at the outset. Of course, it is but a conjecture.

It is better to explain what need not be explained than to leave unexplained what needs explaining. I will therefore make some notes on my own work in logic, in which, as I myself think, my strong point is my power of logical analysis, or definition. I began working intelligently, or as I generally phrase it, "reasonably," that is to say, under the dominance of a well-matured purpose so that my work has been, ever since, what I mean by "self-controlled" that is under a control analogous to moral self-control (just the opposite to what some writers mean by "self-controlled," they ~~taking~~ so calling a process of thought which is under no control other than what issues from this procedure *itself*, which I mean controlled by a "self," or person who *[E, whose]* deliberates, and makes *resolves*, which make or denote, "determinations" or real dispositions causing conduct to agree with the resolves,) I say my work became self-controlled early in the year 1867, when I already had in my mind the substance of my central achievement, the paper of May 14 of that year, "On a New List of Categories." But in some respects, the doctrine of that paper *[begin page 2, draft A]* has greatly grown and developed in the intervening years. It was not until the next year, I think, that I took up seriously the logic of relations. I wanted here to refer to my journal of logical analyses which is contained in a quarto blank book of six or seven hundred pages.*[Peirce's Logic Notebook, R 339]* But my letter has been interrupted \for two days/ by a vain search for that book. Very essential things are contained therein. But the other sex *will* interest itself in arranging my books by separating volumes of the same work putting half the books upside down, and I have to smile and appear enchanted because of the devoted affection that prompts it all

But that I am forbid
To tell the secrets of my prison-house,
I could a tale unfold like quills
Upon the fretful porcupine! *[Hamlet*, Act 1, Scene 5*]*

3–4 I have … installment.] *This paragraph in red pencil on verso of last page.*

I myself began with defining logic as the science of the general relations of *symbols* to their objects. But on the whole, it certainly is concerned with all signs and though the relation to their objects is what logic has specially in view yet I think it should begin with general semeiotic. A sign may be defined as something which mediates between an object and a mind, by being itself in some way influenced or affected or determined by that object and then in its turn producing an effect upon the mind, which effect I call the *interpretant* of the sign, of such a nature that the mind is thereby and therein itself affected indirectly by the object. This wording can be improved; but I think it a valuable definition of a sign. Now we have at once to make certain distinctions both with reference to the *Object* and with reference to the *Interpretant*. As to the Object, we have to distinguish between the *Immediate Object* i.e. the object as it is represented in the sign, and the *Real Object* [Dont forget my definition of *real*. That is real which has characters sufficient to distinguish it from all other things, which characters are true of it whether or not any person or actual collection of persons thinks them to be so or not; (although they may not be independent of what persons inquiring sufficiently would ultimately come to think of their truth.)] As to the Interpretant, we have to distinguish first the *Immediate Interpretant* i.e., that which the sign itself represents as such, second the *Dynamical Interpretant*, or the effect the sign actually produces on a given mind; and third, the *Ultimate Interpretant*, which is the effect that will be attributed to the sign as its reasonable effect, as the result of sufficient inquiry. Now signs are to be divided in ten ways

---

Well, I could not then see that my "ground," "correlate," and "interpretant" were not concepts but merely *elements* of concepts—what *fluorine* was among chemical substances until Moissan isolated it. Or better like *ions*. Only instead of being elements of matter, they are elements of thought. My "ground,"—is ~~that~~ the *univalent* element of thought, to use a chemical term; my "correlate" of 1867 is the *bivalent* element; and my "interpretant" is *[…]*
*[*begin page 2, draft B*]* has ~~been~~ largely grown up and developed in the intervening years. What I then called a "ground," I now call an *Idea*, that is to say, a pure predicate, which cannot stand pure in thought but must be referred to some subject, however indefinite. In itself it is a mere capacity for making such an attribution. What I call there a "correlate" is an ordinary experiential correlate, reference to which is forced upon the mind. We may call it an "occurrence", meaning a *thing* or *fact*, single and definite. Since such an element of thought is due to an experience and that experience is an event, I like the word "occurrence" as reminding the thinker that a thing never is thoroughly singular, but the only ~~th~~ object that is so is an instantaneous event. If you say there is no such thing as an instantaneous event, I reply, I suppose not; but these categories are not concepts but only the three kinds of elementary stuff out of which concepts are built. "Stuff" is not the word either; for they are *forms* of thought, not *matter* of thought.

---

8 and therein] *inserted above*
15 actual] *inserted above*
20 the^2] *after* ~~wha~~

1st, as to the Nature of the Sign in itself
2nd as to the Nature of its Immediate Object
3rd as to the Nature of the Real Object
4th as to the Relation of the Sign to its Real Object
5th as to the Nature of the Immediate Interpretant
6th as to the Nature of the Dynamical Interpretant
7th as to the Relation of the Sign to the Dynamical Interpretant
8th as to the Nature of the Ultimate Interpretant
9th as to the Relation of the Sign to the Ultimate Interpretant
10th as to the manner in which the Sign represents its Object
to the Ultimate Interpretant.

All those ten divisions appear to be trichotomies, and if they were independent would give 3^{10} different species of signs; although there are still further distinctions to be considered. *[…]*

[…] the *trivalent* element. *<u>Theorem</u>: There can be no more.*

Demonstration. For suppose there were another element. Then, being an *element*, it must be indecomposable by logical analysis; and being an element of *form*, it cannot be distinguished by any material peculiarities. For, mind you, I do not say that these three are the sole categories, but only that they are the sole *formal elements* of thought. If then there were a fourth formal element, it must either (1) have infinite valency, or (2) have a higher finite valency than trivalency, or (3) have *nilvalency*, or (4) there must be two formal elements of the same valency.

In alternative Nº. 1, the formal element assumed must have a valency of the order of multitudinousness of some one of the endless series of infinite multitudes, which series begins with the denumeral multitude (or multitude equal to that of the positive integers), and every multitude, M, of the series is followed by an abnumerable multitude of an order higher by one than the abnumerability of the multitude M itself, the denumeral multitude being the multitude of abnumerability zero; and the relation of the multitude of abnumerability $X + 1$ to the multitude of abnumerability X is that the former is the multitude of the collection of all possible collections composed exclusively of individuals of a collection of abnumerability X. If there be any other way of definitely conceiving an abnumerable multitude of any order greater than 0, it is at least as complicated as this. But this is a clearly defined concept, and in that definition it is analyzed. Consequently it is not an *element* of thought whether formal or material; for an *element* is a concept

---

12 appear] apper *[E]*
15 <u>Theorem:</u>] *inserted above*
18 cannot] *after* ~~mus~~
23 alternative] *after* ~~the case~~
23 a] *after* ~~one~~
26 , *M*,] *inserted above*
34 in] *after* ~~therefore a~~

not susceptible of logical analysis. As to the denumeral multitude I have defined it in the *Monist* for July 1908 pp. 432 *et seq.*[132] Its concept, again is not an element; and it is impossible to conceive definitely an infinite valency without introducing the idea of an infinite multitude. It is not necessarily so with ideas of finite valency. We can conceive of them without resort to the general idea of multitude.

Having thus refuted alternative No. 1, I will interrupt my demonstration with a sort of scholium, or remark to clear up possible objection. There is no doubt that our indirect concepts of, 1st, the Universe of all possible pure Ideas, or predicates, 2nd of the Universe of all actual Occurrences of Experience, past and future, and 3rd, of the Universe of all living courses of thinking and psychical life in general, animal and vegetable, are concepts of imaginary concepts of infinite valency. For were all the valencies of either direct concept of such a Universe satisfied, thought of that Universe would be complete and incapable of further progress, an idea contrary to our notion not only of such Universe but also of that of Thought itself. But we have no direct concept of any one of the three Universes. We only have a concept of such an imaginary concept. Now this latter direct concept, though it is imagined to have an infinite valency, is not an *element* of thought. On the contrary, it is built up by thinking of an infinite collection which is a highly complex definable concept. I must also notice a question which is pretty sure to arise, namely, why need thought come to a stop when all valencies are satisfied? Why may we not add on another thought entirely disconnected with the one already completed? For what is a unit of valency but a call for a correlate to a concept of a relative term,—or say, taking my idea of a *rheme*, or blank form of proposition, such that if each blank were filled with a name of a definitely recognized individual it would become a proposition,—why, one may well ask, after all the blanks of a rheme have been filled and a proposition results, may we not go on to form another independent proposition? This difficulty is best resolved by thinking of the matter from the point of view of the system of Existential Graphs, of which I will frankly say that, whatever its imperfections, and however thorough an overhauling it may be found to require, yet when I come to show what uses it can be put to, in my opinion every competent judge must admit that it is the greatest illuminator of logic that ever has been made yet. I say this, not out of vanity,—for I suspect some great

132 "Some Amazing Mazes: Explanation of Curiosity the First."

3 an] *above* ~~a valency of~~
8 of[1]] *before* ~~the~~
12 direct concept] *after* ~~Universe~~
16 direct concept, though it is] \direct concept,/ though it ~~has an im~~ is
22 For] *A*For
23–24 if each blank were] ~~when~~ \if/ each blank ~~is~~ \were/
30 yet] *after* ~~that~~

changes will have to be made in the system,—but to urge you, as a student of logic, to get the most entire mastery of this system. To work it you must simplify the diagrams by cutting them up into bits, and representing the parts not relevant at the moment by some vague signs. Now then, when you look at the above question in addition to the "*pegs*" [for the term "peg," see Monist XVI, p 530, 1 3/4 inches from the bottom.][133] that occur around the periphery of any "spot" [Ibid. 2 inches from bottom.] we must imagine that there is invariably an extra peg by which the "spot" is held down to the "Phemic Sheet" or to such other "area" as it may be "scribed" upon. [For "Phemic Sheet" see p. 526 7/8 inch from top and p. 528 first 2 1/4 inches. For "area" p 528 last 3 lines and first 2 lines of p 529. And writing "down" to mean *from the top*, and "up" to mean "from the bottom," for "scribe" see p. 506 3 in. *up*.] The Phemic Sheet itself is the Graph-Instance [see p. 506 3 in down] of the logical universe. [134] It is a continuum of Pegs. They are essentially *inexhaustible* and therefore exceed all multitude; and therefore they are not actually there at all, but only potentially until any of them are occupied that is *satisfied* (considered as valencies,) when those become actualized. They *cannot* be all actualized because they exceed all multitude. But were they so, there would be no room on the sheet to scribe any proposition.

I now take up alternative Nº. 2. As to this, I say that an idea representable by a graph of more than 3 pegs cannot be an element because it can always be an-

133 Page references here and below are to "Prolegomena to an Apology for Pragmaticism."
134 universe. It is a continuum … p. 531, 1/2 in down.]] *another draft of this MS page reads as follows (variants recorded in-line):*
[..] universe. It is a \true/ continuum of Pegs. They are *essentially* inexhaustible, i.e. not only inexhaustible one by one, but inexhaustible by the process of passing from the collection of all the S's to the collection of all possible collections of S's. They thus exceed all multitude. Consequently, they do not form a collection. Therefore they are not single objects. Therefore they do not actually exist, but only have a potential being, until any one of them is marked, whereby it becomes actualized.

I now take up alternative Nº 2. As to this, the proposition I shall endeavour to prove is that any ~~theme~~ \proposition/ of more than 3 ~~blanks~~ \subjects/ ~~can~~ expresses a state of things which can be expressed in Existential Graphs by a single connected Graph none of whose spots has more than three pegs. If this be true, it analyzes the meaning of the assertion, and consequently such an assertion is not elementary.

Before undertaking a general proof, I had better give illustrations which shall familiarize you with the meaning of what I seek to prove and also with the more intricate of the concepts required in the proof.

There are no better examples of relations of more than three correlates than geometrical examples, espe- *[…]*

1 as] *after* ~~if~~   16 those] *above* ~~they~~

alyzed into a trivalent idea and an idea of valency lower by one than that of the idea to be analyzed. Take for example the idea of selling, where the graph would have 4 pegs to denote respectively the seller, the buyer, the thing whose ownership is transferred and the price. Take any two of these you like, say the seller and the buyer. Then the idea that the seller transfers to the buyer the ownership of the subject of the sale in consideration of the price may be analyzed thus: You take two triadic graphs expressing these rhemes: "___ is the transferrer in a legal act ___ to the transferee ___" and "In a certain legal act ___ a transfer of ___ is made in consideration of a reciprocal payment of ___". The first blank of the first rheme is filled with the seller's designation, the second with the selective, X, and the third with the proper name of the buyer. In the second graph the first blank is filled with X, the second with the thing sold, the third with the price. It is evident that thus every "spot" of more than three pegs can be replaced by "spots" of three "pegs" each, among which "spots" there may be a number of instances of the graph of coidentity. [Called "teridentity" on p. 531, 1/2 in down.] By introducing instances of the graph of co-identity, the analysis of the sale can be carried further. Thus suppose the graph to be analyzed to be

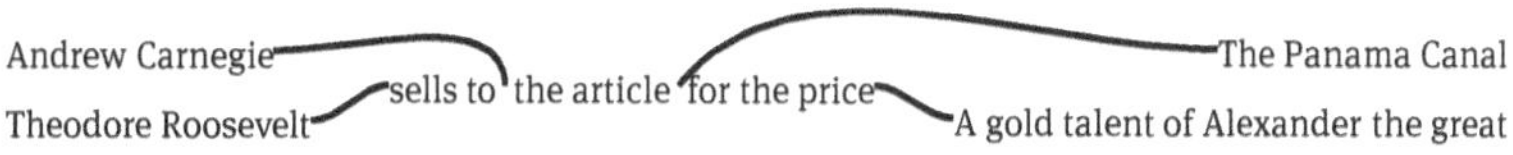

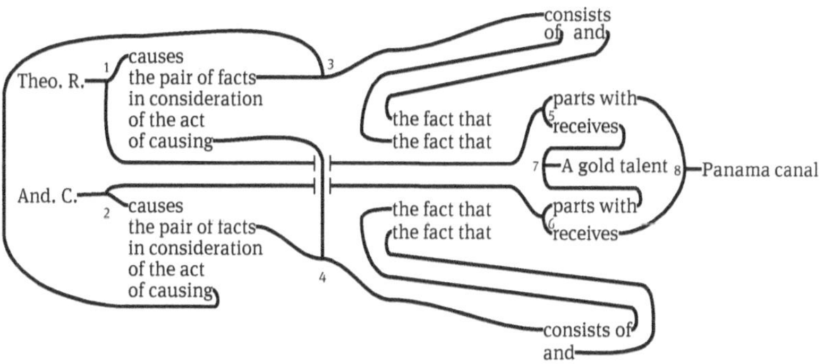

Here *eight* graph-instances of co-identity, enable us to analyze the sale into eight trivalent parts—2 beginning "causes" etc., 2 beginning "consists of" etc.,

1 lower] *after* ~~one~~
6–7 You take … rhemes:] ~~The seller~~ You take *[. .]* rhemes: ~~"___ is the first party in a transfer ___ in which the second party is ___" and "___ is a transfer~~
13 by] *after* ~~with~~
14 a] *above* ~~any~~
17] *[In large diagram, first* "consists of" *near top right]*] ~~parts with~~ /consis*t of*ts /of\ \
21 trivalent parts] *above* ~~elements~~

3 beginning "parts with," and 2 beginning "receives." Making 16 trivalent parts in all. In general, a relation between 4 correlates can be expressed by six relations each among two of the original correlates and one more new correlate in each. Two of these correlates can then be connected with a new one, making 3 correlates of the last set and thus the relation will be analyzed into *ten* triadic relations.

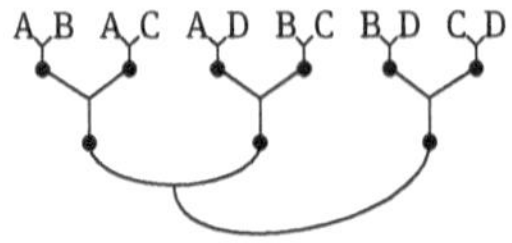

In like manner a pentadic relation can be expressed by $10 = \frac{5\cdot 4\cdot 3}{2\cdot 3} = \frac{5!}{3!2!}$ triadic relations each between 3 of the original and 1 introduced correlates.[135] And these 3 can be connected with the one left over of the first set of introduced correlates. Thus the pentadic relation will be analyzed into 14 tetradic relations and since each of those can be expressed by 10 triadic relations, every pentadic relation can be expressed by 140 triadic relations, at most.

135 correlates. And these 3 … triadic relation etc. etc.] *The MS page before this one ends,* introduced correlates and ~~9 of these 10 <corr> new correlates can be connected into 3tetradic relations introducing 3 new and between these~~

An alternative draft of this page, shown below, continues directly from the above variant from the preceding page, but is less complete than that included in the text.

[… introduced correlates and *]* between these 10 new correlates, there will be 5 more triadic relations introducing 5 more new correlates, and between 4 of these there will be 2 triadic relations with 2 new correlates and between these and the one left over before there will be one more triadic relation. So that the pentadic relation can always be analyzed into 18 triadic relations at most, with 17 instances of the graph of coidentity.

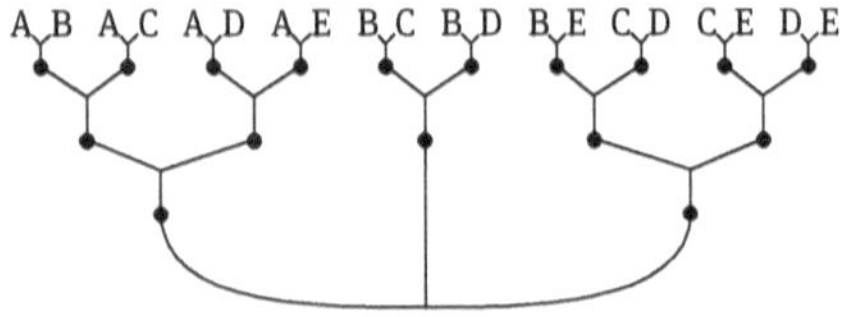

So between the correlates of a hexadic relation *[ ]*

3 new] *inserted above*

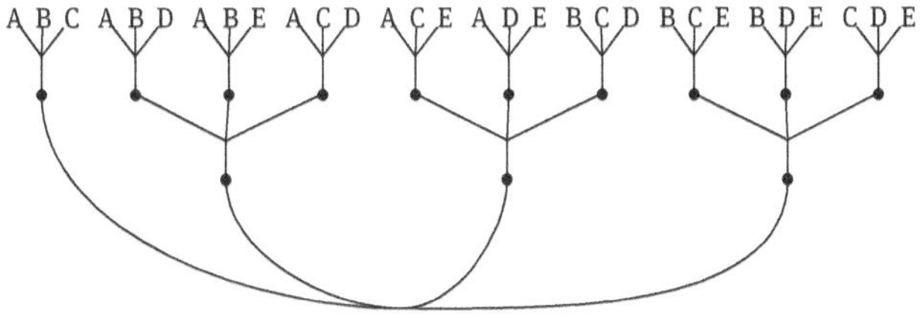

So a hexadic relation can be expressed by $20 = \frac{6!}{3!3!}$ pentadic each between 4 of the original and one introduced correlate; and the 20 introduced correlates being united into 5 pentadic relations each between 4 of the first introduced and 1 freshly introduced correlates, these 5 will be united in one pentadic relation; so that the hexadic relation will be expressed by $20 + 5 + 1 = 26$ pentadic relations; and since each of these can be expressed by 140 triadic relations, any hexadic relation can be analyzed into 3 640 triadic relations at most.

A hexadic relation will give $\frac{7!}{4!3!} = 35$ hexadic with 35 first introduced correlates. Between these 35 there will be 7 hexadic relations with 7 introduced correlates. Of these 5 will make a hexadic relation introducing 1 new correlate. Thus the heptadic relation will be expressed by $35 + 7 + 1 = 43$ hexadic and 1 triadic relation etc. etc. *[ ]*

---

2 $= \frac{6!}{3!3!}$] *inserted above*

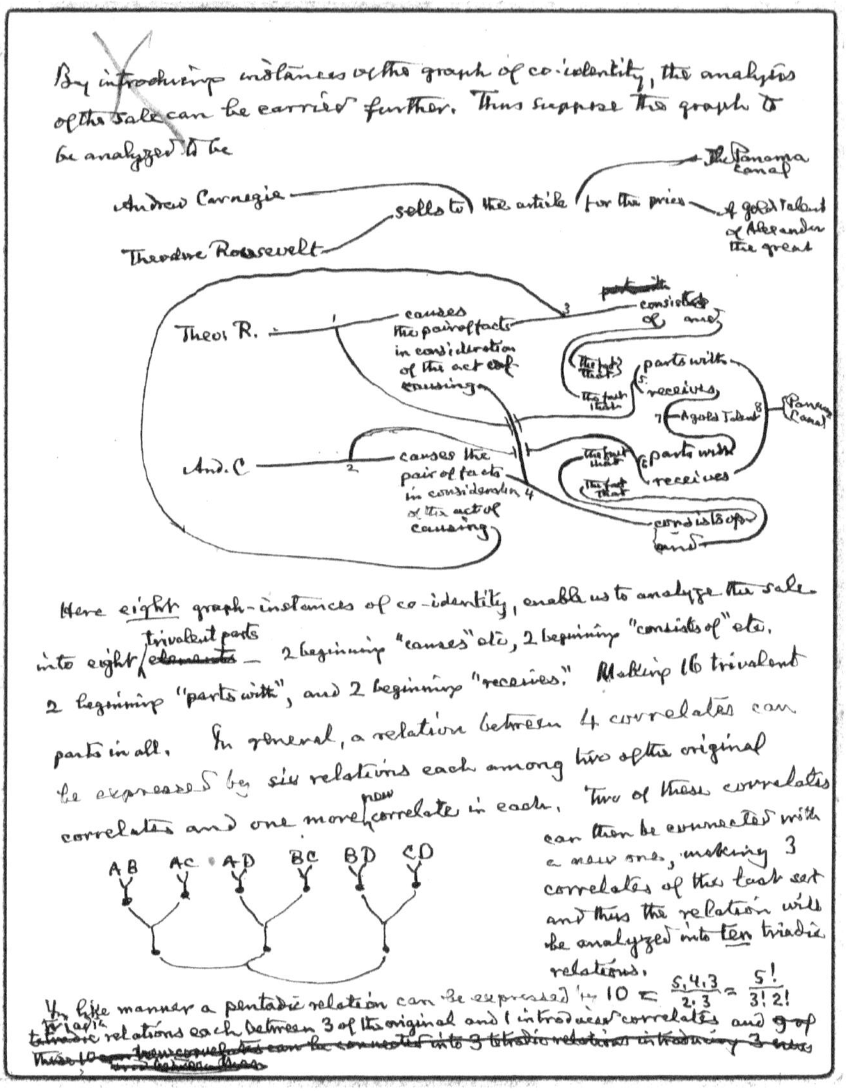

Figure 11: Peirce to Russell 10 July 1908, page of omitted set 1. Existential graphs of co-identity.

[Omitted set 2 of 2: Pages 4–9 of a previous draft]

[…] a perfectly inert, amorphous, insoluble very fine powder) shaken up with gum arabic water, so that it covers and protects raw places, I was treated by various physicians & I never saw one that after a fortnight of my attack did not think I was going off the hooks. But the wonderful part is the sudden disappearance of the inflammation; and since I made the acquaintance of John Wyeth & Brother's Beef-juice, I am left weak but not in a state of inanition. When your letter came I was in the midst of an attack & last night suddenly got well so that I could swallow a couple of eggs this morning. But I am in that state of feebleness of mind which must excuse garrulity if anything can.

A propos of inverse probabilities, I found I had a copy of "Studies in Logic" that I could send you & in case the copy "on your table" does not belong to you, I should desire to do so. For I want you when you get leisure to read carefully from the bottom of p. 146 to the bottom of p. 181.[136] You will then see what my objection to Laplace's practice is.

But before you read that, you had better read what I will here set forth. For there are some points which, though not incorrect in my doctrine of 1882, were then by no means as clear to my mind as they are now, and some important points I did not then know about at all.

Logical research was prompted by a desire to distinguish between good arguments and bad ones, meaning by an *argument* a process of thought tending to produce a state of belief, and by a *good* argument meaning one capable of maintaining its habit in the long run, and meaning by the long run an endless series of experiences. [And by the way the only breed of troglodyte animals that ever originated a logical impulse was that of the Ionian Greeks.] Now since it is plain that what thought a given belief can reasonably spring from or reasonably give rise to depends solely upon the state of things which that thought represents as real and not at all upon the form of the thinking, so that e.g. whether we think that among animals that stand erect some have a pale skin or whether we think that among pale-skinned animals some can stand erect, is perfectly indifferent, it follows that *the form of thinking is quite irrelevant to logic*, and that the consideration of it can

136 "A Theory of Probable Inference," section 6 to the end.

2 very] *after* ~~f~~
3 was] *after* ~~h~~
6 made] *after* ~~learned that~~
11 *A propos* of inverse probabilities,] *inserted above*
13 when you get leisure] *inserted above*
26 thought] *after* ~~process of~~
26 reasonably] *after* ~~give~~

only give rise to confusion. Moreover, it was early recognized that self-observation shows that thinking always takes place in a dialogue, or series of signs, addressed by the self of one moment to the self of another moment, and since a perfectly equivalent *outward* sign of any thought can be found, much more readily identified than the form of thinking itself, Aristotle perceived the great advantage of a canonical system of forms of signs to be used instead of forms of thinking as the object of logical study. Of course, his canonical system was very imperfect. It was

1. πᾶς ἄνθρωπος λευκός. Any man (there may be) would be white. This translation is strictly his own definition (and he was very exact in important definitions) viz: “λέγομεν *[δὲ]* τὸ κατὰ παντὸς κατηγορεῖσθαι, ὅταν μηδὲν ᾖ λαβεῖν τῶν τοῦ ὑποκειμένου, καθ' οὗ θάτερον οὐ λεχθήσεται.”[137]
2. οὐδεὶς ἄνθρωπος λευκός, No man would be white.
3 ἔστι λευκός ἄνθρωπος, Some man is white.
4 οὐκ ἔστι λευκός ἄνθρωπος, Some man is not white.

These are Aristotle’s own examples in cap 7 of the Peri hermeneias. I count 5 systems of Greek logic:

1. The inductive logical analysis of Socrates. I don’t think Plato’s numerous logical remarks amount to a system, and far less do the catches of Megara and Velia.
2. Aristotle’s Analytics etc.
3. The mathematical logic of the first book of the Elements of Euclid, further perfected by Archimedes.
4. The system of Chrysippus, which really didn’t contain anything not in Aristotle
5. The Epicurean system of Philodemus and others, which was curiously like that of J. S. Mill.

---

137 The *Dictum de omni*, from *Prior Analytics*, A, 1, 24b29–32 (Loeb Classical Library, 1934). Peirce provides the following expanded translation in “What Pragmatism Is” (*Monist*, 1905, p. 179): “We call a predication (be it affirmative or negative) *universal*, when, and only when, there is nothing among the existent individuals to which the subject affirmatively belongs, but to which the predicate will not likewise be referred (affirmatively or negatively, according as the universal predication is affirmative or negative).”

---

5 form] *after* ~~thin~~
13 3 ἔστι … is white.] *below* ~~3~~ ~~τὶς ἄνθρωπος λευκός, Some man is white.~~ ~~4 τὶς ἄνθρωπος οὐκ ἔστι λευκός~~
17 inductive] *after* ~~lo~~
17 don’t] don’t dont *[E]*
18 Megara] *after* ~~the~~

The only very important divergence of the Logic of Chrysippus from that of Aristotle is its utter denial of the validity of Induction. It was natural that this should be the logical doctrine of a philosophy expressive of the sternest morality amid a corrupt state of society. Such a morality maintains itself by a sharp demarcation between the sheep and the goats. It can see no colors but black and white. Nothing between necessity and impossibility.

But after the works of Aristotle as we know them (or did, before the recovery of the work on the constitution of Athens) were brought out in Cicero's time, their inherent power made them the standard of the Imperial schools; and in Boethius about A.D. 500 we find nothing but Aristotelian logic developed in the direction of the "hypotheticals" which were regarded as compounded of categoricals instead of being as they really are categoricals with the omission of identity.

If there is a man there is a white thing

instead of

If there is a man *that same* is a white thing.

Of all the impudent charlatans, Carl Prantl takes the cake. He has looked over a great number of old *printed* books and never consulted a MS except in the case of the Summulae Logicales which by the most inconceivable absurdity he calls a "Byzantine Logic" simply because one MS of a Greek version attributes it to "Michael Psellus." But anybody can see that *this* "Michael Psellus" no more writes the Greek of either of the two known authors of that name than the Dr. Thomas Browne who wrote the lectures on the Philosophy of the Human Mind was either the author of the Religio Medici or the flippant Tom Browne. His Greek is full of Latinisms and is influenced by Priscian; while the author of the Summulae Logicales knew so little Greek that he begins by saying (or rather it is his fourth sentence) Dicitur enim dyalectica a dya, quod est duo, et logos quod est sermo.[138] Yet Prantl will have it that this is a translation from the Greek. Why? Because the Middle Ages were incapable of any original thought! Yes for him, because he does not understand the books into which he dips here and there! I for my part have no doubt that the Summulae present a development of logic (of no very great importance, yet of decided originality) which had grown up in the Roman and afterward in the diocesan schools. This is the result of my very careful study of the remains of the Logic and Grammar of the XIIth century, which are very considerable. This was the time of the great discussions about what we call Realism and Nominalism, though the

138 "It is said that 'dialectic' comes from 'dia', which is 'two', and 'logos', which is a discourse ..." (For more on the debate here referenced, see Rijk, *Logica Modernorum*, Part II, Vol. II p. 462 ll. 17–18.)

1 divergence] *after* ~~differe~~ 20 no] *after* ~~is~~

word *realis*, except in the legal sense of real property, was not yet invented. "Res," however, not seldom occurs for what would be called today an *objective* concept. Scholastic realism came out of the long and temperate (for the most part) disputations triumphant. Of course it was not the nonsense which such fools as Prantl suppose it to have been. What is the "real," anyway? It is that which is not the mere figment of individual minds, but that which the full development of thought under experience destines to universal acceptance by *whomsoever may* (not any group of individuals) sufficiently consider the matter. Since *experience* means that belief which is forced upon us willy-nilly and since experience begins with perception, the most certainly *real* of all things are colors, sounds, smells, tastes, and resistances to pressure. Next come those ideas that are founded in human nature, Love, στοργή, honor, justice, truth, etc. Then, ideas as ideas, I mean their substance, or that which we think, such as $\sqrt{-1}$. True, this has the lowest kind of being, since it is being in what it is thought to be, the logically possible. Next, comes whatever science may ultimately discover, electrons, as they may ultimately come to be conceived, and such like. *Matter* and *motion* are far less certainly real; for they are merely the first inferences of science.

Next, I ask, what are the different kinds of *reals*? They are 1st those whose being lies in the substance of the thought itself, mere ideas, objects logically possible, the objects of pure mathematical thought for example. 2nd those whose being consists in their connexions with other things, existents, reacting things. 3rd those whose being consists in their connecting two or more other things;—laws, generals, signs, etc. In short the real is the ultimately undeniable.

The logic of the Summulae was fully adequate to the main current of medieval thought. *[ ]*

Russell to Peirce. MS: RL 387.

1268. Jackson Boul | Chicago Sept 16th 1908

Dear Prof. Peirce.

Before I go very far in the preparation of any logical text I feel that I must clear up some points that I am fearing I may be in serious error about. That is to say I fear so because (*and because only*) I have seemed to draw from divers passages in your papers that you were of a different opinion.

One instance of such points I will dwell upon to day to wit. It has been and is a great hobby of mine that the dependence of our mental estate upon the sign-

---

7 group] *above* ~~number~~
12 truth,] *inserted above*
18 whose] *after* ~~which~~

systems we have in service has been and is persistently ignored.[139] A strong hint of my state of mind on this topic will be furnished by the enclosed M.S. - A - which is a part of what I wrote up while I was trying to work out a *brief* preface to your M.S. - B. -[140] Lordy! Lordy! what a time I did have in trying to explain the regenerated logic "*in a word*". [That reminds me,—In Mrs Sargent's Sketches (&c) of the Boston Radical Club (Osgood & Co, 1880) pp 376–380 is the report of your father's lecture before the club on "The Impossible in Mathematics" and of an inquiry by Minot Savage during the (brief I guess) after interlocution, if he (i.e. your father) could tell him #*in a word*# what was space of four dimensions (!!!!!!!&c).][141]

My own proclivity of mind is to anticipate the *very greatest* advances in the mental competency of man by and through his discoveries and inventions in the (to be) *science of Signification.*[142]—The evolution of mentality proceeds *pari passu* with the evolution of the means of *true thinking.*

I am almost inclined to define Logic as the science of the *means* of correct and successful thinking,[143] and it is just because the *calculative* logic promises to be at least a *schoolmaster* to lead us towards a *general science* of *signification* that I esteem it so highly.[144] In my glowing moments I am prone to project a time to come when by the growth of this science of signification man will be as much above his present mental estate in scope and competence as his present mental estate is above that of the Hottentot (and a boundless horizon)[145]

Better *means* of thinking will generate better thought and better habits of thought and these will produce better constitutional structures for thinking. Talk about the *Laws of Thought.* If anything is "free and libertine" it is thought. It is subject only as it has not (as yet) discovered and invented for itself better *means.*

---

139 At top of page Peirce notes: "I dare say you're right; but I am not in the habit of paying attention to reasoning as grossly bad as that."

140 Meaning, part of what he wrote for his "Hints" accompanying Peirce's "Explanation of Curiosity the First," *Monist* (July 1908). For details about the enclosed MSS, see footnote to postscript below.

141 At p. 379; Minot Judson Savage (1841–1918), Unitarian minister and writer.

142 In left margin Peirce notes: "I regard logic as the general science of signs"

143 In left margin Peirce notes: " 'Not at all bad' as the English would say."

144 At top of page Peirce notes: "I esteem it just so far as it furnishes diagrams of logical relations."

145 In-line Peirce notes: "That will be perhaps in the other world."

---

1 A strong] I strong *[E]*
2 - A -] *in blue pencil*
5–9 [That … &c).]] *brackets in blue pencil*
14 as] is *[E]*
15 *calculative*] *after* ~~so called~~
20 Hottentot] *after* ~~and~~
21 habits] *after* ~~thought~~
22 constitutional] *inserted above*
23 it] at *[E]*

What is the *algebra* of thought? Is it a multiple algebra? If so how many independent units has it?.—And what are the complications and explications (operations?) to which these independent units are amenable?, such are some of the questions that *open up* to my musings.[146]—

As aids and suggestions in this line there are a number of productions that ought to be made more accessible than they now are and I have almost persuaded the Dr. to print and bring out an annotated volume of such. Just what papers should be selected is an open question. One should be, Grassman's paper on "*The different kinds of multiplication*" perhaps. Riemann's Hypothesis &c should be one.[147] I notice that Weierstrass is much mentioned as a philosophical mathematician but as to what essay or essays of his are most significant I am as yet mostly in the dark.[148] Wont you tell me your idea of what such a volume should contain in order to provide the *lay* but *rather competent* reader with information as to what is in gestation among the logical and mathematical competents of yesterday and to day?.—I suppose you have read Martz's chapter in the second volume of his Intellectual History of the 19th century. Have you reviewed it?. If so in what paper *or* magazine? I want to find out your estimate of the work and especially that last chapter.

Very Cordially
Francis C. Russell

146 In left margin Peirce notes: "The logic of dyadic relations is represented by linear associative algebra in general. Deductive logic in general may be exhibited in all cases in the most analytic forms so far attained by Existential Graphs."

147 Hermann Günther Grassmann (1809–1877), German linguist and mathematician; reference is to "Sur les différents genres de multiplication" (1855). Also Georg Friedrich Bernhard Riemann (1826–1866), German mathematician and pioneer in analytic number theory; reference is to the Riemann Hypothesis of non-trivial zeros, introduced in his 1859 "Ueber die Anzahl der Primzahlen unter einer gegebenen Grösse" ("On the Number of Primes Less Than a Given Magnitude").

148 In left margin Peirce notes: "You will get the idea of what Weierstrass insists on from any modern work on the Theory of Functions. He wants more exact logic and less *intuition* in which he is partly wrong, and partly right."

7 accessible] accessable *[E]*
8 annotated] annottated *[E]*
10 Hypothesis] *after* ~~Hip~~
11 Weierstrass] Wierstrass *[E]*
12 or essays] *inserted above*
13 such] *after* ~~sh~~

P.S.—I guess I will send you also another embryonic *part* of my attempt to explain your M.S. - B - I mark it - B - These— - A - & - B - M.S.S. —I would like to have back when you have seen them all you want to.—But the other (Views &c)—is a copy, which you need not return. - Write soon as you feel right for it[149]

F.C.R.

Peirce to Russell. MS: RL 387. Followed by drafts of pages 3 and 6 of letter.

1908 Sep 18

Dear Russell:

I wish you would acknowledge receipt of my letters. I wrote you a long letter which I *infer* you got (I haven't yet finished reading yours of the 16th inst.) But as you didn't acknowledge it, I sent you a reply postal to ask if you got it. It would have been easy to write

Yes

Russell

But it seems to have been too much trouble. You observe that I attend to your letters, though they make some draughts on my supplies of time & energy.

---

149 The enclosed MSS consist of the following, which are not included in this edition:

MS "- A -": 14 MS pages (pp. 2–15) of an expansion of his MS "- B -" below. Describes Calculative Logic and patterns of "higher thinking" as inspired by the "ideas of Prof. Peirce," and discussing many other points of logic related to his and Peirce's thought.

MS "- B -": 2 MS pages (pp. 174–175) of a previous draft of Russell's "Hints" preceding Peirce's "Explanation of Curiosity the First" (Peirce's "MS - B -") in the July 1908 *Monist*. States its purpose "to lead the reader if possible into a very much higher purer and truer structure of thought and intellectual experience."

Various other MSS, perhaps comprising the "Views &c" to which he refers:

3 MS pages (pp. A–C) of an outline headed "Existential Relations"

9 MS pages (pp. A–I) of an outline headed "Genl Nature of Number"

6 MS pages (pp. 6–12) of an undetermined paper on geometry and logic

1 MS page (p. 3) of an undetermined paper or letter dealing with Geometry

5 MS pages (pp. 1–5) headed "Amazing Mazes | The Fourth Curiosity | (Index & Abstract)", containing transcriptions of various sections of Peirce's article.

11 MS pages (pp. 2–12) headed "M.S. = E—Fourth Curiosity", describing a proposed 186-page "thorough essay at the solution of the question as to which of the two Cardinal Numbers or Ordinal Numbers are the first in logical primacy." Also provides an outline similar to the "Index & Abstract" of the Fourth Curiosity above.

---

1 embryonic] *inserted above*

2 - B - These— - A - & - B - M.S.S.] - B - *and* A - & - B - *in blue pencil*

As to the book you say you want to persuade Carus to get out, the idea *has merits*. That does not say that it is on the whole good as it stands. To judge of it, it is necessary to consider what mathematics is. Now to get an idea of what any science is *in its present condition*—which is all that a man of sense will attempt, for any science but his own,—it is necessary to study the men at work on it; for what it is, is what they make it. The mathematicians are men with great powers of embracing in one thought great complications of very simple elements, and of thinking about them with great exactness. They also have extraordinary ingenuity in finding points of view from which the properties of their complexes may be discerned. Having these exceptional powers, of course they devote themselves to the exercise of them. The result is that mathematics is nothing but a collection of deductive reasonings. Of course these group themselves in certain ways. They have *external* groupings of very little value, but commonly noticed; and they have *essential* groupings according to the nature of the leading concepts. It is the rarest thing in the world to find a mathematician who is capable of logical analysis, so that he really understands what it is that he has been doing; the literature and history of mathematics are filled with proofs of the incompetence of the greatest masters in that. Except Leibniz, I should be puzzled to name another. Take Fermat who discovered the *substance* of the calculus, though it took Leibniz to invent the proper notation for it. Well Fermat made a discovery of that method of reasoning about integers whose essence is that whatever is true of zero and is true of the number next higher than any number of which it is true is true of all integers; because any integer whatever can be reached by successive additions of 1 starting from 0. But Fermat though he showed extraordinary skill in using his method, never seems to have been able to formulate it or give a reason for it. He calls it a kind of *reductio ad absurdum*. This simply shows that he didn't know that the *reductio ad absurdum* differs from direct reasoning in no essential particular but only in the order in which the ideas that have to be put together are taken.

The whole Weierstrassian mathematics,—that is to say the way mathematicians reason since Weierstrass showed how loose most of the geometrical reasoning is, showing for example that it does not follow because a real function is continuous that it must have a differential coëfficient, which he did by simply giving an obvious instance of its falsity,—well, the whole Weierstrassian mathematics is characterized by a *distrust of intuition*. Therein it betrays ignorance of a principle

12 group] *after* ~~are~~
13 but] *after* ~~com~~
16–17 the literature and history of] *inserted above*
17 mathematics] mathemas *[E]*
18 Except] {*e*Except
22 integers] *over* n
23 additions] *after* ~~add~~
31 does] *over* i
31 real] *inserted above*

of logic of the utmost practical importance; namely that every deductive inference is performed, and can only be performed, by imagining an instance in which the premisses are true and *observing* by contemplation of the image that the conclusion is true. The image, as *singular*, must of course have determinations that the premisses, as *general*, have nothing to do with. But we satisfy ourselves that the particular determinations of the image chosen, so far as they go beyond the premisses, could make no difference. There is no other way of making a deductive inference than that, however simple it may be. Take for example the conversion of E the universal negative proposition. To ascertain whether or not this follows, I take the example "No bird is a phoenix." To be sure this is not the only universal negative proposition but it does not differ from any other in any respect which could make any difference as to what might logically follow from it or what might be consistent with it. Now I imagine a cage in which are contained all the birds but no phoenices. To be sure all the birds could not in fact be contained in the cage; but we can imagine that they were so compressible that they could, and that would not prevent phoenices from being birds, because there is plenty of room for the phoenices, if there are any outside the cage, and there are none in it. Now then if there are any phoenices they are either in the cage or outside of it. But they are not in it so if there are any they are outside. So then none of them is a bird since every bird is inside and not outside the cage. So then if no bird is a phoenix, no phoenix is a bird. It is true that whatever phoenix there is is a bird. But that does not prevent its being true that whatever phoenix there is is not a bird.

Let us take a different example. Given the two equations

$$ax + by + c = 0$$
$$\alpha x + \beta y + \gamma = 0$$

Required the solution. Add to them the identical equation

$$mx + ny - (mx + ny) = 0$$

where we may give any values we please to m and n without falsifying the equations.

We know from the nature of a determinant that

$$\begin{vmatrix} ax, by, c \\ \alpha x, \beta y, \gamma \\ mx, ny, -(mx + ny) \end{vmatrix} = 0$$

4 determinations] *after* ~~countless~~
8 however ... be.] *inserted above*
8 conversion] converson *[E]*
540.32–541.28 giving an obvious instance ... m and n without] *An aborted draft of this MS page (p. 3) follows this letter.*
30 that] *above* ~~ax + by~~

$$\text{whence}\begin{vmatrix} a, b, c \\ \alpha, \beta, \gamma \\ m, n, -(mx+ny) \end{vmatrix} = 0$$

or $-(\alpha\beta - \alpha b)(mx + ny) + (\alpha c - a\gamma)n + (b\gamma - c\alpha)m = 0$

or $mx + ny = \dfrac{(\alpha c - a\gamma)n + (b\gamma - c\alpha)m}{a\beta - \alpha b}$

when by putting $m = 1 \quad n = 0 \quad x = \dfrac{b\gamma - c\alpha}{a\beta - \alpha b}$ and putting $\begin{matrix} m = 0 \\ n = 0 \end{matrix} \; y = \dfrac{\alpha c - a\gamma}{a\beta - \alpha b}$

The point is that we have to *observe* the equations to see that they do permit the first equation on this page.

Now as to the book you have in mind, its design as I understand it is, by specimens of mathematical reasoning to improve the reasoning of non-mathematicians. For that purpose, it would not often be advisable to give the whole mathematical memoir, but only the logically instructive parts; and those need not be given in the words of the original writer, though his mathematical *forms* should be conserved. The annotations would require a very strong logician, not the kind of man that the average editor or publisher takes to. Small men like small men: they have an instinctive antipathy to great minds; and it would be a pity the thing should be so done that its weakness should be shown up at once. The poor annotator would not be in an enviable situation, nor anybody else concerned.

If the ground were fully covered, the book could not be a very small one. It would have to be at least as large as Russell's Phil. Of Math.,[150]—a weak performance, by the way. Existential graphs should begin the work, because there is no organ of definition and logical analysis that is at all equal to that. I have not yet brought out this feature of the system, though I have much in MS. When I do bring it out, those few who go through the exposition will be amazed. Then should be taken up integers. First the Cantorian *ordinals* though he himself puts multitudes first. But it is wrong. For ordinals are the general, multitudes the special. That is, the different multitudes of plurals (which Cantor badly denominates *Cardinal-Zahlen*, and *Mächtigkeiten* von *Mengen*) are simply the successive grades of plurals; and there are (if you like so to modify the meaning of the term) "cardinal numbers," or as I have called them, 'arithms,' (which is better) to designate those grades. But *ordinals*, in general, are designations of *grades*, each grade being pos-

150 For *Principles of Mathematics* (1903).

18 Existential] *after* ~~Th~~
25 *Mächtigkeiten*] *Mächtigeiten [E]*
26 modify] *above* ~~extend~~
27 or] *after* ~~to desig~~
27 (which is better)] *inserted above*

sibly occupied by what I will call *numerates*, or subjects of numeration. In the special cases in which the numerates are *plurals*, the grades are *multitudes*, and their ordinals become cardinals, or *arithms*. I shall show that all real rational numbers are ordinals, and that irrational reals are ordinals likewise; while imaginaries and quaternions are a special kind of complexes of ordinals. I may mention here that I fully recognize that integer arithms have a special interest owing to the fact that each plural is composed of singular members, while rational arithms denote grades of ratios of two multitudes, and irrational arithms are the arithms of ratios between first-abnumerable multitudes. I have completed or at any rate extended the system of fundamental concepts employed by Cantor, in a very important respect, which I will, if I have time, explain further along in this letter.

Between Grades, there are two highly important dyadic relations, whose corresponding relative terms I call "after," i.e. *counted later*, and "next," i.e. *immediately after*. It is obvious that "next" can be expressed in terms of after. In fact the graph N—next—M is equivalent to 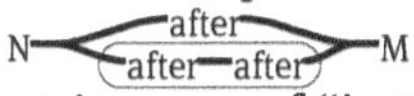. But "after" can also be expressed in terms of "next," (in a certain sense of "in terms of"); and this I am inclined to think is the truer analysis. At any rate, it lets us into the secret connexions between the numerates.

If we confine ourselves to the finite ordinals it is easy to define 'after' in terms of next. It is a part of the definition of ordinals that every Grade has a Grade next to it and that every grade but one is 'next' to a Grade; and and no Grade is 'next' to a Grade other than a Grade to which it is 'next' nor has a Grade 'next' to it which is other than another 'next' to it. Moreover every Grade lacks a certain character which is possessed by a Grade next to it, and which is further possessed by whatever Grade is next to a Grade that possesses it. Now to say that a Grade is 'after' another is to say that it possesses every character of that kind that is not possessed by thc Grade that it is 'after' but is possessed by the Grade that is 'next' to the latter. But this is not every kind of afterness, but is only that kind of afterness that alone occurs between finite Grades.

The higher kinds of afterness were introduced by G. Cantor. They are described by him in Vol XLIX of the Mathematische Annalen[151] and in a form supposed to be adapted to philosophical readers in a little book Zum Lehre vom

151 "Beiträge zur Begründung der transfiniten Mengenlehre" (1897).

1 what] *after* ~~num~~
3 real] *inserted above*
5 I] *after* ~~But Cantor~~
6 integer arithms have] ~~Cantor~~ \integer/ arithms ~~are a~~ have
8 multitudes,] *after* ~~f~~
10 system of] *inserted above*
21 and no] ˅and˅ ~~but one Grade is next to~~ ˅~~any~~˅ ˅no˅
23 lacks] *after* ~~po~~

Transfiniten. Halle: Verlag von C. E. M. Pfeffer. 1890. There is also a very important critical report on the whole of Cantor's work on numbers by Arthur Schönfliess which forms the whole of Heft 2 of Vol. VIII of the Jahresbericht der deutschen Mathematiker-Vereinigung. It is called "Die Entwickelung der Lehre von den Punktmannigfältigkeiten" There has also been some fundamental work by Borel.[152] I am not altogether satisfied with Cantor's work, greatly as I admire it.

I ought to have said above that it is essential to the idea of Ordinals that it is a linear system i.e. in *some* sense of any two Grades one is 'after' the other. In the sense just defined the finite system is *teres totus atque rotundus.*Nothing can be added to it. But there is another kind of *nextness* which can be introduced without modifying any of the above definitions. Namely a Grade may be "next,"—not to any one Grade, but to an endless series of Grades; and Cantor postulates that every endless series of Grades has one, and but one, Grade 'next' to it which is not 'next' to any other Series nor to any Grade. For after Achilles has run through his geometrical series after the Tortoise, time still runs on and he with it, and the very next instant he is supposed to catch up. So this new grade the infinitesimal (i.e. the infinity-eth) is in a new sense 'after' all the finite grades. I will scribe the finitely after -after-_{0}, and this new after -after-_{1}, then

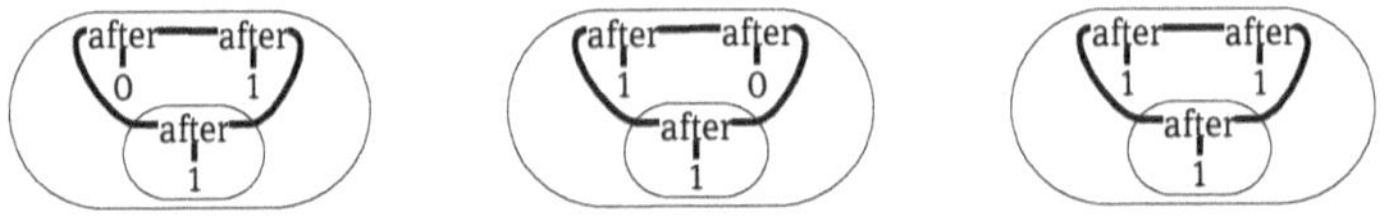

Cantor denotes that which is 'next' to the whole series of finites by the ordinal ω. He notes that the *count* now depends on the order of counting; so that $1 + \omega$ is no longer the same as $\omega + 1$. I consider the *addend* as operating on the augend; and since, in my algebra of dyadic relatives, as well as in all my earlier algebras of logic I always wrote the operator to the left of the operand I write $\omega + 1 = \omega$ and $1 + \omega$ as *next* to ω. But Cantor(and mathematicians generally) write the *addend* after the *augend*, which, in my view, is wrong. Of course it still remains true that every

152 Félix Émile Borel (1871–1956), French mathematician and pioneer in measure theory and probability.

9 *teres totus atque rotundus.*] *For* totus teres atque rotundus, *"entirely smooth and round" or "well-rounded"*
16 grade] *above* ~~numbe~~
17 scribe] *above* ~~call~~
543.18–544.23 between the numerates … as well as in all] *An aborted draft of this MS page (p. 6) follows this letter.*
23 algebra] *after* ~~af~~
24 wrote] *above* ~~put~~
24 write] *after* ~~con~~

Grade has a Grade 'next' to it and therefore we run on to $\omega + \omega$, which Cantor writes 2ω. Then comes $2\omega + 1 \ldots 2\omega + \omega = 3\omega$ and so on, until he has an endless series of endless series of ordinals. To say that there is a *next* Grade to this seems to me a new extension of the idea of nextness, since *this* next, is neither *next* to any Grade nor *next* to any endless series of Grades, but is only next to an endless series of endless grades. Here therefore it seems to me clearly that we have a new kind of afterness, which I write $\overline{\text{after}}_2$. But Cantor simply speaks of ω^2 without regarding himself as introducing any new conception; and his reason is that the *multitude* is unchanged. That is to say there is an order of counting according to which $\omega^2 = \omega$. But it seems to me the better way of expressing this would be to say that the multitude of infinites, or transfinites, as he calls them, depends on the order of the count, and that the minimum count of ω^2 equals that of ω. It is of course easy to distribute the finite integers into any *finite* number of endless series.

Of course there will be a next to any endless series of numerates which is not next to any numerate, and there will be a next to an endless series of endless series of numerates which is neither next to an endless series of numerates each having next to another, (nor is next to any numerate), and there will be a next to an endless series of endless series of endless series of numerates and so on indefinitely. And then (what we can only express indirectly by a sign describing a sign,) the whole lot of numerates expressible by "series of series" etc. is supposed itself to have a next; though it is not evident that that does not land itself in an absurdity. This method Cantor supposes to be carried out endlessly; and he assumes (I can find no attempt to prove it) that in this way his Grades will attain every possible multitude. But that is an entirely illogical assumption. For I have proved that every multitude has a higher multitude, the plural of all plurals of M's being always greater than the plural of all Ms. If Cantor has really proved, as he says he has that by his method, starting from ω he reaches 2^ω (or as he expresses it, ω^ω, which is the same multitude) and 2^{2^ω} and $2^{2^{2^\omega}}$ and so on indefinitely, then he certainly has ordinals that count a series greater than any collection. But his multitudes being themselves a collection, and not skipping but always advancing to the next, it is impossible they should count a series more multitudinous than themselves; so that there must be an error somewhere. It may be the fallacy lies in not observing that the multitude of a collection is not the last ordinal in this or that count of it, but is the earliest ordinal that can count it, the count varying

1 Grade] *after* ~~ordinal~~
2 until] *after* ~~to~~
3 Grade] *after* ~~ordin~~
16–17 each … (nor] each \having/ next to ~~one~~ another, ^(^nor
22–23 assumes] *before* ~~that~~
25 every] *above* ~~there is a~~
25 plural of all plurals] *above* ~~collection of collections~~
29 ordinals] *after* ~~mul~~

according to the order of counting. Or it may be owing to his not having proved that none of his descriptions of ordinals involve any absurdity. Schönfliess agrees with me that Cantor has not proved all he thinks he has.

Cantor does not start with the relation of "next" but with that of "after" and consequently fails to recognize any distinction between one kind of "after" and another, which is a great blemish.

In referring to his paper in Vol XLIX of the Annalen, I omitted to say that it cannot be read without first mastering in minute details a previous paper in Vol XLVI where he treats of multitudes. I should be much disposed to write a work myself on this subject if my logic were not far more important.

I don't know as I ever showed you how rational fractions are fundamentally ordinals. Start with the assumption that $\frac{N}{1}$ where N is an integer represents that integer, and with the assumption that $\frac{A+M}{B+N}$ where A, B, M, N are integers is *after* either $\frac{A}{B}$ or $\frac{M}{N}$ and that the one of these it is not after is *after* it. Of course, the same would be true of $\frac{A}{N}$ and $\frac{M}{B}$. Next assume that $\frac{1}{0}$ is after $\frac{0}{1}$. It follows that using Cantor's sign $P \prec Q$ for Q is after P, $\frac{1}{1} = 1$ is after $\frac{0}{1} = 0$ and that $\frac{1}{1} \prec \frac{1}{0}$. Then having $\frac{0}{1} \prec \frac{1}{1} \prec \frac{1}{0}$ we interpose $\frac{0}{1} \prec \frac{1}{2} \prec \frac{1}{1}$ and $\frac{1}{1} \prec \frac{2}{1}$ $(= 2) \prec \frac{1}{0}$. Next $\frac{0}{1} \prec \frac{1}{3} \prec \frac{1}{2} \prec \frac{2}{3} \prec \frac{1}{1}$ etc. Next $\frac{0}{1} \prec \frac{1}{4} \prec \frac{2}{5} \prec \frac{1}{2} \prec \frac{3}{5} \prec \frac{2}{3} \prec \frac{3}{4} \prec \frac{1}{1}$; and it is easy to prove that we shall then obtain every positive rational value *expressed in its lowest terms*. You can work out the proof for yourself, perhaps; though it is quite a theorem. Now we have supposed nothing at all about the values of these fractions nor have we supposed the intervals between them to be anything in particular, yet we have deduced all that is needed to establish the whole doctrine of fractions. Consequently, that doctrine has nothing to do with *measure* or exact equality, but only with ordinal sequence. You are at liberty to suppose that $\frac{1}{2}$ stands for two thirds or one tenth, and all the consequences will be true. The line between the numerator and denominator is not a sign of division. The fraction only express such a function that it has the properties stated. Of course you must recognize as a corollary that $\frac{P+P}{Q+Q}$ is intermediate between $\frac{P}{Q}$ and $\frac{P}{Q}$ that is equals $\frac{P}{Q}$. So $\frac{7}{13}$ is between $\frac{1}{1}$ and $\frac{6}{12}$ i.e. is $\succ \frac{1}{2}$, between $\frac{1}{4}$ and $\frac{6}{9} = \frac{2}{3}$ Between $\frac{2}{3}$ and $\frac{5}{10} = \frac{1}{2}$ Between $\frac{2}{4} = \frac{1}{2}$ and $\frac{5}{9}$ etc. The assumption usually made (often with a puerile attempt at proof of it) that between all the values of a convergent series and the nearest rational value there is but a single irrational value, is perfectly gratuitous. There may be any multitude of such values of course "infinitely near" one another; and we have a perfect right to assume that there are (i.e. may be, are possibly) and so revert to Leibniz's differential calculus. For since rational quantities are mere

19 positive] *inserted above*
24 or exact equality] *inserted above*
27 The fraction only express] ~~It is~~ \The fraction/ only \expresses/

ordinals, the assumption that because there is no "assignable difference" i.e. no rational difference between two values therefore there is no difference at all, is perfectly arbitrary.

Passing to the subject of true *continua*. The true definition of a continuum should be given; namely a continuum is an object all of whose parts are alike in their relations to *their* parts. So Kant. Then Listing's great memoir, I mean the one on the Census Theorem in the Göttinger Abhandlungen.[153] But this should be annotated and brought into connection with projective geometry on the one hand & quaternions on the other. Besides which, the doctrine of Topical Singularities should be treated. And reference should be made to space of more than three dimensions. I haven't time to write more at present.

Very faithfully
C. S. Peirce

Various things in the Theory of Functions should be brought in; ideas from the Galois Theory of Equation

I will return your MSS A & B on Sep 22 or 23.

[Aborted draft of p. 3]

[...] giving an obvious instance of its falsity,—well, the whole Weierstrassian mathematics maintains its vogue by a *distrust of intuition*. Therein the modern mathematical world with Weierstrass at its head betrays an ignorance of a fundamental fact in the logic of deductive inference; namely, that *every* such inference is performed by making a diagram of the state of things represented by the premisses—which diagram may be an array of letters or it may be composed chiefly of lines,—and having made sure that, though the diagram, as a single image, is necessarily far more determinate than the premisses, since these are general, nevertheless none of the additional determinations can affect the conclusion; the reasoner then *observes* this diagram and notices some relation between parts of it which had not been expressly introduced in the construction of the diagram. Take, for example, the conversion of the Universal Negative. Supposing I know that No protestant adores the Virgin Mary, I imagine a circle with a lot of red sand in it.

153 "Der Census räumlicher Complexe" (1862).

14–15 Various ... Equation] *This line up left margin.*
16 I Will ... or 23] *This line on verso of last page. Refers to Russell's enclosures of 27 Nov 1906.*
19 maintains] *after* ~~are~~
25 the premisses, since] the ~~stat~~ premisses, ~~which~~ since
26 the^3] *after* ~~and th~~
30 I imagine] *after* ~~I imagine~~

I say let every grain of red sand be a protestant & suppose every damned protestant is in that circle; to be sure in point of fact, protestants are not grains of red sand nor in that circle either. But those circumstances can have no effect on the character of those who adore the Virgin, if any such there be.[154] *[ ]*

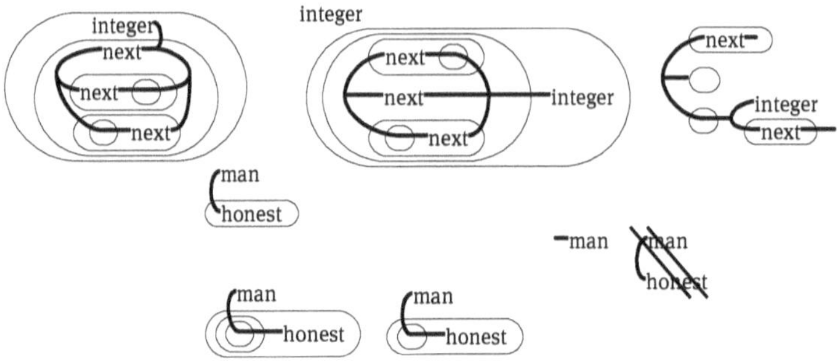

[Aborted draft of p. 6, numbered p. 5 in this draft]

[…] between the numerates. If we confine ourselves to *finite* integer ordinals, our problem is practically solved by the substance of pp 443–446 in the July *Monist* of this year[155]; which is the true kernel of that paper, and I need not tell you that its "sum and substance" is *not* what, on p 407 7cm to 12cm from the top of the page,[156] you tell those who won't read it anyway that it is; for your own remarks further on in your paper show clearly that you yourself fully understand and appreciate the reasoning. But your experience in talking to juries induced you to put off the stupids with that statement; and I have no doubt it was the best that could be done.

Namely in the system of finite integers, the form of "next" is sufficiently defined by saying that "next" is such a relation that if any one grade *N* is "next" to any grade, *M*, then, 1st, *N* is not *M*; 2nd, *N* is identical with whatever is "next" to *M*; and 3rd, *M* is identical with whatever *N* is "next" to *M*. In this system, "after" is reduced to identify with "finitely after" or "after in the 0 degree" which may be defined thus: If any grade *Q* is "after$_0$" to any grade *P*, then there is a certain plural

154 The following Existential Graph notes are found at the bottom of the MS as shown and further exemplify the diagrams discussed in this aborted draft and in the sent letter above.
155 "Explanation of Curiosity the First."
156 Russell's "Hints" in the same number.

18 any] *inserted above*
18 to] *below* ~~to~~ \\~~after~~/
19 to] *above* ~~after~~
20 *M*] *over* m
20 to] *after* ~~after~~
22 any grade] *inserted above*
22 "after$_0$" … *P*,] "after$^{\theta\vee}{}_0{}^{\vee}$" to \any grade/ *P*,

such that Q is a singular of it and P is not a singular of it, and if any grade G is a singular of this plural then any grade H, that is next to G is a singular of it, (and consequently if any grade T is not a singular of it any grade S to which T is next is not a singular of it;) and whatever is a singular of this plural is "after_1" to whatever is not a singular of it. You will easily scribe the graphs for yourself. But in the Cantorian system of ordinals, not only has every *single* grade a grade next to it, but every *plural* of grades such that there is one of its singulars of which every other singular of it is after_n has a single grade that is next after that *plural* of grades. I would call this kind of nextness "nextness in the $Nn^{\underline{th}}$ degree," where Nn means "next to n." And given any plural of grades such that there is one of its singulars of which every other of its singulars is after_n then there is a grade which is next_{Nn} to that plural. *[ ]*

Russell to Peirce. MS: RL 387.

1268. Jackson Boul | Chicago Oct 2^d 1908

Dear Prof. Peirce

You are altogether just in chiding me for my neglect in failing to acknowledge your letters, and now and here I am found in fault as to your last letter written Sept 18^h. I could attempt to palliate for I labor under a number of embarrassments but remembering that *qui excusé se accusé* I will just throw myself upon your charity and promise to be wary and prompt in the future. I cannot tell you how much I appreciate your letters and how sensible I am of your gracious attention to my solicitations. If you suppose that I think myself any "great shaker" in science and philosophy you are in error. I am vividly aware of my inferiority naturally and as a matter of information. I only account myself fortunate in having thirty years ago been awakened and illuminated by your series of papers in the Pop. Sci. Monthly. They have given to my whole subsequent intellectual life its controlling ingredient and there has never been any occasion since when I have not been able to see the advantageous information I thereby gained, especially in absorbing thor-

1 singular[1]] *above undeleted* member
1–2 *G* is a singular[1] of] *above* ~~belongs to~~
2–4 *G* is a singular of it, (and consequently … of it;)] *G* ~~belong~~ \is a singular/ ~~to~~ of it, ~~while~~ \(and consequently/ *[. .]* of it; ^v)^v
4 is a singular of] *above* ~~possesses~~
4 "after₁"] 1 *over* 0
7 its singulars] *above* ~~them~~
9 in] *above* ~~of~~
18 embarrassments] embarassments *[E]*
19 *qui excusé se accusé*] *For* qui s'excuse s'accuse, *"He who excuses himself accuses himself."*
28 advantageous] advantagious *[E]*

oughly that chapter on "How to make our Ideas Clear." I would essay to get the *Dr.* to bring out a reprint of those papers (with your annotations) if I dared.[157]

Now I guess I had better tell you just how the situation is here and how you and I are related to it.

The *Open Court* essay began really through the occasion that Mr. Hegeler felt himself duty bound, *richesse oblige*, to bestow a part of his fortune for the benefit of his fellow men. At the same time he regarded himself as a scientific man and philosopher of a rank worthy of considerable (if not more) deference. As I diagnose his case he is a man of a deeply religious temperament at bottom but having been brought up in a university atmosphere of crass Büchnerian materialism his *mental system* (?) was pervadingly constructed after the model of that *ism*

Accordingly in casting about for the avenue through which his contribution to the welfare of his fellows should have course he resolved on the establishment of a periodical publication. At first he expected to utter through such a publication a good deal of his own authorship. He had conceived an idea of *Monism* which he deemed of vast and far reaching consequence although his idea had (as I guess) not been much elaborated if indeed it was at all clear in his own conception. In pursuit of his purpose he made overtures to Underwood then of the then languishing *Index* with the result that Underwood became installed as the editor of the new publication which was launched under the title of "The Open Court" and published by "The Open Court Publishing Company" a corporation organized for that office.[158] Well I suppose you have long ago "sized up" Underwood. He was simply a disciple and follower of Herbert Spencer and clave to the company of his master and his *ilk*, in all his (U's) opinions, conceptions, and proclivities. Underwood did'nt and could'nt "tumble to" the real situation and from the very first the intercourse demanded the lubrication of much charity and patience on both sides. Hegeler did'nt see how he could go on and do any better and Underwood had embarked pretty much all he had in the venture. But Hegeler was aching to deliver

157 Coincidentally or perhaps under Russell's influence, Carus shortly after wrote Peirce 9 Dec 1908 upon returning from Europe proposing again to take up the "old and cherished plan" of the "Illustrations" reprint proposed 23 Jan 1907. Despite earnest efforts on both sides from this point, the book was never completed due primarily to Peirce's inability to settle on the revisions. For Peirce's proposed preface, which references Russell's involvement in the project, and for Carus's proposed editor's preface, see Appendix.

158 For details beyond what is hereafter explained regarding the dramatic inception of the OCP involving the Underwoods and Carus, see Henderson, 27–44; also "Carus" and "Underwood" in Biographical Register.

5 the] the the *[E]*
9 temperament] temprement *[E]*
15 idea] *after* ~~as~~

a *Monistic* gospel which Underwood couldnt see anything of consequence "*in it*" and filled "The Open Court" with the lucubrations of the old Index bunch, so matters went on for a while. Dr. Carus was then in and about New York and Boston only a year or two or three emigrated from Germany and struggling to exist as a back writer. By some chance, or perhaps by his own instance, his little booklet "Ursach Grund and Zweck" came or was put into the hands of Mr. Hegeler. Upon this occasion Mr. Hegeler sought out Dr. Carus and the latter was invited to visit Mr Hegeler at La Salle.

The result was that Dr Carus was installed as sub-editor of "The Open Court" charged with the office of giving vast to the *Monistic* gospel and as the special *organ* of Mr. Hegeler as to that. Mr. Hegeler as you know is not fluent of discourse anyway, and his command of English is, (or at least he felt it so) inadequate for the illumination of his ideas (*thought* out of course in his vernacular) to Underwood so that the latter even had he felt so disposed could *[not]* render them faithfully in "The Open Court". I have no doubt that Mr. Hegeler found great comfort in being able to utter himself in German to some one of the same vernacular.

Now Dr. Carus, besides being naturally tactful and accommodating in spirit had felt the buffeting of fortune and of course he was swift to discover the opening that circumstances had brought about and he governed himself, as well he might, accordingly. Moreover Mr. Hegeler had a daughter, a *favorite* daughter who became Mrs. Carus.

Now do'nt take me as saying or even as hinting that Dr. Carus played the hypocrite or the flattering courtier. I do'nt believe any such thing. He simply basked in the sunshine that he found provided and took care not to injure his favor.

Well the installation of Dr Carus as sub-editor and the evident special favor with which he was regarded proved very unwelcome to Underwood who with that combativeness that so marks his ilk soon found means to stir up the issue that he himself must be "boss and all hands" as regards the editorship or that he would throw up the whole job. He thought no doubt that the whole project would collapse in default of his support. He had better have known Mr. Hegeler better whose paramount characteristic is a dogged and unconquerable persistence in whatever he once undertakes.

---

4 emigrated] immegrated *[E]*
7 Carus] Caurus *[E]*
23 the] *after* ~~as~~
26 with[1]] \with/ with *[E]*
27 marks] markes *[E]*
30 known Mr. Hegeler better] known ~~better.~~ Mr. Hegeler \better/

So Mr Underwood went, and in sympathy with him the body of the Indexers went also and "The Open Court" proceeded under the auspices of Dr. Carus.[159] The *essay* has grown and grown and grown and is still growing. First, "The Open Court" then "The Monist" and then the publication of books until it has become a publishing house of no contemptible consequence. But it is distinguished from the ordinary publishing house by the fact that it is not "run" for the "*money there is in it.*" It has never paid any profit. Indeed it has never paid expenses by a long shot. It has gone on for over twenty years suffering a deficit of somewhere about $ 20,000.$\underline{00}$ annually. It would be erroneous to say that its promoters plan to thus suffer always. They hope and believe that in due time it will pay at least all expenses and it is organized on a business model on that behalf. But it will go on whether it pays expenses or not for Mr. Hegeler has endowed it with a fund of $ 750,000.$\underline{00}$ put in trust in the hands of Dr. Carus and Mrs. Carus as Trustees and devoted to the perpetual maintenance of its well known projects. In other words "The Open Court" concern instead of being a mere business enterprise is in reality a well endowed *education foundation*, devoted to be conducted not primarily for gain but in a missionary spirit and at the expense of sacrifice so far as may prove needful. If on some aft occasion you could advertise in and about the Eastern metropolis this peculiar and commendable plan, purpose, and embodiment you would gratify me (and perhaps others here and about here).

159 For the correspondence of Hegeler and Underwood leading up to their unsavory separation, see *Open Court* (22 Dec 1887), pp. 621–640. Hegeler published this extensive communication in an effort to "submit the evidence without argument to the readers of The Open Court" regarding the dramatic resignation of the Underwoods, announced in the previous number (24 Nov 1887, pp. 592–593), and to include the following departing remarks from the Underwoods:

> The immediate cause of the editors' resignation is Mr. Hegeler's expressed desire *to make a place* on The Open Court for Dr. Paul Carus, who never had, it should here be said, any editorial connection with the paper, who never wrote a line for it except as a contributor and as Mr. Hegeler's secretary, and who was unknown to Mr. Hegeler when his contract with the editors was made. To the request that Dr. Carus be accepted as an editor, the present editors, for good and sufficient reasons, have unhesitatingly refused to accede, and although always willing to make concessions when required in the interest of the paper, a point is now reached where they feel compelled by self-respect to sever all relations with this journal rather than yield to Mr. Hegeler's latest requirements.

1 in sympathy] *inserted above*
8 shot] *after* ~~shop~~
20 here and about here] *inserted above*

Now when Dr. Carus first started in as editor he was to a degree unfamiliar with the calls and responsibilities of editorship and further he was hardly acclimated as an American citizen. It was at this time that I first connected with him and a mutual liking ensued. Gen. M. M. Trumbull, (now dead) a close friend of mine became also one of a coterie in and about The Open Court office and *I think* that at that time leaned a good deal towards and upon us. I think I may say that you became known to Dr Carus through my offices. All went well enough until your series of articles.

Now Dr. Carus had elaborated or was elaborating a system of philosophy. Now it would be officious for me to say what I think about the Dr's "system." Except as mere tentatives I dont subscribe to any "system". As such a tentative his "system" has *points* worthy of attention although I often find myself asking myself if the Dr. really sees very clearly the full import and implications of the theses he utters. But at the time of the controversy between you and he,[160] he had committed himself pretty deeply to a certain set of doctrines and he had not enough of the true scientific spirit to hold them in deference to instruction that he might possibly acquire in the evolution of his intellectual life. He thought himself called upon to hold the ground he had taken against all infringers and so you and he had your "*scrap*." Now I must say that I dont think you was as tactful as you ought to have been. T'is glorious to have a giant's strength, but tis godlike to forbear from any needless exercise of that strength and I am sure that you deeply wounded the Dr's self contentment. I did my best to help the Dr. out but my support could not have been very efficient for I felt the force of your considerations and beyond a certain amount of allowable accommodation I could not and would not go, so although I in no wise took up cudgels on behalf of your contentions I was not able to conceal from the Dr. my lack of any hearty faith that he had the best of the scrap. This led to a coolness between us and for several years we stood at arm's length although we shook hands with all apparent heartiness whenever we met. But he did not seek me and my converse as he had been wont to do previously. I suppose he thought I would (metaphoricaly) go on my knees to him in the matter of our want of concord. Every once in a while he would drop a remark showing that he regarded me as taking sides with you as against him. Now I take sides with nobody except so far as I am compelled.

160 Carus's and Peirce's Necessity–Tychism feud in *The Monist* April 1892–July 1893, the "scrap" referenced in the paragraphs that follow. For details about this debate, see the general Introduction to *W* 8 (especially at p. lxxiii), individual headnotes to Peirce's "*Monist* Metaphysical Series" articles in *W* 8 and *EP* 1, and Corrington, *An Introduction to C. S. Peirce* (1993), pp. 167–198.

29 converse] convers *[E]*   30 (metaphoricaly)] *inserted above*

For the last five years I have been out of law practice and have had no office down-town. Although I have my books and meditations, besides household chores to occupy my time still my time, a good deal of it, hangs rather heavy on my hands. Besides my means are very limited. They are a good deal like the corselet reported upon by the traveller, that bared the feet whenever he pulled it up to cover his shoulders. So about a year ago I made up my mind to connect again with the Dr. if it was a possible thing to do so. So I wrote him a letter of proposal in aid of the work I knew he had in hand that I would devote my spare time *at home* in reasonable measure on any work of his that was within my competence and to make sure he could make no excuse on the score of expense I told him that $ 10.00 per week would satisfy me and rather to my surprise he closed with me. So I have become installed as a kind of *taster* and general utility man. But he is still very sensitive on the score of my esteem for your work over his own deliveries. I came marvelously near a soreness on his part towards me in connection with the "Amazing Mazes". He do'nt see or sees very obscurely what such matters have to do with philosophy and in trying to explain I have to exercise the utmost of my tact not to excite his inclinations to take issue according to the old score between you and him.

Another thing. He has for a right hand assistant at La Salle a mighty bright and well informed (for a woman) young lady, Miss. Lydia G. Robinson, college bred and naturally very competent. He hired her first for a translation from the German and the French. But she became in a multitude of other ways so useful and relieving to him that she is now almost "*the whole thing*" so far as editorship is concerned. *She has great influence with him* and you know the old saying about the better estate of whoever "*stands in*" with the king's barber over one who only stands well with the king

Now Miss Robinson is *not* favorably inclined towards you, rather otherwise. This disinclination is of course the reflection of what she knows about you as she gathers it by the Dr's manners of utterance. It is only the old story of "like master like man" only the "*like man*" is usually found in grossly exaggerated measure.

Now Miss Robinson is really a very nice amiable lovely woman and so far as opportunity offers I find it well to cultivate her good will and opinion and I *guess* I have been quite successful in so doing.

I could "*tip*" you, perhaps on other points that do'nt just now occur to me. This letter is already a long one. I will only add that it is one of my heartiest wishes that you and the Dr. could come to a mutual co-appreciation and hearty regard.

---

15 or] *after* ~~what~~
17 inclinations] incli-*[line break]*tions *[E]*
28 disinclination] disinclina *[E]*
30 exaggerated] exagerated *[E]*

Cordially
Francis C. Russell

P.S.—

I forgot to tell you that I duly received your long letter in answer to my proposal about writing a Logic. I see its date was July 12, Ny!Ny![161] What a shame to neglect acknowledgment so long.

I remember now the proximate occasion of that incipient "scrap" between the Dr. & me in connection with the "Amazing Mazes." He in a protesting letter to me was exclaiming over the amount of copy you had sent him and over his inability to compass what he could do with it and *especially* over what he regarded as your discordant clamors for money and I wrote back to him a rather (I guess) passionate letter voicing my sense of the shame it was that such a man as you should have to shin around and almost beg for money adequate for subsistence. I virtually hinted to him that the more money he sent you (within reason) the more he honored himself and The Open Court foundation and the better that foundation would be mentioned in history &c &c. I guess it was a pretty "hot one" although I dont remember the text very particularly. I know I felt strongly and let myself out "*some*."

F.C.R

Peirce to Russell. MS: RL 387.

P.O. Milford Pa | 1908 Oct 7

Dear Russell:

Thank you very much for yours of the 2nd. I have not received the Monist of October but I corrected and returned the proofs of my article.[162]

This will be the third of mine that has been printed since I receive last any money from Carus which was $100 on Oct 21 of last year. I have an extremely important article showing the proper method of making logical analyses & definitions.[163] It is only partly written because I have no paper nor money wherewith to purchase any & I owe ten dollars for paper for a long time.

161 Peirce to Russell 10 July 1908.

162 Likely "Some Amazing Mazes: A Second Curiosity," *Monist* (Jan 1909).

163 Undetermined, possibly part of or a precursor to his ca. Dec 1909 "Studies of Logical Analysis" and related material (R 643–650)—see Peirce to Carus 26 Aug 1910, note 1. Also described in more detail in the first unsent draft of Peirce to Carus 7 Jan 1909.

8 the] *after* ~~the first installment of~~ 26 on] *i*on

If you can I wish you would get him to send me some money. The last time he wrote he said I owed him money. I wrote back and asked him to send me a state of the account, but he never did so. I wish he would so that I can find out how many months of hard work a hundred dollars is supposed to pay for.

I have been asked to contribute to the Hibbert Journal & propose to send them my system of Monism in four articles,[164] since Carus complains of my writing too much.

I have always been friendly to Carus and have never lost an opportunity of forwarding his interests. I simply cannot understand the state of mind of a person who is vexed because another does not share his opinions,—*especially* opinions about metaphysics!!! Good God!

I have my taxbill of a hundred dollars to pay for and ninety before long for interest on Mortgage at 6%. I was swindled out of three hundred in the spring; & am in a bad way. Twenty-two boxes of books from my later brother's library are on their way to me & I haven't five dollars in the world, to pay the freight. *Do try to persuade Carus* to send some cash & also to explain how I came to owe him so much. If he will pay, I should be glad to write for him & the Hibbert Journal both.

Peirce to Russell. MS: RL 387.

P.O. Milford Pa | 1908 Dec 8

My dear Russell:

I haven't heard from you for a long time. How is the Logic-book coming on? Do let me know: I am greatly interested in it. The last letter I had from you,[165] by the way, announced that you were to help on the Monist. I was in hopes you would be able to prevent Carus from making bad breaks in mathematics & logic. Yet in the Open Court that I received yesterday the first thing that met my eye was this marvellous statement from Carus's pen, which is bad English and has one error in the history of mathematics and another downright absurdity of the most glaring kind. Here it is "... but only about thirty years ago has Professor Ferdinand Lindemann, of Munich, succeeded in proving that *since π is equivalent to an infinite series* it can never be expressed in a proportion of whole numbers." He does not often write so bad a sentence. Of course he ought to have written (though it

164 "A Neglected Argument for the Reality of God" appeared that same month, but no other articles followed.
165 2 Oct 1908.

1 he] *above* ~~I~~

wouldn't be true,) "_ _ but it was not until about thirty years ago that Prof. Ferd. Lindemann of Munich first succeeded in proving that π cannot be expressed as a *ratio* [not "proportion"] of whole numbers, because it is expressible as the sum of an infinite series of rational terms." The absurdity is the implication that the sum of such a series cannot be rational. I wonder how much less than one he supposes the decimal

0.999999999 *etc. ad inf.*

to be. Has he never heard of a circulating decimal? Of course, that π is an irrational or incommensurable number, in the modern sense, was proved a quarter of a millennium ago by James Gregory and a century later by the Frenchman Lambert (though he was born on German soil) & what Lindemann proved after a fashion, & has since been more clearly proved, is that π is not an "algebraical number," that is, is not a root of a rational algebraic equation.[166]

Perhaps you may have noticed that I had an article in the last Hibbert Journal.[167] I should have sent you a copy but I only received *one*; and that I must keep in order to answer letters about it. I mentioned to you that *I had been asked* to write for that journal; but I did not say I had done so, because I thought my article had been rejected, & I had not made up my mind whether to try again or not. It is true that I had received proofs of a part of it. Yet certain circumstances made me think that it had subsequently been decided not to use it; and in point of fact I never did see all the proofs. I was completely surprised to find that it was already in readers' hands, which I learned by some one who had read it writing to me about it. I was very dubious as to the reception it would meet with; but it has in fact met more favor than anything else I ever wrote, & from the men whose opinion would weigh the most with me. There are some scraps of my theory of scientific methodeutic in it that I think you might be interested in. No doubt you can find it in the office of the Monist. Do let me hear from you

ever faithfully
C. S. Peirce

166 James Gregory (1638–1675), Scottish mathematician and astronomer, also known for his reflecting telescope, the Gregorian telescope; Johann Heinrich Lambert (1728–1777), Swiss mathematician, physicist (specializing in optics), and astronomer; and Carl Louis Ferdinand von Lindemann (1852–1939), German mathematician known largely for his 1882 discovery with π that Peirce here describes.
167 "Neglected Argument."

5 how] *after* ~~what~~
13 a[1]] *above* ~~the~~
17 my] *after* ~~the~~
22 who had read it] *inserted above*

Russell to Peirce. MS: RL 387. Pages 1 and 3 only of letter.

#1268. Jackson Boul | Chicago Dec. 10$^{\underline{th}}$ 1908

Dear Prof. Peirce.

Got your Dec 8$^{\underline{h}}$ letter just now and was mighty glad to hear from you. I sometimes want to say as I observe the goings-on, Why Dr Carus "never opens his mouth but he puts his foot into it" but of course that would be only an extravagant utterance of vexation for I most heartily like the Dr. and moreover he is not without his points as a philosopher. In his Fundamental Problems he makes "Form & Formal Thought" the core of his philosophy (I sometimes think it was I who "put him onto" even that) and however wide and vague and in part indeterminate that text is, if he would only stick to and study out and develop its capabilities and suggestions he would be pretty near on the right track to say the least. But the Dr. wants to figure as a leader, itches for such a leadership with an intense and persistent irritation. He has struck out a *system* and he feels himself committed to it and he is bitterly disappointed that it cuts so little figure in the philosophic world. So he contents and comforts himself with the persuasion that his system will obtain currency "after many days" "as all more than ordinarily profound systems must". In the *[…]*

*[*Page 2 of letter missing*]*

[…] is a sister (?) of Prof. James) and I spoke to him about you and in the course of the short conversation I had with him he said in effect that you did'nt "put yourself on paper often enough and persistently enough that you ought to *push*" yourself more, and that in default of these you were liable to live not long enough to get much of the personal benefit and glory that the great merits of your thinking entitled you to.

I have had in my possession for several months your sections D. & E of "Amazing Mazes"[168] and they confirm my profound admiration especially section E. The Dr. is afraid (so he protests) to put so much mathematics into the Monist. He says that a good many of his patrons *kick* at so much matter that they cannot compass. Nevertheless he says he guesses he will try to "work them in" from time to time perhaps in small parcels. (My Logic is not commenced yet) You may rest assured

168 "Third Curiosity" and "Fourth Curiosity" (R 299 and 300).

24 of the] *inserted above*
27 and they confirm] and ~~that~~ they ~~excite~~ confirm
28 afraid] *after* ~~so~~
30–31 Nevertheless … parcels.] *A short arrow drawn here points to the beginning of the next sentence,* You may *[. .]*

that whenever I get a chance to prod the Dr. on the money question on your behalf I shall not fail to do so. If I were anywhere near in the affluence of the La Salle folks I would just say to you Here! How much of a pension will put you so that you need think of nothing but *[. . .]*

Russell to Peirce ca. 24 Dec 1908 (missing): Sends a "pretty card" for Christmas or New Year.

1909

Peirce to Russell. MS: RL 387.

P.O. Milford Pa | 1909 Jan 1

Thanks for your surprisingly pretty card.

My dear Russell:

A Happy and Prosperous New Year to you! I hope you don't read the reports from Sicily & Calabria. To me, to whom these places were so very familiar a generation ago, all this is enough to shake my reason.

You must think it very strange that, after my protestations, I should not have answered yours of Dec 10$^{\underline{h}}$. But I will tell you how it is. My wife is very ill, and very energetic, conscientious, and the most particular housewife I ever knew by far. Even if I were not in the condition of penury in which I am, it would not (short of *wealth*) be possible for me to provide the servants that her state of nerves and of overwork require; and her devotion to me and acceptance of her share in my lot, when she might roll in luxury if she would leave me, tears at my heart strings. We have *no* servant. For it is better to have none than such as I could afford. Consequently a great part of my energy has to go to *chores*. Besides that my Hibbert article[169] and some other things have caused me to be flooded with letters from people of consequence & which cannot be answered without much labour. In addition some matters of business consequent upon the death of one brother[170] and

---

169 "Neglected Argument."

170 Likely James Mills (Jem), d. 1906. The "matters of business" not determined.

---

9 Thanks . . . card.] *In red pencil at top-left corner.*

18 provide] *after* ~~have~~

19 her^{1}] *after* ~~d~~

the fact that the family take the attitude toward me that families generally do,—though I think *more so* than usual have thrown an additional mass of work upon me. Then there are my own logical studies which it is impossible for me to dismiss from my mind, and which become all the more urgent owing to my saying to myself that the time is brief in which I can be of service now. The result of all is that I have been so overworked that for more than a month I have not averaged 4 hours daily in bed & have worked all the time at the very top of my powers. Not only your letter but many other things of the greatest urgency have been utterly neglected.

As to the straight line, I think very differently from you. In the first place like the geometers of our time I think that projective geometry has nothing at all to do with measurement. Now the straight line is the chief thing in projective geometry & therefore I do not think it ought to be metrically defined. Your definition is metric. If I were to admit a metric I do not see any objection to resorting, as most geometers do, to the definition of the straight line (let me say *Ray*—meaning a limitless straight line) as the shortest *[*distance*]* between two points. (Your definition introduces the idea of a circle. But a *circle* is a metrical concept.) In particular, in 1873 Klein's paper on the *nicht-euclidische Geometrie* in the Annalen der Mathematik seemed to make it luminously clear to me,—even more so than it already was,—that measurement involves something clearly *additional* to the properties of space itself, something dependent entirely upon mechanics, & if there were no such thing as an approximately rigid body to suggest absolute rigidity & (as at present) no way of measuring time apart from measuring space,—what would become of the concept of measurement? Yet we could sight along a range and determine whether it were straight or not. But even *straightness* & *flatness* to my mind have nothing to do with space itself. If you read Listing's paper in the Göttinger *Abhandlungen* on the Census-Theorem,[171] you will see that there is a geometry which knows no difference between a straight line and a curved line.

Imagine all space to be filled with an absolutely rigid glass in which three systems of equidistant parallel lines would be visible the 3 being at right angles to one another and the lines to be either indefinitely close together or so that intermediate-visible lines could be produced at will. Then everything in metrics could be seen to be true if only that glass could be moved about interpenetrating

171 "Der Census räumlicher Complexe" (1862).

4 become] *over* a
19 involves] *above* ~~of~~
26 on the Census-Theorem] *inserted above*
28 in] *inserted above*
29 equidistant] *above* ~~equilat~~

other bodies without resistance. Every proposition in metrics could so be illustrated & if the glass were instead capable of moving like optical rays & shadows, all the propositions of projective geometry could be seen to be true in any special case. But suppose that the glass could be rendered plastic, though we will say incompressible, and distort it in any manner whatsoever. Then render it rigid again, and you will perceive that though the lines are no longer straight, there will be a perfect analogue to every proposition about straight lines.

It will be better however to suppose that in place of the glass being movable like a rigid body, that it is capable of motion of a projective kind. You will thus see that

Any family of surfaces, each in one piece, of cyclosyone and of periphraxy one, which are such that any two of them have in common only one line in one piece and of cyclosy one, and any three of which have in common a single point, or else the three pairs have the same line in common and such that through any three points of space there is one such surface and but one unless a onefold infinity of them all having the same line in common,—any such family of surfaces are in all respects (not metric) related exactly as the planes of space are related to one another, and every non-metrical proposition of geometry has its exact analogue, concerning these surfaces.

Now if it be true that measurement is something quite additional to straightness, you will perceive that it follows that there *cannot be* any purely geometrical property of a straight line, unless of course you make assumptions concerning some other undefined line and define a straight line in terms of that.

All you can do, then, is to define a plane as one of *a* family of surfaces having the above properties. If you wish to render metric geometry true, you will simply add further that there is a certain one of these surfaces which every other of them cuts at right angles and there are some other conditions that I am not equal to working out tonight. I did once write a geometry in which the whole thing was worked out;[172] but I could not get a publisher for the reason that I found the publishers all thought *they* knew something about geometry, whereas in fact they were in the

172 Likely "New Elements of Geometry" or "Elements of Mathematics" (1894–1895; *NEM* 2:1–474).

1 Every] *after* ~~But~~
1–2 illustrated] *after* ~~made~~
2 instead] *above* ~~also~~
11 *one*] above ~~*zero*~~
12 *have ... line*] *have* ~~*a line*~~ *in common* *only one line*/
13 *or else*] ~~*and*~~ *or else* ~~*(if*~~
14 *through*] after ~~*any*~~
15 *one*[1]] after ~~*but*~~
15 *them*] *over suc*
17 *as*] above ~~*like*~~
21 purely] *inserted above*
22 make assumptions] \make/ assum*e*ptions
24 one of] *inserted above*
24 surfaces] *after* ~~lines~~

deepest ignorance of it, knew nothing at all for instance of the 3 kinds of geometry and had notions in conflict with those of all the higher mathematicians' views. The MS was eventually lost. It was rather a pity as it was worked out very thoroughly and clearly. I defined a straight line as the path of a ray of light (though a ray of light is a fiction) and measurement by the properties of a rigid body and its looking glass image, though there is no rigid body known. However, there seems to be *room* for a ray of light were there such a thing, & for a rigid body.

As for Carus, he seems to me to have much ability in the way of language. His English is remarkable. But anything less like a scientific man, it would be difficult to *[*find*]*. In philosophy, I should say there were half a dozen men in this country who are his superiors. Take Lester Ward for example, Royce, and several men less known.[173] What he is ambitious to achieve is *reputation*; but reputation in philosophy in this country without a professor's chair cannot be achieved without great real power; and that is not *born* in a man; it has to be worked out; and the first condition is that the man's soul should be filled with the desire to make out the truth, to the exclusion or almost to the exclusion of other aims; but Carus is full of himself; and it stands irremovably in the way of a thorough devotion to truth. He ought not to try to combine two aims so disparate & incompatible.

As to myself, I have had the misfortune to be interested in a science which nobody but me really cares for or even wants badly to know about. Now to pursue researches alone is so dangerous, one is in such constant peril of taking one's private opinions for the truth, that the greatest caution is necessary. I go over everything many times before I feel that I can venture to be confident of being right. Of course I go slow; and with the detestation of all exact thought requiring painful effort to apprehend it, which is so nearly universal, it naturally cannot happen that I should produce more than one book that is at once widely regarded as important and that really is so. Whether or not if I succeed in writing such a book it would be recognized before intellectual development has gone so far that it is no longer useful, I cannot know. I think it exceedingly dubious. At any rate, however, it cannot do me any good personally; and that is the brightest and most consoling thing about it. I can always say to myself that my motive was as white as the driven snow; and after all one's own good opinion is worth ten times that of all the rest of the world put together.

---

173 Lester Frank Ward (1841–1913), renowned American sociologist and first president of the American Sociological Association. (For Royce, see Biographical Register.)

---

8 His] He *[E]*
15 to] to to *[E]*
26 widely] *after* ~~reg~~
31 about] *after* ~~I~~

Why don't you get on with your logic?

With warm wishes
C. S. Peirce

P.S. Carus has offered me a hundred dollars for the copyright of my 6 articles on "Illustrations of the Logic of Science," I to revise as I please. I should print the first two with merely clerical corrections, but with an addition correcting the errors of doctrine which are implicit rather than explicit. The next on Probability I should not greatly alter but should *define* probability which has never been done correctly without a vicious circle. I should also state & take sides in the great controversy of Boole vs. Laplace. The one about the Order of Nature could stand substantially, but with an important addition. In place of the remaining two, I should give a brief detailed account of the method of scientific reasoning & the grounds of its validity, and also a long series of examples of historical scientific reasonings, very long. The whole would be polished & made as clear as possible; but readers would have to think. It would take me the better part of a year & a hundred dollars would be less than a starvation price. I shall see what another publisher will say. Anything from you in the way of faultfinding with definite passages of those articles will be valued by me.

Peirce to Russell. MS: RL 387.

Milford Pa 1909 Jan 23

Dear Mr. Russell:

I wonder whether everybody does not find himself subject to unaccountable fluctuations in his power of expressing himself. I do, especially perhaps in French. One naturally would in a foreign language, and though I have been studying French diligently for sixty years, I have doubtless less command over it than I have over English. But as to that I am different on different days. Sometimes one language seems a misfit, sometimes t'other. My natural mode of utterance, which I always practiced before knowing that it was a system having rules of its own, is Existential Graphs; and by the way, everybody who sits down & attentively stud-

---

5 "Illustrations] *over* ~~Lo~~

9–10 I should also … Laplace] *inserted above*

13 historical] ~~natural~~ \histori*es*cal/ [*(first revision)* natural \histories/ *(second revision shown here)*

16 less] *over* pr

16–18 Anything … by me.] *up left margin*

25 doubtless … than] ~~no more~~ \doubtless less/ command over it ~~, perhaps less,~~ than

26–27 But as … t'other.] *below* ~~My~~

28 before] *above* ~~without~~

29 by the way,] *inserted above*

ies my next paper on pragmatism[174] will have a surprise to see how that language aids Logical Analysis. Not that I recommend it as a pasigraphy, because it is far too analytic for that. But it might be made the *basis* for a universal language to fulfill the functions of Esperanto, should that language prove disappointing.

I have often expressed to you my astonishment at Carus's mastery of English. Well, I have just been reading one of the most interesting of his papers.[175] papers. It is one that I could not find time to read before; and which in the latter half of it, shows his usual command of our language (except for a few of his usual Chicago-isms, detestable to lovers of genuine good English). The first few pages might have been written by any German with an ordinary good acquaintance with our tongue. It is about Samson and was published 2 years ago.

I will give some examples that have struck me with surprise:

174 Undetermined, possibly part of or a precursor to his ca. Dec 1909 "Studies of Logical Analysis" and other related material (R 643–650)—see Peirce to Carus 26 Aug 1910, note 1. For other estimations, see Peirce to Russell 7 Oct 1908.
175 *The Story of Samson* (1907).

4 fulfill] *after* ~~ful~~
6–7 papers. It is one] papers$^{\vee}$.$^{\vee}$ \It is one/
7 in] *inserted above*
8 of his usual] *inserted above*

What he writes	*What I should say*
p. 33 I was quite astonished to be called to account by Mr. George W. Shaw	I was really astonished at finding myself called to account by so good a Hebrew scholar well versed in biblical criticism as our fellow contributor Dr Geo W Shaw.
The best secular works, such as the *Encyclopaedia Britannica*, repudiate the idea that the story of Samson should be a myth.	Oh, I can hardly believe they do that; although, as far as I can see, it borders on insanity to say that the story of Samson *should be* a myth. Still, there is nothing in it that outrages decency. If a man cannot get along at all with his wife, he nowadays seeks a separation; but he does not repudiate her unless she has fallen into the habit of sleeping with every pretty fellow that comes along. Repudiation is the last degree of censure. But I guess this was a piece of newspaper-lingo, and Chicago newspaper-lingo at that, and that he only meant The best works *reject* the theory that the story of Samson *is* a myth. or They reject the notion of Samson's life & exploits being a solar myth.

10 *What he writes / What I should say*] *in red pencil*
11–12 at finding myself] *in blue pencil*
13 well] *inserted above*
17–18 as far as I can see, it] *above* ~~it certainly~~
18 insanity] *before* ~~(as far~~
35–36 solar myth.] solar ~~or [?heroism]~~ *s*myth.

He calls his paper, which is quite elaborate, a "*treatise*"

But a treatise is a book that presents an extensive subject, especially a branch of theoretic science, systematically, with an attempt at thoroughness, and in formal, explicit and general terms. A separate publication too small and too unsystematic to be called a treatise is called a *tract*. But this use of the word is perhaps obsolete; for I do not remember having come across any instance of it of later date than 1879, when the last of the Revd W. J. Wright's "Tracts relating to the Modern Higher Mathematics" was published.

The last, I remember, bore the motto

τοῦ γὰρ ἀεὶ ὄντος ἡ γεωμετρικὴ γνῶσίς ἐστιν,[176]

which illustrates Plato's materialism, which I am much inclined to share.

---

176 *Republic*, VII, 527 b, on cover of Wright. Often translated as, "the real object of the entire study [of geometry] is pure knowledge," although translations vary widely.

---

3 But] *above* ~~Now~~
5 branch] *after* ~~theory~~
6 systematically,] *after* ~~in a~~
7 in] *after* ~~with~~
9 and] *after* ~~to be called~~
11 is] *after* ~~is~~
20 γεωμετρικὴ] or ρια *noted below as alternate ending*

p. 34

I only wish that the perusal of it will be as interesting and instructive to my readers as the writing of it wasto me.

I only wish something *may* be

For since I *only* wish, I am by no means confident that it will be. In fact, nobody whose ideas never dream of presenting themselves at the front door of his consciousness until they have arrayed themselves in English dress, of their Sunday go-to-meeting best, ever says or thinks

I *wish* it *will* be.

As for the "as the writing of it *was* to me" which he uses as he completes the *first page* (!) of his article, he probably had written a good part of that page on the very day he uses the preterit *was* in reference to the writing. Under these circumstances he should have said "has been." But I suspect he refers to the pleasure of reading up on the subject and of turning over the arguments in his mind.

1 p. 34] *in green pencil*

3 will] *in blue pencil*
5 was] *in blue pencil*

An account which is decked with mythological arguments should not for that reason be regarded as absolutely unhistorical.

This leaves the writer's meaning uncertain in three respects:

1st, I don't know whether by "decked" he means *decked over* (which would be nearer to good English,) or whether he means *decked out* (which seems better to suit the context.) I think the latter is the more probable meaning. 2nd, I cannot imagine what he can mean by "arguments" or how he was led to use the word or what German word he could have had in mind. In fact I do not know what is the German for "Argument" But I suppose "embroidery" in the sense of fictional embellishments would come somewhere near the meaning. 3rdly "Absolutely" may mean entirely, or it may mean certainly, or it may mean unhistorical on all points, though perhaps not utterly so on any one. Absolute is a capital word when one does not himself know just what one means.

17–19 In fact I do not know what is the German for "Argument"] *A note in red pencil to* GO TO THE LEFT COLUMN *inserted here at column-end, with a red arrow pointing to the left-column note* Continued from Right Side *preceding the text that follows.*

What he says	*What he may mean*
"It is quite natural that myth enters into the fabric of history"	Does myth put on a fabric woven of history, or is the fabric an edifice into which myth enters by the front door? Never mind the figure, what is the interpretation of it? The "enters" should be replaced by a subjunctive It is quite natural that myth *should enter*, or else

Myth enters quite naturally. The difference is obvious, but the construction seems to attempt to seat itself on both stools at once. Perhaps the "*enters*" means that the fabric is fabricated chiefly of history but of history fraudulently mixed with myth. In that case, he might have said, It is natural that myth should adulterate the fabric of history, or should pollute the stream of history, or should interlard the page of history. But *entering the fabric* of history simply makes one stop and wonder what the writer meant to say. Perhaps that it is natural that myth, like a stray cat, should creep into the edifice of history.

---

4 *What he says* | *What he may mean*] *in red pencil*

10 should] *after* ~~sh~~

15 fabricated] *above* ~~woven~~

17 or[2]] *after* ~~shou~~

As I have pointed out in my introduction to Mr H R Evans's book on the Napoleonic myth	Since that introduction appeared some four or five years ago, he should have said "As I pointed out"
If a myth has crystallized in a definite form and localized in well-known places etc	Better, either If a myth has crystallized … and has *been* localized– or else If a myth has been crystallized– and (been) localized–
we must not jump to the conclusion that its historicity is well-established	A theory is not *established* until it has been accepted by the general mass of competent critics. But he means one should not jump to the conclusion that its historicity has been proven (or proved.)
The next two pages are just as bad or worse Page 37 has "illumined" for "illuminated" and Might not one literary critic rightly say	
– – – while another would deny this proposition and claim that the hero of the romance is historical though the account is overlaid with mythical ornamentation	This "would" should be "might" Here it would be better to say "in spite of the account being overlaid"

23 four or] *inserted above*

31 (been)] *above* ~~been~~

35 one] *inserted above*

All the rest is free from such faults. I forgot to mention that he translates "faktisch" by *factic*, as if German *-isch* and English *-ic* at all corresponded. There is no such word in English as "factic." It is opposed to the recognized principles of the language, while "factual" is a word used by the best writers of the XIX century.

very truly,
C. S. Peirce

Peirce to Russell. Postcard: RL 387. Recto headed, "Postal Card with Paid Reply | United States of America" with portrait at right.

[recto]

Port Jervis | N.Y. | APR 9 | 430 PM | *[1909]* *[stamped]*

Francis C. Russell, Esq.
1268 Jackson Boulevard
Chicago
Ill.

[verso]

My dear Russell

On p. 294 of the April Monist[177] you take an isosceles right triangle and letting fall a perpendicular from the right angle upon the hypotheneuse, declare that it would be almost an imputation on the reader to prove that the two half triangles are isosceles. Never mind the imputation! *I defy you to prove it*. If the sum of the angles of the original triangle is $\pi - 2x$, then its 3 angles are $\frac{1}{2}\pi$, $\frac{1}{4}\pi x$, and $\frac{1}{4}\pi - x$ and the 3 angles of one of the half-triangles are $\frac{1}{4}\pi$, $\frac{1}{4}\pi - x$, and $\frac{1}{2}\pi$ Therefore by the *Pons Asinorum* such half triangle is *not* isosceles. I hope you dont feel that to be an imputation!

very faithfully
C. S. Peirce

Peirce to Russell. MS: RL 387.

1909 Apr. 15 Milford Pa

My dear Mr. Russell:

177 Russell's "A Modern Zeno."

1 translates] translate*d*s

I hope I am right in thinking your article in the April Monist utterly unworthy of you: to think otherwise I should be obliged to rank you with the circle-squarers.

There are two questions or aspects in regard to the non-Euclidean geometry. The purely mathematical question is whether or not the assumption that the sum of the three angles of a triangle is less or greater than two right angles does or does not lead to a perfectly self-consistent and mathematically interesting system of geometry. As to that *all* mathematicians worthy of the name are at one. You may not be aware of this. But you *do* know that in putting yourself in opposition to it, you are opposing the carefully formed conviction of Gauss, of Riemann, of Cayley and Sylvester and Clifford, of Helmholtz, of Klein, of W. E. Story, and many other profound mathematicians. Even my father, whose "Geometry"[178] was his darling idea and was contrary to Non-euclidean idea, was forced to acknowledge that the non-Euclideans were correct.

The other question was whether experiential space may be in accordance with the non-Euclidean theory and contrary to the Euclidean postulate. In 1870,[179] I put forth, perhaps before anybody else, the *suggestion*—without guaranteeing it,—that it might be possible rationally to consider space as Euclidean whatever discoveries might be made looking to the truth of one or other of the non-Euclidean theories, and the Nancy man whose name escapes me takes that ground strongly.[180] As to how it may be permissible or reasonable to conceive of *real* experiential space is an open question. But it has nothing to do with your position.

For your sake, I must HOPE you have not put yourself into the mortifying position of having read all these great minds the lecture you have read them about over-haste and over-confidence without having read and duly studied at least the greatest of the works they have written in defence of their views; and therefore I am bound to assume that you are familiar with the following:

Cayley's Sixth Memoir on Quantics

Clifford's work on Metrics[181]

My own few words about the division of geometry into Topics, Graphics, & Metrics, observing that the cross-ratio is not a question of metrics at all but only of Graphics[182]

---

178 *Elementary Treatise on Plane and Solid Geometry* (1837).
179 "Description of a Notation for the Logic of Relatives."
180 Possibly Eugenio Beltrami (1835–1900), Italian mathematician similarly mentioned in Peirce's "Description of a Notation" at p. 368n.
181 Likely "Instruments Used in Measurement" (1876).
182 "Topical Geometry" (R 137).

---

17 rationally] *inserted above*
20 or] *after* ~~to~~
24 and duly] *above* ~~at leas~~

Riemann's celebrated contribution Ueber die Hypothetese etc.

Klein's two luminous papers in Vols IV and VI of the Mathematische Annalen[183] (though they must have slipped your mind when you said that nobody had *proved* that the Non-euclidean Geometry involves no contradiction; since otherwise you must have *read* those remarkably clear papers without fully comprehending them;) and also Klein's *two volumes of Lectures on the Nicht Euclidische Geometrie*

W. E. Story's account of the Non-euclidean Geometry, its trigonometry, etc. which are particularly valuable, not only as showing its Graphical relations to the Absolute, but also as enumerating all the kinds of Non-Euclidean geometry which admit the truth of the usual Graphics.[184]

With these works before you, it was indeed a lawyer-like proceeding to confine yourself to a criticism of a translation of Lobatchewski's little book,[185] which though a brilliant and elementary exposition and (barring his admitting the truth of the propositions that he numbers as 4, 5, 7, & 9,) involves, I believe, no real error, unless a mere slip of the pen, which is rendered perfectly obvious by noticing what it is that his demonstration really demonstrates, be reckoned as an error,—which nobody but a lawyer presuming upon the gullibility of his jury would reckon as an error.

You seem to be that lawyer, for otherwise, why in the world should you have given that wordy proof that in the non-euclidean geometry two lines may be parallel to one third without being parallel to each other, when Lobatchewski's original explanation of what he means by "parallel" shows that, in the geometry, every straight line has *two* lines parallel to it through *each point of space* that does not lie in the line itself? You cannot have failed to see that these two lines cutting each other are not parallel, and his defining them as parallel to the third was in obvious contradiction to the proposition that two straight lines both parallel to a third are necessarily parallel to each other.

I press the Question: Why did you not content yourself with this obvious proof of the incorrectness of his proposition №. 25??[186]

183 "Über die sogenannte Nicht-Euklidische Geometrie" (1871 and 1873).
184 "On the Non-Euclidean Trigonometry" and "On the Non-Euclidean Geometry" (1881 and 1882).
185 Nikolai Ivanovich Lobachevsky (or Lobatchewsy; 1792–1856), Russian mathematician and geometer, famous for his hyperbolic geometry, later named "Lobachevskian" geometry. Reference to *Geometrical Researches on the Theory of Parallels*, trans. George Bruce Halsted (1891).
186 "Two straight lines which are parallel to a third are also parallel to each other." (p. 22)

6–7 and also … *Geometrie*] *inserted line-end and between paragraphs*
16 noticing] *after* ~~his~~
17 really] *over* sho

The answer seems to me obvious. If you had done that, your readers would have at once perceived that Lobatchewski merely made a slip of the pen and *meant* that two straight lines parallel to a third *toward the same side* are parallel to each other & then they would have seen that *this* is precisely what his demonstration of Proposition 25 correctly demonstrates & would have reduced your whole argument to a *quibble*.

Your calling Lobatchewski a "Zeno" because he acknowledges a line between all straight lines through any given point that cut a given straight line at a real finite distance and those that cut it at imaginary distances (I need not remind you of Klein's papers) is to my mind down-right inappropriate,—not to use an uncivil expression. All lines in the plane of a circle that pass through any point in that plane may be divided into those that cut the circle and those that dont, unless you deny the principle of Excluded Middle. But are there not two lines that are on the boundary between these two classes? Namely the two tangents to the circle? Do you call everybody who uses that very natural way of speaking, whether it be logically correct or not (and I think it *is*) a modern Zeno? If so, I should think it to be a compliment to be called *by you* a modern Zeno or by any other opprobrious & contemptuous epithet! For Zeno was not the man who held there was a limit between two classes of loci, but on the contrary was the man who did out rightly comprehend the idea of the *Limit*. To my mind, it is you yourself who do not see that two parallels in the elliptically non-Euclidean geometry of Lobachewski (or maybe it is called "hyperbolic": I have forgotten, at the moment) cut one another on the Absolute, or surface at infinity, at an angle equal to *zero*, just as they do in Euclidean geometry. The only difference is that in the latter they cut at two coincident points of the Absolute, since the latter consists in that case of two coincident planes placed, as it were, *back to back*; i.e. the two planes form a sphere of infinite radius having all real points in its interior; while in Lobatchewski's geometry the Absolute is a real quadric surface, an ellipsoid and may be called a sphere, since, without having any definite properties different from any other kind of ellipsoid so placed yet owing to the Absolute serving to define distance (apart from its unit) it must necessarily *be called* a sphere.

---

4 his] *above* ~~the~~
7–8 a line … through] a ~~point~~ \line/ between all \straight/ *h*lines ~~that~~ through
8 straight] *after* ~~line~~
11–12 All lines … may] ~~Through~~ \All lines *[. .]* through/ any point in th*e*at plane ~~of a circle, all lin~~ may
19 but] *before* \~~on the contrary was the man~~/ ~~who could not rightly see that there was such a limit.~~
21 two … geometry] *thes*two ~~line~~ parallels *[?*~~found~~*]* in the ~~hyperbolic~~ \elliptically non-Euclidean/ geometry
29 other kind of] *inserted above*

As to a straight line not having any *definition proper*, it is demonstrable that it cannot be, properly speaking, defined.

You define a straight line metrically. Every modern geometer will object that a straight line is a purely *graphic* conception and graphics knows no such conception as distance or length and does not distinguish a circle from any other ellipse nor from a parabola or hyperbola. Besides, if a *metric* definition were admissible unquestionably the proper definition is that a straight line is the shortest distance between two points.

And after all your talk about the error of the non-euclidean geometry being due to the absence of a definition of the straight line (which is absurd, since the straight line is a graphical concept and the non-euclidean geometry purely metrical, leaving graphics untouched,) you produce a definition of the straight line from which it is demonstrably if not *manifestly impossible* to prove the sum of the angles of a triangle to be equal to two right angles!

And don't you think your remarks at the bottom of p. 289 are rather arrogantly applied to Gauss, Riemann, Klein, Helmholtz etc. rather than to yourself?[187]

Faithful are the wounds of a friend.

your friend
CSPeirce

Peirce to Russell. MS: RL 387. Unsent, aborted letter.

[Unsent]
Milford Pa 1909 Apr 21

My dear Mr. Russell:

You don't seem to lift my gage to prove that a perpendicular let fall from the right angle of a right isosceles triangle divides it into two isosceles triangles.

If I wanted to write a treatise on Elementary Geometry quite ignoring the Non-Euclidean hypotheses, I should proceed as follows:

---

187 "Among men there is no habit more inveterate than the persuasion of each individual that he personally is immune from slips in reasoning. All around him during almost every day of his life he takes notice how badly other people reason without ever saying to himself that probably he is like other people in the same regard."

---

3 metrically] *above* ~~graphically~~
9 all] *over* d
13 demonstrably if not] *inserted above*
22 Milford Pa 1909 Apr 21] *above* ~~My dear Mr~~

I should first treat of Topical Geometry defining a *point* as an indivisible place, and a *particle* as a movable thing which at any one instant of time occupies a single point; A *line* as a place that a particle can occupy, not at an instant but in a lapse of time, and A *filament* as movable thing which at any one instant occupies a line; A *surface* as a place that a filament can occupy, not instantaneously, but in a lapse of time during every instant of which it totally leaves its line, and a *Film* as a movable thing that at any one instance occupies precisely a surface; and a *space* as a place that a film can occupy in a lapse of time during each instant of which it is totally leaving its surface, and a *body* as a thing that at any one instant exactly fills a *solid space*. A *point-figure* is a place consisting of a collection of *points*. A *line figure* is a place containing lines and it may be points, but no surface nor solid space. A *surface figure* as a place containing a surface but no solid space. A *solid figure* as a place containing a solid space.

Any object, A, stands to any object, B, in the relation of being a *part* of B, if and only if a sufficient account of what is true of B, regardless of anything else, logically determines what is true of A regardless of anything else, while an account of what is true only of A does not necessarily determine all that is true of B.

Any object, A, stands to any object, B, in the relation of being a *spatial part* of B, if and only if, the full statement of what places B occupies determines what places A occupies, while the statement of what places A occupies does not suffice to determine what places precisely B occupies. *[ ]*

Russell to Peirce 22 April 1909 (missing): Responds to Peirce's critical letter of 15 April.

Peirce to Russell. MS: RL 387.

Milford Pa | 1909 Apr 24

My dear Russell

I just receive yours of 22 inst. & can only reply to it very hastily. I am glad you're as thick-skinned as you say. Of course it is the only attitude that becomes a logician, if I can still continue to think of you as such. There is nothing to *study* in my letter. Your article struck me as very lawyer-like indeed. My conception of a lawyer

---

6 every] *after* ~~whi~~
11 nor] $^{\vee}$n$^{\vee}$or
12 a place … but] a ~~figure~~ place containing \a surface but/
15 B] *over* A
16 A] *over* B
17 only] *after* ~~of~~
19 full] *inserted above*
20 places] *over* A
26 you're] your *[E]*
29 Your article] *I*You$^{\vee}$r$^{\vee}$ ~~stru~~ article

is the usual one. That is that, in the first place, he is a man who has *studied* law, and therefore one whose ideas of justice between man & man deserve respect,—if we can only be sure that there is no darky in the woodpile, as there are 99 chances to 1 there are; for the lawyer dislikes talking of justice in the abstract or in general, & it is not likely that he has, in any given case, nothing more on his mind. But, in the second place, a lawyer is a man insured to *practice* law, that is to accept pay for defending a given position before he knows anything about it; which is *essentially* corrupting, & which is rendered still more so by the boasted Common Law of England, which I doubt not was the best makeshift practicable once, & perhaps may be so now, but is awful, a system iniquitous beyond words to tell, yet which practice has made the lawyer look upon as ideal. That the Law is the Perfection of Human Reason, considered as a maxim for ascertaining what is the law is sublime. But as characterizing what our courts will uphold is the abomination of desolation to every right-minded layman. Your article is lawyer-like in the sense of showing the degree of love of truth that a cross-examining lawyer trying to confuse a witness does show, to the delight of the presiding judge and the resentful disgust of the jury.

What your article & this letter both show is your utter failure to master a matter so simple as the Non-Euclidean Geometry. Both bristle with evidences of this. Thus, in the letter, you underscore, as highly pertinent, the remark that I have taken no notice of the last ¶of p. 300 and the first of p 301. This shows that to you it is not an evident matter of course that there can be no hesitation, in the mind of anybody who really grasps the Non-Euclidean business, in regard to either of those paragraphs. It did not strike you that an intelligent man has a feeling of resentment against a writer who has so little regard for him as to occupy ten lines with saying that on a plane (or, for that matter, any other uniformly isotropic surface) on which there *can be* a finite rectilateral or geodetic triangle the sum of whose angles is not exactly 180°, there cannot be a rectangle. Your saying it as if it communicated something startling, puts me, and will everybody worth considering, into grave doubt whether you have any clear notion of non-euclidean geometry or whether you have understandingly read anything not non-sensical about it. And as to the paragraph that follows that one its utter inconsequence is beneath discussion.

---

3 darky] *after* ~~monke~~
13 abomination] *[?]*abomination \abomination/ *[E]*
21 to] *after* ~~y~~
27 finite] *inserted above*
28 cannot be a] *above* ~~can be no~~
32 the paragraph that follows that one] the ~~following~~ paragraph \that follows that one/

I began to write for you an account of a way in which, regarding geometry as a science of observation, you could establish the truth of Euclidean geometry, as nearly as anybody really knows it is true, without any doubtful postulate. But I give up; for I am confident you would not begin to understand it. You don't seem to me to understand the logical nature of the non-Euclidean geometry. For you say that ordinary intelligent men look upon the matter, as you do, and upon the mathematicians who assert its truth as apparently cracked. Now they certainly do not, and cannot, think so. For that would imply that they knew what the non-Euclidean position is. An intelligent man has no opinion at all about a doctrine the nature of which he does not know; further than that if the general body of well-informed and competent judges are known to be in accord about it, he presumes they are right, unless it be one of these subjects like metaphysics which has not yet been reduced to any universally acknowledged methods. Now mathematics is of all sciences the one about whose reasonings there is the least question among competent men, especially since Weierstrass

The doctrine of non-Euclidean geometry is that if Euclid's $5^{\underline{h}}$ Postulate (erroneously called by some former writers the $11^{\underline{h}}$ Axiom, though it is not so in the good mss. nor according to the statements of ancient geometers, and certainly it is not an *Axiom* and never has been generally thought to be so) may be supposed false either in excess or in defect, and the new postulate substituted for it will lead to a system of geometry which involves no self-contradiction.

Now everybody knows that geometry must rest upon postulates, and the logical nature of a postulate was correctly described by the ancient mathematicians as a proposition that the mathematician does not pretend to know is true, but that he must assume to be true in order to develope a mathematical theory, because he needs it as a premiss to support his reasonings. For it is not within the province of the mathematician to study the facts of nature and to pronounce upon such matters. An *axiom*, on the other hand, is a proposition supposed to have been already settled before the study of the branch of mathematics in which it figures as an axiom has been taken up. Such, for example, is Euclid's erroneous axiom that any part is less that its whole, which may, however, be freed from *error* by regarding it as meaning that geometry is restricted to *finite* spaces; although modern geometers have conclusively shown (and especially Cayley in his $6^{\underline{h}}$ Memoir on Quantics) that that view renders any comprehensive understanding of geometry as impossi-

3 nearly] *above* ~~well~~
10–11 well-informed] *after* ~~intelligent and~~
15 , especially since Weierstrass] *inserted line-end and between paragraphs*
22 the] *after* ~~a postulate~~
25 because] *after* ~~as a~~
29 study] *after* ~~branch~~
578.34–579.1 impossible] *comma deleted*

ble as the exclusion of from algebra would maim and disfigure that science and the whole theory of functions.

Now ordinary people, far from thinking Euclid's $5^{\underline{h}}$ Postulate satisfactory, have almost universally and in all ages, protested that it was *not* so. Therefore, whether they are aware of it or not, they have really been on the side of the non-Euclideans, and if they think the latter hold a position contrary to common sense, it is because they do not know what their real position is, and imagine that, *I*, for example, and people who occupy such a ground as I do are typical Non-Euclideans. That is not so. I agree entirely with the non-euclideans on the *mathematical* question so far as that can be called a "question"; but on the field of logic I hold a position that non-Euclideans generally would repudiate. Namely, while mathematicians generally hold Laplace's *Théorie des Probabilités*, as a sort of bible, on account of its great *mathematical* power, and I do not dispute this last merit, I maintain that the logical foundation of the work, the explanation of what probability is, is puerile in the extreme, and that no observation of facts can impart any *probability* whatever to any generalization so far as it goes beyond observation. The validity of induction is indeed closely *connected with* a correct theory of probabilities, but does not *consist in* any definite probability attaching to the inductive conclusion.

I cant give much more time to the matter; but if you purpose to write a book about logic, I should say that you ought by all means to begin by *thoroughly mastering* Cayley's Sixth *Memoir on Quantics*,—though if, when you say you have read Story's papers, you mean that you have thoroughly understood him, you must be mistaken, because if youhad, you would understand not only Cayley but the whole ground of certainty of the non-Euclidean Geometry. Therefore, if you have any difficulty about that (as I presume you will) you have only to begin by reading the first two chapters of Lindemann's redaction of Clebsch's lectures on Geometry, a very luminous & beautiful treatise.[188] I have not the original but only Benoist's French translation, so I can't give the German title nor say what the "Chapters" are called in the German Edition. They make about a quarter of the whole work. Any little difficulties you can't surmount, I can aid you in by referring you to other books or otherwise.

188 *Vorlesungen uüber geometrie* (1876).

10 that] *above* ~~it~~
16 so] *after* ~~that~~
24 had] *above* ~~did~~
28 I] *after* ~~Any~~

Some time I will endeavour to give you my own way of imagining the non-Euclidean geometry which enables me to answer any simple question about it instantly. After you have completely digested that immortal memoir of Cayley's,—perhaps the greatest luminary, it was, of all my mathematical life,—By all means get to the very bottom of Klein's papers on the Non-Euclidean Geometry in Vols IV and VI of the Mathematische Annalen[189] which wonderfully clear up the whole matter. Don't satisfy yourself with half-understanding—*Master everything* as you go on.

C. S. Peirce

There are besides two volumes of Lectures by Klein on the Non-Euclidean Geometry.

Russell to Peirce ca. May 1909 (missing): Responds to Peirce's critical letter of 24 April. Also refers Peirce to a "friend's theological talk," possibly enclosed.

Peirce to Russell. MS: RL 387.

1909 June 5 | Milford Pa

My dear Mr. Russell:

I can see no sense at all in trying to associate non-Euclidean geometry with the word "virtual." All I see is that Lobatchewsky's little book is perfectly elementary and if a person goes through it, so as to follow each step of the elementary reasoning, he ought to have sufficient acquaintance with the subject as to answer the question you put me; and you assured me that you had studied the book. But now it appears that you have not read it. Your eye may have passed over every line, but though the reasoning is as plain as a pike staff you have not taken it in. According to Lobatchewsky's kind of geometry the locus at infinity in a plane is an ellipse, which can be called a circle if you like. Therefore the straight line between two points both at infinity will itself be at infinity at those points only. A triangle inscribed in infinity will have each angle = 0 but the area of the triangle will be some finite quantity which may be different in different spaces. Call this

189 "Über die sogenannte Nicht-Euklidische Geometrie" (1871 and 1873).

3 Cayley's] *after* ~~Cy~~

10–11 There are … Geometry] *This line up left margin.*

28 may] *after* ~~will~~

area A. Cut the triangle by a line from one angle so as to divide the area into two equal parts, and plainly the area of each will be $\frac{1}{2}A$ and the sum of the angles will be 90°. Bisect that again and the area will be $\frac{1}{4}A$ and the sum of the angles will be 135°. Bisect that again and the area will be $\frac{1}{8}A$ and the sum of the angles $157\frac{1}{2}°$. Of course A is a finite quantity.

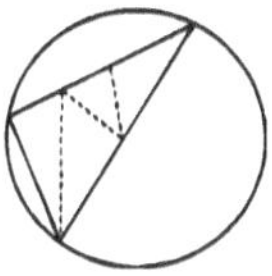

Before you came out and contradicted all the real mathematicians in the world you should have read deeply the works of the great masters, in which number Lobatchewski is not.

As for your friend's theological talk, I have skimmed it pretty well. The best parts seem to me questionable; and I do not think that men have made God. I hold God to be a reality. But anyway, it is theology, and I think that kind of loose theological speculation weakens religion, to which all that is utterly foreign. [See my paper in the Hibbert Journal for October 1908.][190] If it were put forward as a mere idle fancy it would do no harm and no church would persecute the man.

Very truly
C. S. Peirce

P.S. I scribbled on the back of my sheet,[191]—having finished my letter,—without noticing that it was a letter. Instead of atoning for this incivility by copying it, I will do better, by adding as much more matter, which by the way costs time of which I am just now very short.

You speak in your last as if you supposed there were only two kinds of non-Euclidean geometry. This is not so. A non-euclidean geometry consists in the hypothesis of a space of 3 dimensions with the same *topical* properties as Euclidean space; that is having its Chorisy = its Cyclosy = its Periphraxy = its Apeiry = 1 but differing in having a different locus at infinity[192] (which locus Cayley named the "Ab-

190 "A Neglected Argument for the Reality of God."

191 See Figure 12 below.

192 The names Peirce assigns to the four Listing or census numbers. See *EP* 2:542n7 for more information.

2 will[2]] wi*th*ll 3 Bisect] *after* ~~Cut~~

solute") in Euclidean Geometry it is a plane or rather a pair of coincident planes with an *imaginary circular* line of nodes, so that any line that cuts this circumference has a simple contact with the Absolute & owing to this circular nodal line being imaginary any real curve that passes through it once must pass through it twice i.e. in the conjugate point. A circle is a conic which does this. A sphere is a quadric surface on which the Absolute lies.

Some of the non-Euclidean spaces evidently have properties different from our space. But that does not in the least diminish their mathematical Importance, since Mathematics makes no pretense that any of its hypotheses are true. If any of them are so, which I doubt as far as geometry goes, that is an extra-mathematical fact of not the smallest interest to the mathematician, as such.

One very interesting non-Euclidean Space (interesting particularly, however, to the non mathematician) is a space whose Absolute (or place that no movable body or part of a body can ever either reach or if there can leave) instead of having an imaginary circle for nodal line has a *real circle*, so that no plane can ever be rotated so far as to touch it and a plane that is touched by the Absolute can only be turned round this ring, either round from being nearly in its plane on one side to being nearly in the same plane on the other side, or else turned from one part of the circle to another. The interest of this sort of Space is that all its properties must be exactly the same as those of Euclidean space except that what is real in one will generally be imaginary in the other. Thus it enables us to solve a great number of intricate problems in a jiffy or less.

In the next place there is a Space in which the Absolute consists of two planes, with a line of intersection which may be real or imaginary. Then nothing can ever be turned through such an angle that if sufficiently prolonged it would collide with that ray of intersection.

In the next place the two planes may be imaginary in which case one could travel all round space in a straight path without going infinitely far. How great the circumference of space would be we can't say. To some animals it would seem a long journey & to others a mere step.

Next the Absolute might be a double cone in which case nothing could be so turned that any straight line in it should at one time lie on a tangent to the cone and at another time not. For such a turn would be through an *infinite angle*. And there are two varieties of this Space. In one of them it is the inside that is inhabited. In the other it is the outside.

---

7 Some] *after* ~~One interest~~
14 or part of a body] *inserted above*
15 for] *above* ~~of~~
16 rotated] *after* ~~turned~~
17 either round from] ~~the same~~ \either round/ *b*from
31 Absolute] *after* ~~ins~~

Then there is a Space in which the Absolute has a Dice-box shape, or developable hyperboloid. The inside and outside would be *exactly alike* there being certain directions in which one could go all round space while in others you could not without passing through an infinite distance. Round those directions of axes there would be no infinite rotation while round the directions of axes along which the circuit of space is finite, you never could turn a wheel round so as to bring it back to its former position. Anything travelling round space in any of those spaces where such a thing is ever done is (I *suppose* so, for I am writing very hastily for lack of time) turned inside out.

There is also a space in which the Absolute is a surface of the second order & the second class and people living outside of it find some planes in which geometry is Lobatchewski's while other planes have all triangles in them *[and]* have the sum of the angles *greater* than two right angles.

In all *cases whatsoever* there is one and the same rule viz

the distance between two points equals the logarithm of the cross ratio of those points, relative to the points where the ray through the first pair of points cuts the absolute

while the angle between two planes equals the logarithm of the cross ratio of those planes, relative to the planes through the line of intersection of the first pair of planes, which (second pair of planes) are tangent to the absolute.

This was better than copying the first sheet.

C. S. P.

None of this is original with me. It is Cayley x Klein. KC which dont stand for Knot correct.

Of course you dont want to know the arithmetical value of the cross ratio. But you can take a perspective view of the four points or the four planes & then you can calculate the base of the system of logarithms in different in different spaces & maybe any values.

---

4 Round … axes] ~~In~~ \Round/ those directions \of axes/
5 no] *above* ~~some~~
5 round … axes] ~~in~~ \round/ the directions \of axes/
6 so] *after* ~~roll that~~
7 in] *after* ~~and~~
12 all triangles] *after* ~~the su~~
19 through] through through *[E]*
25–28 Of course … any values] *This paragraph down left margin.*

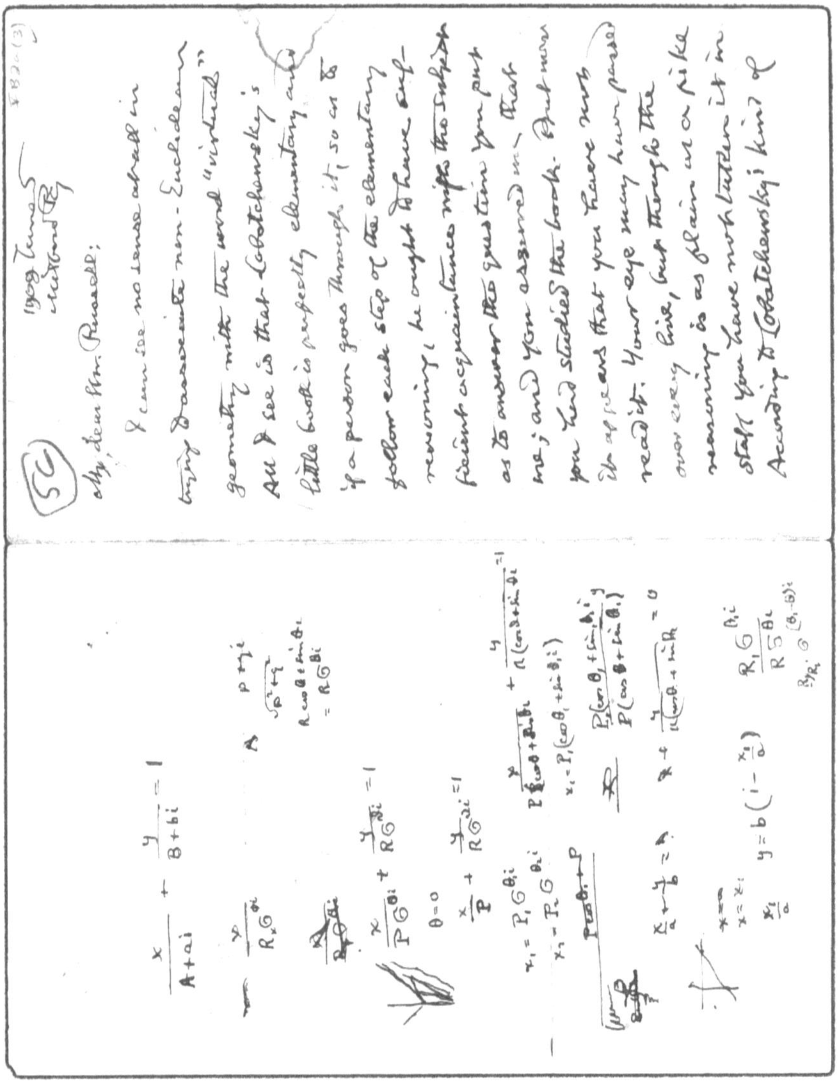

My dear Mr. Russell:
I can see no sense at all in
trying to assimilate non-Euclidean
geometry with the word "virtual"
All I see is that Lobatchewsky's
little book is perfectly elementary and
if a person goes through it, so as to
follow each step of the elementary
reasoning, he ought to have sufficient acquaintance with the subject
as to answer the question you put
me; and you assumed, in that
you had studied the book. But now
it appears that you have not
read it. Your eye may have passed
over every line, but through the
reasoning is as plain as a pike
staff you have not taken it in.
According to Lobatchewsky's kind of

Figure 12: The "scribbled on" final sheet of Peirce to Russell 5 June 1909, referenced in post-script.

Peirce to Russell. MS: RL 387. Likely unsent, entire letter crossed out in blue pencil. The postcard that follows possibly sent in its place.

1909 Aug 23

Dear Russel

I reënclose your “Russell’s Theorem.”[193] It would make a good example in my logic; because it is not easy to find a piece of actual reasoning that involves so obvious a fallacy. So if you will return it to me, I will use it in my book, either with or without your name, as you may prefer.

yours very truly
C. S. Peirce

Peirce to Russell. Postcard: RL 387. Recto headed, “Postal Card” with portrait at right.

[recto]

Matamoras | PA | AUG | 24 | 7 AM | 1909 *[stamped]*

Francis C. Russell, Esq[re]
1268 Jackson Boulevard
Chicago
Ill.

[verso]

Would you mind my using “Russell’s Theorem” in the logic-book I am writing to illustrate a variety of the *petitio-principii* which I would entitle “Russell’s Inference,” showing how it could be explained as arising from a defective love of truth, common among lawyers?

Because it is not known what magnitude the angles at the base of a right-angled isosceles triangle have, you propose to call them “x-angles,” which seems to mean angles whose magnitudes are unknown. Thereupon you produce this immediate inference:

If an isosceles triangle has the angle at the vertex a right angle, then the angles at the base are of unknown magnitude *Ergo*, any isosceles triangle whose angles at the base are of unknown magnitude has a right-angle at the vertex.

193 Undetermined.

24 “x-angles,”] “\[?sense *or* ?seuil]/ x-angles,” *Insertion unclear and not repeated in second reference to “x-angle”later in the message; omitted from main text until better determined.*

Of course, it is admitted on all hands that the difference between 2 right angles and the sum of the angles of any triangle is proportional to the area of that triangle. It cannot therefore be assumed that an x-angle has any constant magnitude for triangles of different areas.

Altogether it is a mighty pretty piece of sophistry from a man who is writing a book on logic.

C. S. P.

Why does not your reasoning apply to spherical triangles? I know you wont answer this, because it would force you to see the truth, which you dont want to see.

1910–1914

The remainder of the correspondence is not found, but based on Peirce and Carus 1910–1914, it dealt mostly with the revising and reprinting of Peirce's "Illustrations of the Logic of Science" papers in book form for OCP, as follows. (For more details about this exchange or other context during this period, see the Carus and Peirce correspondence and Chronology of this edition; see also the Introduction to de Waal, 2014.)

Aug–Sept 1910: After 18 months of working through revisions to his "Illustrations," Peirce sends Carus 26 Aug 1910 a long letter explaining the difficulties he has had in revising the work and agrees to simply add a preface to the original articles stating the alterations he would have made, which he there describes in detail. Carus and Russell 1–3 Sept 1910 (LPCs and MS: SIU 91.7) correspond regarding Peirce's long instructive letter and agree to have Russell edit the "Illustrations" papers accordingly, without waiting for any more input from Peirce for fear "that Peirce may die at any time."

Russell to Peirce ca. 25 Oct 1910 (missing): Confirms the new plan for Peirce's "Illustrations," Russell to annotate and Peirce to produce the preface explaining how he would have altered the text in the new book. (Mentioned in Russell to Carus 22 Nov 1910, MS: SIU 91.7.)

Russell and Carus Nov 1910: Russell explains to Carus 22 Nov (MS: SIU 91.7) that he has been preparing footnotes throughout Peirce's papers and attempting to collaborate with Peirce to that end but with no reply. Feels that Peirce's suggested alterations and additions, especially dealing with his Existential Graphs, are troublesome and out of scope for the book. Believes he and Carus will have "a time" in getting together Peirce's contributions and suggests that they "could make up a better book without him and still not underrepresent nor mis-represent him." Carus agrees with Russell 28 Nov (LPC: SIU 91.7) to move forward with the work with little revision and without overly involving Peirce. Reiterates his eagerness to publish this soon for "Peirce is old and might break down" and Carus will shortly leave for Europe.

1 between] *before* ~~two is~~

Russell to Peirce postcard ca. 24 Dec 1910 (missing): Relays that Carus plans to avoid substantial alterations and annotations in the "Illustrations" book and to note only Peirce's dissents as described in Peirce to Carus 26 Aug 1910.

Carus to Russell 30 Dec 1910 (LPC: SIU 91.7): Relays the general sentiments of Peirce's confidential letter to him 24 Dec (see Peirce and Carus correspondence) and suggests that since Peirce, as discussed in his letter, does not wish to have Russell's footnotes that Russell instead collect the notes and publish them as a comment in *The Monist*.

Jan 1911–Aug 1913: Peirce loses contact with OCP and engages in other endeavors (see Chronology). Meanwhile, Carus and Russell continue collaborating and preparing for the "Illustrations" book, especially during spring 1911 (LPCs and MSS: SIU 91.7). By summer 1913, Carus had written a proposed preface to the work, which he sent to Peirce 23 Aug (included in Appendix, "Peirce and Carus").

April 1914: Peirce dies of cancer 19 April at Arisbe. Carus invites Russell 22 April to write Peirce's obituary. Russell accepts and sends 24 April his "In Memoriam Charles S. Peirce," which appeared in the July *Monist* (also included in Appendix, "Peirce and Russell").

Carus never published Peirce's "Illustrations."

3 Appendix: Enclosures and Other Related Material

Items are dated by the letters in which they were enclosed or referenced, unless MSS are dated independently.

3.1 Peirce and Carus

ca. March 1909. (TS: SIU 91.25) Statement of payments made to Peirce by OCP. Requested by Peirce ca. 30 Jan 1909. This version was procured by Carus after discarding a "rather complicated" version sent him ca. 26 Feb 1909. Incomplete, lists payments from Chicago only, not La Salle.

Remaining payments made to *Charles S. Peirce* for articles in THE MONIST.

Architecture of Theories, I, 161. Jan. 1891 15 pages;	paid	9/2/90	$ 140.
Doctrine of Necessity Examined, II, 321. April 1892, 12pp.	"	11/16/91	$ 160.
The Law of Mind, II, 533. July 1892, 28 pp.	"	6/6/92	$ 200.
Man's Glassy Essence, III, 1. October 1892 22 p.	"	6/14/92	$ 100.
Evolutionary Love, III, 176. January 1893 25 pp.	"	7/21/92	$ 100.
Reply to the Necessitarians, III, 526. July 1893 46 pp.	"		
The Regenerated Logic VII, 19. Oct. 1896 21 pp.	"	5/8/96	$ 210.

During the years 1892 and 1893 there also appeared the following articles in THE OPEN COURT for which some of the above payments may apply. In any event it would indicate that payments must have been made from La Salle of which no record has been made here.

The Critic of Arguments, VI, 3391, 3415.

Dmesis, VI, 3399.

Pythagorics, VI, 3375

The Marriage of Religion and Science, VII, 3559.

What is Christian Faith? VII, 3743.

7 The Law of Mind,] The Law of the Mind *[E]*

https://doi.org/10.1515/9783110768756-003

Carus to Peirce 8 Dec 1909, enclosure. (TS: SIU 91.25) Correspondence extracts enclosed in Carus's letter of same date.

Extracts from Correspondence between Dr. Paul Carus to Mr. Chas. S. Peirce.

January 11, 1909.

As to the payment I am now willing to add another $ 100, and so I will pay you for the whole $ 250. I think very much of this article of yours and shall be very glad to have the revised copy *within about three months*, payment to be made on receipt of copy. I should be very glad to have those papers of yours on logic brought out and have them widely known among all kinds of readers. You have a certain limited number of admirers.

February 26, 1909.

But since I note you are waiting for money, I send you at once $ 150 as per enclosed check.

March 15, 1909.

I handed your letter to Mrs. Carus and I will say concerning the money question, that I will have a check made out at once to the amount of $ 250, accepting the conditions under which you desire that sum.

Extracts from Peirce letters.

March 9, 1909. "thanking you for your kind attention in sending me the check for a hundred and fifty dollars, which I duly received and which squares us up."

March 9, 1909. "In view of all this, if I could persuade you to advance the whole $ 250 *on my solemn promise* to devote *all* my energies, ... exclusively to those *two jobs* for you; namely the *completion of the revision of the Popl Sci. articles* and the article *on my method of analyzing concepts* (i.e. of defining them) a theory completely worked out now—then I would close with you with the understanding that your holding these articles should not be a reason for your not accepting *new* articles, etc."

(Check for $ 250 sent to Peirce).

April 5, 1909. "Your letter and check which I received Saturday afternoon have fired me with the desire to pitch in and make my "Essays on Meaning of Thought" as useful to a relatively large circle of readers as I possibly can."

Carus to Peirce 17 Dec 1909, enclosure. (MS: SIU 91.2) Peirce's edits to the proof, enclosed in Carus's letter this date, of the excerpt of his letter to Russell ca. early Oct 1895 included in Carus's "The Nature of Logical and Mathematical Thought" *Monist*, Jan 1910, quoted at p. 45). The edits, bolded here, were made in the margins and all but the first were honored in the published version. Only the edited portion of the proof is transcribed below.

[. .] With the permission of the writer I ~~quote it in this connection:~~ **give the following statement, which has been revised by him:**

"~~I studied these [writings of Schroeder and B. Russell] and others b~~**B**efore I took up the general study of relatives, ~~in the course of~~ **I made some** investigation into the consequences of supposing the laws of logic to be different from what they are. *[. .]* ~~I only mention it because a~~ **Another** writer ~~of another character~~ afterward suggested such a false logic, as if it were the wildest lunacy, instead of being a plain and natural hypothesis worth looking into **[notwithstanding its falsity]** ~~and which, as the investigation turned out, lacked little of being worth printing~~."

[At the bottom of the same proof, Peirce added the following explanation, the quoted portion of which appeared at p. 158 of the same Monist *issue. Peirce's variants are bolded in-line.]*

Mr. Peirce ~~**He**~~ now adds: "It does not seem to me to have been a lunatic study. On the contrary, perhaps if I had pursued it further, it might have drawn my attention to features of logic <,> that had been overlooked. However, I came to the conclusion that it was not worth my while to pursue that line of thought further. In order to show what sort of false hypotheses \they were that/ I traced out **<,>** to their consequences, I will mention tha***n***t one of them was that instead of the form of necessary inference being, as it is, that from A being in a certain relation to B, and B in the same relation to C, it necessarily follows that A is in the same relation to C, ~~**it is**~~ **\I/** supposed, in one case, that the nature of Reason were such that the fundamental form of inference was, A is in a certain relation to B and B in the same relation to C, whence it ~~**would**~~ necessarily follows that C ~~**was**~~ **\is/** in the same relation to A; and I followed out various other similar modifications of logic."

Carus to Peirce 1 Aug 1910, enclosure. (LPC: SIU 91.26) Carus's copyright permission statement for Peirce's "Illustrations" book. Enclosed in Carus's letter of the same date.

This is to certify that for a consideration of more than One Dollar ($ 1.00) in value received, I, Charles S. Peirce, author of "Illustrations of the Logic of Science" published in the *Popular Science Monthly* for 1877–78, having reserved my copyright to said articles by special arrangement with Mr. Youmans, at that time editor and manager of the *Popular Science Monthly*, hereby authorize the Open Court Publishing Company to publish said articles in such form as they may deem advisable.

Peirce to Carus 26 Aug 1910, enclosure. (MS: RL 77) Peirce's own copyright permission statement for his "Illustrations." Enclosed in his long letter of the same date and proposed to be used instead of Carus's sent 1 Aug. Two unsent drafts of the enclosure also follow.

Milford Pa | 1910 Aug 26

I acknowledge the right of the Open Court Company (for which full payment in lawful money of the U.S. has already been received by me, and for which I have received the courteous consent of the head of the Appleton firm,) to publish in the form of a separate volume *all* the articles that I contributed in 1877 and 1878 to the Popular Science Monthly, and I should be glad if they would include with them the article that appeared in that same Monthly in the number for January, 1901,[1] either in whole, or omitting the first four and a half or five pages as they may prefer. Along with this permission as essential parts of it and conditions of it, there have been all along 3 understandings. The first is that, in case the Open Court Company does publish the volume, it shall be thereby authorized, within two years previous to such publication, to insert any or all of the articles in any number or numbers of the Monist. The Second Understanding is that there be appended to any republication of the whole or any part of the Series of Articles such statement of a change of views on my part as I may furnish upon notice that the printing is to commence at once, my Appendix to be mailed by me within a month of my receiving such notice. [I have ventured to state this somewhat more in detail that it was explicitly made, but so as not to depart from my understanding of what was meant by both parties.] The third Understanding is that I be given an opportunity to make such corrections in the copy to be used by the printer as I can write clearly in the side margins. [I entirely abandon by original wish to make further alterations, since I

---

1 "Pearson's *Grammar of Science*."

---

10 full] *inserted in-line*
15 the[2]] *after* ~~Jan.~~
18 understandings] *after* ~~und~~
19 does] *above* ~~do~~
20 insert] *after* ~~an~~
20 any[2]] *after* ~~num~~
22 such] *above* ~~a~~

find it far more feasible to state how I have changed my mind in an Appendix or Appendices.]

CSPeirce

[Unsent draft 1 of 2, 26 August 1910]

I acknowledge the right of the Open Court Company (for which payment in lawful money of the U.S. has already been made,) to publish in the form of a separate book all the articles that I contributed in 1877 and 1878 to the Popular Science Monthly to which I should be glad to have them add the one that appeared in the number for January 1901 (or, if they prefer it, the major part of that article, omitting the enumeration of motives etc.) This permission, in case the said company carries it into effect, is to be understood as including the right to printing any of the articles separately (within two years before their collective appearance) in the Monist. But there have been two understandings essential to the arrangement that along with any article printed separately as well as along with the whole collectively there appear such statement of my subsequent changes of view as I may mail to Dr. Carus within six weeks of my receiving notice that the volume is to be immediately printed. Such statement may be appended at the end of the article or book. The second understanding is that I have the opportunity to make such small alterations in the text as can be written in the side margins of the copy to be used by the compositor, and that I have facilities for correcting the proofs of the appended statement or statements.

I should be much obliged *[if]* a dozen copies of the volume in cloth, may be sent to me to send to persons whose favorable judgments I may desire.

5–6 (for … made,)] *inserted above*
6 publish] *above* ~~print~~
8 to] *over* (
10 This] ~~But it is~~ *P*T̲his
11 any of] *inserted in-line*
12 (within … appearance)] (within a ~~reasonable time~~ \\~~eighteen~~ two years/ before their collective appearance ~~, \so that it shall not be/~~) *(omitted* "a" *mine)*
13 But … arrangement] But ~~the~~ \there have been two/ understanding˅s˅ ~~is~~ \\~~are 1st~~/ /essential to the arrangement\\
14 along[2]] *inserted above*
14–15 collectively] *before* ~~) I be able~~
16 Dr. Carus] *after* ~~the company~~
16 six] *over* e
17 printed] *before* ~~)~~
18 opportunity to make such] opportunity ~~to correct small errors~~ make \such/ *(cancelled* "to" *reverted)*
22 I … obliged] *above* ~~I trust that~~
23 desire] *after* ~~submit~~

[Unsent draft 2 of 2]

Milford Pa | 1910 Aug 26

I acknowledge the right of the Open Court Company (for which full payment in lawful money of the U.S. has already been received by me, and for which I have received the courteous consent of the head of the Appleton firm,) to publish in the form of a separate volume *all* the articles that I contributed in 1877 and 1878 to the Popular Science Monthly, and I should be glad if they would include with them the article that appeared in that same Monthly in the Number for January 1901 either in whole or omitting the first four and a half or five pages, as the Company may prefer. Along with this permission as essential parts and conditions of it there have been three understandings. The first is that in case the Open Court Company does publish the volume, it shall be thereby authorized, within two years previous to such publication to insert any or all the articles in any number or numbers of the Monist.

ca. 26 Aug 1910. (MS: RL 387) Draft of Peirce's proposed preface to his "Illustrations" book, requested by Carus 1 Aug. Estimated to be from around the time of Peirce's letter of this date. The text comes from an incomplete copy (first two pages only) made by Russell and containing edits by him.

Preface

The papers that are reprinted in this book appeared first in The Popular Science Monthly as a series of six articles beginning in the November 1877 number and following in January March April June and August 1878. The author has very much desired to revise and partially to rewrite as well as to supplement the papers before their reprinting but that has been found to be impracticable without delays that would be indefinite and irksome for the publisher. They are therefore issued in their original text which in spite of the dissent of the author seems to the publisher to have an array of merits that far outweighs the prospect of a more complete expression of such doctrines of the author as over thirty years ago he embodied in the articles. For, for one thing, the articles as originally printed contain the origi-

---

3 full] *inserted above*
4 been received by me,] been ~~made)~~ received by me~~)~~ $^{\vee}$,$^{\vee}$
10 essential] *after* ~~an~~
10–11 there ... understandings.] ~~I acknowledge and claim~~ there have been three ~~conditions~~ understandings.
11 in case] *inserted above*
19 *Preface*] below ~~*Introduction by Editor*~~
22 in] *after* ~~as fo~~
27 more complete] *above* ~~better~~
28 such] *above* ~~the~~
28 over] *after* ~~he~~

nal format by and upon which the voluminous literature of Pragmatism has arisen and have hence an historical interest and value besides their conspicuous merits in themselves.

In order however to make the reader to supplement the text here reprinted by recourse to the other writings of the author as well as to enable the author to present in fuller measure and with better satisfaction to him, perhaps, his doctrines cognate with the doctrines of this text the same has been annotated by Mr. Francis C. Russell who for over thirty years has been a close student and hearty admirer of the writings of the author.[2] While the author has for over *[…]*

Carus to Peirce 23 Aug 1913, enclosure. (TS: SIU 91.40) Draft of Carus's proposed preface to Peirce's "Illustrations". Carus later settled on another draft that follows (also included in de Waal, ix–x), which he bound with a table of contents to the set of "Illustrations" articles that Russell extracted from the *PSM* volumes, bound, and sent to Carus ca. 1888 (see Russell to Peirce 22 Jan 1889). Peirce's responding, and final, letter to Carus of 28 Aug 1913 does not, however, address the proposed preface nor indicate any editorial emendation.

*[*Draft 1 of 2, sent to Peirce*]*

ILLUSTRATIONS OF THE LOGIC OF SCIENCE.
By C. S. Peirce
PUBLISHER'S PREFACE

Mr. Charles S. Peirce is one of the most prominent logicians not only of the present generation but of all ages. He is the son of Benjamin Peirce, professor of Mathematics at Harvard, and known all over the world as one of the keenest mathematicians. With the freshness and enthusiasm of youth, Charles S. Peirce started out to expound the nature of logic in a series of essays which he called by the simple name "Papers on Logic". In later years he justified the expectations of his friends by discussing several logical problems in abstruse essays almost taking pride in being quite recondite and unintelligible except to the select few. Among all his writings this earlier series remains the most popular exposition of his thoughts and it well deserves to be better known. For this reason the publishers have deemed it advisable to acquire from the author the right of their republication.

2 For Russell's own words of profound admiration for Peirce's "Illustrations" series, see his letter to Peirce 2 Oct 1908.

7 annotated] *above* ~~edited~~   9 has] *before* ~~been~~

These six articles appeared in the *Popular Science Monthly* for 1877, 1878, and 1879.[3] The first deals with "The Fixation of Belief", and contrasts "the method of tenacity", "the method of authority", "the *a priori* method", and the scientific method, used in the struggle to attain a state of belief. The second deals with the problem "How to Make Our Ideas Clear", in which the thesis is maintained that our beliefs are rules for action, and is the place from which the "pragmatic" movement started. To attain clearness of apprehension, we must "consider what effects, which might conceivably have practical bearings, we conceive the object of our conception to have. Then, our conception of these effects is the whole of our conception of the object". Chapter III. is on "The Doctrine of Chances", and in it the idea of probability is made clear. Chapter IV. is on "The Probability of Induction"; Chapter V. is on "The Order of Nature"; and Chapter VI. is on "Deduction, Induction and Hypothesis."

[Draft 2 of 2, bound to printed PSM articles]

PUBLISHER'S PREFACE

Charles Santiago[4] Peirce's six papers of "Illustrations of the Logic of Science" first appeared in *The Popular Science Monthly* for 1877, 1878, and 1879[5] (Vols. XII. and XIII.) They stand outside his chief contributions to logic—which consist principally of papers on what he called "the logic of relatives"—and are not mentioned by Schröder in his bibliography of Peirce's logical works.[6]

For some, the chief interest of these *Illustrations* will lie in their heralding the beginnings of Pragmatism. The term "Pragmatism" was "first introduced into philosophy," says William James,[7] "by Mr. Charles Peirce in 1878. In an article entitled 'How to Make Our Ideas Clear', in the *Popular Science Monthly* for January of that year, Mr. Peirce, after pointing out that our beliefs are really rules for action, said that, to develop a thought's meaning, we need only determine what conduct it is fitted to produce: that conduct is for us its sole significance. And the tangible

---

3 For "1877 and 1878", though Carus might also have had in mind the first two articles in the series that also appeared in the *Revue Philosophique* Dec 1878 and Jan 1879.

4 An alternate middle name given him by Schröder in the bibliography of his 1890 *Algebra der Logik* and soon after used by others. Peirce would later embrace the name, presumably in honor of William James ("Saint James"), and beginning ca. 1906 went by "Charles Santiago Sanders Peirce" in his *Monist* articles and other works. (*W* 8:xlvii n35)

5 For "1877 and 1878" (though see note to draft 1).

6 *[Carus's footnote] Vorlesungen* über *die Algebra der Logik (exakte Logik)*, Vol. I, Leipsic, 1890, pp. 710–711; Vol. II. Part II., Leipsic, 1905, p. 603.

7 *[Carus's footnote] Pragmatism, a new name for some old ways of thinking*, New York, 1907, p. 46.

---

27 conduct] confuct *[E]*

fact at the root of all our thought-distinctions, however subtle, is that there is no one of them so fine as to consist in anything but a possible difference of practice. To attain perfect clearness in our thoughts of an object, then, we need only consider what conceivable effects of a practical kind the object may involve—what sensations we are to expect from it, and what reactions we must prepare. Our conception of these objects, whether immediate or remote, is then for us the whole of our conception of the object, so far as that concept has positive significance at all." However, Peirce's statement that our beliefs are really rules for action "is", as Dr. Carus[8] has remarked, and as we can see from the second Chapter below, "an explanation, not a principle, and the explanation is made so that we may rightly understand the nature of belief. Beliefs are never held at random; they serve a purpose and the purpose of a belief is ultimately to ensure a definite line of conduct. It is not probable that anyone would take exception to this. Professor James, however, goes beyond the original meaning of the term by changing the statement of fact into a principle, and he applies it to his conception of truth." For a criticism of pragmatism, we may refer to Mr. Bertrand Russell's *Philosophical Essays*[9] and the work of Dr. Carus already cited. But a criticism of pragmatism is rather a criticism of the doctrine held by William James and, under the name of "Humanism", by Dr. F.C.S. Schiller, rather than of that held by C.S. Peirce. And it will surprise many to see that the term "pragmatism" was not used in Peirce's paper of 1878 which is reprinted in the second Chapter below. Yet Peirce[10] has himself informed us that in the paper of 1878 he had set in motion the subject although not the word of pragmatism. He had only used this expression in oral conversation until James, who was not acquainted with him when he wrote *The Will to Believe* of 1897,[11] had appropriated it and put his stamp upon it as a philosophical term.

Peirce's own logical work is hardly touched upon in these essays, and his enduring work—which ranks next to De Morgan—on the logic of relations is not touched upon at all. And yet anyone who is even slightly acquainted with Peirce's later logical work will recognize in these essays the very characteristic thought and mode of expressing it of our author. The sixth chapter, in particular, contains much that is purely logical—such as a discussion of syllogisms; but, in the main, it is probability and induction to which these chapters on the logic of science are devoted.

Any notes not due to the author are put in square brackets.

---

8 *[*Carus's footnote*]* *Truth on Trial*, Chicago, 1911, p. 5.

9 *[*Carus's footnote*]* London, 1910, articles "Pragmatism" and "William James's Conception of Truth", pp. 87–149.

10 *[*Carus's footnote*]* "What Pragmatism Is", *Monist*, April, 1905; Carus, *op. cit.*, p. 116.

11 As de Waal, 2014, also notes at this point, James and Peirce had been friends since 1861 (Brent, 364, rev. ed.), so it is unclear why Carus thought otherwise.

3.2 Peirce and Hegeler

Late Jan–early Feb 1893. (MS: SIU 91.7) Outline of Peirce's proposed "A Quest for a Method" book (or "Search for a Method"), delivered to Hegeler with a memorandum of their publishing agreement in their late Jan–early Feb 1893 meeting in Chicago. On 27 Feb Hegeler asked Peirce to send another memorandum, explaining that the initial memorandum was with Sacksteder, and on the same day asked Sacksteder to forward this outline. On 1 March, Sacksteder enclosed the outline in a letter to Hegeler, explaining, "Replying to your favor of Feb. 27 enclosed is the list of Prof. Peirce's articles marked in lead pencil by Mr. Russell as to when many of the articles can be found." (Comments in smaller type are OCP glosses.) On 7 March Peirce sent a revised book outline included in that letter. For publication details of these works, see the Bibliography of this edition. Some clarifying footnotes have been added.

To be included

1 Boole's Logic
2 New List of Categories
3 Classification of Arguments
4. Logical Extension & Comprehension
5 Logic of Mathematics (To be rewritten)
6. Questions concerning certain Faculties
7 Consequences of Four Incapacities
8 Validity of the Laws of Logic
9 Review of James's Secret of Swedenborg
(To be in part rewritten)
10. Berkeley and the English Philosophy[2]
(With additions)
Nov 1877
11. Fixation of Belief.
12 How to make our Ideas Clear
13–16. Other Popular Science Monthly Articles.
Pop Sci Monthly (Popular Science Monthly)
17. Theory of Probable Inference —In Studies in Logic
18.–21. On the Logic of Relatives.
Three old papers and *one new one.*
Memoirs of Amer. Academy of Arts & Sciences
—Mes Am Ac Arts & Sc
Am Jour of Math
" " " "
" " " "
American Journal of Mathematics
22. Logic of Number
23. Economy of Research Smth Inst (Smithsonian Institute)
24. Spiritism (Never Published.)[3]
25–26 Phantasms of the Living

Not Included

Think Again
Philosophy of History[1]
Shakespearian Pronunciation
Sensation of Color
Include in an appendix
if there is room

1 Undetermined, possibly review of Hegel ca. Peirce–Harris exchange on Hegel 1867–1868 (*W* 2:132–161).

2 Review of *The Works of George Berkeley* and Harvard "Lectures on British Logicians."

3 For "Spiritualism," as in his 1890 article prepared for (though never published in) Lorettus Metcalf's spiritualism series in *The Forum*. Article not titled, but included as "Logic and Spiritualism" in *W* 6:380–394.

20 Mes] *over [?Pres]*

27, 28, 29, 30, 31 *Monist Articles*
32-33. New Articles of the Same Series
34. Critic of Arguments
35. Comparative Biography of Great Men (New)

Small Articles

1 Pythagorics
(Add full life of Pythagoras)
2. Dmesis
3. John Locke[4]
4. Johann Keppler (New)
5. Comte's Calendar of Great Men
6. Lombroso's Man of Genius.
7. Criticism of H Spencer
From *New York Times* with changes and additions.

The whole will be 700 or 800 octavo pages.

ca. 7 March 1893. (MS: SIU 91.12) Undated MS by Diesterweg summarizing the agreement proposed by Peirce in his 7 March letter to Hegeler, perhaps for records.

Professor Peirce in his *letter of March 7·93* states the arrangement between himself and Mr. Hegeler to be in substance as follows:

1. Prof. Peirce sells his Library to Mr. Hegeler for \$ 2 000$^{\underline{00}}$.
2. Prof Peirce is to offer to buy the Library back when he sells his place.
3. Mr. Hegeler is to pay Prof. Peirce \$ 250$^{\underline{00}}$ a month for seven months at least, for the arithmetic book which he is to write during that time, but says "at least" to mean that there is no express agreement in regard to any longer time, although, if the arithmetic is in the hands of the publisher he is led to hope that Mr. Hegeler may then be willing to help him a little further if he is in need of it.
4. It is understood that the money so advanced will be repaid from the first five years profit of the Book.

4 Review of Fraser's *Locke* in *The Nation* (25 Sept 1890). Articles 5 and 6 also refer to *Nation* reviews.

6 Pythagorics] Pythagor*ean*<u>ics</u>

12 Man of Genius.] *above* ~~Genius & Insanity~~

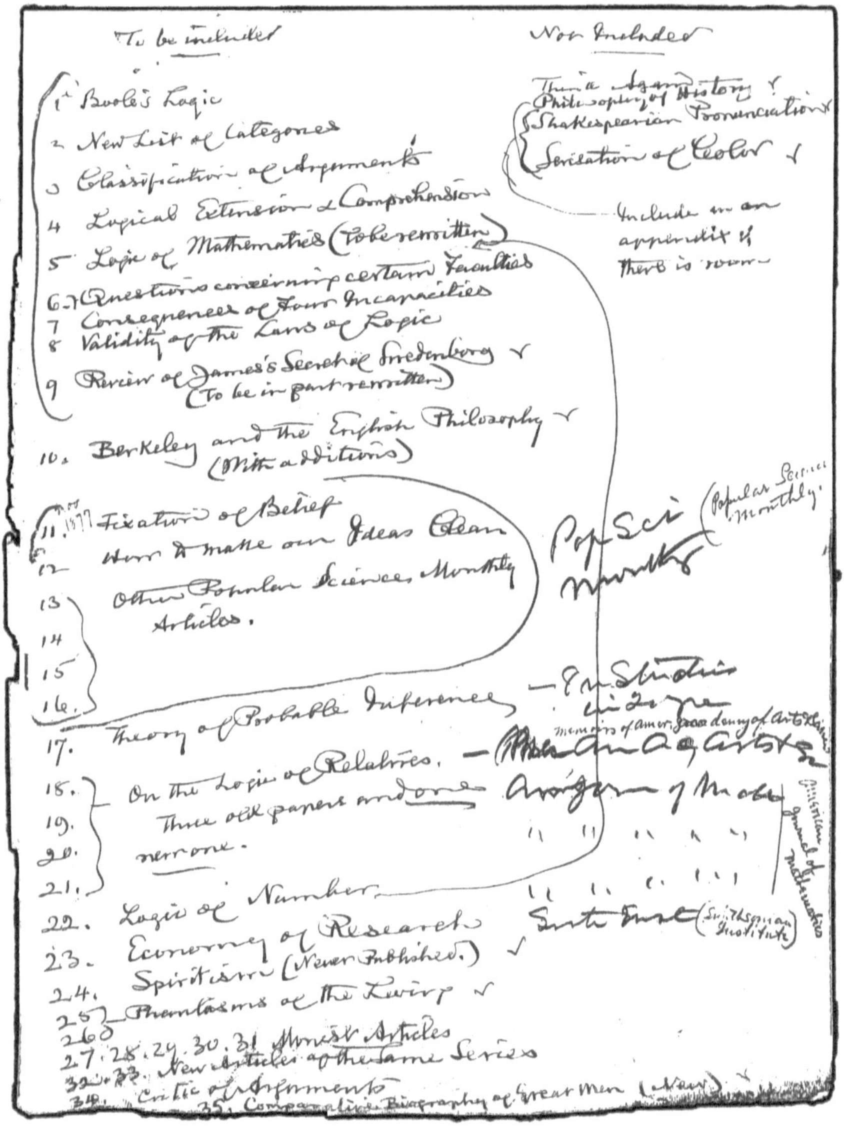

To be included

1. Boole's Logic
2 New List of Categories
3 Classification of Arguments
4 Logical Extension & Comprehension
5 Logic of Mathematics (To be rewritten)
6-7 Questions concerning certain Faculties
7 Consequences of Four Incapacities
8 Validity of the Laws of Logic
9 Review of James's Secret of Swedenborg ✓
(To be in part rewritten)

10. Berkeley and the English Philosophy ✓
(With additions)

11. Nov 1877 Fixation of Belief
12 How to make our Ideas Clear
13 Other Popular Science Monthly
14 Articles.
15
16.

17. Theory of Probable Inference
18. On the Logic of Relatives.
19. Three old papers and one
20. new one.
21.
22. Logic of Number
23. Economy of Research ✓
24. Spiritism (Never Published.)
25 Phantasms of the Living ✓
26
27. 28. 29. 30. 31 Monist Articles
32 + 33. New Articles of the same Series
34. Critic of Arguments
35. Comparative Biography of Great Men (New) ✓

Not Included

This Age and History ?
Philosophy of History
Shakespearian Pronunciation ✓
Serialation of Color ✓

Include in an appendix if there is room

Pop Sci Monthly (Popular Science Monthly)

— Studies in Logic

— Mem Am Ac Arts & Sci (Memoirs of Amer. Academy of Arts & Sciences)

Am Jour of Math (Journal of Mathematics)

" " " " "

" " " " "

Smith Inst (Smithsonian Institute)

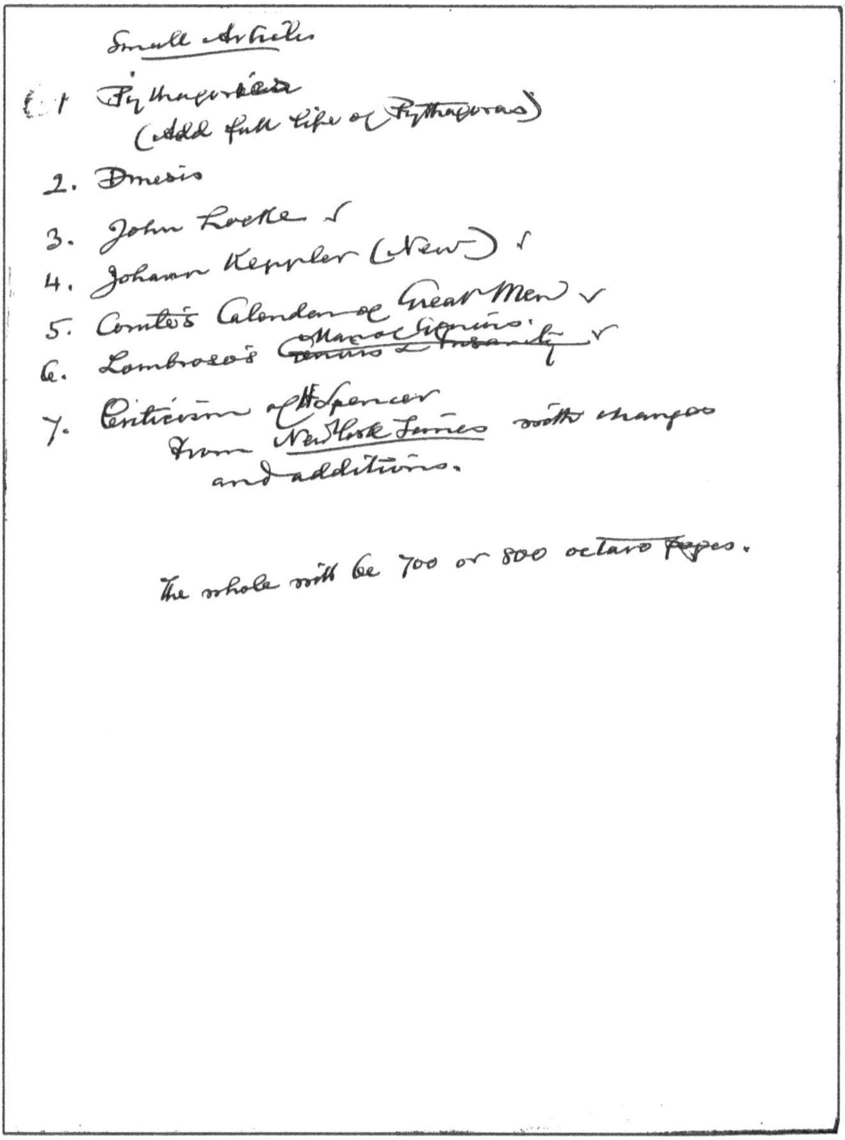

Small Articles

1 Pythagorica
(Add full life of Pythagoras)

2. Dmesis

3. John Locke ✓

4. Johann Keppler (New) ✓

5. Comte's Calendar of Great Men ✓

6. Lombroso's ~~Genius & Insanity~~ Man of Genius ✓

7. Criticism of H Spencer
from New York Times with changes
and additions.

The whole will be 700 or 800 octavo ~~pages~~.

Figure 13: Outline of Peirce's proposed "A Quest for a Method" book (or "Search for a Method"), 2 pages, late Jan–early Feb 1893.

5. Prof. Peirce understands this to be all the positive arrangement there is between himself and *Mr. Hegeler*.
6. Upon a request from Mr. Hegeler (5/15·93) for a full settlement for work already done by him, Mr. Peirce says in his letter dated 5/19·93 "As far as I know there is nothing due except for reading proof of Mach, for that I said I would take whatever Dr. Carus saw fit to give".

In this letter[5] Prof. Peirce seems also to ask permission to use his spare time, while writing the arithmetic to hold a Summer School at Milford.

ca. mid-April 1893.(LPC (MS): SIU 91.12) Undated annotated list by Diesterweg of several of Peirce's letters to Hegeler. Likely in mid-April because the quoted entries stop at 7 April and because Hegeler expressed frustration around this time, as in his letter to Peirce 3 April, with Peirce's unreliable postal addresses—calling for a list such as this. Variants between the portions shown and Peirce's original letters from which the quotes were taken are minimal and minor—such as differences in punctuation and capitalization, as well as slight variations as *to morrow* for *tomorrow* or *continues* for *continues to be* —and have not been emended or identified in the text.

Extract from Letters of Prof C. S. Peirce.

Letter of Feb 24. 93 written from *Port Jervis* N.Y.
"My address continues Milford Pa."

Letter of Feb 28. 93. written from *Port Jervis* N.Y.
Margin note on top of letter "Please address me Port Jervis N.Y. till March 3. 93."

Letter of March $2^{\underline{nd}}$ *93* written from *Port Jervis* N.Y.
"I will go to New York with her (Mrs. Peirce) to morrow and return, if I can, Saturday."

Letter of March $7^{\underline{th}}$ *93* written from *Port Jervis* N.Y.
"Let me begin by acknowledging receipt of Bill exchange for \$ $250^{\underline{00}}$ It is made to order of C. L. Peirce instead of C. S Peirce but I do not think this will matter. The letter was addressed Port Jervis Pa. instead of Port Jervis N.Y. but the Post Office officials discovered where it was meant to go. This shows the advantage of such

---

5 19 May 1893.

a name as Port Jervis. Now there are 3 or 4 Milfords in Pennsylvania alone, and numbers in other states."

Telegram from New York March 5th *93*
"Will try and see book through press, return Monday to Port Jervis."

Letter of April 7. 93 from *Milford Pa*
"I hasten to say that my address is Milford Pa. till further notice."

C.D.

1 Pennsylvania] Pensylvania *[E]* 4 try] *after* ~~see~~

3.3 Peirce and McCormack

Peirce to McCormack ca. Jan–Feb 1893. (MS: SIU 91.1) Peirce's statement of the mechanical units in use in the United States and Great Britain included in McCormack's Mach edition (Chapter III, Section III, Sub-section 8), the proofs of which were enclosed in McCormack to Peirce 11 March 1893. The section was included in the 1893 edition almost exactly as Peirce submitted it—with the exception of McCormack's minor edits standardizing spelling and punctuation, and correcting various calculations and measurements (not reflected in this transcription)—and was honorably mentioned in McCormack's "Translator's Preface," as shown in the next entry of this Appendix. The corrected text as it appeared in the edition is also available in HP 1:538–544.

Also included here is an alternate draft of the first page not sent to McCormack (MS: RL 77). Whether the rest of the draft is missing or was ever completed is unknown.

[8. The following statement of the mechanical units at present in use in the United States and Great Britain is substituted for the statement by Dr. Mach of the units formerly in use on the continent of Europe. All the civilized governments have united in establishing an International Bureau of Weights and Measures in the Pavillon de Breteuil, in the Parc of St. Cloud, at Sèvres, near Paris. In some countries, the standards emanating from this office are exclusively legal; in others, as the United States and Great Britain, they are optional in contracts, and are usual with physicists. These standards are a standard of length and a standard of *mass* (not *weight*.)

The unit of length is the International Metre, which is defined as the distance at the melting point of ice between the centres of two lines engraved upon the polished surface of a platiniridium bar, of a nearly X-shaped section, called the International Prototype Metre. Copies of this, called National Prototype Metres, are distributed to the different governments. The international metre is authoritatively declared to be identical with the former French metre, used until the adoption of the international standard; and it is impossible to ascertain any error in this statement, because of doubt as to the precise length of the latter, owing partly to the imperfections of the standard, and partly to obstacles now intentionally put in the way of such ascertainment. The French metre was defined as the distance, at the melting-point of ice, between the ends of a platinum bar, called the *mètre des archives*. It was against the law to touch the ends, which made it diffi-

---

15 at] *above* ~~near~~
16 office are exclusively] *bur*office are ~~the~~ exclusively
20–21 The unit … point of ice] The ~~standard~~ \unit/ of length is the ~~International Prototype Metre. It is~~ \International Metre, which is defined as/ the distance \at the melting point of ice/
22 a nearly] *an*a nearly
24 international] *after* ~~International~~
27 doubt] *after* ~~the~~
30 between] *above* ~~of~~

cult to ascertain the distance between them. Nevertheless, there was a strong suspicion they had been dented. The *mètre des archives* was intended to be one ten-millionth of a quadrant of a terrestrial meridian. In point of fact such a quadrant is 3 937 622 720 958 175 inches, which is 1 000 147 205 metres, according to Clarke.[1]

The international unit of mass is the kilogram, which is the mass of a certain cylinder of platiniridium called the International Prototype Kilogram. Each government has copies of it called National Prototype Kilograms. This mass was intended to be identical with the former French Kilogram, which was defined as the mass of a certain platinum cylinder called the *Kilogramme des archives*. The platinum being somewhat spongy contained a variable amount of occluded gases, and had perhaps suffered some abrasion. The kilogram is 1 000 grams; and a gram was intended to be the mass of a cubic centimetre of water at its temperature of maximum density, about 3·93° C. It is not known with a high degree of precision how nearly this is so, owing to the difficulty of the determination.

The regular British unit of length is the Imperial Yard which is the distance at 62° F between the centres of two lines engraved on gold plugs inserted in a bronze bar usually kept walled up in the Houses of Parliament in Westminster. These lines are cut relatively deep, and the burr is rubbed off and the surface rendered mat, by rubbing with charcoal. The centre of such a line can easily be displaced by rubbing; which is probably not true of the lines on the Prototype metres. The temperature is, by law, ascertained by a mercurial thermometer; but it was not known, at the time of the construction of the standard, that such thermometers may give quite different readings, according to the mode of their manufacture. The quality of glass makes considerable difference, and the mode of determining the fixed points makes still more. The best way of marking these points is first to expose the thermometer for several hours to wet aqueous vapor at a known pressure, and mark on its stem the height of the column of mercury. The thermometer is then brought down to the temperature of melting ice, as rapidly as possible, and is immersed in pounded ice which is melting and from which the water is not allowed to drain off. The mercury being watched with a magnifying glass is seen to

1 In his 11 March 1893 letter enclosing these proofs, McCormack notes: "Clarke's statement of the length of the earth's meridian quadrant is in metres (Ency. Brit Vol VII, p 607), and not in inches (see Galley 71, A). If you want the statement in inches to stand, please mark it."

2–3 ten-millionth] ten~~th~~-millionth
3 meridian] me\ri/dian
5 international] *inserted above*
11 is 1 000 grams] ~~was~~ is 1 000 gram~~mes~~
16 at 62° F] *inserted above*
18 are] a *[E]*
26 wet] *inserted above*
27 height] *after* ~~high~~

fall, to come to rest, and to commence to rise, owing to the lagging contraction of the glass. Its lowest point is marked on the stem. The interval between the two marks is then divided into equal degrees. When such a thermometer is used, it is kept at the temperature to be determined for as long a time as possible, and immediately after is cooled as rapidly as it is safe to cool it, and its zero is redetermined. Thermometers, so made and treated, will give very constant indications. But the thermometers made at the Kew observatory, which are used for determining the temperature of the yard, are otherwise constructed. Namely the melting-point is determined first and the boiling-point afterward; and the thermometers are exposed to both temperatures for many hours. The point which upon such a thermometer will appear as 62° will really be considerably hotter (perhaps a third of a centigrade degree) than if its melting-point were marked in the other way. If this circumstance is not attended to in making comparisons, there is danger of getting the yard too short by perhaps one two-hundred-thousandth part. General Comstock finds the metre equal to 39·369 85 inches, Genl Walker to 39·369 90 inches. The mean between these makes the inch 2·540 013 centimetres.

At the time the United States separated from England, no precise standard of length was legal; and none has ever been established. We are, therefore, without any precise legal yard; but the United States office of weights and measures, in the absence of any legal authorization, refers standards to the British Imperial Yard.

The regular British unit of mass is the Pound, defined as the mass of a certain platinum weight, called the Imperial Pound. This was intended to be so constructed as to be equal to 7 000 grains, each the 5 760 th part of a former Imperial Troy pound. This would be within 3 grains, perhaps closer, of the old avoirdupois pound. The British pound has been determined by Miller to be .4536 kilogram; that is the kilogram is 2.204 621 pounds.

At the time the United States separated from Great Britain, there were two incommensurable units of weight, the *avoirdupois pound* and the *Troy pound*. Congress has since established a standard Troy pound, which is kept in the Mint in Philadelphia. It was a copy of the old Imperial Troy pound which had been

---

1 fall] *above* ~~descend~~
2 Its] *after* ~~At~~
3 divided] ~~sub~~divided
5 safe] *after* ~~p~~
7 at the Kew observatory,] ~~in~~ at the \Kew/ observatory,
8 melting-point] *after* ~~po~~
9 boiling-point] *after* ~~p~~
10 many] *after* ~~a long~~
12 melting-point] ~~zero~~ \meltingpoint/ *(hyphen mine)*
15 metre] *after* ~~yard~~
17–18 At the time … was legal*] *[Peirce's footnote]* *The so-called standard of 1758 had not been legalized.
17–18 of length] *inserted above*
22 Imperial] *after* ~~Britis~~
23 as] *after* ~~that~~
27 there] *above* ~~no~~
28 units] *after* ~~standa~~

adopted in England after American independence. It is a hollow brass weight of unknown volume; and no accurate comparisons of it with modern standards have ever been published. Its mass is, therefore, unknown. The mint ought by law to use this as the standard of gold and silver. In fact, they use weights furnished by the office of weights and measures, and no doubt derived from the British unit; though the Mint officers profess to compare these with the Troy pound of the United States, as well as they are able to do. The old avoirdupois pound, which is legal for most purposes, differed without much doubt quite appreciably from the British Imperial pound; but as the Office of Weights and Measures has long been, without warrant of law, standardizing pounds according to this latter, the legal avoirdupois pound has nearly disappeared from use of late years. The makers of weights could easily detect the change of practice of the Washington Office.

Measures of capacity are not spoken of here, because they are not used in mechanics. It may, however, be well to mention that they are defined by the weight of water at a given temperature which they measure.

The universal unit of time is the mean solar day or its one 86 400 th part, which is called a second. Sidereal time is only employed by astronomers for special purposes.

Whether the International or the British units are employed, there are two methods of measurement of mechanical quantities, the *absolute* and the *gravitational*. The *absolute* is so called because it is not relative to the acceleration of gravity at any station. This method was introduced by Gauss.

The special absolute system, widely used by physicists in the United States and Great Britain, is called the Centimetre-Gram-Second system. In this system, writing C for centrimetre, G for gram mass, and S for second,

the unit of length is .... C;
the unit of mass is.......................................G;
the unit of time is .... S;
the unit of velocity is .... C/S;
the unit of acceleration (which might be called a "galileo," because Galileo Galilei first measured an acceleration) is .... C/S^2;
the unit of density is .... G/C^3;

3 ever] *inserted above*
3 , therefore,] *inserted in-line and above*
6 Troy] *after* ~~Unite~~
9–10 has long ... of law,] has \long/ been, \without warrant of law,/
14 It may,] *above* ~~I may~~
22 station.] *above* ~~po~~
23 widely used by physicists] ~~com~~ widely used by ~~scienti~~ physicists
25 mass] *inserted above*
32 acceleration] *before* ~~, as such~~

the unit of momentum is . . . . GC/S;
the unit of force (called a *dyne*) is . . . . GC/S^2;
the unit of pressure (called one millionth
of an absolute atmosphere) is. . . . . G/CS^2;
the unit of energy (*vis viva*, or
work, called an *erg*) is. . . . . $\frac{1}{2}\text{GC}^2/\text{S}^2$;
etc.

The gravitational system of measurement of mechanical quantities, takes the kilogram or pound, or rather the attraction of these towards the earth, compounded with the centrifugal force,—which is the acceleration called gravity, and denoted by *g*, and is different at different places,—as the unit of force, and the foot-pound or kilogram-metre, being the amount of gravitational energy transformed in the descent of a pound through a foot or of a kilogramme through a metre, as the unit of energy. Two ways of reconciling these convenient units with the adherence to the usual standard of length naturally suggest themselves, namely, 1st, to use the pound weight or the kilogram weight divided by *g* as the unit of mass, and, 2nd, to adopt such a unit of time as will make the acceleration of *g*, at an initial station, unity. Thus, at Washington, the acceleration of gravity is 980·07 galileos. If, then, we take the centimetre as the unit of length, and the 0·031943 second as the unit of time, the acceleration of gravity will be 1 centimetre for such unit of time squared. The latter system would be for most purposes the more convenient; but the former is the more familiar.

In either system, the formula $p = mg$ is retained; but in the former *g* retains its absolute value, while in the latter it becomes unity for the initial station. In Paris, *g* is 980·8 galileos; in Washington it is 980·1 galileos. Adopting the more familiar system, and taking Paris for the initial station, if the unit of force is a kilogram's weight, the unit of length a centimetre, and the unit of time a second, then the unit of mass will be $\frac{1}{980{\cdot}8}$ kilogram, and the unit of energy will be a kilogramme-centimetre, or $\frac{1}{2}\cdot\frac{1000}{980{\cdot}8}\,\text{GC}^2/\text{S}^2$. Then, at Washington the gravity of a kilogram will be, not 1, as at Paris, but $\frac{980{\cdot}1}{980{\cdot}8} = 0{\cdot}9993$ units, or Paris kilogram-weights. Consequently, to produce a force of one Paris kilogram-weight we must allow Washington gravity to act upon $\frac{980{\cdot}8}{980{\cdot}1} = 1{\cdot}0007$ kilograms.]

6 , called an *erg*] *inserted in-line*
12 being] *after* ~~as~~
14–15 adherence to] *above* ~~adoption of~~
16 weight] *inserted above*
16 weight] *inserted above*
17 2nd, to adopt such a] 2nd, ~~to use the second divided \multiplied/ by *g* as the unit of time, that is, to <tak> adopt such a unit of time~~ \to adopt such a/
19 0·031943] *above* ~~second~~
24 absolute] *inserted above*
26 if the unit of force] \if/ the unit of ~~mass w~~ force

[Unsent draft, first page only]

[8. The translator substitutes, in place of Dr. Mach's statement about weights and measures, the following which refers to the present state of things and to the usage of this country. The unit of length generally employed for scientific purposes is the *metre*. The present metre is the distance between the centres of two lines engraved upon a certain bar of platiniridium, measured at the melting-point of ice. This bar, which is known as the International Prototype Metre, has been adopted by most civilized governments, and is preserved and compared with other bars, at the International Bureau in the Pavillon de Breteuil, in the Park of St. Cloud, near Paris. The different governments are furnished with copies of it, which are compared with copies at their several offices of weights and measures. The International Metre is, as nearly as possible, equal to the former French mètre, now abandoned. That metre was defined as the length at the melting-point of ice between the ends of a certain platinum bar, known as the mètre des archives. Comparisons with this bar could not be made with all the precision desired, and there was always a slight uncertainty as to its length. The metre was adopted, *[…]*

McCormack to Peirce 10 May 1893, enclosure. (TS: RL 77) Draft of McCormack's "Translator's Preface" to Mach containing Peirce's edits, highlighted in-line. McCormack later settled on another draft that follows (TS: SIU 92.1). McCormack's edits in that version are noted.

[Draft 1 of 2, sent to and edited by Peirce]

TRANSLATOR'S PREFACE.

This ~~book~~ **work** originally appeared in the *Internationale Wissenschaftliche Bibliothek* **(The International Scientific Series)** ***? Is it not?.*** ~~and~~ The Open Court Publishing Company has acquired the sole right of English translation.

A considerable number of textual errors and irregularities ~~in the original work~~ have been corrected; marginal notes of reference have been supplied; and a more complete index has been added: in this way, **it is believed that** the ~~work~~ **book** has been rendered much more serviceable.

The Translator is indebted to Mr. Charles S. Peirce, ~~our foremost authority in this branch~~ **who has read the proofs**, for a ~~great~~ number of valuable suggestions~~,~~ ~~modifications and additions.~~ **and notes.** Mr. Peirce~~'s independent work are the footnotes on pages …………………, and~~ **has rewritten** Paragraph Eight in the

6 measured] *after* ~~this bar being~~
7 has] *after* ~~was~~
13 between] *after* ~~of~~
15 could] *above* ~~were~~
15 all the] *above* ~~one~~

Chapter on Units and Measures~~.~~**, the original being inapplicable to this country and out of date.**

La Salle,

*[*Draft 2 of 2, edited by McCormack and published*]*

TRANSLATOR'S PREFACE.

The Open Court Publishing Company has acquired the sole right of English translation of this work, which in its German original formed a volume of the *Internationale Wissenschaftliche Bibliothek* of F. A. Brockhaus of Leipsic.

In the reproduction, many textual errors and irregularities have been corrected, marginal titles have been inserted, and the index has been amplified. It is hoped that the usefulness of the book has thus been increased.

No pains have been spared to render the author's meaning clearly and faithfully. In this, it has often been necessary to depart widely from the form of the original, but never it is hoped from its spirit.

The thanks of the translator are due to Mr. C. S. Peirce, well known for his studies both of analytical mechanics and of the history and logic of physics, for numerous suggestions and notes. Mr. Peirce has read all the proofs and has rewritten § 8 in the chapter on Units and Measures, where the original was inapplicable to this country and slightly out of date.

*[*signed*]* T. J. McC.

La Salle, Ill., June 28, 1893

McCormack to Peirce 23 May 1893, enclosure. (TS: SIU 92.1) McCormack's translation of Mach's "Preface to the Translation," published in the edition. McCormack's minor edits are noted.

AUTHOR'S PREFACE TO THE TRANSLATION.

Having read the proofs of the present translation of my work, *Die Mechanik in ihrer Entwickelung*, I can testify that the publishers have supplied an excellent, accurate, and faithful rendering of it, as their previous translations of essays of

---

12–14 No pains … its spirit.] No pains ha*s*<u>ve</u> been spared to render the author's meaning clearly and faithfully*,*<u>.</u> ~~but the difficulties of this task \the preservation of the author's precise tone & spirit/ have often made~~ ˅In this,˅ it \has often been/ necessary to depart widely from the form of the original˅,˅ \but never it is hoped from its spirit./ ~~in order to preserve its precise tone and spirit. To those acquainted with the nature of such work, the methods resorted to need no justification.~~

21 La Salle, … Ĩ893] La Salle, \Ill.,/ ˅June 28, 1893˅

27 as their … of essays] as \their/ previous translations ~~which they had made \by them/~~ of essays

mine gave me every reason to expect. My thanks are due to all concerned, and especially to Mr. McCormack, whose intelligent care in the conduct of the translation has led to the discovery of many errors, heretofore overlooked. I may, thus, confidently hope, that the rise and growth of the ideas of the great inquirers, which it was my task to portray, will appear to my new public in distinct and sharp outlines.

E. Mach.

Prague, April 8th, 1893.

3.4 Peirce and Russell

ca. Oct–Dec 1894. (Publication: SIU 91.27) Advertisement by Henry Holt & Co. (see *W* and *EP* Chronology) for Peirce's proposed 12-volume "The Principles of Philosophy," discussed in Peirce to Russell 1 Sept 1894. A similar advertisement consisting of a more-detailed outline of the work is found in *CBPW* P00552 and in SIU 91.27.

AN EXTENSIVE WORK BY C. S. PEIRCE

MEMBER OF THE NATIONAL ACADEMY OF SCIENCES

The Principles of Philosophy: or, Logic, Physics, and Psychics, considered as a unity, in the Light of the Nineteenth Century

IN TWELVE VOLUMES

EACH VOLUME A DISTINCT WORK

Price, to subscribers only, $2.50 per volume

Address: MR. C. S. PEIRCE, "Arisbe," MILFORD, PA.

The first volume, entitled "A REVIEW OF THE LEADING IDEAS OF THE NINETEENTH CENTURY," will go to press shortly.

THIS philosophy, the elaboration of which has been the chief labor of the author for thirty years, is of the nature of a Working Hypothesis for use in all branches of experiential inquiry. Unmistakable consequences can be deduced from it, whose truth is not yet known but can be ascertained by observation, so as to put the theory to the test. It is thus at once a philosophy and a scientific explanation of observed facts.

The actual comparison of its consequences with

Christianity, than any attempt to make a christianoidal metaphysics serve in lieu of religion. Still less could it accept a theology of phrases which should label an abstraction "God" and influence with posterity "A future life." It distinctly upholds a *Christian Sentimentalism*, as contra-distinguished from a gospel of salvation through intelligent greed.

The following synopsis of the volumes is liable to modification:

Vol. I.—*Review of the Leading Ideas of the Nineteenth Century.*

Vol. II.—*The Theory of Demonstrative Reasoning.*

Vol. III.—*The Philosophy of Probability.*

Vol. IV.—*Plato's World: An Elucidation of the Ideas of Modern Mathematics.*

Vol. V.—*Scientific Metaphysics.*

Vol. VI.—*Soul and Body.*

Vol. VII.—*Evolutionary Chemistry.* (The title may probably be changed.)

Vol. VIII.—*Continuity in the Psychological and Moral Sciences.*

Vol. IX.—*Studies in Comparative Biography.*

Vol. X.—*The Regeneration of the Church.*

Vol. XI.—*A Philosophical Encyclopædia.*

Vol. XII.—*Index* raisonné *of Ideas and Words.*

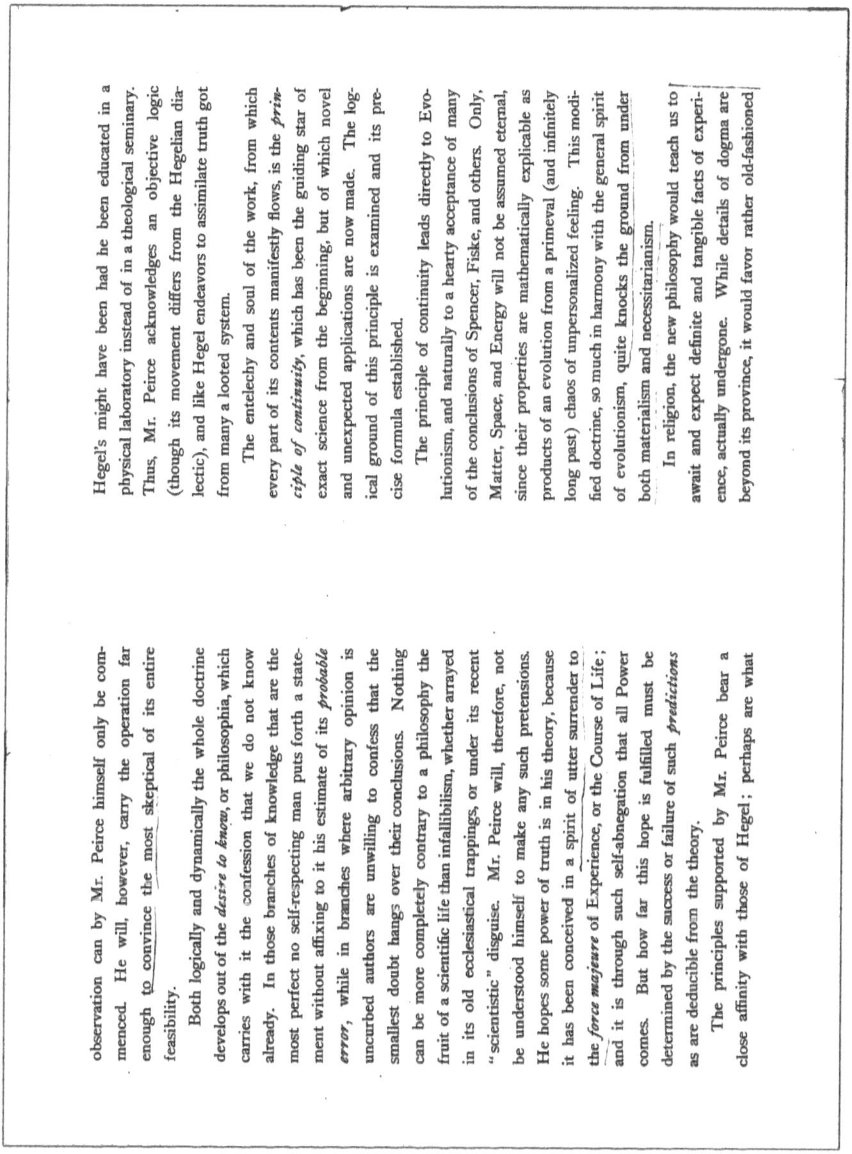

observation can by Mr. Peirce himself only be commenced. He will, however, carry the operation far enough to convince the most skeptical of its entire feasibility.

Both logically and dynamically the whole doctrine develops out of the *desire to know*, or philosophia, which carries with it the confession that we do not know already. In those branches of knowledge that are the most perfect no self-respecting man puts forth a statement without affixing to it his estimate of its *probable error*, while in branches where arbitrary opinion is uncurbed authors are unwilling to confess that the smallest doubt hangs over their conclusions. Nothing can be more completely contrary to a philosophy the fruit of a scientific life than infallibilism, whether arrayed in its old ecclesiastical trappings, or under its recent "scientistic" disguise. Mr. Peirce will, therefore, not be understood himself to make any such pretensions. He hopes some power of truth is in his theory, because it has been conceived in a spirit of utter surrender to the *force majeure* of Experience, or the Course of Life; and it is through such self-abnegation that all Power comes. But how far this hope is fulfilled must be determined by the success or failure of such *predictions* as are deducible from the theory.

The principles supported by Mr. Peirce bear a close affinity with those of Hegel; perhaps are what Hegel's might have been had he been educated in a physical laboratory instead of in a theological seminary. Thus, Mr. Peirce acknowledges an objective logic (though its movement differs from the Hegelian dialectic), and like Hegel endeavors to assimilate truth got from many a looted system.

The entelechy and soul of the work, from which every part of its contents manifestly flows, is the *principle of continuity*, which has been the guiding star of exact science from the beginning, but of which novel and unexpected applications are now made. The logical ground of this principle is examined and its precise formula established.

The principle of continuity leads directly to Evolutionism, and naturally to a hearty acceptance of many of the conclusions of Spencer, Fiske, and others. Only, Matter, Space, and Energy will not be assumed eternal, since their properties are mathematically explicable as products of an evolution from a primeval (and infinitely long past) chaos of unpersonalized feeling. This modified doctrine, so much in harmony with the general spirit of evolutionism, quite knocks the ground from under both materialism and necessitarianism.

In religion, the new philosophy would teach us to await and expect definite and tangible facts of experience, actually undergone. While details of dogma are beyond its province, it would favor rather old-fashioned

Figure 14: Advertisement by Henry Holt & Co. for Peirce's proposed 12-volume "The Principles of Philosophy," 2-sided quarto brochure, ca. fall 1894.

7 Oct 1904. (MS: RL 387) Peirce's review of Hippolyte Taine's 1870 *De L'Intelligence*, which Carus, Russell, and Peirce agreed to have submitted for the April 1905 *Monist* under Russell's name, and in his "own language" and "own point of view" as Peirce recommends in his letter to Russell 3 Oct 1904. Of the three drafts of this article extant in RL 77 and RL 387, the one included here is most complete, is dated, and appears to be the rewrite Peirce mentions in the same letter to Russell.

1904 Oct. 7.

Substitution in Logic

To the Editor of The Monist:

Just now, I accidentally come upon the following particularly positive assertion in a recent periodical:

"In Taine's brilliant book on 'Intelligence,' substitution was for the first time named as a cardinal logical function, though the facts had always been familiar enough."[1]

Taine, by substitution, meant the substitution of a sign for an object signified or one sign for another. He illustrated it copiously, but whether he made any explicit formulation of its being "a cardinal logical function," I do not know. Probably not, his work being psychological. Ockham, Hobbes, Leibniz, and all the other logicians who looked upon thought as a sign made it perfectly apparent that *substitution* was very important. In a paper "On the Classification of Arguments," published in 1867 (a copy with a letter calling attention to the doctrine were sent to M. Taine but probably did not receive the attention they called for) made substitution not '*a* cardinal logical function,' but the sole hinge on which thought turns in all its movements. For it was there evident, and in the paper of the same year on a list of categories,[2] that the term "argument" was meant to cover every movement of thought proper. Near the beginning of that paper one can read as follows:

"Every conclusion may be regarded as a statement substituted for either of its premisses. Nothing is relevant to the other premisses, except what is requisite to justify the substitution.... Every substitution of one proposition for another must consist in the substitution of term for term. Such substitution can be justified only so far as the first term represents what is represented in the second." Upon this idea the whole thought of the paper turns; and the writer of it goes on to show demonstratively *that*, and *how*, every kind of necessary reasoning, not omitting

1 William James, "A World of Pure Experience," *Journal of Philosophy*, vol. 1 (1904), p. 541.
2 "On a New List of Categories."

9 particularly] *above* ~~statement~~
16–17 Probably ... psychological.] *inserted above*
21 Taine] *before* ~~)~~
22 hinge] *after* ~~cardinal~~

those involving relative terms, and even including every kind of immediate inference, together with scientific induction, the problematic adoption of hypothesis, and reasoning from analogy, can all be considered as substitutions (not "substitutions of similars," which Jevons shortly after fell in love with[3]).

Taine's book was a detailed application of the logical principle to psychology, —a very wearisome book to a logician,—a wall-paper book in its endless repetitions of the old pattern.

But the bottom has been knocked out of the logical theory of substitution as a fundamental logical principle, since it has been demonstrated that every logical substitution is valid only in so far, first, as it is a valid operation to insert, the substitute and, secondly, as it is a valid operation, following that, to omit the substituend; so that the operations of insertion and omission must take the place of substitution. No doubt an insertion or omission may be reduced to the formula of substitution, but only by a complication to no purpose, while the substitution is resolved into simpler elements depending on simpler logical conditions, by being analyzed into an insertion followed by an omission.

Russell to Carus 24 April 1914, enclosure. Russell's obituary of Peirce, published in *The Monist*, July 1914.

IN MEMORIAM CHARLES S. PEIRCE.

(Born 1839, died 1914.)

Concerning genius, its advent discovery and nurture, history informs us that with rare exceptions its worldly case is one of the utmost austerity. On reflection this appears not at all strange. *Pro re natâ*, genius issues as an outlaw. It breaks over and through the accustomed rules and conceptions to the confusions and perplexity of a world otherwise comfortable in conventions regarded by it as settled possessions. Hence it is unwelcome. Hence the futility of all extant provisions in its favor. Had any Nobel foundation been in existence is 1841, would any of its benefits have found its way to Hermann Grassmann? Not in a thousand years. His case is typical of the general case of genius. Neglect and poverty are its portion

---

3 *The Substitution of Similars* (1869).

---

2 induction,] *before* ~~and~~
3 all] *after* ~~only~~
5 logical] *inserted above*
10–11 valid … secondly,] valid only ~~becau~~ in so far, \first,/ as it is \a/ valid operation~~, first,~~ to substitute and ~~then~~, secondly
14 complication] *after* ~~purp~~

in life. Then afterwards lapse of time reveals to a stupid, jealous and oftentimes spiteful world that it has conspired for the suffocation of a divine messenger.

In the late sixties the distinguished Prof. Benjamin Peirce of Harvard, lecturing before the Boston "Radical Club" on "The Impossible in Mathematics," spoke of his son Charles and of his expectations that the latter would develop and fertilize the vistas he had been able only to glimpse. On April 19, 1914, after at least half a century of assiduous probings into the most recondite and the most consequential of all human concerns, in a mountain hut overlooking the serene Delaware, in privation and obscurity, in pain and forsakenness, that son, Charles S. Peirce, left this world and left also a volume of product the eminent value of which will sooner or later be discovered, perhaps only after it has been rediscovered. For his issues have so far anticipated the ordinary scope of even professional intellectual exercise that most of them are still only in manuscript. Publishers want "best sellers." At least they want sellers that will pay the expenses of publication, and buyers of printing that calls for laborious mental application are scarce. Let me here with the utmost solicitude beg all to whom it falls to handle his books and papers to beware how they venture to cast away any script left by him.

Is this panegyric unwarranted? If so, then why should Professor James in his *Varieties of Religious Experience* call Mr. Peirce "our greatest American Philosopher"?[4] Why should Professor Schroeder base his great work "Exact Logic" on the prior work of Mr. Peirce?[5] Why should the editors of the great *Century Dictionary* employ Mr. Peirce to write so many of its logical, mathematical and scientific definitions?[6] Why should the editors of Baldwin's Dictionary[7] make a similar draft? Why should the editors of the *New York Evening Post* and of *The Nation* for years refer their books of serious import to Mr. Peirce for examination and review?[8] Why should Dr. Carus recognize in Mr. Peirce a foeman worthy of his incisive steel on the fundamental problem of necessity?

---

4 p. 444. Text actually identifies Peirce as "an American philosopher of eminent originality."

5 Schröder's *Algebra der Logik* series (1890–1905), and Peirce's work on the logic of relatives and algebra of logic especially 1870–1886. (See Bibliography; also "Schröder" in Biographical Register.)

6 Peirce contributed definitions to *The Century Dictionary* throughout much of his career, particularly leading up to the 1889–1891 edition. His *Century* writings between 1883 and 1909 are to comprise the core of *W* 7.

7 J. M. Baldwin's *Dictionary of Philosophy and Psychology*, for which Peirce wrote extensively 1901–1902 (see *CBPW*).

8 Peirce wrote over 300 reviews 1869–1908 for *The Nation* (many referenced throughout this edition), renamed *The New York Evening Post*, which provided critical income for him, especially beginning 1890, and kept him in touch with his intellectual environment.

Of course genius is uncomfortable. "'T is its nature to." It is often very hard to get along with. It tries the patience to the limit. It is so immersed in and so saturated with the inspiration of non-conformity that it often neglects to observe what is really and plainly only a merely defensive right on the part of the world of conformity. There ought to prevail a mutual spirit of forgiveness. If much is to be forgiven because of much love why should not much be forgiven to much promising and well directed power?

Mr. Peirce died a faithful man. His earlier studies led him far towards the goal of materialism, but in the course of those studies he was led to the discovery of that touchstone of values, that at first until the conception and word became mangled and aborted out of its true intent and utility he called Pragmatism, the principle that all rational significance of conceptions and of the terms embodying the same lies between the four corners of their *conceived* consequences in and to actual practice mental and otherwise. Since all logic is only a comparison and criticism of conceptions, this principle affects and effects our whole rational life and conduct. He was thus led to his conception of *reality* as that which has the natural prerogative of persistence as a possession forever. He perceived that intellectual entities, like, say, the law of gravitation or the ratio of the radius to the circumference of a circle, have just as abiding persistence as any material entity and hence just as *real* an obtaining. Hence actual medieval realism, when better introduced and explained, is more pragmatically valuable than any case of nominalism or conceptualism can possibly be. The recognition of ideal realities opens out into the recognition that all existence is grounded in and upon that ideal substance the best names for which are *Form*, alias *Reason*, alias *Mind*, alias *Truth*, alias *the Good*, alias *Beauty*. The perception of Reason immanently in and throughout the universe and identical in nature with human reason solves at once the vexed question of the relation of body and mind, invites the soul to faith and repose and at the same time stimulates the soul to a vivid aspiration after cooperation with the Universal Spirit in accordance with it course of procession.

So lives Charles S. Peirce. The Universal Spirit has him and the world that neglected him will care for him—after many days perhaps, but most assuredly.

Francis C. Russell.

Chicago.

Biographical Register

Listed below are many of Peirce's contemporaries mentioned repeatedly in the letters. Persons mentioned incidentally or those deemed to be well-known historical figures (e.g., Kepler, Kant, and the like) are identified as needed in individual letters. This list is broadened to also include several organizations and one place, "Arisbe," Peirce's Milford estate. Sources used for these entries include *W* 1–8, Brent (1993), Henderson (2009), Sheriden (1957), "Stories from the Big House" (online), and the General Reference materials listed in the Bibliography.

OCP Correspondents

Carus, Paul (1852–1919). German philosopher, prolific scholar and writer, and managing editor at OCP 1887–1919. Received his PhD from Tübingen, and was initially an educator in Germany, England, and, by 1884, New York and Boston. Several publications 1885–1886, including *Monism and Meliorism*, contributions to the Boston *Index*, and a book of German poems (*Leben in Liedern*), drew the attention of Hegeler who in turn invited Carus in Jan 1887 to join OCP as co-editor (of sorts) with the newly hired Underwood. After a year-long dispute between Hegeler, Carus, and Underwood, the latter and his wife resigned, leaving Carus the post Dec 1887, where he remained until his death. During his 32-year tenure he wrote 74 books, nearly 1 500 articles, produced 732 issues of *The Open Court*, and 113 issues of *The Monist*. Among the myriad of authors with whom he corresponded while at OCP, one of his first and longest-standing was Peirce. He held an appreciation of Peirce's intellect and writing not generally shared by OCP, and ultimately published more of Peirce's philosophical articles during Peirce's lifetime than any other editor.

Wife: Marie Henriette Hegeler (1861–1936), daughter of Edward C. Hegeler.

Children: Edward Hegeler, Gustav Kruger, Laura Camilla Paula Holler, Marie Elizabeth Weisbach, and Hermann Dietrich

Diesterweg, Charles. Secretary at OCP La Salle office. Typed letters to Peirce largely on behalf of Hegeler and maintained related internal records. Also a shoe and boot merchant in La Salle.

Garbutt, W. J. Editorial assistant at OCP La Salle office. Corresponded with Peirce briefly 1892 regarding his first *Open Court* and *Monist* series.

Hegeler, Edward Carl (1835–1910). German miner and founder of OCP. Immigrated to the U.S. in 1856 with schoolmate and mining associate Frederick

https://doi.org/10.1515/9783110768756-004

Matthiessen, with whom he co-founded in 1858 the Matthiessen & Hegeler Zinc Company of La Salle, IL, by 1880 the largest zinc company in the U.S. Founded OCP in 1887 to provide a forum for discussing philosophy, science, and religion, and to make philosophical classics more affordable and widely available. Corresponded with Peirce primarily during 1893 to publish his proposed books on arithmetic and geometry, as well as his "A Quest for a Method" (or "Search for a Method"). By 1894, due to consistent complications and misunderstandings with agreements, Hegeler no longer worked with Peirce and none of the books were published.

Wife: Maria Camilla Weisbach (1835–1908), daughter of Julius Ludwig Weisbach (1806–1871), Hegeler's professor at Freiberg University of Mining and Technology 1853–1856

Children: Marie Henriette (Carus), Helene Emma, Camilla (Bucherer), Meta Rosalie, Julius Weisbach, Gisela Lina, Herman, Annie, Lina Zuleikha, and Catharina Olga

McCormack, Thomas Joseph (1865–1932). Assistant editor of *The Open Court* and translator (German and French) at OCP La Salle office ca. 1890–1903, considered by OCP compositor Fred Sigrist to be the "best man and intellectual backbone of the enterprise" (Sigrist to Peirce 26 June 1903). Worked with Peirce mostly 1892–1893 on a translation of Ernst Mach's 1889 edition of *The Science of Mechanics* (*Die Mechanik In Ihrer Entwickelung*) published 1893 by OCP. Other translations by him include Schubert, *Mathematical Essays and Recreations* (1898); Lagrange, *Lectures on Elementary Mathematics* (1901); Berkeley, *Principles of Human Knowledge* (1904); Delitzsch, *Babel and Bible* (1906); and Koerner, *Memoirs of Gustave Koerner* (1909).

Nattinger, J. G. Secretary at OCP La Salle office. Typed letters to Peirce largely on behalf of Carus and maintained related internal records.

Robinson, Lydia Gillingham (b. 1875). Assistant editor and translator (German, Dutch, and French) at OCP La Salle office beginning ca. 1905. By 1908 considered by Russell to be Carus's right hand with great influence over him, and "'*the whole thing*' so far as editorship is concerned" (Russell to Peirce 2 Oct 1908). Corresponded with Peirce 1905–1910 on various editorial matters, especially Peirce's series on pragmaticism and Amazing Mazes. Other translations by her include Delitzsch, *Babel and Bible* (1906, with McCormack); Archimedes, *Geometrical Solutions* (1909); and Couturat, *The Algebra of Logic* (1914).

Russell, Francis C. Chicago attorney and judge, amateur philosopher, logician, and mathematician, and avid admirer of Peirce. Friend of Carus and Hegeler and periodically a contract editor for OCP. Persuaded Carus to invite Peirce to contribute to *The Monist* (Carus to Peirce 2 July 1890) and served as Peirce's ongoing confidant and liaison with OCP. At Carus's request, reviewed and edited several of Peirce's *Monist* publications, including "The Regenerated Logic" (1896), "Substitution in Logic" (1905, published under Russell's name), and Peirce's 1877–1878 "Illustrations of the Logic of Science" series to be republished in book form (never published, Russell's work mostly 1910–1911). His own OCP contributions include "Postulates vs. Axioms" (*Open Court*, Feb 1889), "Prison or Citadel—Which?" (*Open Court*, July 1894), "Hints for the Elucidation of Mr. Peirce's Logical Work" (*Monist*, July 1908), "Minos and Niemand Again" (*Open Court*, Nov 1908), and Peirce's obituary, "In Memoriam Charles S. Peirce" (*Monist*, July 1914).

Sacksteder, Martin A. Clerk and sales manager at OCP Chicago office. Corresponded periodically with Peirce on administrative matters.

Sigrist, Fred. Compositor at OCP beginning ca. 1897. Despised Carus and corresponded with Peirce briefly 1903 and 1907 to recruit him for a scheme to expose Carus for alleged plagiarism and other pretensions. Although no letters from Peirce are found, he seems not to have gotten involved for fear of anything "squinting at blackmail" (Sigrist to Peirce 9 Jan 1903). The veracity of the allegations is undetermined (though this editor doubts it very much).

Others

Abbot, Francis Ellingwood (1836–1903). Peirce's classmate at Harvard (class 1859), helped found the Free Religious Association of Boston 1867 and edited its periodical *The Index* ca. 1870–1880 (succeeded by Benjamin Underwood; see below). Member with Peirce of the Metaphysical Club at Cambridge (formed 1872). Later theologian and Unitarian minister. Author of *Scientific Theism* (1885) and *The Way Out of Agnosticism* (1890), the latter attacked by Royce the same year in the *International Journal of Ethics* for philosophical fraud and allegedly plagiarizing Hegel. Peirce came to Abbot's defense in *The Nation* 12 Nov 1891 (*CN* 1:115–117) with his "Abbot against Royce." In Peirce's obituary for Abbot in *The Nation* 5 Nov 1903 (*CN* 3:147–148), he said fondly of Abbot that he gave "a new conception to the saying, 'The pure in heart shall see God' " and praised the "unsophisticated purity of his love of and apprehension of truth, oblivious of the tide of opinion."

Adler, Felix (1851–1933). German-American philosopher and professor of social and political ethics, whose ethics Carus mentions differ from his own (Carus to Peirce 31 May 1892). Considered an "ethical idealist with great practical reforming zeal" who promoted an idealistic moral perfectionism. Also appealed to a balance between universal principles and moral particularism or individualism in determining ethically sound choices. ("Felix, Adler," *Encyclopedia of Ethics*, 2001)

Agassiz, Alexander Emmanuel Rodolphe (1835–1910). American scientist and engineer, and friend and correspondent of Peirce (RL 5). Son of Peirce family friend Louis Agassiz, under whom Peirce privately studied zoological classifications at Harvard 1860.

Appleton & Co. Publishing company founded 1831 by Daniel Appleton (1785–1849), succeeded 1848 by son William Henry (1814–1899) in partnership with brothers John Adams, George Swett, Daniel Sidney, and Samuel Francis. Publisher of Youmans's *Popular Science Monthly* 1872–1900 and co-founder of the American Book Company 1890. William became acquainted with Peirce en route to Europe 1875 when the two discussed Peirce's "Illustrations" series to appear in *PSM* 1877–1878. They met again (and with Henry Vail of American Book Company) early Feb 1893 in consideration of his proposed arithmetic book and for copyright permission to republish with OCP his "Illustrations" series. Appleton et al. maintained periodic correspondence with Peirce 1893–1900 (RL 21).

Arisbe. Peirce's estate in Milford, PA, purchased May 1888 from Quick family, expanded to 2,000 acres 1889–1890, and re-named "Arisbe" 1891 (formerly "Quicktown"). The origin of the name is unclear and a source of much speculation, but a popular theory is that it was named after the Milesian colony northeast of Troy along the Hellespont, home of the first philosophers of Greece (*W* 8:xlvii). Its initial renovations and expansions (detailed in Peirce to Russell 17 Sept 1892) reputedly served as an impetus for Peirce's maiden *Monist* article, "The Architecture of Theories" (Jan 1891), and as a type for his philosophy in general. In recruiting Hegeler 19 May 1893 for a proposed boarding school to be held here, Peirce described it as the "most ravishing pieces of country that are still left just as it was turned out of God's studio, ... To look out my window is to praise God." Here the Peirces would become involved with a community of prominence and affluence, especially via their neighbors James W. and Mary E. Pinchot, that would provide much-needed social and financial support. The property would soon become a source of unrelenting financial strain, as is evidenced by his desperate pleas to Carus (and many others) for employment and loans throughout this correspondence. Nevertheless, Peirce remained here until his death 1914, as did Juliette until hers 1934. Arisbe is now the home of the Research and Resource Planning Division of the Delaware Water Gap National Recreation Area.

Barnard, Lydia Augusta Allen (1827–1911). Family friend of the Peirces. Married to James Munson Barnard (1819–1904), philanthropist, youth mentor, and mental health advocate who became acquainted with Benjamin Peirce, Alex Agassiz, and others of prominence as a student at Lawrence Scientific School of Harvard 1854–1858. Lydia corresponded with Charles and Juliette 1896–1900 (RL 35).

Becker, George Ferdinand (1847–1919). Geologist, miner, and metallurgist, serving in that specialization at UC–Berkeley 1875–1879, at the U.S. Geological Survey beginning 1879 through most of his life, and in the Spanish-American War 1898. Long-standing friend of Peirce, arranged for him to deliver his 1892–1893 "History of Science" Lowell Institute lectures, provided temporary employment for Peirce late 1897 with the Geological Survey. Maintained correspondence with Peirce 1891–1910 (RL 39).

Bierstadt, Albert (1830–1902). German-born American painter specializing in landscapes, especially western, grouped with the Hudson River School and later Rocky Mountain School movements. Acquainted with Peirce via The Century Club and partnered with him (and Edmund Stedman, below) in a failed entrepreneurial endeavor ca. 1896–1898 to develop affordable domestic lighting from acetylene gas (RL 45). Briefly housed Peirce in New York early November 1896 (see Peirce to Russell ca. 5 Nov 1896, postscript).

Calderoni, Mario (1879–1914). Italian philosopher and pragmatist considered "more Peircean than Peirce" (Peirce to Russell 5 July 1905). Student of Italian pragmatist Giovanni Vailati (see below) and mutual correspondent of Peirce and Peirce's semiotic confidant Victoria Lady Welby (York University, Welby, Box 2). Peirce wrote to Calderoni late summer or fall 1905 a now popular letter describing his pragmaticism (RL 67; partly in *CP* 8.205–213, in full in Fisch and Kloesel, "Peirce and the Florentine Pragmatists," *Topoi*, 1982).

Cayley, Arthur (1821–1895). Renowned British mathematician appointed Sadleirian professor of pure mathematics at Cambridge University 1863, admired and well read by Peirce. Met Peirce at Johns Hopkins University as visiting lecturer Jan–June 1882 during Peirce's short tenure there (1879–1884), where Peirce presented a brief paper "On the Relative Forms of Quaternions" (*W* 4:334–335) in a series of Mathematical Seminary lectures in honor of Cayley's visit. Around this time, Peirce also discovered that Cayley's 1858 "Memoir on the

Theory of Matrices" substantially anticipated his own algebra of dual relatives, as Peirce noted in the postscript to his 1882 "Brief Description of the Algebra of Relatives" and reputedly perpetuated in his JHU lectures. Peirce later wrote Cayley's obituary for *The New York Evening Post* 28 Jan 1895 (*NEM* 2:642–649).

Century Club (or Century Association). Private New York club founded 1847 by editor and poet William C. Bryant (1794–1878) to promote literature and the fine arts. Attracted affluence and intellect among bankers, investors, scientists, writers, publishers, politicians, and eccentrics. Peirce was a member 1876–1895 and often used it as his mailing address. There he formed relationships with prominent men and became involved with some in extravagant (though failed) entrepreneurial endeavors (e.g., see Albert Bierstadt and Edmund Stedman in this register). Peirce's first reading of his "Excursion into Thessaly" was held at the Club April 1892 (Peirce to Russell 14 May 1892), indicating that it was likely an outlet for his work. Partly due to Peirce's failed business affairs with his fellow members, the Club was also a source of food and occasional charity during financial difficulties near the end of his membership, which ended with his expulsion.

Cushing & Co., J. S. Boston printer founded ca. 1808 by Joshua S. Cushing (1775–1832), later with locations in Salem, Lunenburg, and Norwood, MA. One of the printers Peirce considered early 1893 to contract with OCP for his proposed arithmetic book (never published).

d'Aulby del Borghetto, Count John Edward (b. 1868). Wealthy French nobleman, married to the daughter of Adelaide Lunt, a Peirce family friend, partnered with Peirce (as well as Bierstadt, above, and Stedman, below) in a failed entrepreneurial endeavor ca. 1896–1898 to develop affordable domestic lighting from acetylene gas. Took control of their European prospects for the business ca. July 1898 and cut Peirce out altogether.

De Morgan, Augustus (1806–1871). Renowned British mathematician and logician, professor at London University (now University College London) 1828–1866, and pioneer in the study and notation of the logic of relatives. Presumed to have influenced Peirce's own work with the logic of relatives, particularly Peirce's "Description of a Notation for the Logic of Relatives" (1870), about which he and De Morgan met in London summer 1870 following their exchange of logical papers late 1867–1868 (RL 119). Among De Morgan's chief logical works are his *Essay on Probabilities* (1838), *Formal Logic* (1847), and *Syllabus of a Proposed System of Logic* (1860). His 1836 *On the Study and Difficulties of Mathematics* was republished by OCP 1898 and reviewed by Peirce in *The Nation* 21 Sept 1899 (*CN* 2:214–216). His obituary in *The Nation* 13 and 20 April 1871 has been attributed to Peirce (*CN* 1:41–42).

Delbœuf, Joseph Rémi Léopold (1831–1896). Belgian philosopher and psychologist who Peirce believed to be one of the "world-renowned thinkers," along with Montgomery and Renouvier, who shared Peirce's opinion of *tychism* in "The Doctrine of Necessity Examined" (reference identified in *EP* 2:334n1 as Delbœuf's "Déterminism et liberté," *Revue Philosophique* 13 and 14, 1882), and for whom Peirce requested a copy of that issue to be sent by Carus (3 June 1892). See Peirce's footnote at p. 1 of Peirce's "Man's Glassy Essence," Oct 1892.

DeVinne Press. New York printer founded ca. 1885 by Theodore Low De Vinne (1828–1914). Printed Peirce's "Prospectus" ca. 1892–1893 for his translation of the 1269 *Epistle of Petrus Peregrinus on the Lodestone* to Sygerus de Foucaucourt, to be printed by DeVinne (never published). Also one of the printers Peirce considered early 1893 to contract with OCP for his proposed arithmetic book (never published).

Edmund Duncan Montgomery (1835–1911), Scottish philosopher, physician, and opponent to the monistic objectives of Hegeler and the *Open Court*. Carus to Peirce 3 Aug 1890 refers to "the Montgomery controversy of recent date," meaning his controversial article "The Monism of 'The Open Court' Critically Examined" to appear in *The Open Court* 21 August 1890. He was also a correspondent and an admirer of Peirce on grounds of their mutual rejection of mechanical or necessitarian evolutionary theory, and of Carus's tolerance of it (e.g., see Peirce and Carus's exchanges between April and June 1892). For a discussion of the Peirce–Montgomery correspondence, see *W* 8:402–403 and Keeton, *The Philosophy of Edmund Montgomery*, p. 306.

Faye, Hervé Auguste Étienne Albans (1814–1902). French astronomer and president of the Bureau of Longitudes in Paris while at the International Geodetic Association conference in Stuttgart where Peirce was in attendance for the Coast Survey. There Faye proposed a double pendulum method to reduce flexure and inaccuracy in recording gravitational data, which Peirce further investigated and confirmed, publishing his results in "On a method of swinging Pendulums for the determination of Gravity, proposed by M. Faye" (*American Journal of Science*, Aug 1879; *W* 4:12–20) and by way of a letter to Faye July 1880, included in the Association's proceedings (*W* 4:157–160).

Fiske, John (1842–1901). Harvard Law graduate (class 1865) turned philosopher and historian. Lecturer at Harvard 1869–1871 and a member with Peirce of the Cambridge Metaphysical Club (though a less-frequent attender). Later held various university lectureships on American history in the U.S. and U.K., notably Washington University of St. Louis, University College London, and the Royal Institution of Great Britain. Author of several historical and theological works, including his 1899 *Through Nature to God* and *A Century of Science and Other Essays*, reviewed by Peirce in *The Nation* 10 Aug 1899 and 4 Jan 1900, respectively. Corresponded with Peirce 1891–1894 regarding Peirce's inquiries for employment following his dismissal from the Coast Survey, including a proposed lecture circuit to be given by Peirce (RL 146).

Garrison, Wendell Phillips (1840–1907). Son of prominent abolitionist and journalist William Lloyd Garrison (1805–1879) and editor of *The Nation* 1865–1906. Published nearly 300 book reviews by Peirce 1869–1906, providing critical income for Peirce, especially beginning 1890, and keeping Peirce in touch with his intellectual environment. Became a close friend and financial support to Peirce until his retirement from *The Nation* 1906 (RL 159).

Gibbs, Josiah Willard (1839–1903). Renowned mathematician and physicist, received the first American doctorate in engineering from Yale 1863, where he served as professor of mathematical physics 1871 until his death. Corresponded briefly with Peirce ca. 1881–1882 (RL 162).

Gilman, Daniel Coit (1831–1908). First president of the Johns Hopkins University during Peirce's short tenure there (1879–1884). Dismissed Peirce from the university as a result of Peirce's instability and growing reputation of immorality. Also first president of the Carnegie Institution 1902–1904, where he oversaw Peirce's expansive and heavily endorsed 1902 grant application for his "Memoirs on Minute Logic," which was ultimately (perhaps not surprisingly) rejected. Reputedly influenced in his opinion of and actions against Peirce by Simon Newcomb, described below.

Ginn & Co. Boston textbook publisher founded 1867 by Edwin Ginn (1838–1914), named Ginn Brothers until 1885. Under Jem's instigation (see below), worked with Peirce 1894–1895 in consideration of several proposed math books, including his arithmetic, his "New Elements of Geometry," and later his "Elements of Mathematics" (none published).

Hall, Granville Stanley (1846–1924). Served with Peirce in the newly formed philosophy faculty at Johns Hopkins University during Peirce's short tenure there (1879–1884), specializing in experimental psychology, a passion he and Peirce shared. Held a deep appreciation for Peirce's work, especially his 1877–1878 "Illustrations of the Logic of Science" series. Was awarded the JHU philosophy professorship over Peirce (and over their mutual colleague George S. Morris) in 1882. In 1889 became president of Clark University, which he modeled after JHU and where Peirce was twice turned down by Hall for a lectureship there. The two nevertheless remained close friends.

Halsted, George Bruce (1853–1922). American mathematician and pioneer in the field of non-Euclidean geometry in the U.S. Renowned JHU professor J. J. Sylvester's first PhD student graduating 1879, appointed professor at Princeton that year and later at University of Texas–Austin 1884. While under J. J. Sylvester, corresponded with Peirce 1878 (the year before his JHU appointment) describing Peirce's poor reputation at JHU, and again 1891–1895 (RL 181) initially out of interest in Peirce's ideas on non-Euclidean geometry and soon after in regard to Halsted's translations of Nicholai Lobachevsky's *Geometrical Investigations on the Theory of Parallels* (1891) and János Bolyai's *The Science of Absolute Space* (1896), which Peirce reviewed in *The Nation* 11 Feb 1892 and 6 Feb 1896, respectively.

Harris, William Torrey (1835–1909). American philosopher and founding editor of the *Journal of Speculative Philosophy* 1867, in which Peirce published his 1868–1869 cognition series. During the same period corresponded with Peirce on Hegel (*W* 2:132–161; RL 183). Editor in chief of the 1900 *Webster's International Dictionary*, reviewed by Peirce in *The Nation* 24 Jan 1901 (*CN* 3:23–25).

Holden, Edward Singleton (1846–1914). Astronomer and director of Washburn Observatory at University of Wisconsin–Madison 1881–1885, where he befriended Peirce during Peirce's 1885 visit on assignment for the Coast Survey. Became an admirer of Peirce and a mutual friend of Peirce and Carus. President of University of California 1885–1888 and the first director of the Lick Observatory 1888–1897, founder of the Astronomical Society of the Pacific 1889, and later librarian at West Point 1901–1914.

Huntington, Capt. Edward Stanton (Ned) (d. 1895). Peirce's cousin, U.S. Army 1861–1869. Author of *Dreams of the Dead* (1892), which, per Huntington's request 3 Aug 1892 (RL 209), Peirce reviewed in *The Nation* 8 Sept that year (*CN* 1:165–166).

Huntington, Rev. William Reed (1838–1909). Episcopal priest and author, fellow classmate of Peirce at Harvard (class 1859) where he taught chemistry 1859–1860. Possibly one of the Huntington relatives to the Peirce family (see Huntington, Edward). Rector at All Saints Church (Worcester, MA) 1862–1883 and at Grace Church (New York) 1883–1909. Authored several theological works, notably his 1870 *The Church-Idea*, which established the four-point statement of Anglican identity later known as the Lambeth Quadrilateral. Corresponded briefly with Peirce 1893 and 1897 (RL 213).

Jacobi, Abraham (1830–1919). German physician and active communist, immigrated to New York 1854, held various university posts 1861–1902 in New York, mostly at Columbia, specializing in childhood diseases and general pediatrics. Established the first department of pediatrics in the U.S. at Mount Sinai Hospital. Considered later as the father of American pediatrics. Married prominent New York physician Mary Putnam Jacobi 1873 (below), likely through whom he became acquainted with Peirce.

Jacobi, Mary Corinna Putnam (1842–1906). Prominent New York physician, author, and suffragist. Married pediatric pioneer Dr. Abraham Jacobi (above). Daughter of New York publisher George Palmer Putnam (1814–1872) of Wiley & Putnam (1838), G. P. Putnam & Co. (1848), and later G. P. Putnam's Sons (1872), the publisher with whom Peirce collaborated 1898–1903 regarding his proposed "History of Science" volumes (RL 78, 364). Peirce also corresponded with Jacobi herself around this time, presumably under related pretexts, and again 1902 to engage her influence with his Carnegie Institution grant application for "Memoirs on Minute Logic" (RL 75, 221).

James, William (1842–1910). Harvard-trained physician turned eminent philosopher and psychologist, and one of Peirce's most loyal friends and proponents. Entered the Lawrence Scientific School at Harvard with Peirce 1861 and for the rest of his life showed unfailing admiration for Peirce and his work. Instigated the formation of the Cambridge Metaphysical Club 1872, considered the birthplace of American pragmatism fathered by Peirce, James, Nicholas St. John Green (1830–1876), and others—although James would later credit Peirce as the founder of pragmatism in his 1898 "Philosophical Conceptions and Practical Results" (Philosophical Union address, UC–Berkeley). Dedicated his *Will to Believe* (1897) to Peirce, "to whose philosophic comradeship in old times and to whose writings in more recent years I owe more incitement and help than I can express or repay." Arranged for Peirce to deliver his 1898 Cambridge Conferences lectures on logic and later his 1903 Harvard lectures on pragmatism. Provided other financial support for Charles and Juliette as early as 1898 and even organized an unofficial Peirce fund with an estimated 20 contributors ca. 1902–1908. Peirce's alternate middle name "Santiago" given him by Schröder 1890 was later embraced by Peirce, especially beginning 1906, reputedly in honor of James ("Saint James"). (More details in "A Note on James's Aid of Peirce," *Transactions of the Charles S. Peirce Society*, vol. 12, no. 1 (Winter 1976): 71–76.)

Jastrow, Joseph (1863–1944). Prominent psychologist and former student of Peirce during Peirce's short tenure at Johns Hopkins University (1879–1884). Together they conducted psychological experiments (a first at the time) and co-authored "On Minimum Differences of Sensibility," presented by Peirce Oct 1884 to the National Academy of Arts and Sciences (*W* 5:122–135). Professor at University of Wisconsin–Madison 1888–1927 and New School of Social Research (NY) 1927–1933, where he continued his experiments in perception, color sensation, and optical illusions (e.g., his famous Rabbit-Duck Illusion). After Peirce's death, contributed in his honor "The Passing of Master Mind" in *The Nation* (1914) and "Charles Sanders Peirce as a Teacher" in the *Journal of Philosophy* (1916). Added in his autobiography that Peirce was "one of the most exceptional minds that America has produced, who stimulated me most directly" ("Joseph Jastrow," *History of Psychology in Autobiography*, 1930).

Johnson, William Ernest (1858–1931). British logician and professor at University of Cambridge 1893–1931. Known widely for his 1921–1924 *Logic* in three volumes, in which he introduced logical exchangeability. Estimated to have been greatly influenced by Peirce's formal logic, manifest as early as Johnson's 1892 "The Logical Calculus" in *Mind*.

Kempe, Alfred Bray (1849–1922). English mathematician, barrister by trade but widely known for his work with four-color theory, linkages, and graphical relations. Corresponded with Peirce 1886 (RL 232a) regarding Kempe's "Memoir on the Theory of Mathematical Form" that year, which profoundly impacted Peirce, particularly its graphical notation of spots and lines on chemical diagrams, and would later play a role in Peirce's Existential Graphs. In Peirce's "The Logic of Relatives" (*Monist*, Jan 1897), for instance, he credits Kempe's "masterly" article for inspiring his graphical system that Peirce there lays out—though Kempe responded in the April 1897 *Monist* with "A Correction and Explanation" to Peirce's flattering citations. Peirce prepared a reply to Kempe (R 708–715), but it was never published.

King, Clarence (1842–1901). American geologist and mountaineer, leading explorer of the Fortieth Parallel Survey for eastern California and much of the Mountain West region 1867–1872, and first director of the U.S. Geological Survey 1878–1881. Acquainted with Peirce via The Century Club in New York and presumed to be the anonymous respondent "Kappa" to Peirce's March 1890 "Herbert Spencer's Philosopher" in *The New York Times*. Published several geological and evolutionary works, notably *Mountaineering in the Sierra Nevada* (1872) and "Catastrophism and Evolution" (*American Naturalist*, 1877), the latter discussed in Peirce's 1891 "Architecture of Theories" (*EP* 2:290).

Ladd-Franklin, Christine (1847–1930). First female PhD student at Johns Hopkins University 1878–1882, one of Peirce's prized students during his short tenure there (1879–1884), married to fellow Peirce classmate Fabian Franklin (1853–1839). Her dissertation "On the Algebra of Logic" published by Peirce, her thesis advisor, in his *Studies in Logic* (1883), along with work from fellow students Allan Marquand, O. H. Mitchell, and B. I. Gilman. Member of the JHU Metaphysical Club instigated by Peirce (formed 1879). Remained Peirce's life-long friend and periodic correspondent (RL 237). After Peirce's death, contributed in his honor "Charles Peirce at the Johns Hopkins" in the *Journal of Philosophy* (1916). Other publications by her include "On Some Characteristics of Symbolic Logic" (*American Journal of Psychology*, 1889), "Epistemology for the Logician" (*Kongresses fur Philosophie*, 1908), and *Colour and Colour Theories* (1929).

Langley, Samuel Pierpont (1834–1906). American astronomer, physicist, and aeronautics engineer. Professor at the Western University of Pennsylvania (now University of Pittsburgh) and the first director of Allegheny Observatory 1867–1891, where he met Peirce early 1879 on assignment with the Coast Survey to oversee the Observatory during Langley's impending trip to Mt. Etna. Later, as secretary of the Smithsonian Institution 1887–1906, provided periodic (and much-needed) employment for Peirce as translator of French and German scientific publications, especially 1901 (RL 409; also *CBPW*).

Listing, Johann Benedict (1808–1882). German mathematician and topologist. Discovered topological invariants later known as "Listing numbers," a classification presumably coined by Peirce (e.g., see Peirce's "Logic of Continuity" Cambridge Conferences lecture of 1898, *RLT*, 242–270).

Little and Co., J. J. New York printer founded 1867 by Joseph James Little (1841–1913) and associates, named Little, Rennie & Co. until 1876. One of the printers Peirce considered early 1893 to contract with OCP for his proposed arithmetic book (never published).

Mach, Ernst Waldfried Josef Wenzel (1838–1916). Austrian philosopher and physicist, inventor of the Mach number, and among "the very best of names" Carus recruited as an early contributor to *The Monist* (Carus to Peirce 2 July 1890)—"The Analysis of the Sensations," Oct 1890. Peirce worked with McCormack June 1892–May 1893 on a translation of Mach's 1889 edition of *The Science of Mechanics* (*Die Mechanik In Ihrer Entwickelung*), published by OCP 1893. Peirce contributed a lengthy statement on measurements included in the edition as Chapter III, Section III, Sub-section 8 (see Appendix, "Peirce and McCormack"). Peirce also reviewed the book in *The Nation* 4 Oct 1893.

Marquand, Allan (1853–1924). American logician and one of Peirce's prized students during his short tenure at Johns Hopkins University (1879–1884). His

"The Logic of the Epicureans" and "A Machine for Producing Syllogistic Variations" published by Peirce in his *Studies in Logic* (1883), along with work from fellow students O. H. Mitchell, Christine Ladd-Franklin, and B. I. Gilman. Member of the JHU Metaphysical Club instigated by Peirce (formed 1879). Professor of Latin and logic at Princeton 1881–1883, during which time he built his logical machine referenced in *Studies in Logic* and inspired by English logician William Jevons (1835–1882). Transitioned his career at Princeton to teach art history 1883–1922. Peirce's letter to him Dec 1886 (*W* 5:421–424) included circuit diagrams for an electric version of the machine that provided prophetic insight into modern electrical computing. Marquand's attempts to pursue Peirce's suggestions were ultimately to no avail.

Mendenhall, Thomas Corwin (1841–1924). American physicist, meteorologist, and educator. Held various professorial posts 1873–1889 in Ohio, Tokyo, and Indiana until his appointment as superintendent and Peirce's superior at the U.S. Coast and Geodetic Survey. Corresponded with Peirce 1889–1890 (RL 91) regarding his long-overdue report of extensive field operations from 1882–1886, but due to Mendenhall's differing opinions about gravitational studies and to the negative influence of his mentor and Peirce-adversary Simon Newcomb, rejected Peirce's report and soon after asked for his resignation from the Survey effective December 1891.

Mitchell, Oscar Howard (1851–1889). American mathematician, logician, and one of Peirce's prized students during his short tenure at Johns Hopkins University (1879–1884). His "On a New Algebra of Logic" published by Peirce in his *Studies in Logic* (1883), along with work from fellow students Allan Marquand, Christine Ladd-Franklin, and B. I. Gilman. Admired by Peirce for his work in logic and with whom Peirce corresponded 1882 on the logic of relatives (RL 294; *W* 4:394–400).

Morison, George Shattuck (1842–1903). Harvard Law graduate (class 1866) turned civil engineer and bridge builder. Built bridges mostly over the Missouri, Ohio, and Mississippi Rivers, his largest and most renowned being the Memphis Bridge (now Frisco Bridge) of the lower Mississippi. Hired Peirce 1895 (per recommendation, possibly of George Becker of U.S. Geological Survey) to prepare calculations for a suspension bridge over the Hudson River, which under other auspices fifty years later became the George Washington Bridge. Remained a friend to Peirce and maintained correspondence 1895–1903 (RL 300). Supported Peirce financially, occasionally paying his mortgage and small debts.

Müller, Friedrich Max (1823–1900). German philologist and orientalist at Oxford (1850–1900) and among "the very best of names" Carus recruited as an early contributor to *The Monist* (Carus to Peirce 2 July 1890)—"On Thought and Language," July 1891.

Newcomb, Simon (1835–1909). Renowned astronomer and mathematician and an early acquaintance of the Peirce family as a computer for the Nautical Almanac Office at Cambridge 1857–1861 and as a student of Benjamin Peirce while at Lawrence Scientific School 1857–1858. Appointed professor of mathematics at U.S. Naval Observatory 1861, director of the Nautical Almanac 1877, and professor of mathematics at JHU 1884. Although a friend to the Peirce family, apparently hated Charles and played a destructive role in several episodes of Peirce's professional life, notably Peirce's dismissal from JHU 1884, his "resignation" from the U.S. Coast Survey 1891, and the rejection of Peirce's 1902–1903 grant application to the Carnegie Institute for his "Memoirs on Minute Logic." Corresponded with Peirce 1871–1908 (RL 314), published in "The Correspondence with Simon Newcomb," Eisele, 1979.

Oliver, James Edward (1829–1895). Student of Benjamin Peirce at Harvard (class 1849), mathematical assistant with Nautical Almanac Office at Cambridge 1850–1867 (per B. Peirce's recommendation), and later mathematics professor at Cornell ca. 1870–1895 (Hill, 1896). Corresponded with Peirce periodically 1871–1893 (RL 322).

Otis, Dr. William K. (1860–1906). Prominent New York urologist and genitourinary surgeon, fellow of The New York Academy of Medicine, and son of also prominent Dr. Fessenden Nott Otis (1828–1900). Presumably one of Juliette's doctor for her gynecological health issues. Corresponded with her and Peirce periodically 1891–1902 (RL 326).

Palmer, George Herbert (1842–1933). Alford professor of natural religion, moral philosophy, and civil polity at Harvard (1889–1913). Believed to have been met by Peirce in Cambridge May 1892 regarding a professorship at the newly formed University of Chicago that Peirce was then pursuing, Palmer having recently declined an offer as chair of Philosophy from the same university (see Russell to Peirce 19 May 1892). After discovering that William James had recommended Peirce for the position, Palmer wrote the university president William R. Harper warning him against Peirce on account of his "broken and dissolute character" that had elsewhere impeded his prospects (*W* 4:lxix–lxv). Peirce was ultimately denied the position.

Peirce, Benjamin (1809–1880). Peirce's father and renowned Harvard mathematician and astronomer 1831–1880. Played a key role in establishing scientific institutions in the U.S., including the Lawrence Scientific School at Harvard (1847), the National Academy of Sciences (1863), and the U.S. Coast and Geodetic Survey (1807), which he reorganized and greatly improved as superintendent 1867–1874. His clout and position provided academic and professional opportunities for Charles he might otherwise not have had, particularly his long tenure and high rank (Assistant to the Superintendent) with the Coast Survey. A year after Benjamin's death, Charles revised and republished in his honor his 1870 *Linear Associative Algebra*, and ca. 1894–1895 also attempted to republish Benjamin's 1837 *Elementary Treatise on Plane and Solid Geometry* in his proposed "New Elements of Geometry" (*NEM* 2:233–474; see also Peirce to Russell 1 Sept 1894).

Peirce, Herbert Henry Davis (t) (1849–1916). Peirce's younger brother and U.S. diplomat. First Secretary of U.S. Legation in St. Petersburg 1894–1901, appointed by President Roosevelt as Third Secretary of State 1901 and U.S. Minister to Norway 1906. Like his older brother Jem, Bert supported Peirce during his financial difficulties, especially beginning ca. 1890, and also played a role in gaining President Roosevelt's endorsement of Peirce's 1902–1903 Carnegie Institute grant application for "Memoirs on Minute Logic" (ultimately rejected nonetheless, due mostly to Newcomb—see above).

Peirce, James Mills (Jem) (1834–1906). Peirce's oldest brother and Harvard professor of mathematics and astronomy 1859–1906, appointed dean of the newly formed Harvard Graduate School 1890. It was in Jem's room and with his copy of Whatley's *Elements of Logic* (1826) that young Peirce in 1851 was introduced to his lifelong passion for logic. Jem provided frequent financial support for Charles, especially 1890–1905, although apparently opposed him during a lawsuit against Charles by a house servant, Laura Walters (see Peirce to Russell ca. 10 March 1896). Worked with Ginn & Co. 1894 and 1895 to publish Peirce's proposed arithmetic and "New Elements of Geometry" books (neither published), and continually tried to find employment for Peirce at Harvard and elsewhere.

Peirce, Juliette Froissy Pourtalais (or Juliette Annette Pourtalai) (d. 1934). Peirce's second wife of undetermined origins, his travel companion and mistress ca. 1877–1883 during the final years of his first marriage to Harriet Melusina Fay, whom he divorced two days before marrying Juliette 1883. Their scandalous relationship created strife in Peirce's familial and professional life, and reputedly was partly responsible for his dismissal from JHU 1884. She suffered throughout their marriage from tuberculosis and gynecological problems, the latter requiring an emergency hysterectomy April 1897. Her illness often took her away from Peirce to Europe, New York, and elsewhere to be with family, to seek medical attention, or to convalesce in better climates. Her devotion to Peirce and his work was unrelenting during their marriage and continued after his death. Arranged with Josiah Royce and James Haughton Woods of Harvard to donate the late Peirce's manuscripts to the philosophy department ca. Christmas 1914.

Players Club (or The Players). Private New York club founded 1888 by Shakespearean actor Edwin T. Booth (1833–1893, brother of John Wilkes Booth) to engage actors and artists with men in industrial, financial, or other professions. Details about Peirce's affiliation are undetermined, though he likely became affiliated with the Club via his wife Juliette, previously an actress with strong New York ties.

Powell, Major John Wesley (1834–1902). Civil War veteran, geologist, and explorer. Professor at Illinois Wesleyan University beginning 1865 and leader of the Powell Geographic Expedition 1869 along the Green and Colorado rivers, the first thorough geological investigation of the region. Later director of U.S. Geological Survey 1881–1894, where in 1885 he received moral support from Peirce at the Coast Survey during government funding difficulties affecting both the Geological and Coast Surveys. They corresponded periodically thereafter until ca. 1895 (RL 360).

Renouvier, Charles Bernard (1815–1903). French philosopher who Peirce believed to be one of the "world-renowned thinkers," along with Montgomery and Delbœuf, who shared Peirce's opinion of *tychism* in "The Doctrine of Necessity Examined" (reference identified in *EP* 2:334n1 as Renouvier's *Essais de critique générale*, 1875), and for whom Peirce requested a copy of that issue to be sent by Carus (3 June 1892). See Peirce's footnote at p. 1 of Peirce's "Man's Glassy Essence," Oct 1892.

Risteen, Allan Douglas (1866–1932). Peirce's computing assistant late 1885–1887 in U.S. Coast and Geodetic Survey, consultant spring–summer 1891 for Peirce's proposed arithmetic book and for his mathematical and scientific contributions to the upcoming volume of *Century Dictionary*. Researcher for Hartford Steam Boiler Inspection & Insurance Co. 1888–1911 and safety engineer with Traveler's Insurance Co. beginning 1911. Author of *Molecules and the Molecular Theory of Matter* (1895), reviewed by Peirce in *The Nation* 13 Feb 1896 (*CN* 2:128–130). Maintained correspondence with Peirce 1887–1913 (RL 376).

Romanes, George John (1848–1894). English evolutionary biologist and comparative psychologist, friend and colleague of Darwin, and pioneer in neo-Darwinism. Among "the very best of names" Carus recruited as an early contributor to *The Monist* (Carus to Peirce 2 July 1890)—"Mr. A. R. Wallace on Physiological Selection," leading article in maiden issue Oct 1890. His *Mind and Motion, and Monism* (1895) reviewed by Peirce in *The Nation* 26 March 1896 (*CN* 2:130–131).

Rood, Ogden Nicholas (1831–1902). American physicist, color theorist, and physics department chair at Columbia University 1863–1902. Personal friend of Peirce as early as 1878, when he provided a flattering letter of recommendation for Peirce in his request for a raise at the Coast Survey (RL 91). His 1879 *Modern Chromatics, with Applications to Art and Industry* broke new ground in color theory and was reviewed by Peirce in *The Nation* 16 Oct 1879 (*CN* 1:58–61). The two remained in correspondence until ca. 1894 (RL 382).

Royce, Josiah (1855–1916). Prominent American philosopher and objective idealist, Harvard professor 1884–1916. Respected Peirce's work and advocated for its success, though frequently opposed Peirce in many areas of philosophy, including religion, God, realism, and Peirce's tychism (or absolute chance). Was also well-read by Peirce. His 1885 *The Religious Aspect of Philosophy*, for instance, reputedly influenced Peirce's return to philosophy after his extended focus on logic at Johns Hopkins University (1879–1884). Peirce treated his work often in *The Nation*, including his *The World and the Individual* (1899–1901), reviewed 5 April 1900; *Outlines of Psychology* (1903), reviewed 29 Sept 1904; and *Herbert Spencer: An Estimate and Review* (1904), reviewed 26 Jan 1905. Shortly after Peirce's death, Royce organized the effort of Dec 1914 to retrieve Peirce's manuscripts from Milford to be housed at Harvard. Later said of Peirce's work that "the two works, which, if they could ever have been completed, were intended by Peirce to be the proper fruits of his studies, were a 'History of Science' and a 'Comprehensive Treatise on Logic.'... No more erudition has ever existed amongst us regarding the topics which were here in question." ("Charles Sanders Peirce," *Journal of Philosophy*, 1916; for more on these two works, see note preceding 1898 Peirce and Carus letters, and Peirce to Carus 6 and 7 Jan 1909.)

Schiller, Ferdinand Canning Scott (1864–1937). German-British philosopher and pragmatist, preferred the term "humanism" to "pragmatism." Professor at Cornell 1893–1897 and from 1897 divided his professorship between Oxford and UCLA the remainder of his life. Disagreed with the distinctions of Peirce's "pragmaticism" and found his insistent pragmatic disentanglements unnecessary.

Schröder, Friedrich Wilhelm Karl Ernst (1841–1902). German mathematician famous for his work in algebraic logic and one of Peirce's most influential followers. Corresponded with Peirce at Johns Hopkins ca. 1879–1884 regarding Peirce's work at the time on the algebra of logic (see Bibliography), which impressed and influenced Schröder's own work. Their correspondence resumed 1890–1898 (see Houser, 1991), centering primarily around Schröder's masterwork *Vorlesungen über die Algebra der Logik* (3 vols., 1890–1905) and Peirce's rekindled work on his own algebra of logic, specifically the algebra of the copula (see *W* 8:208–228). Schröder's third installment of the series, *Algebra und Logik der Relative* (1895), was reviewed by Peirce in *The Nation* (April 1896) and treated at great length in *The Monist* in his "The Regenerated Logic" (Oct 1896) and "The Logic of Relatives" (Jan 1897)—the latter two per Schröder's insistence with Carus that Peirce was the only one suited to review his work (Brent, 256). The Peirce–OCP correspondence, particularly Peirce to Russell, is a rich source of Peirce's commentary on Schröder's logical work.

Searle, Father George Mary (1839–1918). Astronomer and Catholic priest, assistant at Dudley Observatory 1858–1859 where he discovered asteroid 55 Pandora. Worked with Peirce at the U.S. Coast Survey ca. 1860–1862 and again at Harvard Observatory 1866–1868. Left Harvard to enter the Paulist Order, was appointed 1871 as lecturer at the Catholic University of America, and became superior general of the Paulist Order 1904–1909. His 1895 *Plain Facts for Fair Minds* on papal infallibility and aimed at non-Catholics elicited a letter of criticism from Peirce 9 Aug that year stating, "I would with all my heart join the ancient church of Rome if I could. But your book is an awful warning against doing so." Peirce proceeded with a defense of the inevitability of fallibilism in searching for truth. (RL 397) Searle's response to Peirce is not found.

Sigwart, Christoph von (1830–1904). German logician known widely for his 1873–1878 *Logik* in two volumes. Intended to be among "the very best of names" Carus recruited as an early contributor to *The Monist* (Carus to Peirce 2 July 1890), though nothing from him appeared—only a review of his *Logik* July 1894.

Smith, Benjamin Eli (1857–1913). Graduate student at Johns Hopkins University during Peirce's short tenure there (1879–1884) and on the staff of the *Century Dictionary*, later managing editor and editor in chief. Presented his "Wundt's Theory of Volition" and "On Brown's 'Metaphysics'" Feb–March 1882 to the JHU Metaphysical Club (instigated by Peirce), around which time he invited Peirce to contribute definitions to the *Century* project. Peirce's assignment included terms in logic, metaphysics, mathematics, mechanics, astronomy, and weights and measures. Peirce's work with Smith and the *Century Dictionary* 1883–1891 provided much-needed financial and intellectual support during a time Brent considers to be "the destruction of Peirce's career and the beginning of his descent into ruin and poverty" (139). In their remaining years that followed, the two remained in contact and Peirce contributed sporadically to the *Century* project.

Spencer, Herbert (1820–1903). Renowned British philosopher, sociologist, and evolutionary theorist. Popularized the phrase "survival of the fittest" in his *Principles of Biology* (1864–1867), one of five works composing his ten-volume *A System of Synthetic Philosophy*—the other four including *First Principles* (1862), *Principles of Psychology* (1870–1872), *Principles of Sociology* (1882–1898), and *Principles of Ethics* (1892). Peirce instigated a collective criticism of Spencer's evolutionary philosophy in *The New York Times* March–April 1890, his own contributions (signed "Outsider") including "Herbert Spencer's Philosophy. Is it Unscientific and Unsound?" and "'Outsider' Wants More Light" (23 March and 13 April). Spencer's mechanical notion of the universe had also elicited Peirce's early criticisms of the doctrine of necessity in his 1887 "Science and Immortality" (Christian Register Symposium; *W* 6:61–64) that would set the stage for his wider-read (and much-debated between him and Carus) "The Doctrine of Necessity Examined" in *The Monist* April 1892. Peirce also reviewed in *The Nation* Spencer's 1891 *Essays: Scientific, Political, and Speculative* 8 Oct 1891 and his *Ethics* in part 19 Oct 1893. Carus too was highly critical of Spencer, especially in his "Herbert Spencer on the Ethics of Kant" (*Open Court*, 16 and 23 Aug 1888).

Stedman, Edmund Clarence (1833–1908). New York stockbroker, poet, and editor. Became acquainted with Peirce via The Century Club ca. 1876 and reputedly introduced him to the area of Milford, PA, where Stedman owned property and where Charles and Juliette settled 1887. Also partnered with Peirce (and Albert Bierstadt, above) in a failed entrepreneurial endeavor ca. 1896–1898 to develop affordable domestic lighting from acetylene gas. Some of his published poems included *Poems, Lyrical and Idyllic* (1860), *Alice of Monmouth* (1864), and *The Blameless Prince* (1869), and his larger edited works included *A Library of American Literature* (11 vols., 1888–1890) and *The Works of Edgar Allan Poe* (10 vols., 1895).

Stickney, Albert (d. 1908). Peirce's former classmate at Harvard (class 1859) and later prominent New York attorney. Had little contact with Peirce until ca. Sept 1891 when Peirce invited him to visit his home in Milford and soon after recruited his help (and wealthy friends) in renting his home for additional profit. Stickney remained in contact with Peirce on the matter and served as Peirce's legal counsel for several years thereafter (RL 421). Also housed Peirce in New

York July–Sept 1896. Authored several jurisprudential books, such as *Organized Democracy* (1906) which Peirce reviewed in *The Nation* 12 Sept 1907 (*CN* 3:290).

Stokes, George Gabriel (1819–1903). English mathematician and Lucasian professor of mathematics at Cambridge University 1849–1903. Also a major contender in the field of fluid dynamics, which was the pretext for his correspondence with Peirce 1886 (RL 423). Peirce was at that time preparing a report for the Coast Survey on his extensive field operations conducted since 1882 (*W* 6:275–353), and in the report he would propose an improvement to Stokes's hydrodynamical theory and an emended formula for calculating the effect of the viscosity of air on the motion of a pendulum. This was likely a major topic of their correspondence, in addition to Stokes's desire to have Peirce take part in the new gravitational operations underway in England that summer (Brent, 178).

Sylvester, James Joseph (1814–1897). English mathematician, chair of mathematics at the newly formed Johns Hopkins University 1876–1883 during Peirce's short tenure there, founder of the *American Journal of Mathematics* 1878, and Savilian professor of geometry at Oxford 1883–1897. Involved in a controversy with Peirce 1882–1883 initiated by Peirce's belief that Sylvester's universal algebra was derived from his own logic of relatives and escalated by an explicit reference to that effect added by Peirce in the proofs of a published abstract by Sylvester (JHU *Circulars*, Aug 1882), who apparently overlooked the addition in the proof. A correction was made but not without dispute. (For details, see *W* 4:liii–lviii; also Brent, 140–141.) Peirce later wrote Sylvester's obituary for *The New York Evening Post* 16 March 1897 (*HP* 1:502–505).

Trumbull, Matthew Mark (1826–1894). London-born American lawyer, politician, and Civil War Brigadier-General. Later Chicago resident and contributor to *The Monist*, *The Open Court*, and other periodicals on topics of politics, economics, and philosophy. Close friend to Russell and "one of a coterie in and about The Open Court office" (Russell to Peirce 2 Oct 1908).

Underwood, Benjamin Franklin (1839–1914). Virginian freethought lecturer and writer, succeeded Francis E. Abbot as editor of the Free Religious Association's periodical *The Index* in Boston 1880–1886, recruited by Hegeler Jan 1887 as the first editor and manager of *The Open Court*, with his wife and associate editor Sara A. Francis. Hegeler had shortly after invited Carus to be his secretary and tutor for his children, and to play some role in the publication of *The Open Court*. Carus's uncertain appointment with the magazine threatened the Underwoods and caused a year-long dispute that ended with their resignation

Dec 1887. Despite his brief tenure, Underwood successfully initiated the magazine and set up a printing office in Chicago.

University Club. Private New York club established 1865 by a group of Yale alumni to promote the active study of literature and art, providing reading rooms and art galleries to that end. Details about Peirce's affiliation are undetermined.

Vail, Henry Hobbart (1839–1925). Editor in chief at the American Book Company (co-founded by D. Appleton & Co) 1890–1907, and vice president beginning ca. 1904. Author of *A History of the McGuffey Readers: The Bookish Books—Vol. IV* (1910). Peirce met with him (as well as William Appleton) early Feb 1893 in consideration of his proposed arithmetic book and an Elementary Geometry book.

Vailati, Giovanni (1863–1909). Italian philosopher, mathematician, historian of science, and pragmatist. Greatly influenced by Peirce and James, between whose thought he was among the first to distinguish. Presumably in response to Peirce's 1905 *Monist* articles on pragmaticism, addressed in his Oct 1906 *Monist* article "Pragmatism and Mathematical Logic" the contributions of Italian philosophy to the logico-pragmatic thought promoted by Peirce—the article was fittingly followed in the same issue by Peirce's "Prolegomena to an Apology for Pragmaticism."

Wilson & Son, John. Cambridge printer founded ca. 1855 in Boston (moved 1865) by Scottish printers John Wilson (1802–1868) and son John Wilson, Jr. (1826–1903). One of the printers Peirce considered early 1893 to contract with OCP for his proposed arithmetic book (never published).

Winslow, Dr. John (1836–1900). Ithaca physician originally from Massachusetts. Relationship to Peirce undetermined. Possibly a family friend of the Peirces or a mutual acquaintance of Peirce and James E. Oliver of Cornell. Among the individuals Peirce requested from Carus to have sent a copy of the July 1892 *Monist* containing his "Law of Mind" (Peirce to Carus 15 July 1892).

Wright, Chauncey (1830–1875). American philosopher, mathematician, and friend of Peirce beginning 1857. Appointed computer for the Nautical Almanac Office at Cambridge 1852, lecturer at Harvard in psychology 1870–1871 and later 1874–1875 in mathematics (before his untimely death), and member with Peirce of the Metaphysical Club at Cambridge (formed 1872). Also a valued contributor to biology and evolutionary theory. His "The Genesis of Species" in the *North American Review* July 1871, for instance, was reprinted in England at the request of Darwin, who shortly after befriended Wright. Peirce's review of Fraser's *Works of George Berkeley* in the *North American Review* Oct 1871 was critiqued by Wright in *The Nation* 30 Nov 1871 (*CN* 1:43–45).

Youmans, Edward Livingston (1821–1887). Scientific writer, editor for *Appleton's Journal* ca. 1869–1872, and founding editor of *Popular Science Monthly* magazine 1872, in which Peirce published his 1877–1878 "Illustrations of the Logic of Science" series and his 1901 "Pearson's *Grammar of Science*" review.

Bibliography

For Peirce's published works, the original publisher is identified. For his unpublished works, an authoritative edition containing the work (such as *Writings*, *Essential Peirce*, or *Collected Papers*) is cited where possible; otherwise the manuscript location corresponding to the Robin Catalogue is provided. Sources in the abbreviations list in this edition are excluded. Multiple sources by the same author are listed chronologically (in some cases roughly). Peirce's *Monist* and *Open Court* articles are grouped together for ease of reference, as are other article series. Entries with an asterisk (*) are citations by the editor (in notes or other matter) not found in the letters. General reference materials are listed separately.

*Abbot, Francis E. *The Way Out of Agnosticism, Or the Philosophy of Free Religion*. Boston: Little, Brown, and Co., 1890.

Anonymous. "Charles Sanders Peirce." *Sun and Shade*, vol. 4, no. 12 (August 1892), photogravure numbered XC, with short biographical sketch.

Anonymous. "Recent Philosophical Works." *The Nation*, vol. 52 (25 June 1891): 521–522. Paragraph about Carus's *Soul of Man* at p. 522.

*Bisanz, Elize, ed. *The Logic of Interdisciplinarity: The Monist-series*. Berlin: Akademie Verlag (now de Gruyter), 2009.

*Blacoe, William Victor. *Understanding the New Testament: 1st and 2nd Timothy, Titus, and Philemon*. Springville, UT: Cedar Fort, 2011.

Boëthius, Anicius Manlius Severinus. *De Consolatione Philosophæ*. Ghent, 1485.

Boltzmann, Ludwig. "The Recent Development of Method in Theoretical Physics." *The Monist*, vol. 11 (January 1901): 226–257.

Boole, George. "The Calculus of Logic." *Cambridge and Dublin Mathematical Journal*, vol. 3 (1848): 183–198.

*Brent, Joseph. *Charles Sanders Peirce: A Life*. Indianapolis: IUP, 1993.

Browne, Sir Thomas. *Pseudodoxia Epidemica: Or, Enquiries into Very Many Received Tenets and Commonly Presumed Truths*. London: Harper, 1646.

Bunyan, John. *The Pilgrim's Progress from This World to That Which Is to Come; Delivered under the Similitude of a Dream*. London: Nath. Ponder, 1678.

Cantor, Georg. *Zur Lehre Vom Transfiniten: Gesammelte Abhandlungen Aus Der Zeitschrift Für Philosophie Und Philosophische Kritik*. Halle: C.F.M. Pfeffer, 1890.

Cantor, Georg. "Beiträge zur Begründung der transfiniten Mengenlehre." *Mathematische Annalen*, vol. 49 (1897): 207–246.

*Carus, Paul. *Monism and Meliorism: A Philosophical Essay on Causality and Ethics*. New York: F. W. Christern, 1885.

*Carus, Paul. *Ein Leben in Liedern: Gedichte eines Heimathlosen*. Milwaukee: Freidenker, 1886.

Carus, Paul. *Ursache, Grund und Zweck: eine philosophische Untersuchung zur Klärung der Begriffe*. Dresden: von Grumbkow, 1887.

Carus, Paul. "Herbert Spencer on the Ethics of Kant." *The Open Court*, vol. 2 (16 Aug 1888): 1155–1160.

Carus, Paul. "Herbert Spencer on the Ethics of Kant. (Concluded.)." *The Open Court*, vol. 2 (23 Aug 1888): 1165–1169.

Carus, Paul. "Is Nature Alive?" *The Open Court*, vol. 2, in four parts (17 October 1888): 1264–1267.

https://doi.org/10.1515/9783110768756-005

Carus, Paul. "Form and Formal Thought." *The Open Court*, vol. 2 (November–December 1888): 1310–1312, 1336–1339, 1349–1351, and 1369–1372.
Carus, Paul. *Fundamental Problems: The Method of Philosophy as a Systematic Arrangement of Knowledge*. Chicago: OCP, 1889.
Carus, Paul. "Feeling and Motion." *The Open Court*, vol. 4 (31 July 1890): 2424–2426.
Carus, Paul. "Feeling and Motion. (Concluded.)." *The Open Court*, vol. 4 (7 Aug 1890): 2435–2437.
Carus, Paul. "Kant on Evolution. In Criticism of Mr. Herbert Spencer's Presentation of Kantism." *The Open Court*, vol. 4 (4 September 1890): 2492–2497; revised and reprinted (quoting in a note at p. 44 Peirce to Carus 6 Sept 1890) in *The Monist*, vol. 2 (July 1892): 33–53.
Carus, Paul. "The Unity of Truth." *The Open Court*, vol. 4 (11 September 1890): 2501–2502.
*Carus, Paul. "The Superscientific and Pure Reason." *The Open Court*, vol. 4 (11 September 1890): 2509–2511.
Carus, Paul. "The Criterion of Truth: A Dissertation on the Method of Verification." *The Monist*, vol. 1 (January 1891): 229–244.
Carus, Paul. *The Soul of Man: An Investigation of the Facts of Physiological and Experimental Psychology*. Chicago: OCP, 1891.
*Carus, Paul. "Mr. Charles S. Peirce on Necessity." *The Monist*, vol. 2 (April 1892): 442.
Carus, Paul. "Mr. Charles S. Peirce's Onslaught on the Doctrine of Necessity." *The Monist*, vol. 2 (July 1892): 560–582.
Carus, Paul. "The Idea of Necessity, Its Basis and Its Scope." *The Monist*, vol. 3 (October 1892): 68–96.
*Carus, Paul. "Religion Inseparable from Science." *The Open Court*, vol. 7 (16 February 1893): 3560.
Carus, Paul. "The Founder of Tychism, His Methods, Philosophy, and Criticisms: In Reply to Mr. Charles S. Peirce." *The Monist*, vol. 3 (July 1893): 571–622.
*Carus, Paul. "Notes" [on Peirce's article "What is Christian Faith?"]. *The Open Court*, vol. 7 (27 July 1893): 3750.
Carus, Paul. *The Philosophy of the Tool*. Chicago: OCP, 1893. From a lecture delivered 18 July 1893 at the Chicago World's Fair.
Carus,j894): 4121–4125.
Carus, Paul. "The Circle-Squarer. (Concluded.)." *The Open Court*, vol. 8 (28 June 1894): 4130–4133.
Carus, Paul. *The Gospel of Buddha*. Chicago: OCP, 1894.
*Carus, Paul. Review of *Algebra und Logik der Relative*, dritter band. By Ernst Schröder. *The Monist*, vol. 6 (January 1896): 312.
*Carus, Paul. "Buddhism in Its Contrast with Christianity, as Viewed by Sir Montier-Williams." *The Open Court*, vol. 10 (23 January 1896): 4783–4789.
Carus, Paul. *The Dharma, or The Religion of Enlightenment: An Exposition of Buddhism*. Chicago: OCP, 1896.
Carus, Paul. *Buddhism and Its Christian Critics*. Chicago: OCP, 1897.
Carus, Paul, trans. and ed. *Lao-Tze's Tao-Teh-King*. Chicago: OCP, 1898.
Carus, Paul. *Kant and Spencer: A Study of the Fallacies of Agnosticism*. Chicago: OCP, 1899.
Carus, Paul. "The Philosophical Foundations of Mathematics." *The Monist*, vol. 13 (January 1903): 273–294.
Carus, Paul. "The Foundations of Geometry." *The Monist*, vol. 13 (April 1903): 370–397.
Carus, Paul. "The Foundations of Geometry. (Concluded)." *The Monist*, vol. 13 (July 1903): 493–522.
Carus, Paul. "Esperanto." *The Monist*, vol. 16 (July 1906): 450–455.

Carus, Paul. *The Story of Samson and Its Place in the Religious Development of Mankind.* Chicago: OCP, 1907.
*Carus, Paul. "Pragmatism." *The Monist*, vol. 18 (July 1908): 321–362.
Carus, Paul. "The Nature of Logical and Mathematical Thought." *The Monist*, vol. 20 (January 1910): 33–75. Peirce quoted at pp. 45 and 158–159. For Peirce's edits to this excerpt, see Appendix of this edition under "Peirce and Carus."
Carus, Paul. "The Monism of 'The Monist,' Compared with Professor Haeckel's Monism." *The Monist*, vol. 23 (July 1913): 435–439.
Cayley, Arthur. "Memoir on the Theory of Matrices." *Philosophical Transactions of the Royal Society of London*, vol. 148 (January 1858): 17–37.
Cayley, Arthur. "A Sixth Memoir upon Quantics." Ibid., vol. 149 (January 1859): 61–90.
The Century Dictionary. 6 vols. Ed. William Dwight Whitney. New York: The Century Company, 1889–1891. Digitized at http://www.global-language.com/century/, accessed 22 August 2016.
Chase, Charles H. "A Strange Attack on Some Physical Theories." *The Monist*, vol. 8 (April 1900): 463–465.
*Corrington, Robert S. *An Introduction to C. S. Peirce: Philosopher, Semiotician, and Ecstatic Naturalist.* Lanham, MD: Rowman and Littlefield, 1993.
Clebsch, Alfred. *Vorlesungen uüber geometrie*, ed. Ferdinand Lindemann. Leipzig: Teubnar, 1876. Also as *Leccons sur la geéomeétrie*, trans. Adolphe Benoist. Paris: Gauthier-Villars, 1879.
Clifford, William K. "Instruments Used in Measurement." *Handbook to the Scientific Loan Collection of Scientific Apparatus.* London: Chapman and Hall, 1876.
d'Alembert, Jean Le Rond. *Traité de dynamique*. Paris: Coignard, 1743.
Dedekind, Richard. *Was sind und was sollen die Zahlen?* Braunschweig: F. Vieweg, 1888. Trans. Wooster Beman, "The Nature and Meaning of Numbers" in *Essays on the Theory of Numbers.* Chicago: OCP, 1901.
*Degen, Nicholas. "History of the schlieren method," in *An Overview on Schlieren Optics and Its Applications*. Zurich: ETH, 2012.
De Morgan, Augustus. "On the Syllogism." *Transactions of the Cambridge Philosophical Society*, 5-part series, vols. 8–10 (1846–1864).
De Morgan, Augustus. *A Budget of Paradoxes*. 2 vols. Chicago: OCP, 1915.
*de Waal, Cornelis, ed. *Illustrations of the Logic of Science*. By Charles S. Peirce. Chicago: OCP, 2014.
*Dewey, John. "The Superstition of Necessity." *The Monist*, vol. 3 (April 1893): 362–379.
Dirichlet, Peter Gustav Lejeune. *Vorlesungen über Zahlentheorie*. Braunschweig: F. Vieweg, 1871.
Dreher, Eugene. "The Basis of Dualism." *The Open Court*, vol. 7 (18 June 1893): 3670.
Dureège, H., George E. Fisher, and Isaac J. Schwatt. *Elements of the Theory of Functions of a Complex Variable: With Especial Reference to the Methods of Riemann*. Philadelphia: G. E. Fisher and I. J. Schwatt, 1894. Reviewed by Peirce in *The Nation*, vol. 63 (3 September 1896): 181–182.
*Eisele, Carolyn. *Studies in the Scientific and Mathematical Philosophy of Charles S. Peirce.* Ed. Richard M. Martin. The Hague: Mouton, 1979.
*Elsenhans, T., ed. "Fortsetzung der Diskussion über den Pragmatismus." *Bericht über den III. internationalen Kongress für Philosophie zu Heidelberg*. Heidelberg, 1909.
Emerson, Ralph Waldo. *Nature*. Boston: James Munroe & Co., 1836.
*Emerson, Ralph Waldo. *Essays*. Boston: James Munroe and Co., 1841.
*Fisch, Max. *Peirce, Semeiotic, and Pragmatism: Essays by Max H. Fisch*. Ed. Kenneth L. Ketner and Christian J. W. Kloesel. Indianapolis: Indiana University Press, 1986.

Freytag, Gustav. *The Lost Manuscript*. Chicago: OCP, 1890.
*Funk, Isaac K, and Francis A. March. *A Standard Dictionary of the English Language*. New York: Funk & Wagnalls, 1894.
Greenhill, William Alexander. *Religio Medici, Letter to a Friend &c. and Christian Morals*. London: Macmillan, 1881.
Goethe, Johann Wolfgang von. *Faust: Eine Tragödie*. Tübingen: J. G. Cotta'schen Buchhandlung, 1808.
Gomperz, T. *Philodem Über Induktions-schlüsse*. Leipzig: Teubner, 1865.
Gould, George Milbry. *The Modern Frankenstein*. Chicago: OCP, 1889.
Gould, George Milbry. *Dreams, Sleep, and Consciousness: A Psychological Study*. Chicago: OCP, 1889.
Grassman, Hermann. "Sur les différents genres de multiplication." *Journal für die reine und angewandte Mathematik*, vol. 49 (1855): 123–141.
Gunkel, Hermann. *The Legends of Genesis*. Trans. W. H. Carruth. Chicago: OCP, 1901.
*Gunlogsen, A. H. "The Philosophy of the Vedanta," *The Open Court*, vol. 4 (6 March 1890): 2131–2133.
Hegel, Georg Wilhelm Friedrich. *The Science of Logic* (*Wissenschaft der Logik*). 2 vols. Nürnberg: Johann Leonhard Schrag, 1812 and 1816.
Hegeler, Edward C. "Religion and Science." *The Open Court*, vol. 4 (28 August 1890): 2473–2474.
*Henderson, Harold. *Catalyst for Controversy: Paul Carus of Open Court*. Carbonale, IL: SIU Press, 2009.
Herschel, John F. W. *Outlines of Astronomy*. London: Longman, 1849.
Hill, G. W. "Memoir of James Edward Oliver." Paper read before National Academy of Sciences, April 1896, accessed 22 August 2016, http://www.nasonline.org/publications/biographical-memoirs/memoir-pdfs/oliver-james.pdf.
Hobbes, Thomas. *Elementorum philosophiae sectio prima De corpore* (*Elements of Philosophy, The First Section, Concerning Body*). Londini: A. Crook, 1655.
Höffding, Harald. *Outlines of Psychology*. London: Macmillan & Co., 1891.
Hodgson, Shadworth. *The Philosophy of Reflection*, vol. 1. London: Longmans, Green, and Co., 1878.
*Houser, Nathan. "The Schröder-Peirce Correspondence." *The Review of Modern Logic*, vol. 1, nos. 2–3 (1991): 206–236.
Hume, David. *Dialogues concerning Natural Religion*. London, 1779.
James, William. *The Principles of Psychology*. 2 vols. New York: Henry Holt and Co., 1890.
*James, William. "Abbot against Royce." *The Nation*, vol. 53 (19 November 1891): 389–390.
James, William. *The Will to Believe and Other Essays*. London: Longmans, Green, and Co., 1897.
James, William. *Varieties of Religious Experience: A Study in Human Nature*. London: Longmans, Green, and Co., 1902.
James, William. "A World of Pure Experience." *Journal of Philosophy, Psychology, and Scientific Methods*, vol. 1 (1904): 533–543 and 561–570.
James, William. *The Meaning of Truth: A Sequel to 'Pragmatism.'* London: Longmans, Green, and Co., 1909.
Jevons, William Stanley. *The Substitution of Similars: The True Principle of Reasoning, Derived from a Modification of Aristotle's Dictum*. London: Macmillan and Co., 1869.
Johnson, William Ernest. "The Logical Calculus." Parts I–III. *Mind*, new series vol. 1 (Jan, April, and July 1892): 3–30, 235–250, and 340–357.

Kant, Immanuel. *Critique of Pure Reason* (*Critik der reinen Vernunft*). Riga, Latvia: Johann Friedrich Hartknoch, 1781.

Kant, Immanuel. *Critique of Judgement* (*Critik der Urtheilskraft*). Berlin and Libau: Lagarde and Friedrich, 1790.

Kant, Immanuel. *Anthropology from a Pragmatic Point of View* (*Anthropologie in pragmatischer Hinsicht abgefaßt*). Königsberg: Friedrich Nicolovius, 1798.

Kant, Immanuel. *Prolegomena to Any Future Metaphysics* (*Prolegomena zu einer jeden künftigen Metaphysik die als Wissenschaft wird auftreten können*). Trans. Paul Carus. Chicago: OCP, 1902.

*Keeton, Morris. *The Philosophy of Edmund Montgomery*. University Press in Dallas, 1950.

*Kegley, Jacquelyn Ann K. "Josiah Royce: Classical American Philosopher; Pragmatist, Phenomenologist, Process Thinker and Advocate for Community." *Pragmatism Today (Journal of the Central-European Pragmatist Forum)*, vol. 4, issue 1. (Summer 2013): 49–55.

Kempe, Alfred Bray. "Memoir on the Theory of Mathematical Form." *Philosophical Transactions of the Royal Society of London*, vol. 177 (1886): 1–70.

Klein, Felix. "On the So-Called Non-Euclidian Geometry" ("Über die sogenannte Nicht-Euklidische Geometrie") *Mathematische Annalen*, vol. 4, no. 4 (1871): 573–625; and vol. 6, no. 2 (1873): 112–145.

Klein, Felix. *Lectures on the Icosahedron and the Solution of Equations of the Fifth Degree* (*Vorlesungen über das Ikosaeder und die Auflösung der Gleichungen vom fünften Grade*). Leipzig: Teubner, 1884.

Klein, Felix. *Nicht-Euklidische Geometri*, 2 vols. Göttingen, 1890–1892.

Lagrange, Joseph-Louis. *Lectures on Elementary Mathematics* (*Leçons élémentaires sur les mathématiques*). Trans. Thomas J. McCormack. Chicago: OCP, 1898.

Langland, William. *Visio Willelmi de Petro Ploughman* (*William's Vision of Piers Plowman*). ca. 1370.

Laplace, Pierre-Simon. *Théorie analytique des probabilités*. Paris: Ve. Courcier, 1812.

Laplace, Pierre-Simon. *Mécanique Céleste*. 4 vols. Trans. Nathaniel Bowditch. Boston: Hillard, Gray, Little, and Wilkins, 1829.

Listing, Johann Benedict. "Der Census räumlicher Complexe, oder Verallgemeinerung des Euler'schen Satzes von den Polyädern." *Göttingen Abhandlungen*, vol. 10 (1862): 97–182.

Lobachevsky, Nikolai Ivanovich. *Geometrical Researches on the Theory of Parallels*. Trans. George Bruce Halsted. Austin: University of Texas, 1891, reprinted by OCP 1914. Reviewed by Peirce in *The Nation*, vol. 54 (11 February 1892): 116.

Louis Couturat. *L'algèbre de la Logique* (*The Algebra of Logic*). Paris: Gauthier-Villars, 1905; trans. OCP, 1914.

Lutosławski, Wincenty. *The Origin and Growth of Plato's Logic*. New York: Longmans, Green & Co., 1897.

Mach, Ernst. "Some Questions of Psycho-Physics. Sensations and the Elements of Reality." *The Monist*, vol. 1 (April 1891): 393–400.

Mach, Ernst. *The Science of Mechanics: A Critical and Historical Exposition of Its Principles*. Trans. T. J. McCormack. Chicago: OCP, 1893 (from *Die Mechanik In Ihrer Entwickelung: Historisch-Kritisch Dargestellt*, Second ed. Leipzig: F. A. Brockhaus, 1889).

McCrie, George M. "The Issues of 'Synechism.' " *The Monist*, vol. 3 (April 1893): 380–401.

Mill, John Stuart. *A System of Logic, Ratiocinative and Inductive*. London: John W. Parker, 1843.

Minchin, George. *A Treatise on Statics*. London: Longmans, Green & Co., 1877.

Moore, A. W. "Humanism." *The Monist*, vol. 14 (October 1904): 747–752.

Morgan, Conway Lloyd. "The Doctrine of Auta." *The Monist*, vol. 3 (January 1893): 161–175.

Muir, Thomas. *The Theory of Determinants in the Historical Order of Development*. 4 vols. London and New York: Macmillan, 1890–1923. First vol. reviewed by Peirce in *The Nation*, vol. 51 (28 August 1890): 177.

*Murphey, Murray. *The Development of Peirce's Philosophy*. Cambridge: Harvard University Press, 1961.

Naville, Ernest. *La définition de la philosophie*. Paris: Genève et Bale, 1894.

Nichols, Herbert. *A Treatise on Cosmology*. Cambridge University Press, 1904.

Oliver, James Edward. "A Mathematical View of Free Will." *The Philosophical Review*, vol. 1 (May 1892).

Palmer, Charles T. *Artificial Persons: A Philosophical View of the Law of Corporations*. Chicago: OCP, 1889.

Peirce, Benjamin. *An Elementary Treatise on Plane and Spherical Trigonometry*. Boston: James Munroe and Co., 1836.

Peirce, Benjamin. *An Elementary Treatise on Sound*. Boston: James Munroe and Co., 1836.

Peirce, Benjamin. *An Elementary Treatise on Algebra*. Boston: James Munroe and Co., 1837

Peirce, Benjamin. *An Elementary Treatise on Plane and Solid Geometry*. Boston: James Munroe and Co., 1837.

Peirce, Benjamin. *An Elementary Treatise on Curves, Functions, and Forces*. 2 vols. Boston: James Munroe and Co., 1841 and 1846.

Peirce, Benjamin. *A System of Analytic Mechanics*. Boston: Little, Brown, and Co., 1855.

Peirce, Benjamin. *Linear Associative Algebra*. Washington, 1870 (lithograph). Later edited by Charles S. Peirce with the subtitle *New Edition, with Addenda and Notes by C. S. Peirce, Son of the Author*. New York: Van Nostrand, 1882.

Peirce, Benjamin. *Ideality in the Physical Sciences*. Ed. James Mills Peirce. Boston: Little, Brown, and Co., 1881.

Peirce, Charles S. "Think Again!" *The Harvard Magazine*, vol. 4 (April 1858): 100–105.

Peirce, Charles S. "Shakespearean Pronunciation." Reviews of *Lectures on the English Language*. By George P. Marsh. *The Works of William Shakespeare*. By Richard Grant White. *The English of Shakespeare Illustrated in a Philological Commentary on his Julius Caesar*. By George L. Craik. *The North American Review*, vol. 98 (April 1864): 342–369. Written in collaboration with John Buttrick Noyes.

Peirce, Charles S. Review of *The Logic of Chance*. By John Venn. *The North American Review*, vol. 105 (July 1867): 317–321.

Peirce, Charles S. Articles for *Proceedings of the American Academy of Arts and Sciences*, vol. 7 (1868):

"On an Improvement in Boole's Calculus of Logic." 250–261. Read before the Academy on 12 March 1867.

"On the Natural Classification of Arguments." 261–287. Read before the Academy on 9 April 1867.

"On a New List of Categories." 287–298. Read before the Academy on 14 May 1867.

"Upon the Logic of Mathematics." 402–412. Read before the Academy on 10 September 1867.

"Upon Logical Comprehension and Extension." 416–432. Read before the Academy on 13 November 1867.

Peirce, Charles S. Articles for *The Journal of Speculative Philosophy*:

"Nominalism versus Realism." vol. 2 (1868): 57–61.

"Questions Concerning Certain Faculties Claimed for Man." vol. 2 (1868): 103–114.
"Some Consequences of Four Incapacities." vol. 2 (1868): 140–157.
"What Is Meant by 'Determined'." vol. 2 (1868): 190–191.
"Grounds of Validity of the Laws of Logic: Further Consequences of Four Incapacities." vol. 2 (1869): 193–208.

Peirce, Charles S. "Lectures on British Logicians." Series of 15 lectures given by Peirce at Harvard, 1869–1870. Published, in part, in *CP* 1.28–29 and 1.30–34, and in *W* 2:310–347.

Peirce, Charles S. Review of *The Secret of Swedenborg: Being an Elucidation of his Doctrine of the Divine Natural Humanity.* By Henry James. *The North American Review*, vol. 110 (April 1870): 463–468.

Peirce, Charles S. "Description of a Notation for the Logic of Relatives, Resulting from an Amplification of the Conceptions of Boole's Calculus of Logic." *Memoirs of the American Academy of Arts and Sciences*, vol. 9 (1870): 317–378. Also published separately as an extraction by Welch, Bigelow, and Company for Harvard University (1870).

Peirce, Charles S. Review of *The Works of George Berkeley, D.D., formerly Bishop of Cloyne: including many of his Writings hitherto unpublished.* Ed. Alexander Campbell Fraser. *The North American Review*, vol. 113 (October 1871): 449–472.

Peirce, Charles S. "On Logical Breadth and Depth." n.p. *W* 3:98–101 (1873).

*Peirce, Charles S. "On Quaternions, as Developed from the General Theory of the Logic of Relatives." Paper read before the Philosophical Society of Washington, Washington, D.C., 14 March 1874. Cited in *Bulletin of the Philosophical Society of Washington*, vol. 1 (1874): 94.

Peirce, Charles S. "Note on the Theory of the Economy of Research." *W* 4:72–78, extracted from *Report of the Superintendent of the United States Coast Survey*, 1876. *CBPW* P00259 places this in Peirce's "Force of Gravity" paper of the Coast Survey Report of 1882.

Peirce, Charles S. "Note on the Sensation of Color." *The American Journal of Science and Arts*, third series 13, whole series 113 (January to June 1877): 247–251. Reprinted in *The London, Edinburgh, and Dublin Philosophical Magazine and Journal of Science*, fifth series, 3 (supplement, same date): 543–547.

Peirce, Charles S. "Illustrations of the Logic of Science" articles:
"First Paper.—The Fixation of Belief." *The Popular Science Monthly*, vol. 12 (November 1877): 1–15.
"Second Paper.—How to Make Our Ideas Clear." *The Popular Science Monthly*, vol. 12 (January 1878): 286–302.
"Third Paper.—The Doctrine of Chances." *The Popular Science Monthly*, vol. 12 (March 1878): 604–615.
"Fourth Paper.—The Probability of Induction." *The Popular Science Monthly*, vol. 12 (April 1878): 705–718.
"Fifth Paper.—The Order of Nature." *The Popular Science Monthly*, vol. 13 (June 1878): 203–217.
"Sixth Paper.—Deduction, Induction, and Hypothesis." *The Popular Science Monthly*, vol. 13 (August 1878): 470–482.
*"La Logique de la Science. Première Partie. Comment se fixe la croyance." *Revue Philosophique de la France et de l'Etranger*, vol. 6 (December 1878): 553–569.
*"La Logique de la Science. Deuxième Partie. Comment rendre nos idées claires." *Revue Philosophique de la France et de l'Etranger*, vol. 7 (January 1879): 39–57.

Peirce, Charles S. Articles for *American Journal of Mathematics*:
"On the Algebra of Logic." vol. 3 (1880): 15–57.

"On the Logic of Number." vol. 4 (1881): 85–95.
"On the Algebra of Logic: A Contribution to the Philosophy of Notation." vol. 7 (January 1885): 180–202.

Peirce, Charles S. "Brief Description of the Algebra of Relatives." Privately printed brochure. Baltimore: 7 January 1882; with a postscript dated 16 January 1882.

Peirce, Charles S. "On the Relative Forms of Quaternions." Paper read before the Mathematical Seminary, Johns Hopkins University, January 1882. Cited in *The Johns Hopkins University Circulars*, vol. 1 (1882): 179.

Peirce, Charles S. "A Theory of Probable Inference." In *Studies in Logic, By Members of the Johns Hopkins University*. Ed. Charles S. Peirce. Boston: Little, Brown, and Company, 1883, 126–181. (Other chapters by Peirce include "Preface" iii-vi, "Note A (On a Limited Universe of Marks)" 182–186, and "Note B (The Logic of Relatives)" 187–203.)

*Peirce, Charles S. "The Logic of Relatives: Qualitative and Quantitative." n.p. *W* 5: 372–378 (1886).

Peirce, Charles S. ["Science and Immortality"]. *Science and Immortality: The Christian Register Symposium, Revised and Enlarged*. Ed. Samuel J. Barrows. Boston: George H. Ellis, (April 1887): 69–76.

Peirce, Charles S. "Criticism on 'Phantasms of the Living.' An Examination of an Argument of Messrs. Gurney, Myers, and Podmore." *Proceedings of the American Society for Psychical Research*, old series vol. 1 (December 1887): 150–157.

*Peirce, Charles S. "A Guess at the Riddle." n.p. *EP* 1:245–279 (1887–1888).

Peirce, Charles S. "Reflections on the Logic of Science." n.p. *W* 6:246–259 (January 1889). Part of a proposed book on the philosophy of physics.

Peirce, Charles S. "On Sensations of Color." Paper read before the National Academy of Sciences, Washington, 16–19 April 1889. Cited in *Report of the National Academy of Sciences for the Year 1889*, Senate Mis. Doc. No. 47, 51st Congress, 2d Session, Washington: Government Printing Office, 1891, p. 6.

Peirce, Charles S. "Herbert Spencer's Philosophy. Is it Unscientific and Unsound?—Its Pretensions Attacked and a Demonstration Called For." *The New-York Times*, vol. 39 (23 March 1890): 4, columns 6–7. Signed "Outsider"

*Peirce, Charles S. "Six Lectures of Hints toward a Theory of the Universe." n.p. R 972 (spring 1890).

Peirce, Charles S. "[Logic and Spiritualism]." n.p. *W* 6:380–394 (summer 1890). Untitled article on spiritualism prepared for (though never published in) Lorettus Metcalf's spiritualism series in *The Forum*.

Peirce, Charles S. Review of *Fundamental Problems: The Method of Philosophy as a Systematic Arrangement of Knowledge*. By Paul Carus. *The Nation*, vol. 51 (7 August 1890): 118–119.

Peirce, Charles S. Review of *Locke*. By Alexander Campbell Fraser. *The Nation*, vol. 51 (25 September 1890): 254–255.

Peirce, Charles S. "Notes" [on the first number of the Monist]. *The Nation*, vol. 51 (23 October 1890): 96–97. Accredited to Peirce in *CBPW*.

Peirce, Charles S. Articles for *The Monist*, published and unpublished:
"The Architecture of Theories." vol. 1 (January 1891): 161–176.
"The Doctrine of Necessity Examined." vol. 2 (April 1892): 321–337.
"The Law of Mind." vol. 2 (July 1892): 533–559.
"Man's Glassy Essence." vol. 3 (October 1892): 1–22.
"Evolutionary Love." vol. 3 (January 1893): 176–200.
"Reply to the Necessitarians: Rejoinder to Dr. Carus." vol. 3 (July 1893): 526–570.

"The Regenerated Logic." vol. 7 (October 1896): 19–40. Review of Schröder's 1895 *Algebra und Logik der Relative*.
"The Logic of Relatives." vol. 7 (January 1897): 161–217. Partly a review of Schröder's 1895 *Algebra und Logik der Relative*.
"Knotty Points in the Doctrine of Chances." n.p. R 209 (ca. December 1898).
"What Pragmatism Is." vol. 15 (April 1905): 161–181. Two internal dates by Peirce are included: main text as "Milford, Pa., September, 1904" and a postscript as "Feb. 9, 1905."
"Substitution in Logic." vol. 15 (April 1905): 294–295. Submitted by Francis C. Russell and slightly revised, but based on Peirce's article found in Appendix of this edition under "Peirce and Russell."
"Issues of Pragmaticism." vol. 15 (October 1905): 481–499.
*"The Consequences of Pragmaticism." n.p. R 288–289 (ca. April 1905). Partly in Bisanz, 301–306. Initially Peirce's proposed second article of the *Monist* pragmatism series, later replaced by "Issues of Pragmaticism.")
*"The Basis of Pragmaticism." n.p. *EP* 2:360–397 (ca. August 1905–April 1906). Six variations of Peirce's proposed third article of the *Monist* pragmatism series, R 279–284, later replaced by "Prolegomena to an Apology for Pragmaticism." Attempts five and six are published in *EP* and other portions published in Bisanz, 259–286, and *CP* 1 and 5 (see Robin Catalog).
"Mr. Peterson's Proposed Discussion." vol. 16 (January 1906): 147–151.
Review of *Foundations of Sociology*. By Edward Alsworth Ross. vol. 16 (July 1906): 470–473.
"Prolegomena to an Apology for Pragmaticism." vol. 16 (October 1906): 492–546.
*"Phaneroscopy." n.p. R 298 (late 1906–early 1907). One of several variations of Peirce's proposed fourth article of the *Monist* pragmatism series, to follow "Prolegomena." Included in Bisanz, 352–365.
"The First Part of an Apology for Pragmaticism." n.p. R 288–289 (fall 1907). One of several variations of Peirce's proposed fourth article of the *Monist* pragmatism series, to follow "Prolegomena." Included in Bisanz, 366–374.
*"The Bed-Rock Beneath Pragmaticism." n.p. R 300 (ca. 1908). One of several variations of Peirce's proposed fourth article of the *Monist* pragmatism series, to follow "Prolegomena." Included in Bisanz, 375–393.
"Some Amazing Mazes: The First Curiosity." vol. 18 (April 1908): 227–241.
"Some Amazing Mazes: Explanation of Curiosity the First." vol. 18 (July 1908): 416–464.
"Some Amazing Mazes: A Second Curiosity." vol. 19 (January 1909): 36–45.
"The Third Curiosity." n.p. R 199 (ca. December 1907).
"The Fourth Curiosity." n.p. R 200, partly in *CP* 4.647–681 and 6.318–348 (ca. December 1907–January 1908).

Peirce, Charles S. "The Algebra of the Copula." n.p. *W* 8:210–216 (1891).

Peirce, Charles S. "Abbot against Royce." *The Nation*, vol. 53 (12 November 1891): 372.

Peirce, Charles S. "Comparative Biography of Great Men." n.p. Related material in *W* 5:26–106 (ca. 1891–1892, although beginning as early as 1883 at JHU). Peirce's alternative to "The History of Science" as his topic for the 1892–1893 Lowell Institute Lectures. Also precursor to his 1901 "The Century's Great Men in Science."

Peirce, Charles S. "The Comtist Calendar." Review of *The New Calendar of Great Men*. By Frederic Harrison. *The Nation*, vol. 54 (21 January 1892): 54–55.

Peirce, Charles S. Review of *The Man of Genius*. By Cesare Lombroso. *The Nation*, vol. 54 (25 February 1892): 151–153.

Peirce, Charles S. "An Excursion into Thessaly: a Tale" (a.k.a. "Embroidered Thessaly"). n.p. *W* 8: 296–340 (ca. May 1892). Brent, 233, reports the full title as "Embroidered Thessaly, found among the papers of an attorney recently deceased," likely referring to Russell.

Peirce, Charles S. "Keppler." n.p. *W* 8:286–291 (ca. 1892). Part of his proposed lecture series on "Great Men."

Peirce, Charles S. Articles for *The Open Court*, published and unpublished:

[Announcement, correspondence lessons in the Art of Reasoning]. vol. 6 (1 September 1892): 3374.

"Pythagorics." vol. 6 (8 September 1892): 3375–3377.

"The Critic of Arguments. I. Exact Thinking." vol. 6 (22 September 1892): 3391–3394.

"Dmesis." vol. 6 (29 September 1892): 3399–3402.

"The Critic of Arguments. II. The Reader is Introduced to Relatives." vol. 6 (13 October 1892): 3415–3418.

"The Marriage of Religion and Science." vol. 7 (16 February 1893): 3559–3560.

"Immortality in the Light of Synechism," n.p. *EP* 2:1–3 (ca. May 1893).

*"The Critic of Arguments. III. Synthetical propositions a priori." n.p. R 590 (ca. fall 1893). One of several options for Peirce's proposed third article of the *Open Court* "Critic of Arguments" series.

*"Critic of Arguments. That all inferences concerning matters of fact are subject of certain qualifications." n.p. R 590 (ca. May 1893). One of several options for Peirce's proposed third article of the *Open Court* "Critic of Arguments" series.

*"Cogito Ergo Sum." vol. 7 (15 June 1893): 3702. A "letter to the editor."

"What is Christian Faith?" vol. 7 (27 July 1893): 3743–3745.

*"A Letter from Mr. Peirce." vol. 22 (May 1908): 319. The letter endorses Carus's "Problems of Modern Theology" from the previous month, pp. 234–246.

Peirce, Charles S. "The History of Science." 12 lectures delivered at the Lowell Institute in Boston, 28 November 1892 to 5 January 1893. Published in *HP* 2:139–296 and in part ("Concluding Remarks") in *CP* 7.267–275. Also a precursor to a one-volume edition with the same name prepared by invitation for G. P. Putnam's, (n.p. R 1290), 1898.

Peirce, Charles S. "Practical Arithmetic." n.p. R 167–168, 188 (ca. 1893). R 188 also alludes to an "advanced" arithmetic.

Peirce, Charles S. "Elementary Arithmetic." n.p. R 176 (ca. 1893).

Peirce, Charles S. "C. S. Peirce's Vulgar Arithmetic: Its Chief Features." n.p. R 178 (ca. 1893).

Peirce, Charles S. "Primary Arithmetic." n.p. R 180–182 (ca. 1893).

Peirce, Charles S. "The Association of Ideas." n.p. R 400–402 (1893).

Peirce, Charles S. "The Categories." n.p. R 403 (1893).

Peirce, Charles S. "The Materialistic Aspect of Reasoning." n.p. *CP* 6.278–286 (1893).

Peirce, Charles S. "What is the Use of Consciousness?" n.p. *CP* 7.559–564 (1893).

Peirce, Charles S. "Analysis of Propositions." n.p. R 410 (1893).

Peirce, Charles S. "The Aristotelian Syllogistic." n.p. *CP* 2.445–460 and 517–536 (1893).

Peirce, Charles S. "The Boolian Calculus." n.p. *CP* 3.20–41 (1893).

Peirce, Charles S. "The Quantification of the Predicate." n.p. *CP* 2.532–535 (1893). Part of R 414 of "The Aristotelian Syllogistic".

Peirce, Charles S. Chapter III, Section III, Sub-section 8 [Statement of the mechanical units in use in the United States and Great Britain] in *The Science of Mechanics*. By Ernst Mach, translated from the second German edition by Thomas J. McCormack. Chicago: The Open Court Publishing Co., 1893, 280–286.

Peirce, Charles S. "Hale's New England Boyhood." *The Nation*, vol. 57 (17 August 1893): 123–124.
*Peirce, Charles S. "Mach's *Science of Mechanics*." *The Nation*, vol. 57 (5 October 1893): 251–252.
*Peirce, Charles S. "Prospectus. The Treatise of Petrus Peregrinus on the Lodestone: Latin Text, English Version, and Notes. With an Introductory History of Experimental Science in the Middle Ages. By C. S. Peirce. Printed in two colors on hand-made paper. Bound in full Persian Morocco, hand-tooled. 140 pages." New York: DeVinne, ca. 1893. Privately printed prospectus for a translation of the *Epistle of Petrus Peregrinus on the Lodestone* to Sygerus de Foucaucourt (MS 7378 Bibliotheque Nationale). Also announced in *The Nation*, vol. 58 (11 January 1894).
Peirce, Charles S. "The Principles of Philosophy: Or, Logic, Physics and Psychics, Considered as a Unity, In the Light of the Nineteenth Century." Privately printed brochure announcing Peirce's proposed work in 12 volumes, ca. 1893. Planned for sale through subscription. Also announced in *The Nation*, vol. 58 (11 Jan 1894): 30; and by Henry Holt & Co. fall 1894.
Peirce, Charles S. Note [on Beckford's *Vathek*]. *The Nation*, vol. 5 (9 November 1893): 350.
Peirce, Charles S. "Leland's Memoirs." *The Nation*, vol. 57 (30 November 1893): 414–415.
Peirce, Charles S. "Napoléon Intime." Review of *Napoléon Intime*. By Arthur Lévy. *The Independent*, in two parts, vol. 45 (21 and 28 December 1893): 1725–1726, 1760.
Peirce, Charles S. "A Quest for a Method" (a.k.a. "Search for a Method," later "How to Reason"). n.p. R 592–594 (ca. 1893–1894).
Peirce, Charles S. "How to Reason: A Critick of Arguments" (posthumously a.k.a. "Grand Logic"). n.p. *NEM* 4:353–358 (1894).
Peirce, Charles S. Review of *Familiar Letters of Sir Walter Scott*. 2 vols. Ed. David Douglas. *The Nation*, vol. 58 (8 February 1894): 105–107.
Peirce, Charles S. "Rienzi, Last of the Tribunes," n.p. R 1318 (ca. July 1894).
Peirce, Charles S. "What Is a Sign?" n.p. *EP* 2:4–10 (1894).
Peirce, Charles S. "Achilles & Tortoise." n.p. R 814–815 (ca. 1894).
Peirce, Charles S. [Announcement and endorsements of *The Principles of Philosophy*]. n.p. R 1581 (ca. fall 1894). Robin reports that "the endorsements are by William James, Josiah Royce, G. Stanley Hall, Francis Abbot, Simon Newcomb, and O. C. March, one-time President of the National Academy of Science."
Peirce, Charles S. "New Elements of Geometry by Benjamin Peirce, rewritten by his sons, James Mills Peirce and Charles Sanders Peirce." n.p. *NEM* 2:233–474 (ca. 1894–1895). Revision of Benjamin Peirce's 1837 *Elementary Treatise on Plane and Solid Geometry*.
Peirce, Charles S. "Elements of Mathematics." n.p. *NEM* 2:1–232 (ca. 1895).
Peirce, Charles S. "Of Reasoning in General." n.p. *EP* 2:11–26 (ca. 1895).
Peirce, Charles S. "The Logic of Mathematics: An Attempt to Develop My Categories from Within." n.p. *CP* 1.417–520 (ca. 1896).
*Peirce, Charles S. Review of *Algebra und Logik der Relative, der Vorlesungen über die Algebra der Logik*. By Ernst Schröder. *The Nation*, vol. 62 (23 April 1896): 330–331.
Peirce, Charles S. "On Logical Graphs" (a.k.a. "Existential Graphs"). n.p. Related material in *CP* 4.347–584, also R 479–514 (ca. August 1896, though material ranges 1880–1908).
Peirce, Charles S. "Number: A Study of the Methods of Exact Philosophical Thought." Lecture before the Mathematical Department, Bryn Mawr College, PA, ca. 5 April 1897. Cited in *Annual Report of the President of Bryn Mawr College, 1896–1897*. Philadelphia: Alfred J. Ferris, 1898, 35.
Peirce, Charles S. "Note on the Age of Basil Valentine." *Science*, vol. 8 (12 August 1898): 169–176.

*Peirce, Charles S. Review of *Leibniz: The Monadology and Other Philosophical Writings*. Trans. Robert Latta. *The Nation*, vol. 68 (16 March 1899): 210.

Peirce, Charles S. Note [on Carus' Kant and Spencer]. *The Nation*, vol. 70 (8 February 1900): 109.

Peirce, Charles S. "Infinitesimals." *Science*, new series 11 (16 March 1900): 430–433. MS header indicates "Milford, Pa., Feb. 18, 1900."

Peirce, Charles S. "Pearson's *Grammar of Science*. Annotations on the First Three Chapters." *The Popular Science Monthly*, vol. 58 (January 1901): 296–306.

Peirce, Charles S. "The Century's Great Men in Science." *The Evening Post*, vol. 100 (12 January 1901): section three, page 1, columns 1–3. Reprinted in the *Annual Report of the Board of Regents of the Smithsonian Institution for the Year Ending June 30, 1900*, Washington: Government Printing Office, 1901, 693–699.

Peirce, Charles S. "The Classification of the Sciences." Paper read before the National Academy of Sciences, Washington, 15–17 April 1902. Cited in *Report of the National Academy of Sciences for the Year 1902*, Senate Document No. 81, 57th Congress, 2d Session, Washington: Government Printing Office, 1903, 13.

*Peirce, Charles S. "The National Academy of Sciences." *The Nation*, vol. 74 (24 April 1902): 322–324. Review of the NAS meeting of 19 April 1902 in Washington, in which Peirce refers to himself as C. "L." Peirce.

*Peirce, Charles S. "Note" [on Kant's *Prolegomena to Any Future Metaphysics*. Ed. Paul Carus]. *The Nation*, vol. 76 (18 June 1903): 497–498.

*Peirce, Charles S. "A Syllabus of Certain Topics of Logic." n.p. R 478 and partly in *EP* 2:258–299 (ca. November 1903). Pamphlet accompanying his Lowell Institute Lectures of Nov–Dec 1903.

Peirce, Charles S. "On Topical Geometry." Paper read before the National Academy of Sciences, New York City, 15–16 November 1904. Cited in *Report of the National Academy of Sciences for the Year 1904*, Senate Doc. No. 178, Washington: Government Printing Office, 1905, 16.

Peirce, Charles S. "Topical Geometry." n.p. R 137 (ca. 1904). Intended as an article for *The Popular Science Monthly*.

*Peirce, Charles S. "The Relation of Betweenness and Royce's O-collections." Paper on Josiah Royce's *The Relation of the Principles of Logic to the Foundations of Geometry* (1905) read before the National Academy of Sciences, New Haven, 14–15 November 1905. Cited in *Report of the National Academy of Sciences for the Year 1905*, Senate Document No. 144, 59th Congress, 1st Session, Washington: Government Printing Office, 1906, 15.

Peirce, Charles S. "Gosse's Sir Thomas Browne." *The Nation*, vol. 81 (14 December 1905): 486–488.

*Peirce, Charles S. "Recent Developments of Existential Graphs and their Consequences for Logic." Paper read before the National Academy of Sciences, Washington, 16–18 April 1906. Cited in *Report of the National Academy of Sciences for the Year 1906*, Senate Document No. 308, 59th Congress, 2d Session, Washington: Government Printing Office, 1907, 15. R 490 is a set of notes for this presentation.

Peirce, Charles S. "Pragmatism." n.p. *EP* 2:398–433 (ca. March–July 1907). Intended as a "letter to the editor" for *The Nation* and *Atlantic Monthly*.

Peirce, Charles S. "A Neglected Argument for the Reality of God." *The Hibbert Journal*, vol. 7 (October 1908): 90–112.

Peirce, Charles S. "A System of Logic." n.p. Related material in R 723–751 (ca. 1908). Various efforts toward this book during Peirce's life, especially ca. 1908 until death.

*Peirce, Charles S. "A Sketch of Logical Critics." n.p. *EP* 2:451–462, also various drafts in R 673–677 (ca. 1911).
*Peirce, Charles S. "A Method of Computation." Paper read before the National Academy of Sciences, 21–22 November 1911.
*Peirce, Charles S. "The Reasons of Reasoning, or Grounds of Inferring." Paper read before the National Academy of Sciences, 21–22 November 1911.
*Peirce Edition Project. "In the Works." *Peirce Project Newsletter*. vol. 1, no. 3 (December 1994), accessed 22 August 2016, http://www.iupui.edu/~peirce/news/1_3/13_4x.htm.
*Pietarinen, Ahti-Veikko. *Logic of the Future: Peirce's Writings on Existential Graphs, Volume 3/2: Correspondence* (forthcoming).
*Ransdell, Joseph. "MS L75: Logic, Regarded As Semeiotic (The Carnegie application of 1902)," accessed 22 August 2016, www.iupui.edu/~arisbe/menu/ library/bycsp/L75/l75.htm. Peirce's grant proposal for his "Memoirs on Minute Logic."
Renvouvier, Charles, and Loui Prat. *La Nouvelle Monadologie*. Paris: Armand Colin, 1899.
Ribot, Théodule. *The Diseases of Memory: An Essay in the Positive Psychology*. London: Kegan Paul, Trench & Co., 1882.
Ribot, Théodule. *The Diseases of the Will*. London: Kegan Paul, Trench & Co., 1884. Also Chicago: OCP, 1894.
Ribot, Théodule. *The Psychology of Attention*. Chicago: OCP, 1890.
Riemann, Georg Friedrich. "Ueber die Anzahl der Primzahlen unter einer gegebenen Grösse" ("On the Number of Primes Less Than a Given Magnitude"). *Monatsberichte der Königlich Preußischen Akademie der Wissenschaften zu Berlin*. (November 1859).
*De Rijk, L. M., ed. *Dialectic Monacensis, Logica Modernorum*. Assen: Van Gorcum, 1967.
Romanes, George J. *Darwinism Illustrated: Wood Engravings Explanatory of the Theory of Evolution*. Chicago: OCP, 1892.
*Royce, Josiah. "Dr. Abbot's 'Way Out of Agnosticism.' " *International Journal of Ethics*, vol. 1 (Oct 1890): 98–113.
*Royce, Josiah. *The Spirit of Modern Philosophy: An Essay in the Form of Lectures*. Boston: Houghton Mifflin, 1892.
*Royce, Josiah. *Studies of Good and Evil: A Series of Essays upon Problems of Philosophy and Life*. New York: Appelton and Co., 1898.
Royce, Josiah. *The Relation of the Principles of Logic to the Foundations of Geometry*. Lancaster, PA: New Era, 1905. From *Transactions of the American Mathematical Society*, vol. 6, no. 3 (July 1905): 353–415.
Russell, Francis C. "Hints for the Elucidation of Mr. Peirce's Logical Work." *The Monist*, vol. 18 (July 1908): 406–415.
Russell, Francis C. "A Modern Zeno." *The Monist*, vol. 16 (April 1909): 289–308.
*Russell, Francis C. "In Memoriam Charles S. Peirce: (Born 1839, died 1914)." *The Monist*, vol. 24 (July 1914): 469–472.
Russell, Bertrand. *The Principles of Mathematics*. Cambridge, England: CUP, 1903.
Sargent, Mary Elizabeth Fiske, ed. *Sketches and Reminiscences of the Radical Club of Chestnut Street, Boston*. Boston: Osgood & Co., 1880.
Schönflies, Arthur. "Die Entwickelung der Lehre von den Punktmannigfaltigkeiten." *Jahresbericht der Deutschen Mathematiker-Vereinigung*, vol. 8 (1900): 1–250.
Schröder, Ernst. *Vorlesungen über die Algebra der Logik (Exakte Logik)* (*Lectures on the Algebra of Logic*). 3 vols. Leipzig: Teubner, 1890–1905.

Schröder, Ernst. *Algebra und Logik der Relative, der Vorlesungen über die Algebra der Logik* (*Algebra and the Logic of Relatives*). Considered vol. 3 of *Algebra of Logic*. Leipzig: Teubner, 1895.

Schröder, Ernst. "Note über die Algebra der binären Relative." *Mathematische Annalen*, vol. 46 (1895): 144–158.

Schubert, Hermann. "The Squaring of the Circle: An Historical Sketch of the Problem from the Earliest Times to the Present Day." *The Monist*, vol. 1 (January 1891): 197–228.

Scott, Robert Forsyth. *Treatise on the Theory of Determinants and Their Applications in Analysis and Geometry*. Cambridge, England: CUP, 1880.

*Scotus, John Duns. *Ordinatio* II, dist. 3. In *Five Texts on the Medieval Problem of Universals*. Trans. Paul Vincent Spade. Indianapolis: Hackett, 1994.

*Sheridan, James F. *Paul Carus: A Study of the Thought and Work of the Editor of the Open Court Publishing Company*. PhD diss., University of Illinois, 1957.

*Shorey, Paul. Review of *The Origin and Growth of Plato's Logic*. By Wincenty Lutoslawski. *The Monist*, vol. 8 (July 1898): 621–626.

Smith, William Benjamin. "The Pauline Manuscripts F and G." *American Journal of Theology*, vol. 7 (October 1903): 452–482 and 662–688.

Smith, William Benjamin. *Der vorchristliche Jesus*. Giessen: Toepelmann, 1906. Reviewed by Karl Borinski in *The Monist*, vol. 16 (October 1906): 487–497.

Smith, William Benjamin. *Ecce Deus: Studies of Primitive Christianity*. Chicago: OCP, 1913.

Smith, William Benjamin. *Ecce Deus: The Pre-Christian Jesus*. Boston: Roberts, 1894.

Spencer, Herbert. "The Ethics of Kant." *Popular Science Monthly*, vol. 33 (August 1888): 464–479.

*Spenser, Edmund. "An Hymn in Honour of Beauty." *Fowre Hymnes*. London: Printed [by Richard Field] for William Ponsonby, 1596.

Stallo, Johann Bernhard. *The Concepts and Theories of Modern Physics*. New York: Appleton, 1882.

Steele, Robert. *Opera Hactenus Inedita Rogeri Baconi, No. I*. Oxford: Henry Frowde, 1909.

*"Stories from the Big House." Hegeler Carus Mansion website, accessed 22 August 2016, http://hegelercarusmansion.blogspot.com/p/family.html.

Story, William Edward. "On the Non-Euclidean Trigonometry." *American Journal of Mathematics*, vol. 4 (1881): 332–340.

Story, William Edward. "On the Non-Euclidean Geometry." *American Journal of Mathematics*, vol. 5 (1882): 180–211.

Super, Robert H., ed. *The Complete Prose Works of Matthew Arnold*. 11 vols. Ann Arbor: University of Michigan Press, 1960–1977.

Trumbull, Mathew Mark. *Making Bread Dear: Controversy between Wheelbarrow and Sympathizer upon Corners and the Board of Trade*. Chicago: OCP, 1889.

Turner, Jonathan Baldwin. *The Only Good Thing in All the Worlds*. Chicago: OCP, 1891.

*Turrisi, Patricia, ed. *Pragmatism as a Principle and Method of Right Thinking: The 1903 Harvard "Lectures on Pragmatism."* Albany: State University of New York Press, 1997.

Vailati, Giovanni. "Sull'arte d'interrogare." *Rivista di Psicologia*, vol. 1 (1905): 83–89.

Vailati, Giovanni. "Pragmatism and Mathematical Logic (tr. H. D. Austin)." *The Monist*, vol. 16 (October 1906): 481–491.

Vedanta Series, Calcutta: Vedanta Publishing. Including the following:

Dhole, Nandalal, trans. *A Manual of Adwaita Philosophy : The Vedantasara*. 1881.

Dhole, Nandalal, trans. *The Panchadasi. A Handbook of Hindu Pantheism*. 1886.

Sahib, Lala Sree-ram, trans. *The Vicharsagar; or Metaphysics of the Upanishads*. 1885.

Sahib, Lala Sree-ram, trans. *The Vicharmala*. 1886.

Venn, John. *The Logic of Chance: An Essay on the Foundations and Province of the Theory of Probability, with especial Reference to its Application to Moral and Social Science*. London: Macmillan and Co., 1866.
*Vos, Antonie. *The Philosophy of Duns Scotus*. Edinburgh University Press, 2006.
Ward, Lester Frank. *The Psychic Factors of Civilization*. Boston: Ginn & Co., 1893.
Watts, Isaac. *Logick: Or, the Right Use of Reason in the Enquiry After Truth, with a Variety of Rules to Guard against Error in the Affairs of Religion and Human Life, As Well As in the Sciences*. London: [publisher unknown], 1724.
Weisbach, Julius Ludwig. *Lehrbuch der Ingenieur- und Maschinenmechanik* (*Principles of the Mechanics of Machinery and Engineering*). 3 vols. Braunschweig: F. Vieweg und sohn, 1845–1847.
Wilkin, Simon. *The Works of Sir Thomas Browne*. 3 vols. London: H. G. Bohn, 1852.
*Wordsworth, William. *The Excursion: A Poem*. London: Simpkin, Marhsall, and Co., 1814.
Wright, Rev. W. J. *Tracts relating to the Modern Higher Mathematics, Tract No. 3, Invariants*. London: C. F. Hodgson and Son, 1879.
Wundt, Wilhelm M. *Vorlesungen über die Menschen- und Thier-Seele* (*Lectures on the Soul of Man and Animal*). 2 vols. Leipzig: Voss, 1863.
Wundt, Wilhelm M. *System der Philosophie*. 2 vols. Leipzig: Kröner, 1889.
*Zeman, Jay. "Peirce and Philo." *Studies in the Logic of Charles Sanders Peirce*. Ed. Nathan Houser, Don D. Roberts, and James Van Evra. Bloomington: Indiana University Press, 1997, 402–417.

General Reference Materials

The following have been consulted for this edition, though not cited specifically.

Appletons' Cyclopædia of American Biography. 6 vols. New York: D. Appleton & Co., 1887–1889, accessed 22 August 2016, https://en.wikisource.org/wiki/Appletons% 27_Cyclop%C3% A6dia_of_American_Biography.
Encyclopedia Americana. 30 vols. New York: Encyclopedia Americana Corporation, 1918–1920, accessed 22 August 2016, http://onlinebooks.library. upenn.edu/webbin/metabook?id= encyamer.
Encyclopedia Britannica. 15th ed. 32 vols. New York: Encyclopedia Britannica, Inc, 2002.
National Encyclopedia of American Biography. 46 vols. New York: James White and Co., 1891–1963, accessed 22 August 2016, http://onlinebooks.library. upenn.edu/webbin/serial? id=ncycusbio.
New Century Cyclopedia of Names. 3 vols. New York: Appleton Century Crofts, 1954.
Oxford Dictionary of National Biography. 60 vols. New York: OUP, 2004.
Oxford Dictionary of Philosophy. 2nd ed. revised. New York: OUP, 2008.
Oxford German Dictionary. 3rd ed. New York: OUP, 2008.
Oxford Greek Dictionary. American ed. New York: OUP, 2000.
Oxford Latin Dictionary. 1st ed. Oxford: OUP, 2005.
Stanford Encyclopedia of Philosophy. Metaphysics Research Lab, Stanford, accessed 22 August 2016, http://plato.stanford.edu/ .

Index

https://doi.org/10.1515/9783110768756-006

www.ingramcontent.com/pod-product-compliance
Lightning Source LLC
Chambersburg PA
CBHW031717230426
43669CB00007B/166
* 9 7 8 3 1 1 1 5 3 1 9 4 6 *